J. Ford Freick

Studies in Surface Science and Catalysis

Advisory Editors: B. Delmon and J.T. Yates
Series Editor: G. Centi

Vol. 167

NATURAL GAS CONVERSION VIII

Proceedings of the 8th Natural Gas Conversion Symposium, Natal, Brazil, May 27–31, 2007

Edited by

Fábio Bellot Noronha
Instituto Nacional de Tecnologia, Rio de Janeiro, Brazil

Martin Schmal
Universidade Federal do Rio de Janeiro, Brazil

Eduardo Falabella Sousa-Aguiar
CENPES/Petrobras, Rio de Janeiro, Brazil

ELSEVIER

Amsterdam – Boston – Heidelberg – London – New York – Oxford – Paris
San Diego – San Francisco – Singapore – Sydney – Tokyo

Elsevier
Radarweg 29, PO Box 211, 1000 AE Amsterdam, The Netherlands
Linacre House, Jordan Hill, Oxford OX2 8DP, UK

First edition 2007

Library of Congress Cataloging-in-Publication Data
A catalog record for this book is available from the Library of Congress

British Library Cataloguing in Publication Data
A catalogue record for this book is available from the British Library

ISBN: 978-0-444-53078-3
ISSN: 0167-2991

For information on all Elsevier publications visit our website at books.elsevier.com

Printed and bound in The Netherlands

07 08 09 10 11 10 9 8 7 6 5 4 3 2 1

Preface

The present volume contains peer reviewed manuscripts presented at the 8th Natural Gas Conversion Symposium held in Natal, Brazil, on May 27–31, 2007. The Brazilian event continues the tradition of excellence and the status as the premier technical meeting in this area established by the previous meetings: Auckland, New Zealand (1987), Oslo, Norway (1990), Sydney, Australia (1993), Kruger National Park, South Africa (1995), Taormina, Italy (1998), Girdwood, Alaska, USA (2001) and Dalian, China (2004).

The Editors would like to thank the International Advisory Board for choosing Natal, Brazil as the site for the 8th Natural Gas Conversion Symposium. The 8th Natural Gas Conversion Symposium has been stewarded by an Organizing and Programme Committee chaired by Fabio Bellot Noronha (Brazilian National Institute of Technology), Eduardo Falabella Sousa-Aguiar (Petrobras Research Centre) and Martin Schmal (Federal University of Rio de Janeiro). The individual contributions of the members of both Committees to the success of this symposium are gratefully acknowledged. Also, the editors would like to express their gratitude to the generous financial support given by the Sponsors, whose names are listed below.

The Natural Gas Conversion Symposium is arranged every third year with the aim of bringing together scientists and engineers from academia and industry working in the field of natural gas conversion. It has always been regarded as important to keep a balance between different regions and between Industry and Academia. For that reason, the Symposium is for the first time located in South America and Brazil feels honoured to host such distinguished event. The discovery of new off-shore reserves of associated gas in Brazil and the existence of yet unexplored reserves in the Amazon are excellent motivations for the development of gas conversion technologies in this country.

The Editors thank the authors of these manuscripts for their contributions, which have been divided into the following topics: Production of synthesis gas, FT synthesis of hydrocarbons, Production and conversion of methanol and dimethyl ether, Hydrogen production to fuel cells, Production of alternative fuels from natural gas, Natural gas to chemicals, Production and conversion of light paraffin, Catalytic combustion, Direct conversion of methane to oxygenates, olefins and aromatics, Technology demonstration and commercial activities, Industrial process and economics and Energy production from natural gas. Aiming at keeping the balance between Industry and Academia, contributions from both sides have been encouraged. The present issue contains papers from the Academia and the Industry, not to mention papers originated from collaborations between both sides. Furthermore, the same balance has been a major concern of the organizers when choosing the plenary lectures and keynotes.

The Editors do believe that the present volume of Studies in Surface Science and Catalysis represents an outstanding overview of the natural gas conversion area, enclosing not only recent advances in terms of new technologies and technological aspects of traditional chemical routes, but also scientific innovation concerning catalysts, products and chemical paths. We are sure that this publication will attract the interest of those devoted to all aspects of the chemical transformation of natural gas all over the world.

Eduardo Falabella Sousa-Aguiar, Fabio Bellot Noronha and Martin Schmal

Editors

Program Committee

Fábio Bellot Noronha (Co-chair), Instituto Nacional de Tecnologia
Martin Schmal (Co-chair), Universidade Federal do Rio de Janeiro

Carlos Apesteguia – INCAPE (Argentina)
Xinhe Bao – Dalian Institute of Chemical Physics (China)
Fernando Baratelli Jr.- CENPES/Petrobras (Brazil)
Carla Eponina Hori – Universidade Federal de Uberlândia (Brazil)
Anders Holmen – Norwegian University of Science and Technology (Norway)
Graham Hutchings – University of Wales (UK)
Krijn P. de Jong – Utrecht University (The Netherlands)
Lisiane Veiga Mattos – Instituto Nacional de Tecnologia (Brazil)
Domenico Sanfilippo – Snamprogetti (Italy)
James J. Spivey – Louisiana State University (USA)

Organizing Committee

Martin Schmal (Co-chair), Universidade Federal do Rio de Janeiro
Eduardo Falabella Sousa-Aguiar (Co-chair), CENPES/Petrobras

Lucia Gorestin Appel – Instituto Nacional de Tecnologia
Maria Auxiliadora Scaramelo Baldanza – NUCAT/COPPE
Antonio Marcos Fonseca Bidart – CENPES/Petrobras
Henrique Soares Cerqueira – CENPES/Petrobras
Pedro Neto Nogueira Diógenes – CTGas
Marco Andre Fraga – Instituto Nacional de Tecnologia
Ana Guedes – IBP
Paulo Roberto Barreiros Neves – CENPES/Petrobras
Fabio Barboza Passos – Universidade Federal Fluminense
Ana Carlota Belizário dos Santos – CENPES/Petrobras
Antonio Luiz Fernandes dos Santos – Petrobras
Victor Luis Teixeira da Silva – Universidade Federal do Rio de Janeiro
Valéria Vicentini – Oxiteno

International Scientific Advisory Board

Financial Support

Master Level

Ruthenium Level

ExxonMobil
Research and Engineering

Cobalt Level

ConocoPhillips

Sponsor Level

HALDOR TOPSØE A/S

Table of Contents

VI. Natural gas to chemicals

VII. Production of synthesis gas

Natural Gas Conversion VIII
F.B. Noronha, M. Schmal, E.F. Sousa-Aguiar (Editors)

Catalytic combustion of methane over Pd/SBA-15/ Al_2O_3/FeCrAl catalysts

Fengxiang Yin, Shengfu Ji*, Fuzhen Zhao, Zhongliang Zhou, Chengyue Li*

State Key Laboratory of Chemical Resource Engineering, Beijing University of Chemical Technology, 15 Beisanhuan Dong Road, P.O.Box 35, Beijing, 100029, China

Abstract: A series of Pd/SBA-15 with different Pd loadings and 0.5%Pd/SBA-15/Al_2O_3/FeCrAl as well as 0.5%Pd/5%$Ce_{1-x}Zr_xO_2$/SBA-15/Al_2O_3/FeCrAl (x=0, 0.5, 1) catalysts were prepared and characterized. Their catalytic activity and stability for combustion of methane were investigated. The results showed that Pd/SBA-15 catalysts exhibited an excellent activity, and methane could be combusted completely below 400 °C. The 0.5%Pd/SBA-15/Al_2O_3/FeCrAl monolithic catalyst showed a good activity. The activity of 0.5%Pd/5%$Ce_{1-x}Zr_xO_2$/SBA-15/Al_2O_3/FeCrAl monolithic catalysts was enhanced due to the addition of $Ce_{1-x}Zr_xO_2$, and the activity increased with Zr content increasing. The 0.5%Pd/SBA-15 catalyst exhibited a rapid deactivation. However, the 0.5%Pd/SBA-15/Al_2O_3/FeCrAl catalyst and 0.5%Pd/5%$Ce_{1-x}Zr_xO_2$/SBA-15/Al_2O_3/FeCrAl catalyst were more stable than 0.5%Pd/SBA-15 catalyst. The monolithic catalysts still had a well-defined hexagonal pore structure of SBA-15 when supported on FeCrAl supports.

Key words: Pd; SBA-15; FeCrAl; catalytic combustion; methane

1. Introduction

Catalytic combustion of methane was found to be more beneficial than conventional flame combustion due to lower NO_x emission and higher energy efficiency [1-2]. The alumina-supported palladium catalysts exhibit a good activity for catalytic combustion of methane [3-4]. However, they are not stable at high temperature. It has reported that Ce-Zr mixed oxides used as promoter can increase significantly the stability of noble-based combustion catalysts [5-6]. At the same time, the mesoporous SBA-15 silica has been considered to be one of the more promising materials in the field of heterogeneous catalysis due to its

* Corresponding authors. Tel.: +86 10 64412054; Fax: +86 10 64419619. *E-mail addresses*: jisf@mail.buct.edu.cn (S. Ji), licy@mail.buct.edu.cn (C. Li).

relatively high mechanical, thermal and hydrothermal stabilities [7]. However, the fixed-bed reactors used in industry present much higher pressure drops and much higher temperature gradients because they are randomly packed with catalyst particles, easily resulting in deactivation of catalysts and lower energy efficiency. Nowadays, a novel metallic monolithic catalyst/reactor has considered being promising due to the elimination of these drawbacks of fixed-bed reactors [8-9]. In this study, a series of Pd/SBA-15 with different Pd loadings and 0.5%Pd/SBA-15/Al_2O_3/FeCrAl as well as 0.5%Pd/5%$Ce_{1-x}Zr_xO_2$ /SBA-15/Al_2O_3/FeCrAl (x=0, 0.5, 1) metallic monolithic catalysts were prepared. The goal is to understand the structure and the catalytic performance.

2. Experimental

SBA-15 was prepared according to the literature [7]. Pd/SBA-15 samples were prepared by the wetness impregnation method with an aqueous of Pd $(NO_3)_2$ as metal precursors. The 5%$Ce_{1-x}Zr_xO_2$/SBA-15 (x=0, 0.5, 1) samples were prepared by co-impregnation of pure SBA-15 with an aqueous solution of cerium nitrates and zirconium nitrates. 0.5%Pd/5%$Ce_{1-x}Zr_xO_2$/SBA-15 samples were prepared by impregnation method. All samples were dried at room temperature and calcined in air at 500 °C for 4 h. The monolithic samples were prepared using the FeCrAl alloy foils as supports. The foils were rolled into metallic monolithic supports, which were made up of several cylinders in different diameter and 100 mm in length. Then, the supports were calcined at 950 °C for 15 h in air. The heat-treated supports were immersed in a boehmite primer sol, then dried at room temperature and thereafter at 120 °C for 3 h, and then calcined at 500 °C for 4 h. Then, a mixing slurry of 0.5%Pd/SBA-15 or 0.5%Pd/5%$Ce_{1-x}Zr_xO_2$/SBA-15 and the sol was deposited on the supports coated with the sol. Finally, the 0.5%Pd/SBA-15/Al_2O_3/FeCrAl and 0.5%Pd/5%$Ce_{1-x}Zr_xO_2$/SBA-15/Al_2O_3 /FeCrAl (x=0, 0.5, 1) catalysts were obtained via the same dry and calcination procedure above. XRD patterns were obtained on a Rigaku D/Max 2500 VB2+/PC diffractometer operated at 40 kV and 200 mA with a nickel filtered CuKα radiation. Catalytic activity tests were carried out in a conventional quartz fixed-bed reactor at atmospheric pressure. Methane combustion involves a gas mixture of 2 vol.% CH_4 in air, with a GHSV of 6000 $ml/g_{(cat.+\ washcoat)} \cdot h$. The outlet products were measured with gas chromatography with TC detector.

3. Results and discussion

3.1. Catalytic activity

The catalytic activity of Pd/SBA-15 catalysts is shown in Fig. 1. All the catalysts exhibit higher activities, and methane can be combusted completely below 400 °C. Their activity increases gradually with increasing of Pd content.

Therefore, the 0.1%Pd/SBA-15 catalyst shows the lowest activity. The temperature for 10% conversion of methane (T_{10}) is 295 °C and the temperature for 90% conversion of methane (T_{90}) is 395 °C. The 1%Pd/SBA-15 catalyst exhibits the best activity, the conversion of methane is reached 12.9% when the reaction temperature is 275 °C, and the T_{90} is 348 °C. Fig. 2 shows the activity of the metallic monolithic catalysts. The 0.5%Pd/SBA-15/Al_2O_3/FeCrAl catalyst exhibits an excellent activity. When the $Ce_{1-x}Zr_xO_2$ is used as a promoter, the catalysts show the higher activity and the activity increases with increasing of Zr content. The 0.5%Pd/5%ZrO_2/SBA-15/Al_2O_3/FeCrAl catalyst has the best activity, the T_{10} is 276 °C, and the T_{90} is 396 °C. In a word, for these catalysts with the same Pd loading, the activity of the monolithic catalysts is slightly lower than that of Pd/SBA-15 catalytic under the same temperature likely due to low surface area of the monolithic catalysts.

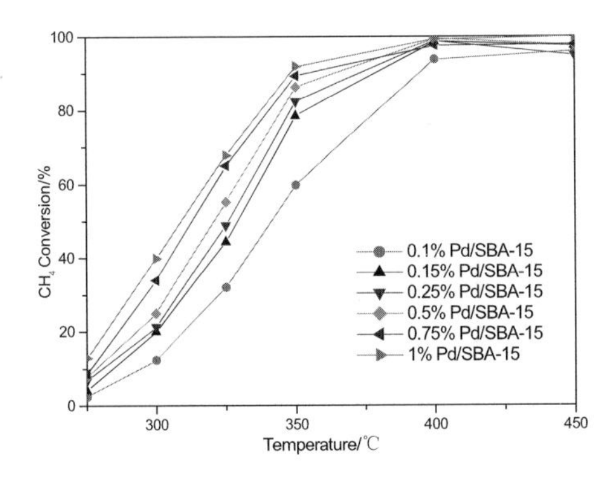

Fig. 1 Methane conversion over the Pd/SBA-15 catalysts

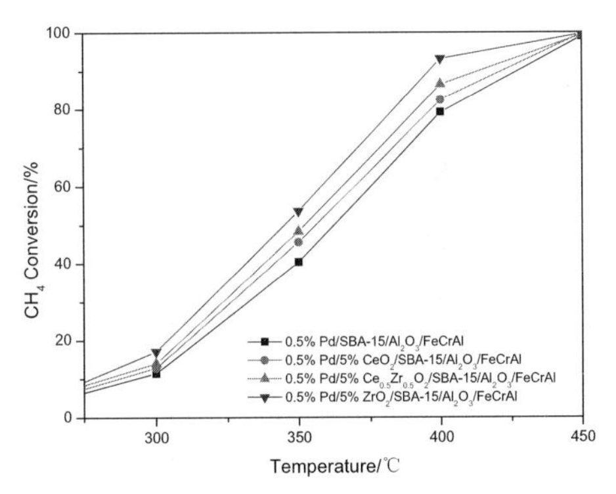

Fig. 2 Methane conversion over the metallic monolithic catalysts

3.2. Stability of catalysts

Fig. 3 shows the results of long-time run test of the prepared catalysts. For 0.5%Pd/SBA-15 catalyst, the conversion of methane begins to decrease when run time is about 20 h, and the conversion is 76.6% after about 92 h. The conversion of methane of 0.5%Pd/SBA-15/Al_2O_3/FeCrAl catalyst begins to decrease when the run time is about 145 h, and the conversion is 75.6% after about 390 h. When $Ce_{0.5}Zr_{0.5}O_2$ is used as promoter, the conversion of methane keeps unchanged during the test. From these results, the 0.5%Pd/SBA-15 catalyst is deactivated rapidly likely related to the decomposition of PdO into metallic Pd [10]. The deactivation of 0.5%Pd/SBA-15/Al_2O_3/FeCrAl catalyst is slower than that of 0.5%Pd/SBA-15 catalyst. This fact is likely related to the addition of FeCrAl supports and of Al_2O_3 with high surface area, which prevents the decomposition of PdO into metallic Pd. However, for 0.5%Pd/5% $Ce_{0.5}Zr_{0.5}O_2$/SBA-15/Al_2O_3/FeCrAl catalyst, the addition of ZrO_2 in CeO_2

forming solid solutions can significantly increase the oxygen storage [11], and Al_2O_3 is also able to stabilize the particle size of the solid solutions [12], hence, this catalyst has an excellent thermal stability.

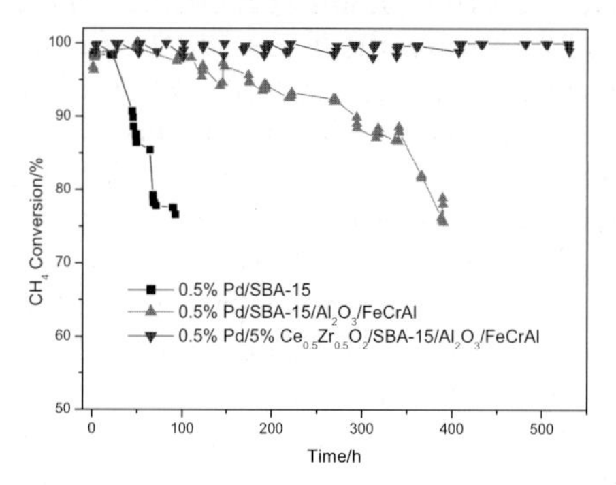

Fig. 3 Methane conversion vs. time on stream at 450 °C

3.3. XRD of catalysts

The small-angle XRD patterns of the Pd/SBA-15 catalysts are shown in Fig. 4A. All the catalysts has a well-defined hexagonal pore structure of SBA-15 [7], however, the (100) diffraction peaks shift to high angle, and their intensity decreases, suggesting that Pd incorporation in SBA-15 has an effect on the pore structure [13]. Fig. 4B shows the wide-angle XRD patterns of Pd/SBA-15 catalysts. There is no PdO species in Pd/SBA-15 catalysts when Pd content is less than 0.25%, suggesting that PdO species is highly dispersed. When Pd content is more than 0.25%, PdO species is detected, and the peak intensity of PdO increases gradually with Pd content increasing further.

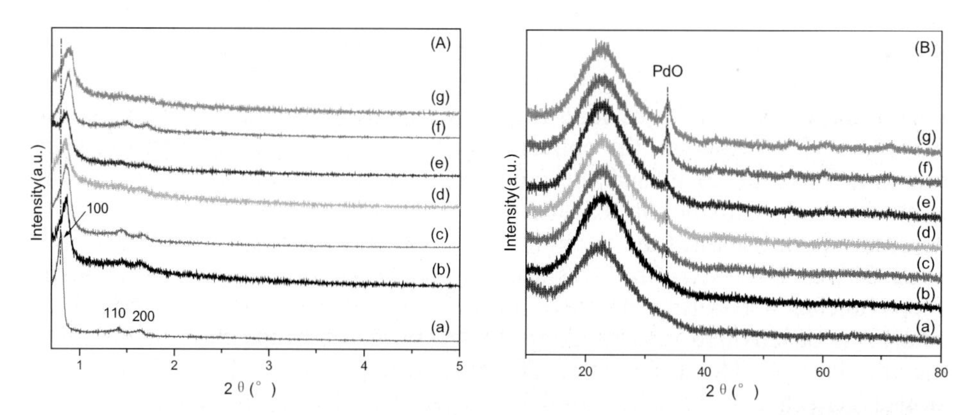

Fig. 4 XRD patterns of Pd/SBA-15 catalysts (A) the small-angle XRD patterns (B) the wide-angle XRD patterns (a) SBA-15; (b) 0.1%Pd/SBA-15; (c) 0.15%Pd/ SBA-15; (d) 0.25%Pd/SBA-15; (e) 0.5%Pd/SBA-15; (f) 0.75%Pd/SBA-15; (g) 1%Pd/SBA-15

The small-angle XRD patterns of the metallic monolithic catalysts are shown in Fig. 5A. All the samples have a well-defined hexagonal pore structure of SBA-15. In comparison with SBA-15/Al_2O_3/FeCrAl, when Pd is supported on the SBA-15/Al_2O_3/FeCrAl, the (100) diffraction peak of the catalyst shifts to high angle, and the intensity decreases likely due to the fact that the corporation of Pd in SBA-15, resulting in a change in pore structure (Fig. 5A-b). However, with the addition of $Ce_{1-x}Zr_xO_2$ the position and the intensity of the (100) diffraction peaks keep almost unchanged (Fig. 5A-c~e). Fig. 5B shows the wide-angle XRD patterns of the samples. The characteristic peaks of FeCr (JCPDS 34-0396) are observed in FeCrAl (Fig. 5B-f). After the heat treatment, besides the characteristic peaks of FeCr, α-Al_2O_3 (JCPDS 88-0826) is formed on the FeCrAl surface (Fig. B-g). The formation of α-Al_2O_3 can improve the combination between SBA-15 washcoat and FeCrAl supports [14]. When 0.5%Pd/SBA-15 is coated on the FeCrAl, PdO is detected (Fig. 5B-b). When CeO_2 is used as a promoter (Fig. 5B-c), the intensity of diffraction peak of PdO keeps almost unchanged, but the diffraction peak of CeO_2 is observed. When $Ce_{0.5}Zr_{0.5}O_2$ is used as a promoter the intensity of peak of PdO keeps almost unchanged, however, the intensity of peak of $Ce_{0.5}Zr_{0.5}O_2$ is weaker (Fig. 5B-d). The addition of Zr forming solid solutions can maintain smaller particle sizes and higher surface area than CeO_2 [11]. Hence, $Ce_{0.5}Zr_{0.5}O_2$ has better dispersion. The diffraction peak of ZrO_2 is not observed when x is 1, suggesting that ZrO_2 is highly dispersed (Fig. 5B-e).

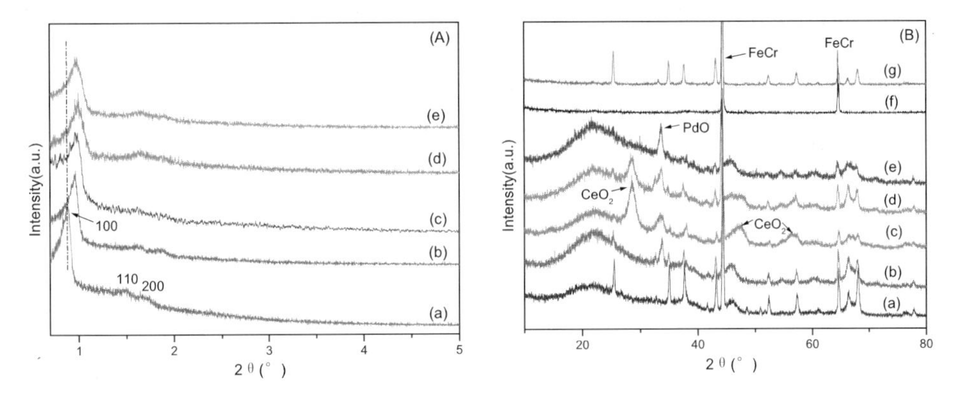

Fig. 5 XRD patterns of the metallic monolithic catalysts (A) the small-angle XRD (B) the wide-angle XRD (a) SBA-15/Al_2O_3/ FeCrAl; (b) 0.5%Pd/SBA-15/Al_2O_3 /FeCrAl; (c)0.5%Pd/5%CeO_2/ SBA-15/Al_2O_3 /FeCrAl; (d) 0.5%Pd/5%$Ce_{0.5}Zr_{0.5}O_2$ /SBA-15/ Al_2O_3/ FeCrAl; (e) 0.5%Pd/5% ZrO_2/SBA-15/Al_2O_3/ FeCrAl; (f) FeCrAl foil; (g) FeCrAl foil pre-oxidized at 950°C for 15 h;

4. Conclusion

The Pd/SBA-15 catalysts exhibited an excellent activity for combustion of methane, and the activity increased with increasing of the Pd loading. The

0.5%Pd/SBA-15/Al_2O_3/FeCrAl monolithic catalyst showed a good activity. When $Ce_{1-x}Zr_xO_2$ is used as a promoter, the activity of the monolithic catalysts increased. Furthermore, the activity increased gradually with increasing of Zr content. The 0.5%Pd/SBA-15 catalyst is deactivated rapidly. The stability of the 0.5%Pd/SBA-15/Al_2O_3/FeCrAl and 0.5%Pd/5%$Ce_{0.5}Zr_{0.5}O_2$/SBA-15/Al_2O_3/ FeCrAl catalyst was much better than that of 0.5%Pd/SBA-15 catalyst. $Ce_{0.5}Zr_{0.5}O_2$ could significantly increase the stability of the catalysts, and the stability of the monolithic catalysts was also likely relative to Al_2O_3 and FeCrAl supports. The prepared Pd/SBA-15 catalysts and the monolithic catalysts had a well-defined hexagonal pore structure of SBA-15.

Acknowledgements

Financial funds from the Chinese Natural Science Foundation (Project Nos. 20376005, 20473009) and the National Basic Research Program of China (Project No. 2005CB221405) are gratefully acknowledged.

References

1. C. Lahousse, A. Bernier, P. Grange, B. Delmon, P. Papaefthimiou, T. Ioannides, X.Verykios, J. Catal. 178 (1998) 214.
2. D. Ciuparu, M. R. Lyubovsky, E. Altman, L. D. Pfefferle, A. Datye, Catal. Rev. 44 (2002) 593.
3. M. Lyubovsky, L. Pfefferle, Appl. Catal. A 173 (1998) 107.
4. C. F. Cullis, B. M. Willatt, J. Catal. 83 (1983) 267.
5. L. F. Liotta, A. Macaluso, A. Longo, G. Pantaleo, A. Martorana, G. Deganello, Appl.Catal. A 240 (2003) 295.
6. B. H. Yue, R. X. Zhou, Y. J. Wang, X. X. Han, X. M. Zheng, Applied Surface Science 246 (2005) 36.
7. D. Zhao, J. Feng, Q. Huo, N.Melosh, G. H. Fredrickson, B. F. Chmelka, G. D. Stucky, Science, 279 (1998) 548.
8. T. A. Nijhuis, A. E. W. Beers, T. Vergunst, I. Hoek, F. Kapteijn, J. A. Moulijn, Catalysis Reviews 43 (2001) 345.
9. S. Roy, A. K. Heibel, W. Liu, T. Boger, Chemical Engineering Science 59 (2004) 957.
10. P. Euzen, J. L. Gal, B. Rebours, G. Martin, Catal. Today, 47 (1999) 19.
11. C. E. Hori, H. Permanna, K. Y. Simon Ng, A. Brenner, K. More, K.M. Rahmoeller, D. Belton, Appl. Catal. B 16 (1998) 105.
12. A. I. Kozlov, D. H. Kim, A. Yezerets, P. Aandersen, H. H. Kung, M.C. Kung, J. Catal. 209 (2002) 417.
13. X. K. Li, W. J. Ji, J. Zhao, Z. B. Zhang, C. T. Au, J. Catal. 238 (2006) 232.
14. X. Wu, D. Weng, S. Zhao, W. Chen, Surface and Coatings Technology, 190 (2005) 434.

Natural Gas Conversion VIII
F.B. Noronha, M. Schmal, E.F. Sousa-Aguiar (Editors)

Catalytic Combustion of Methane over $PdO-CeO_2/Al_2O_3$ and $PdO-CeO_2/ZrO_2$ catalysts

Daniela Domingos, Lílian M. T. Simplício, Genicleide S. Estrela, Marcionila A. G. dos Prazeres and Soraia T. Brandão.

Universidade Federal da Bahia, Instituto de Química. Rua Barão de Geremoabo, s/n, Campus Universitário de Ondina, CEP.: 40170-290, Salvador-Ba.

1. Introduction

Hydrocarbon combustion is one of the most important processes for heat and energy generation; however, it results in polluting emissions of nitrogen oxides and carbon monoxide. Methane is one of the most commonly used fuel, and the development of more efficient and less pollutant processes of its combustion (complete combustion at lower temperatures) is a priority. In this context catalytic combustion appears to offer an alternative to flame combustion due to its ability to promote combustion at lower temperatures than conventional flame combustion, thus reducing residual emissions of NOx, CO and unburned hydrocarbons (1). Noble metal catalysts are considered extremely active for the total oxidation of hydrocarbons and palladium is described in literature as the most active for methane combustion. The active phase of noble metal catalysts is usually dispersed on supports such as SiO_2, Al_2O_3, ZrO_2, TiO_2 and zeolites (2-8). Alumina is the most commonly used because of its high surface area while zirconia presents high thermal stability due its oxygen rich surface (9). This later is very interesting for palladium based catalysts because it promotes PdO thermal stability, which is the active phase for methane combustion. This work describes the preparation, characterization and activity tests in methane combustion using palladium and cerium supported on alumina and zirconia catalysts.

2. Experimental

PdO/Al_2O_3, PdO/ZrO_2, $PdO\text{-}CeO_2/ZrO_2$, $PdO\text{-}CeO_2/Al_2O_3$ and $PdO/Al_2O_3/ZrO_2$ were prepared using wet impregnation method. CeO_2 modified supports were prepared with zirconia and alumina impregnation with a cerium nitrate solution for 24 hours. After impregation, the materials were dried at 110°C for 24 hours and calcined at 600°C for 10 hours. $PdO\text{-}CeO_2/ZrO_2$, $PdO\text{-}CeO_2/Al_2O_3$ were prepared by impregnation of the modified ceria supports with palladium acetylacetonate solution in toluene for 24 hours. After impregnation the materials were dried at 110°C for 24 hours and calcined at 600°C for 10 hours.

Alumina (Pural SB) and zirconia were used as supports. Palladium acetylacetonate (Achros Organics) and cerium nitrate (Fluka) were used in amounts to result in 1% w/w of Pd and 6% w/w of Ce in the catalysts. All the samples were calcined at 600°C for 10 hours under air flow and characterized by X-ray diffraction (XRD), O_2 temperature programmed desorption (TPD), specific surface area (BET) and X- ray fluorescence (XRF).

The X-ray diffraction measurements were carried out on a Shimadzu apparatus (XRD-6000). The Cu Kα radiation (λ = 1.5418 Å) and the following experimental conditions were used: 2θ range = 10-80°, step size = 0.02° and time per step = 4.80 s. The powder samples were analyzed without further treatment.

The palladium content in the samples was obtained by X-ray fluorescence on a Shimadzu WDS (XRF-1800) apparatus.

The reduction-oxidation properties of the catalysts were studied by temperature programmed desorption (TPD) of oxygen in a flow system. 1000 mg of the catalyst was loaded in a quartz reactor and heated at a rate of $5°C.min^{-1}$ from room temperature to 1000°C, under helium flow. Oxygen in the reactor outlet was detected by using a quadrupole mass spectrometer, Balzers QMS-200. The mass-to-charge ratio (m/e) = 32 was used to monitor the oxygen concentration.

Catalytic properties were evaluated using temperature programmed surface reaction (TPSR) feeding a mixture containing 0.5% of CH_4, 2% of O_2 and 97.5% of N_2. Transient flow microreactor measurements were performed in a quartz reactor fixed with 100 mg of catalyst diluted in 100 mg of quartz granules. The samples were heated to 600°C at a rate of $10°C.min^{-1}$, under a flow of 100 $mL.min^{-1}$ of the gas mixture. The reactant gases were fed to the reactor by means of electronic mass flow meter controller MKS 247. The effluent gases were detected by using quadrupole mass spectrometer, Balzers QMS-200, connected in the reactor outlet. The following mass-to-charge ratios (m/e) were used to monitor the concentrations of the products and reactants: 15 and 16 (CH_4), 18 (H_2O), 28 (CO), 32 (O_2), 44 (CO_2).

3. Results and Discussion

XRD patterns of all catalysts (Figure 1) presented the characteristic peaks of the active phase, PdO, in 2θ = 33.8 and 60°. PdO/Al_2O_3 and $PdO\text{-}CeO_2/Al_2O_3$ patterns presented peaks in 2θ = 19, 36, 39, 45, 60 and 67° confirming the presence of the γ-alumina phase of the support.
For PdO/ZrO_2 and $PdO\text{-}CeO_2/ZrO_2$ catalysts, the characteristic peaks of the zirconia monoclinic phase in 2θ = 17, 24, 28, 31, 34, 50 and 55° were observed and another peak at 2θ = 30°, which is characteristic of the tetragonal phase. According to literature, the calcination temperature, 600°C, used here makes the coexistence of these two zirconia phases possible (10). All phases described elsewhere were also found in the $PdO/Al_2O_3/ZrO_2$ catalyst. As well as the tetragonal form of PdO, XRD patterns of $PdO\text{-}CeO_2/ZrO_2$ and $PdO\text{-}CeO_2/Al_2O_3$ catalysts, the presence of cubic CeO_2, with reflection peaks at 2θ = 28° and 56° was detected.

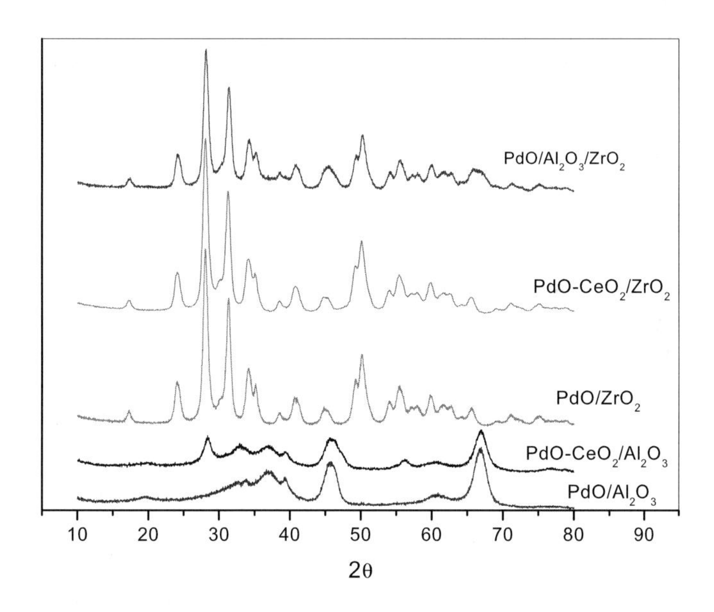

Figure 1. XRD patterns of the catalysts

BET and XRF results analysis are presented in Table 1. The XRF indicated that Pd concentrations are approximately 1% w/w. BET analysis indicated that the catalysts supported on alumina have higher specific surface areas than the zirconia supported catalysts.

Table 1. Palladium content and specific surface area of the catalysts.

Catalysts	Pd (%)	Specific surface area ($m^2.g^{-1}$)
PdO/Al_2O_3	1.00	178.6
PdO/ZrO_2	0.85	67.8
$PdO\text{-}CeO_2/Al_2O_3$	0.84	158.9
$PdO\text{-}CeO_2/ZrO_2$	0.96	68.3
$PdO/Al_2O_3/ZrO_2$	0.91	128.8

TPD analysis (Figure 2) showed that on catalysts containing ZrO_2, PdO decomposition temperatures were higher than those supported on alumina.

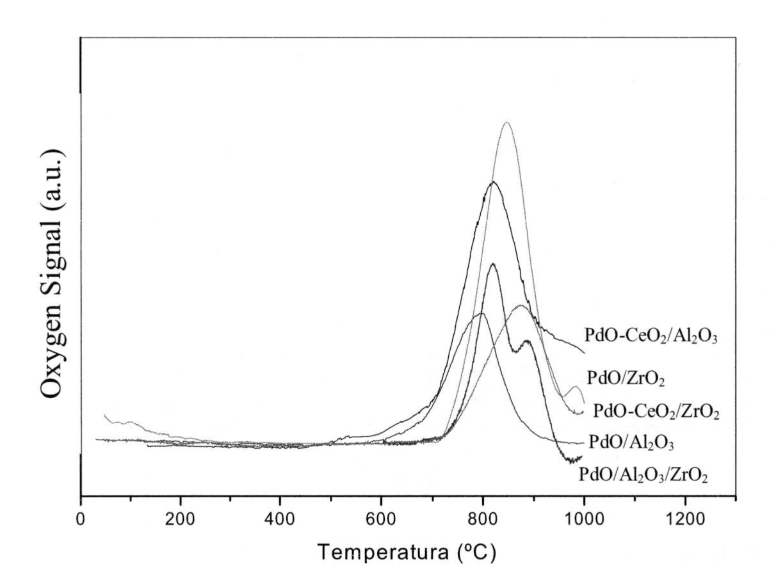

Figure 2. TPD profiles of the catalysts.

The PdO/ZrO_2 catalyst, for example, presented a peak of O_2 desorption at 847°C, while the PdO/Al_2O_3 presented it at 797°C. This behavior is confirmed by comparing catalysts $PdO\text{-}CeO_2/Al_2O_3$ and $PdO\text{-}CeO_2/ZrO_2$ and indicates that at high temperatures zirconia increases PdO stability, which can be attributed to the oxygen storage capacity of zirconia (9).
The cerium effect on PdO decomposition temperatures was also observed. TPD of $PdO\text{-}CeO_2/Al_2O_3$ and $PdO\text{-}CeO_2/ZrO_2$ presented O_2 desorption peaks which shifted to higher temperatures than PdO/Al_2O_3 and PdO/ZrO_2. This indicates that cerium promotes PdO stabilization. This stabilization can be attributed to CeO_2 ability to shift easily between reduced and oxidized state (Ce^{+4}/Ce^{+3}), which provides O_2. This ability and the high mobility of oxygen species contributes to the stabilization of palladium oxide phase.
Figure 3 illustrates the profiles of methane conversion as a function of temperature. The results indicated that all catalysts are active for methane combustion. As can be observed, alumina catalysts were the most active presenting the lowest ignition temperatures.

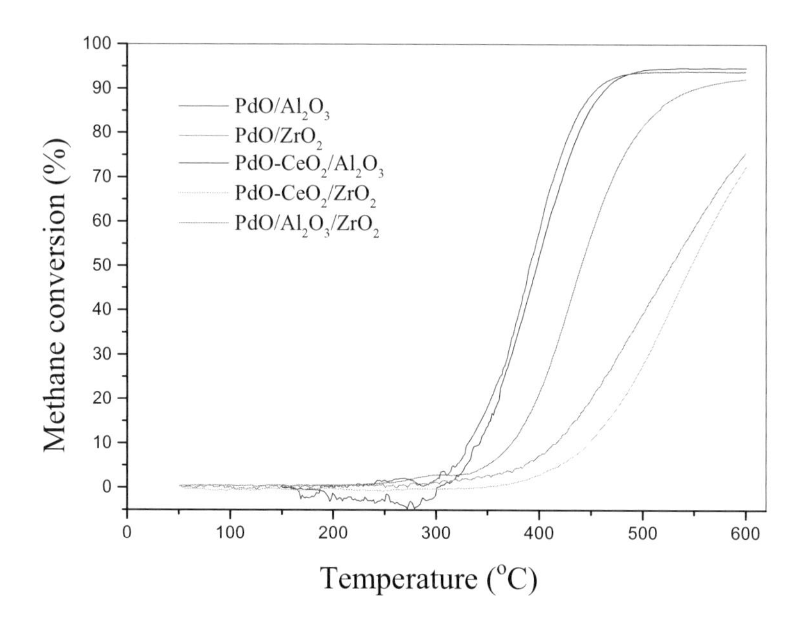

Figure 3. TPSR profiles of the catalysts

Catalysts supported on ZrO_2 were less active, showing the highest ignition temperatures. On the other hand, the catalyst containing zirconia and alumina at a ratio of 1:1 presented an ignition temperature of 372°C, indicating that the presence of alumina promotes methane ignition at lower temperatures. As regards the lower activity, PdO phase on zirconia catalysts are more stable, which suggests that these catalysts are better suited to catalytic combustion of methane at higher temperatures.

4. Conclusion

The results obtained in this work indicated that the catalysts prepared in this work showed to be very active for methane combustion. PdO/Al_2O_3 and $PdO-CeO_2/Al_2O_3$ were the most active ones. Besides its lower activity in this temperature range, zirconia catalysts are promising for high temperature combustion due to its high thermal stability. Combining alumina and zirconia supports can be an alternative to improve activity and PdO thermal stability. The later can also be promoted by the addition of CeO_2 in the catalyst formulation.

5. References

1. P. Forzatti and G. Groppi, Catal. Today 54 (1999) 165.
2. T. R. Baldwin and R. Burch, Appl. Catal. 66 (1990) 359.
3. J. H. Lee and D. L. Trimm, Fuel Processing Technology 42 (1995) 339.
4. K. Muto et al., Appl. Catal. A: Gen. 134 (1996) 203.
5. S. Yang et al., Appl. Catal. B: Env. 28 (2000) 223.
6. K. Eguchi and H. Arai, Appl. Catal. A: Gen. 222 (2001) 359.
7. K. Okumura et al., Appl. Catal. B: Env. 140 (2003) 151.
8. B. Yue et al., Appl. Surf. Sci. (2004) in Press.
9. J. P. P. Porto et al., Levantamento de dados cinéticos de reações de hidrogenação catalítica utilizando planejamento experimental, ANP – Agência Nacional do Petróleo, 2005.
10. B.H.J. DAVIS, Am. Ceram. Soc 67 (1984) 168.

Natural Gas Conversion VIII
F.B. Noronha, M. Schmal, E.F. Sousa-Aguiar (Editors)

Methane conversion to chemicals, carbon and hydrogen (MCCH) over modified molybdenum-NAS catalysts

M.W. Ngobeni[a], M.S. Scurrell[a*] and C.P. Nicolaides[b*]
scurrell@chem.wits.ac.za; chris.nicolaides@sasol.com
[a]*Molecular Sciences Institute, School of Chemistry, University of the Witwatersrand Johannesburg, South Africa 2050; tel.:+27 11 717 6716, fax: +27 11 717 6749*
[b]*Sasol Technology, Research & Development Division, P.O. Box 1, Sasolburg, South Africa 1947*
tel.: +27169604318, fax: +2711522 4142

SUMMARY The catalytic behaviour of Mo-NAS (novel aluminosilicates – zeolite-based) for MCCH reactions has been probed, with focus on the aliphatics/aromatics ratio in the gas-phase products, the aromatics/coke selectivity and the overall activity of the catalysts. In addition, the %XRD crystallinity of the zeolite used, effects of alkali-metal doping and effects of pre-silanation of the parent zeolite on methane dehydroaromatization have been evaluated.

INTRODUCTION The concept of using natural gas for both fuel and chemical feedstock use is not new but the details of suitable process options remain to be examined. One approach involves methane conversion to chemicals (aliphatics and aromatics), carbon and hydrogen (MCCH) as a route to carbon-free fuel (i.e. hydrogen) and carbon-containing non-fuel products. Mo/zeolites offer potential here [1-4] for conversion of methane under anaerobic conditions. The economic sacrifice in free energy available by restricting fuel use to hydrogen, could, in principle, be off-set by the intrinsic value of the carbon-rich products deposited on the catalyst, some of which possess interesting structures at the nano-scale [5]. As an option, some of the carbon, particularly the more active forms could be fed into suitable direct carbon fuel cells [6] so that even though some carbon dioxide would be produced, the overall energy efficiency using both hydrogen and carbon fuel cells would be expected to be higher than conventional natural gas combustion systems. The Mo/ZSM-5 is a well studied system, but most work has considered issues such as catalyst activity and stability, and the drive in most studies has been the production of gaseous aromatic products. Our work has focused on the selectivity aspects of the three major products, aliphatics and aromatics in the gas-phase and the carbon-rich deposits formed on the solid catalyst used. In this paper, the control of selectivity to gas-phase aliphatics versus gas-phase aromatics, and the selectivity gaseous hydrocarbons versus solid carbon-containing deposits are considered. The degree of control on these selectivities exerted by means of catalyst modification is assessed, using variables such as the %XRD crystallinity of the parent zeolite, silanation of the zeolite and doping with alkali metal ions. The aim has been to provide a basis for evaluating the practical viability of such a catalyst system, whilst simultaneously probing further the mode of action of these types of catalysts in natural gas conversion.

EXPERIMENTAL Mo-NAS (novel aluminosilicates - ZSM-5-based [7]) catalysts were prepared with 2 mass% Mo, by impregnation with an aqueous solution of ammonium heptamolybdate. Parent %XRD crystallinity was varied between 5% and 97%. Some NAS samples were impregnated with alkali metal cations (lithium, sodium or potassium) ions before molybdenum was introduced, and some of the NAS preparations were subjected to silanation treatments using 3-aminopropyltriethoxysilane in ethanol, followed by calcination at 500°C, before the molybdenum was introduced. Initial attempts to characterize the chemical state of the Mo present have also been made using uv-vis reflectance, temperature programmed reduction and temperature programmed oxidation of used catalysts. Methane conversion was studied at 750°C and 0.075 MPa at a GHSV at 800 h^{-1}, using a conventional quartz micro-reactor system. Analysis of products focused on the gas-phase components present in the exit stream, with conversion levels being determined using argon as an internal standard (present at 10 vol%) in the inlet methane.

RESULTS AND DISCUSSION. As the %XRD crystallinity increased in the series 5, 17, 47, 57 and 73, the BET surface area rose with values of 177, 217, 333, 374 and 391 m^2g^{-1} respectively, in line with the expected development of the intrazeolitic voids. In general, activity increases slightly with increasing %XRD crystallinity up to a broad maximum at about 20-50% crystallinity and then falls (Figure 1). The cystallinity dependence is not as marked as that observed in other closely related systems such as the Zn- and Ga-NAS systems examined by us earlier [8,9], but some trends are similar, in that low crystallinity samples tend to have lower activity, as do samples having high XRD crystallinity. At low crystallinity, catalysts are usually much less active

because it is known that the both the zeolitic domains are poorly developed and the aluminium content relatively low [7]. The lower activities at high crystallinities are associated with diffusional restrictions due to the increased crystallite size then normally encountered [8,9]. In the case of Mo/NAS, however, the activity of the low crystallinity sample is only slightly lower than those of the medium-crystallinity samples, which is consistent with the fact that a well-developed zeolitic structure is not necessarily required for methane activation [10]. Even poorly crystalline samples of parent zeolite can lead to reasonably active dehydroaromatization catalysts. Crystallinity does, however, have a marked effect on the aromatics versus coke selectivity (Figure 2). Aliphatic C_2 selectivity is only weakly dependent on crystallinity levels. The nature of the coke deposited on these catalysts is revealed by direct observations via transmission electron microscopy (Figure 3) which clearly indicates that nanotubular structures are present in relatively high volume. We use the term "coke" in this work to refer to any carbon-containing residue which reports to the catalyst bed under our conditions and is likely to comprise binuclear and polynuclear aromatics, polyolefinic species, and carbon of various types.
Silanation [11] of the parent zeolite improved the catalytic conversion of methane at the expense of catalyst stability (Figure 4). This result emphasizes that the poisoning of external sites is much more prevalent in this system than in other acid-catalyzed hydrocarbon conversions on zeolites, and that the removal of external hydroxyls can do little to prevent coke formation. Doping with alkali metal ions affects catalytic activity and selectivities but a simple chemically-based trend is not seen for Li^+, Na^+ and K^+ ions added at a Mo/alkali metal ion molar ratio of 0.5 (Table 1). The incorporation of lithium ions in particular renders the catalysts relatively inactive.

The reducibility of Mo (initially in the +6 state, as confirmed by uv-vis reflectance spectrscopy) is affected by the nature of the added alkali metal ion, but in the case of Na^+ we note that the reducibility is very similar to that of the untreated Mo/NAS system. In the case of potassium or lithium ions significant differences in the TPR traces are observed. Turning to the TPO traces, for the types of catalyst under study here, these normally reveal two major peaks associated with the oxidation of the coke deposited on Brønsted acid sites and the re-oxidation of molybdenum carbides phases respectively [12]. The small intensity of the lower temperature component for the Li^+ treated catalyst suggests that the molybdenum carbide may not be present at significant levels [12] and its formation is severely disrupted by the presence of Li^+, ions consistent with the reduced activity of this catalyst A complete understanding of the TPR and TPO data does, however, demand the use of additional characterization techniques. The inhibition of methane dehydroaromatization by lithium has been observed elsewhere [13].

The essential data obtained on MCCH over molybdenum-NAS catalysts shows that the approach is clearly viable provided that process problems (especially coke fouling of active sites and/or catalyst regeneration) can be overcome.

ACKNOWLEDGEMENT We express thanks for the financial support provided by Sasol Technology and the University of the Witwatersrand Johannesburg via the Carnegie-Mellon Foundation

REFERENCES

[1] L. Wang, L. Tao, M. Xie and G. Xu, Catal. Lett., **21** (1993) 35
[2] Y. Xu, X. Bao and L. Lin, J. Catal., **216** (2003) 386
[3] Y. Xu and L. Lin, Appl. Catal., **188** (1999) 53
[4] Y. Shu and M. Ichikawa, Catal. Today, **71** (2001) 55
[5] M.W. Ngobeni, M.S. Scurrell, and C.P. Nicolaides, to be published
[6] S.Zecevic, E.M. Patton and P. Parhami, Direct Carbon Fuel Cell Workshop, NETL-Pittsburgh, PA, July 30, 2003; http://www.netl.doe.gov/publications/proceedings/03/dcfcw/dcfcw03.html, accessed 23 Nov 2006.
[7] K.S Triantafyllides, L. Nalbandian, P.N. Trikalitis, A.K. Ladakos, T. Mavromoustakos and C.P. Nicolaides, Microporous and Mesoporous Mater., **75** (2004) 89
[8] C.P. Nicolaides, N.P. Sincadu and M.S. Scurrell, Stud. Surf. Sci. Catal., **136**, (2001) 333
[9] C.P. Nicolaides, N.P. Sincadu, and M.S.Scurrell, Stud. Surf Sci. Catal.,**154** (2004) 2347
[10] F. Solymosi and A. Szöke, Catal. Lett. **39** (1996) 157
[11] L. Wang, R.Ohnisji and M. Ichikawa, Catal Lett **62** (1999) 29
[12] Y Shu, R. Ohnishi and M. Ichikawa, Catal Lett. **81** (2002) 9
[13] L. Chen, L.Lin, Z.Xu, X.Li and T. Zhang, J. Catal. **157** (1995) 190

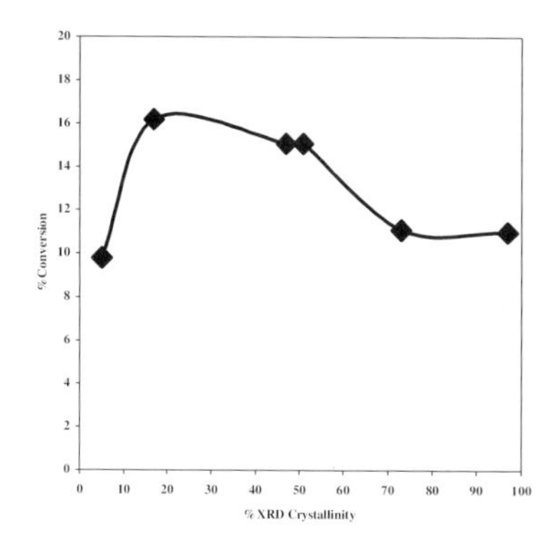

Figure 1: The effect of %XRD crystallinity on the conversion of methane at 750°C on the conversion at 1h time on stream

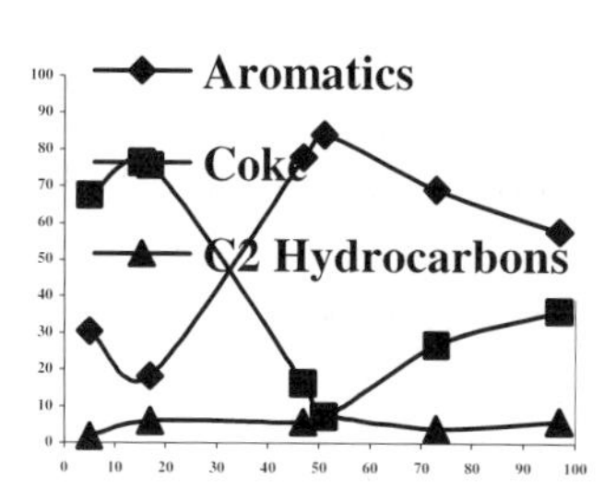

Figure 2: The effect of %XRD crystallinity of H-ZSM-5 on the selectivity to various products over 2%Mo/H-ZSM-5 at 750 °C and at 10% methane conversion.

Table 1: The effect of the addition of alkali metal ions on the catalytic performance of the 2%Mo/H-ZSM-5 for methane conversion at 750 °C and at 1 h on-stream

sample	%conversion	%C selectivity		
		C2's	aromatics	coke
No dopant	16.8	2.2	48.1	49.7
Li	2.8	7.8	5.0	87.2
Na	7.7	10.0	36.2	53.7
K	6.6	3.5	37.5	59.0

Figure 3: A TEM image of the carbon nano structures obtained in the conversion of methane at 750°C

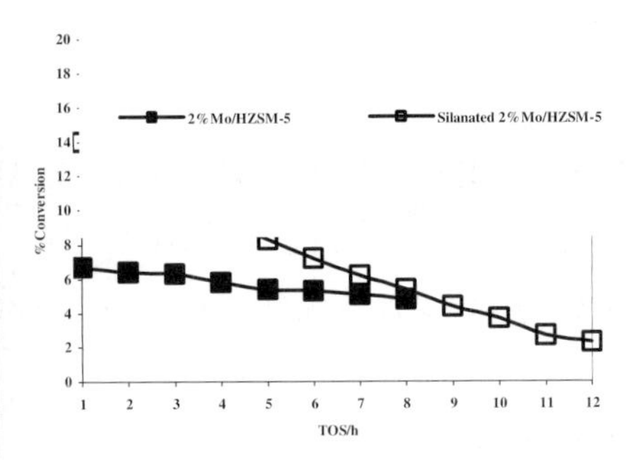

Figure 4: Comparison of the catalytic activity of silanated and non-silanated catalysts

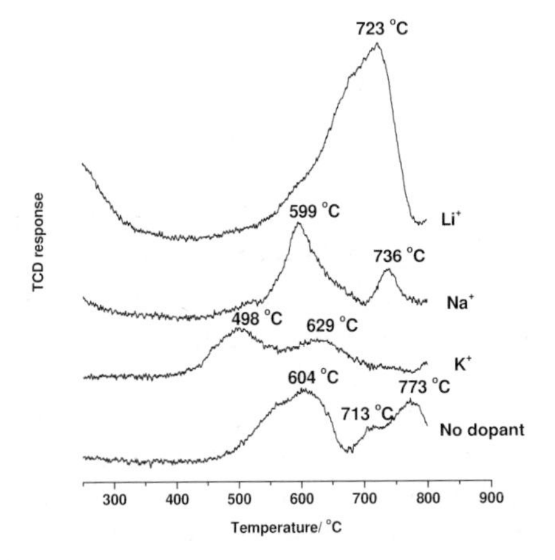

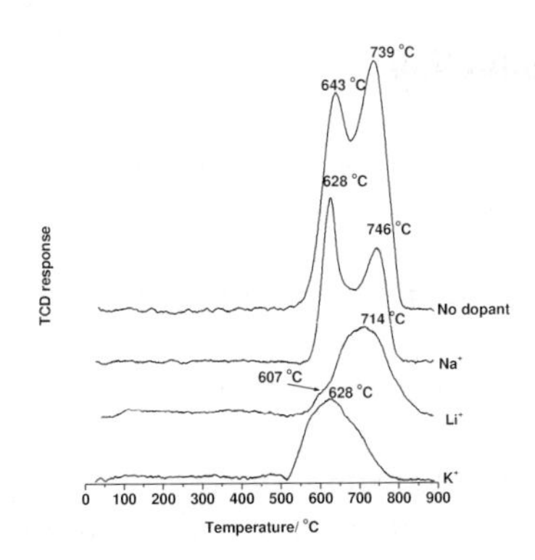

Figure 6: TPO profiles of used 2%Mo/ZSM-5 catalyst with and without alkali metal ion incorporation.

Figure 5: H_2-TPR profiles of 2%Mo/ZSM-5 catalyst with and without alkali metal ions incorporation.

Natural Gas Conversion VIII
F.B. Noronha, M. Schmal, E.F. Sousa-Aguiar (Editors)

Oxidative coupling of methane in a catalytic membrane reactor: Impact of the catalyst-membrane interaction on the reactor performance.

Stephane Haag,[a] Magdalena Bosomoiu,[a] Andre C. van Veen,[a] Claude Mirodatos[a]

[a]*Institut de Recherches sur la Catalyse – CNRS, 2 avenue Albert Einstein, 69626 Villeurbanne Cedex, France*

Results for the oxidative coupling of methane to higher hydrocarbons in dense membrane reactors with surface catalysts are reported. It is demonstrated that the concept is viable, but potential improvements are identified enhancing the activity of the oxidative coupling catalyst for highly permeable membranes.

1. Introduction

Abundant resources of natural gas available at reasonable prices and the foreseeable shortage of petroleum reflected by a recent rise of prices to all-time highs stimulate renewed interest in the oxidative coupling of methane (OCM). Despite past extensive research efforts, suggesting mainly concepts based on fixed bed reactors, the need for sustainable solutions requires reviewing alternative reactor concepts with inherent need for adapted catalysts. One promising design solution could be a catalytic membrane reactor allowing potentially the use of air instead of oxygen. However, published works [1] used either porous membranes with fixed bed catalysts or dense membranes without catalysts having the obvious disadvantages to employ either the membrane only for oxygen distribution (feeding catalysts working in a conventional fashion) or comparably low productivity, respectively. Our previous work [2] indicated the interest to use membranes feeding oxygen directly the catalyst layer without passing via the gas-phase, thus allowing the dense perovskite membrane to provide activated oxygen to the catalyst for the methane oxidation.

2. Experimental

Information on oxygen permeability, key property of dense mixed conducting membranes, was acquired in the temperature range of 800 to 1000°C.

Reconstituted air was fed to the oxygen-rich membrane compartment, while helium was used as a sweep gas on the permeate side. The membrane disk (about 1 mm thick) was sealed with gold rings between two dense alumina tubes (outer diameter: 12 mm, inner diameter: 8 mm) heating to at least 900°C for 48h. The total pressure at the oxygen-rich side was adjusted to 1.1 bar passing a constant total flow rate of mixed O_2 and N_2 streams controlled by mass flow controllers. Gas chromatography (HP 5890 Series II, 13X packed column) allowed complete analysis of gases at both sides of the membrane. For the catalytic tests, an additional mass flow controller injected methane into the helium flow used as carrier gas on the reactant side. Maximum flow rates were 85 $mL.min^{-1}$ for the CH_4/He mixture and 100 $mL.min^{-1}$ for reconstituted air, respectively. Various feed rates, partial pressures of O_2 and CH_4 and a suitable temperature range were explored. Experiments with a non-optimized but temperature stable Pt/MgO catalyst were conducted in the temperature range from 800 to 1000°C. A second gas chromatograph (TCD with HayeSep D packed column) was used to analyze carbon containing gases.

3. Results and Discussion

3.1. Comparison of the permeability for different membranes

Performances of a $Ba_{0.5}Sr_{0.5}Co_{0.8}Fe_{0.2}O_{3-\delta}$ (BSCFO) membrane have been reported [2] using 100 $mL.min^{-1}$ O_2/N_2 mixture as feed and 50 $mL.min^{-1}$ of He as sweep gas on the permeate side. The oxygen fluxes for the activated permeation process and an estimation of the activation energy were presented and indicated a change in the rate determining step, i.e. for $T < 725$°C, the permeation of the oxygen was limited by surface steps and for $T > 725$°C, the diffusion through the bulk becomes the rate limiting step.
In the case of a $BaBi_{0.4}Fe_{0.6}O_x$ (BBFO) membrane [3], the reconstituted air feed was 100 $mL.min^{-1}$, while the He flow rate was 100 $mL.min^{-1}$. The oxygen permeation fluxes (Arrhenius plot) and an estimation of the activation energy are presented in Figure 1 (left side). Oxygen permeation fluxes are significantly lower than those obtained for the BSCFO membrane and are comparable to values reported in literature for a $La_{0.6}Sr_{0.4}Co_{0.6}Fe_{0.4}O_{3-\delta}$ (LSCFO) membrane [4]. The main drawback with this kind of membrane is its lack of thermal stability. In fact, the bulk BBFO membrane is not stable at temperatures above 960°C and cracking or partial melting of the disk is observed when increasing the temperature beyond that point. This is a substantial limitation for its utilization in a membrane reactor especially for the high temperature OCM reaction.
The last membrane tested was a $Ba_{0.5}Sr_{0.5}Mn_{0.8}Fe_{0.2}O_{3-\delta}$ (BSMFO) disk. The reconstituted air feed was 100 $mL.min^{-1}$ and the sweep flow was 100 $mL.min^{-1}$ He. The oxygen permeation fluxes and an estimation of the activation energy

are presented in Figure 1 (right side). The oxygen permeation flux increases with increasing temperature but to a much smaller extend compared to the former described membranes. The comparably high activation energy at high temperatures indicates that the oxygen diffusion through the bulk of the BSMFO membrane is much more difficult compared to the other perovskite samples. This behavior could relate to a low oxygen mobility in the volume of the membrane caused by the relatively small amount of vacancies present in the structure. However, further studies, e.g. the determination of the non-stoichiometry as a function of temperature, are required to confirm.

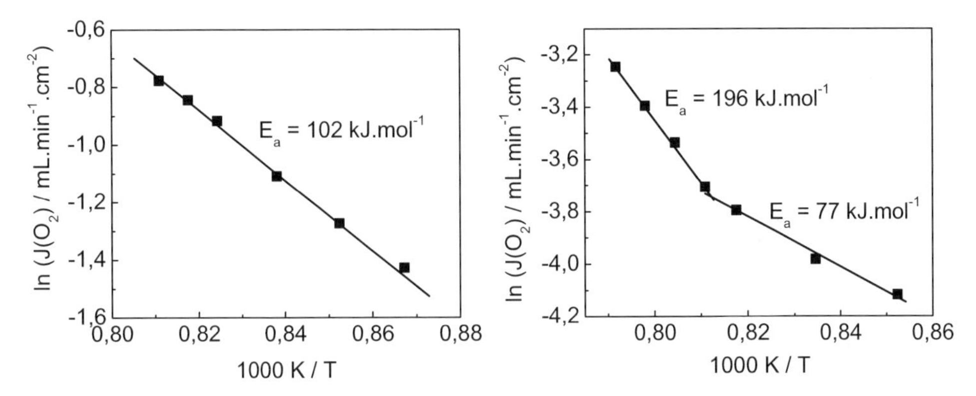

Figure 1: Arrhenius plot of the oxygen permeation for a BBFO (left) and BSMFO (right) membrane using a reconstitued air: 100 mL.min^{-1} / He: 100 mL.min^{-1} gradient

3.2. Comparison of the sol-gel and wash-coat catalyst performance

A catalytic surface modification of the membrane was performed to improve efficiency of the membrane reactor, i.e. to enhance selectively CH_4 conversion to C_2 products. In this work, MgO was chosen as support and platinum was selected as highly stable active species. Two routes were tested for the deposition of the oxide, a sol-gel and a wash-coating method, followed by calcination at 800°C. Scanning electron microscopy (Fig. 2a, left) revealed that about half of the membrane surface is coated by MgO applying the sol-gel method. At one magnitude lower magnification it can be observed that, the sol-gel method yields non-uniform deposition of the catalyst whereas a wash coat layer covers well the whole surface (Fig. 2b, right). Platinum was added by impregnation, using tetraammine platinum(II) nitrate as metal precursor (Sigma-Aldrich).

Permeation tests were performed with both modified membranes and compared to results obtained to a bare membrane. In fact, oxygen fluxes are similar for the bare and the sol-gel modified membranes while the wash-coat modified membrane shows an about 2.5 times higher oxygen permeability (Fig. 3).

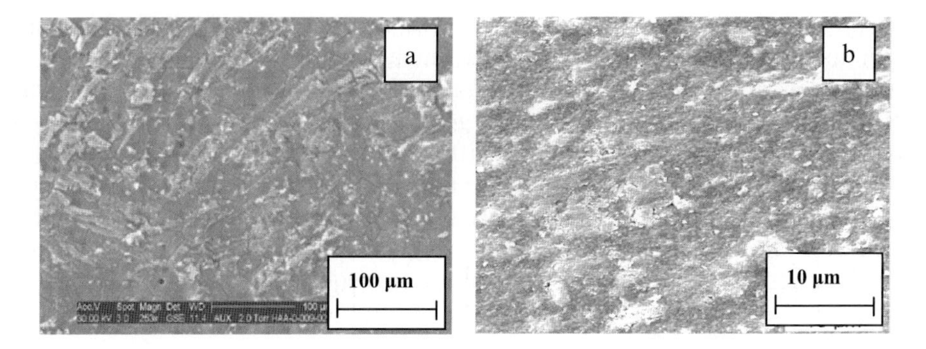

Figure 2: SEM of a surface modified BSCFO membrane by a sol gel method (a, left) and by a wash-coat method (b, right)

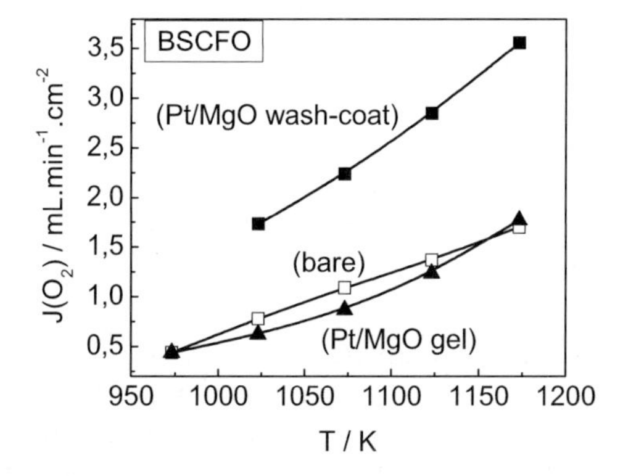

Figure 3: Oxygen permeation through BSCFO-based modified membranes (reconstitued air: 50 $mL.min^{-1}$ / He: 100 $mL.min^{-1}$)

3.3. Comparison of the performance of BSCFO and BSMFO membranes

The catalytic tests for the OCM used both kinds of modified BSCFO membranes. The amount of platinum on the MgO support was estimated to be around 2%. Surprisingly, a non-negligible amount of gaseous oxygen is found on the reactant side for the sol-gel prepared sample while it was not the case for the wash-coat modified one. Given the higher oxygen permeability of the wash-coated sample, the disadvantage of the sol-gel catalyst is less related to an insufficient methane activation but to a shortcoming in contacting oxygen and methane. Obviously, this is caused by the imperfect membrane coverage in the sol-gel case where oxygen desorbs from the membrane material without intimate contact to the catalyst. Conversion of methane was higher with the wash-coat catalyst but selectivity was lower than that obtained for the sol-gel catalyst (Fig. 4). Higher conversions could relate to a better spread of the catalyst on the membrane while the decreased selectivity relates probably to a higher degree of oxidation of the catalyst as demonstrated by the enhanced oxygen permeation.

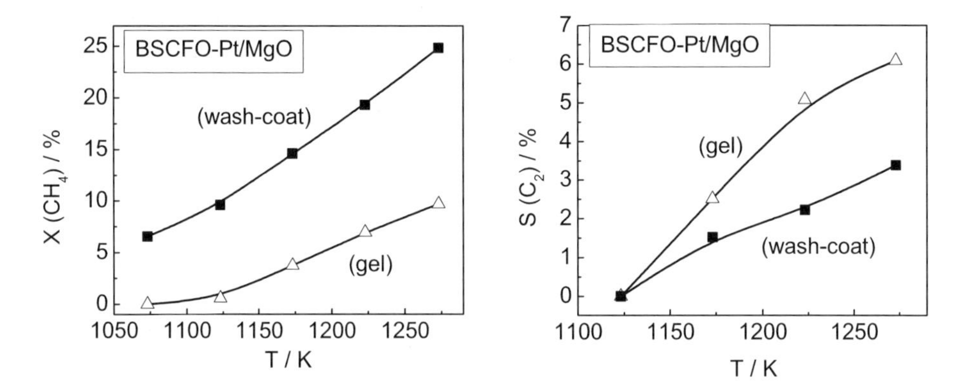

Figure 4: Comparison of CH_4 conversion (left) and C_2 selectivity (right) between for a sol-gel deposited and wash-coated catalysts at low CH_4 conc. in the reactant feed (CH_4 concentration: 10% / reactant flow rate: 85 $mL.min^{-1}$)

Although the BBFO membrane covers an interesting permeability window, its brittleness did not allow obtaining performance data for a sufficient time span. This membrane may be reconsidered when more active catalyst allow operation at lower temperature where the poor thermal stability does not present a significant issue. Finally, the BSMFO membrane with sol-gel catalyst showed reasonable results (Fig 5) with a C_2 selectivity of 46 % and a CH_4 conversion of 6.6 % (yield: 3 %) at 950°C. The C_2 selectivity is much higher than that observed for the equivalent BSCFO membrane exhibiting comparable methane conversions. Unfortunately, the membrane was not stable under the severe OCM conditions and after some cycles, it disintegrated – most probably because of deep reduction of the perovskite, being even that severe for a wash-coated membrane that no clear results may be reported.
The improved selectivity using a BSMFO instead of a BSCFO membrane remains the most significant finding. The obvious difference in oxygen permeability between those membranes (BSCFO, BSMFO) allows comparing the catalytic performance of the catalyst at different oxygen supply rates. On the other hand, the partial surface coverage (sol-gel deposition) might also give rise to side reactions on the bare membrane surface. However, the membrane surface contribution does not seem dominant as methane activation clearly relates to the catalyst demonstrated by increased conversions when improving the catalyst deposition (wash-coat instead of sol-gel).

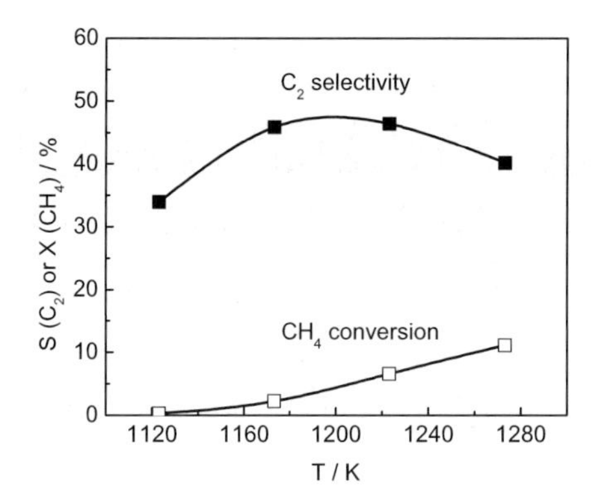

Figure 5: Catalytic activity of the BSMFO membrane (CH_4 concentration: 10% / reactant flow rate: 85 $mL.min^{-1}$)

4. Conclusions

It may be concluded that the BSCFO membrane reactor suffers of decreased selectivity as the oxygen supply rate is too high compared to the conversion rate for the given catalyst loading and conditions. Opting to limit the oxygen supply as demonstrated with the BSMFO membrane might help to increase selectivity, but poor performance and stability present major inconveniences. Thus, aiming on stable reactors with high performance, enhancements in the catalytic activity are obviously required to balance the oxygen permeation when using well performing membrane material like BSCFO, which will give then access to better performing membrane reactors.

Acknowledgements

The research stay of M.B. was supported by the EC Marie Curie program (contract HPMT-2000-00160) and the work of S.H is supported by the European research project "TOPCOMBI" (contract NMP2-CT2005-515792).

References

1. S. Liu, X. Tan, K. Li, R. Hughes; Catal. Rev. **43** (2001) 147.
2. M. Rebeilleau-Dassonneville, S. Rosini, A.C. van Veen, D. Farrusseng, C. Mirodatos; Catal. Today **104** (2005) 131.
3. A.C. van Veen, D. Farrusseng, M. Rebeilleau, T. Decamp, A. Holzwarth, Y. Schuurman, C. Mirodatos; J. Catal. **216** (2003) 135.
4. Y. Zeng, Y.S. Lin, S.L. Swartz; J. Mem. Sci. **150** (1998) 87.

Natural Gas Conversion VIII
F.B. Noronha, M. Schmal, E.F. Sousa-Aguiar (Editors)

Characterization of Mo/ZSM-11 and Mo-Ru/ZSM-11 catalysts. Effect of Mo content and Ru addition on the nature of Mo species.

Renata L. Martins[a,b], Luiz E.P.Borges[a], Fabio B. Noronha[b]

[a]*DEQ, Instituto Militar de Engenharia, Pça. Gal. Tibúrcio 80, Rio de Janeiro, 22290-270, Brazil. luiz@ime.eb.br.*
[b]*Instituto Nacional de Tecnologia, Av. Venezuela 82, Rio de Janeiro20081-312, Brazil. fabiobel@int.gov.br*

1. Abstract

The 2%Mo/HZSM-11 sample contain, basically, isolated molybdenum species while the 2%Mo-0.1%Ru/HZSM-11 sample exhibit dimers species. The increase of Mo content led to the appearance of polymerized structures such as dimers and MoO_3 particles. The Ru addition promoted the reduction of molybdenum oxide due to the presence of more polymerized structures.

2. Introduction

Several studies have been performed in order to evaluate the effect of the nature of support on the dehydroaromatization of methane since Wang et al [1] reported that Mo/ZSM-5 exhibited activity on the reaction. Zhang et al. [2] studied the performance of Mo based catalysts supported over different materials. Mo/ZSM-5 and Mo/ZSM-11 were the most active catalysts. The authors correlated the catalytic performance to the zeolite structure. According to them, the zeolites with two dimensional structures and pore size close to the dynamic diameter of benzene exhibit the higher activity on methane aromatization reaction. Besides the support, the nature, localization and distribution of molybdenum species in the zeolite channels also play an important role on methane dehydroaromatization reaction [3]. Furthermore, the addition of a second metal such as Ru may affect the Mo properties and therefore, the catalyst activity and stability [4]. Hence, the aim of this work was to study the effect of both Mo content and Ru addition on the nature of molybdenum species before and after carburization.

3. Experimental

xMo/HZSM-11 samples (x = 2, 4, and 10 wt.%) were prepared by incipient wetness impregnation of a HZSM-11 (Si/Al ratio (SAR) = 44) with an aqueous solution of $(NH_4)_6Mo_7O_{24}.4H_2O$. 2%Mo-0.3%Ru/HZSM-11 sample was obtained by incipient wetness impregnation of 2%Mo/HZSM-11 sample with an aqueous solution of $RuCl_3$. After impregnation, the samples were dried at 383 K in oven overnight and then calcined at 773 K / 6h under air stream. Diffuse reflectance spectra were recorded between 200 and 800 nm on an UV-Vis NIR spectrometer (Cary 5 – Varian) equipped with an integrating sphere (Harrick), using the HZSM-5 support as reference. The carburization study (TPC) was carried out in a reactor couple to a quadrupole mass spectrometer (MKS PPT). 0.5g of xMo/HZSM-5 was pre-heated at 773 K under He. After this treatment, the reactor was cooled down to room temperature and the sample was contacted with the 20% (v/v) CH_4/H_2 reactant mixture, while the temperature was raised linearly up to the final temperature of 1273 K using a heating rate of 10 K min^{-1}.

4. Results and discussion

Figure 1 shows the diffuse reflectance spectra of xMo/HZSM-11. For comparison, the spectra of reference compounds (Na_2MoO_4; $(NH_4)_6[Mo_7O_{24}].4H_2O$; MoO_3) are also presented in this Figure. All catalysts showed a charge transfer band in the range of 246 – 254 nm. The catalysts containing 4 and 10 wt% Mo exhibited another band at 308 nm. The addition of Ru did not lead to the appearance of a new band around 300 – 500 nm, associated to Ru coordinated to water or chloride ions.

Chen et al [5] also studied Mo/HZSM-5 samples containing different Mo content (1, 2, 5 and 10 wt% Mo) through diffuse reflectance spectroscopy (DRS) analysis. The UV spectra showed only one broad band in the range between 200 – 300 nm on the samples containing up to 5 wt% of Mo. The 10%Mo/HZSM-5 exhibited two bands, which were barely distinguished. The authors attributed these bands to both tetrahedrally and octahedrally coordinated Mo (VI). However, the effect of Mo loading on the distribution of Mo species was not clearly defined.

In order to identify the nature and the coordination of molybdenum species, the energy values of the absorption edges were determined using the methodology proposed by Weber [6]. The positions of the absorption edges were determined by the interception of the straight line fitted through the low energy side of the curve $[F(R_\infty) \cdot h\nu]^2$ versus hν, where $F(R_\infty)$ is the Kubelka-Munk function and hν is the energy of the incident photon. The edge energies obtained for the reference compounds and Mo/HZSM-5 samples are presented in Table 1. A correlation between edge energy and the number of nearest molybdenum neighbors in different molybdenum model compounds was calculated and the estimation of the condensation degree of molybdenum in Mo/HZSM-5 samples was thus done (Table 1). The 2%Mo/HZSM-11 sample presented, basically, isolated molybdenum species while the 2%Mo-0.3%Ru/HZSM-11 sample exhibited polymerized structures such as dimers species. On the other hand, the 4%Mo/ZSM-11 and 10%Mo/ZSM-11 showed two absorption edges values, corresponding to dimers and MoO_3 structures.

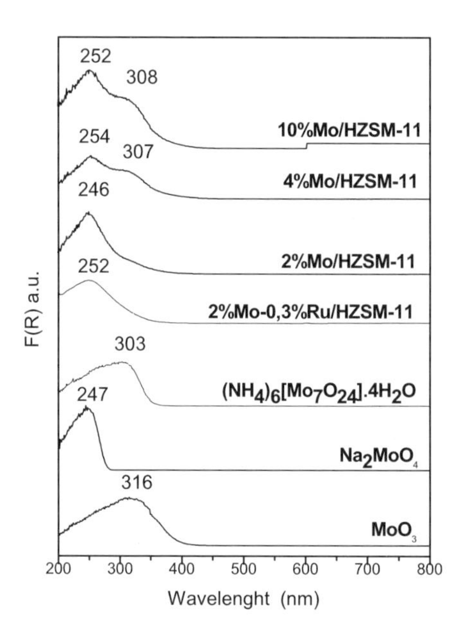

Figure 1- DRS spectra of both reference compounds and Mo/ZSM-11 samples containing different Mo content.

According to the literature, different Mo species are identified on supported Mo catalyst, which depends on Mo loading [6]. At low Mo content, the molybdenum oxide species are likely to be highly dispersed as monomers such as $(MoO_4)^{2-}$ (edge energy: 4.3 eV) in which the metal is in tetrahedral coordination [6]. Increasing the Mo loading causes molybdenum polymerization, first forming dimeric species such as $(Mo_2O_7)^{2-}$ (edge energy of 3.9 eV and tetrahedral coordination), and then more condensed species such as

$(Mo_7O_{24})^{6-}$ (edge energy of 3.3 eV and octahedral coordination) and MoO_3 (edge energy of 3.0 eV and octahedral coordination). Iglesia's research group [7,8] studied the structure of Mo species in Mo/ZSM-5 catalysts containing 1.0, 2.0 and 3.6 wt% Mo. They showed that MoO_3 spread and migrated into the ZSM-5 channels to form MoO_x species. After treatment in air at 973K, MoO_x species exchanged with acid sites to produce $(Mo_2O_5)^{2+}$ dimers connected to two zeolite framework oxygens. The amount of dimer species increased with the Mo content.

Table 1- Edge position and number of nearest molybdenum neighbors of prepared samples and reference compounds.

Samples	Edge position (eV)	Number of neighbors
2%Mo/HZSM-11	4.40	0
4%Mo/HZSM-11	4.16 and 3.53	1 and 5
10%Mo/HZSM-11	4.18 and 3.54	1 and 5
2%Mo-0.3Ru/HZSM-11	4.21	1
MoO_3	3.39	6
$(NH_4)_6[Mo_7O_{24}].4H_2O$	3.65	4
Na_2MoO_4	4.57	0

Our results are in agreement with the literature. When the Mo content is higher than 2 wt%, the appearance of polymerized structures such as dimers and MoO_3 particles is observed. The addition of Ru also favored the Mo polymerization and the formation of dimers.

Figure 2 shows the water and CO profiles during carburization of Mo/ZSM-5 catalysts. The water profiles revealed the presence of two peaks in all catalysts; the first peak at around 796 – 813K and the second one between 910-960K. The addition of Ru shifted both peaks of water formation to lower temperatures. In all samples, CO formation begins at around 900K, and two peaks are observed. It is important to stress that the m/e=28 signal did not return to baseline as temperature increased.

Jiang et al. [11] studied the induction period of methane aromatization on Mo/ZSM-5 catalysts with 2 and 10wt% Mo through TPSR. The H_2O and CO_2 profiles showed two peaks. The intensity of the first peak increased as Mo content increased. The peak at high temperature was followed by a large formation of CO. Ethene and benzene were detected after CO_2 production had finished. The authors suggested that Mo^{6+} is partially reduced in the first step and then the active site for methane aromatization is formed. TPSR of methane over Mo/MCM-22 catalysts containing different Mo loading identified three reaction regions [12]. In the first one, only CO_2 and H_2O were detected while CO and H_2 were also formed at high temperature. Benzene formation occurred only after CO_2 production had ended. The first two steps were described by the following reactions:

$MoO_3 + CH_4 \rightarrow MoO_2 + CO_2 + H_2O$ (1)
$MoO_2 + CH_4 \rightarrow Mo_2C + CO + H_2 + CO_2 + H_2O$ (2)

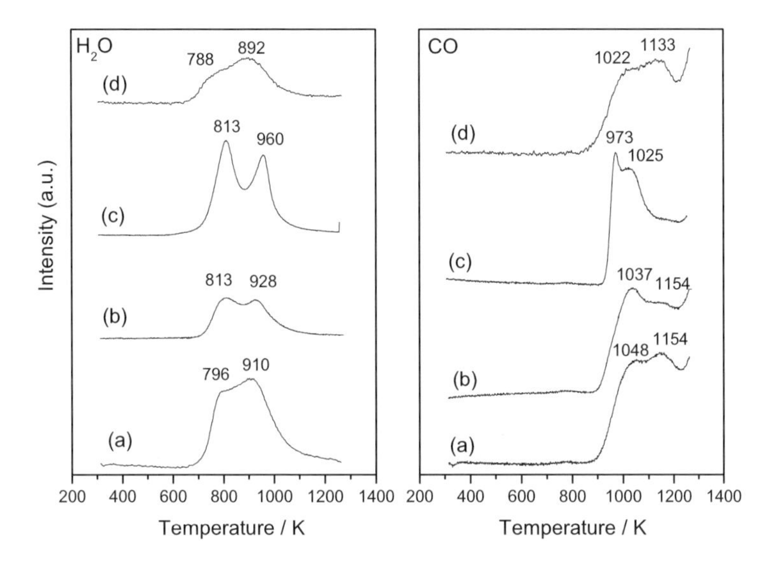

Figure 2- Water and CO formation profiles obtained during carburization of (a) 2%Mo/HZSM-5; (b) 4%Mo/HZSM-5; (c) 10%Mo/HZSM-5; (d) 2%Mo-0.3%Ru/HZSM-11.

In our work, the carburization was carried out with a CH_4/H_2 mixture and then, CO_2 was not produced in the first step. The molybdenum oxide reduction proceeds through the following reaction:

$MoO_3 + H_2 \rightarrow MoO_2 + H_2O$ (3)

The second stage is related to the carbon monoxide evolution and represents the Mo_2C formation which occurs due to MoO_2 carburization according to equation 2. The intensity of the first peak increased as Mo content increased while the second one is shifted to higher temperature. According to Laachen and Iglesia [11], H_2O formed during carburization of exchanged (Mo_2O_5) and unexchanged MoO_3 inhibits and even prevents carburization and lengthens activation periods. This may explain the shift of the maximum of the second peak observed in our work. Ru addition promoted the reduction of molybdenum oxide due to the presence of more polymerized structures (dimers), which reduces at temperatures lower than that related to isolated species. The increase in the CO signals observed in Figure 2 can be associated to methane dehydroaromatization reaction and ethene formation.

5. Conclusions

The increase of Mo content led to the appearance of polymerized structures. H_2/CH_4 mixture converts Mo species to active Mo_2C structures, which takes place at higher temperature on catalyst containing high Mo content. Ru addition promoted the reduction of molybdenum oxide due to the presence of more polymerized structures (dimers).

6. Acknowledgements

The authors wish to acknowledge the financial support from FINEP/PETROBRAS (65.04.1060.17). R.L. Martins acknowledges the scholarship received from CAPES.

7. References

1- Wang L., Tao L., Xie M., Xu G., Catal. Letters, 21 (1993) 35.
2- Zhang C., Li S., Yuan Y., Zhang W., Wu T., Lin L., Catal. Letters 56 (1998) 207.
3- Ohnishi R., Liu S., Dong Q., Wang L. and Ichikawa M., J. Catal,182 (1999) 92.
4- Shu Y., Xu Y., Wong She-Tin, Wang L. and Guo X., J. Catal., 170 (1997) 11.
5- L. Chen, L. Lin, Z. Xu, Z. Xu, X. Li, T. Zhang, J.Catal. 157 (1995) 190.
6- R.S. Weber, J. Catal. 151 (1994) 318.
7- R. W. Borry III, Y. H. Kim, A. Huffsmith, J. A. Reimer, E. Iglesia, J. Phys. Chem. B 103 (1999) 5787.
8- W. Ding, G. D. Meitzner, E. Iglesia, J. Catal. 206 (2002) 14.
9- H. Jiang, L. Wang, W. Cui, Y. Xu, Catal. Letters 57 (1999) 95.
10- D. Ma, Y. Shu, M. Cheng, Y. Xu, X. Bao, J.Catal. 194 (2000) 105.
11- H.S. Lacheen, E. Iglesia, PCCP 7 (2005) 538.

Natural Gas Conversion VIII
F.B. Noronha, M. Schmal, E.F. Sousa-Aguiar (Editors)

Conversion of methane to benzene via oxidative coupling and dehydroaromatization

J. S. Espindola, N. R. Marcilio, O. W. Perez-Lopez*

Department of Chemical Engineering - Federal University of Rio Grande do Sul
Rua Luiz Englert s/n – CEP 90040-040 – Porto Alegre – Brazil

Abstract. The combination of methane oxidative coupling and dehydroaromatization processes by catalysts arranged in a double-layer bed leads to an increase in the benzene productivity under inert atmosphere. This effect is ascribed to an increase in the ethylene production, which plays a role of intermediary.

1. Introduction

Natural gas reserves are considered an important source of energy and raw material for the chemical industries [1]. Therefore, the catalytic conversion of methane, which is the main component of natural gas, is of great interest. Several processes that allow the conversion of methane to products of greater value, have been studied, including the oxidative coupling of methane to ethylene, the selective oxidation of methane to methanol and formaldehyde, and the direct conversion to aromatics [2].
Among the different catalytic processes, the study of the methane aromatization to benzene and toluene in non-oxidative conditions has been of great interest, emerging as a potential route to the aromatic production. In the industrial process, benzene is produced from naphtha through thermal pyrolysis or catalytic reform [3].
Results previously obtained by our group, over ZSM-5 supported catalysts with different loads of molybdenum [4], have shown that the aromatic yield at 700°C is maximized on a catalyst containing 5 wt% Mo. They have also shown that ethylene is an intermediary in the methane aromatization, which is in agreement with results presented by Liu *et al.*[2]. On the other hand, in the study of the oxidative coupling of methane over co-precipitated lanthanum-magnesia catalysts we have found that the higher production of ethylene over a La-Mg catalyst happens when reaction is carried out at 700°C [5].
Considering that ethylene is an intermediary in the methane conversion to aromatic compounds, the objective of this work is the study of the combination of these two catalysts (Mo/ZSM-5 and La-Mg) in the methane aromatization reaction. This work is also motivated because of the scarceness of papers found in literature related to the coupling of these two processes.

2. Experimental

The methane dehydroaromatization catalyst, containing 5 Mo wt.% supported on ZSM-5, was prepared by impregnation of a H-ZSM-5 commercial sample (Si/Al = 15; S_{BET} = 312m^2/g) with an aqueous solution of ammonium heptamolybdate. Impregnation was carried out, under constant agitation, at room temperature for 24 hours. After that, it was dried in an oven at 80°C and later calcined during 6 hours at 600°C under air flow.
The oxidative coupling catalyst based on magnesium promoted by lanthanum was prepared according to the continuous co-precipitation method presented by Perez-Lopez *et al.*[6], from a solution containing nitrates of the metals and a mixture of carbonate and sodium hydroxide as precipitating. The co-precipitation was carried out under constant agitation at 60°C, and constant pH obtained by adjusting the flow of the solutions. The precipitate was washed with de-ionized water, vacuum filtered and dried in an oven at 80°C. The oxides were obtained by thermal treatment in air flow at 600°C. The two catalyst samples used in the activity runs were labeled as presented in Table 1.

Table 1- Composition of the catalyst samples

Sample	Composition
Mo	5wt% Mo/ZSM-5
La	15mol% La, 85mol% Mg

The catalytic activity runs were carried out combining both catalysts, La and Mo (La-Mg and Mo/ZSM-5), in a fixed bed. The runs were performed at 700°C in a quartz tubular reactor. The flow of the fed gases was adjusted through mass flow controllers (Bronkhorst). The products of the reaction were analyzed by on-line gas-chromatography (Varian 3600cx) with thermal conductivity and flame ionization detectors, using nitrogen as carrier gas.
Preliminary runs were carried out in order to determine the effect of the reaction atmosphere and also the adequate disposition of the catalysts in the bed. The effect of the reaction atmosphere was studied through runs carried out in two different environments: a) inert atmosphere; b) oxidative atmosphere with the CH_4/O_2 ratio equal to 10, which is equivalent to the one used in the oxidative coupling reaction [6, 7]. Concerning the arrangement of the bed, the catalysts were arranged in two different ways: a) a mixture of the two catalysts in a single bed; b) catalysts separated in a double-layer bed, with La catalyst placed at the inlet and Mo catalyst at the exit of the reactor.
In order to evaluate the effect of the residence time and ratio between the two catalysts (Mo/La) on the conversion and selectivity to methane aromatization, runs were carried out according to the results obtained in the preliminary runs for catalyst bed arrangements and different reaction atmospheres. The runs carried out for this study are presented in Table 2.

Table 2 – Runs carried out to evaluate the influence of the Mo/La ratio and residence time

Run #	Mo/La ratio	Residence time (g.min/l)
1	1	45
2	2	45
3	4	45
4	1	90
5	2	70

3. Results and Discussion

Influence of the reaction atmosphere

Since the methane aromatization takes place under inert atmosphere, while the oxidative coupling of methane takes place in the presence of air, preliminary runs were carried out to verify the influence of the reaction atmosphere over the dehydroaromatization reaction.

Fig. 1 shows the results for methane conversion and benzene selectivity, obtained at 700°C, with the two catalysts loaded in a double-layer bed. The results obtained under inert atmosphere are identified by full symbols, while the results in the presence of an oxidant agent are represented by empty symbols.

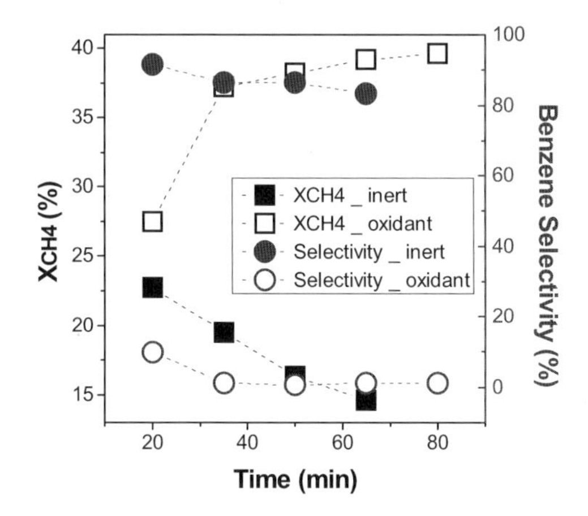

Figure 1 – Methane conversion and benzene selectivity for runs carried out under inert and oxidant atmospheres.

The results under oxidative atmosphere show that, although there is an increase in the methane conversion, the selectivity to benzene decreases drastically, showing that the presence of air influences negatively the benzene production, since this product is easily converted to CO or CO_2. The increase in the methane conversion can be ascribed to the methane oxidation to carbon oxides [5].

Influence of the catalytic bed arrangement

Fig.2 presents the results obtained at 700°C for methane conversion and benzene selectivity by different arrangements of the bed, that is, when both catalysts were loaded in a single bed (Mo_La Single) or for these catalysts loaded in a double bed (Mo_La Double). For comparison, the results obtained using only the aromatization catalyst (Mo), were also included.

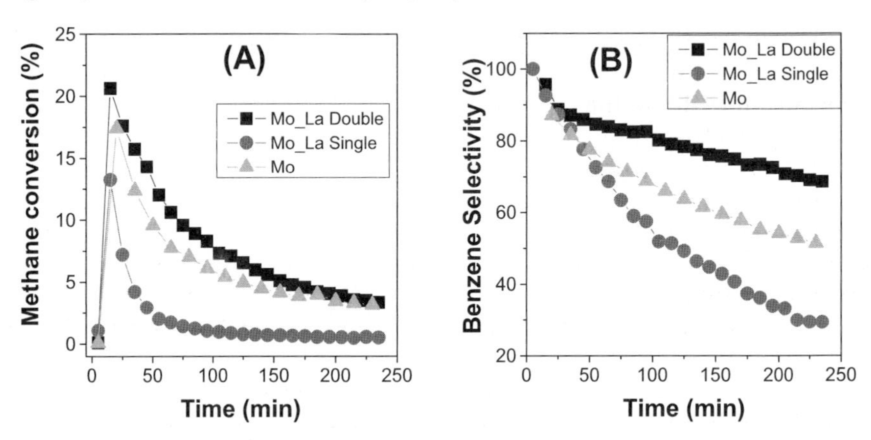

Figure 2 – Methane conversion (A) and benzene selectivity (B) for catalysts loaded in different dispositions in the reactor.

The results relative to the disposition of the two catalysts (single or double-layer bed) demonstrate that there is a significant increase in the benzene productivity when the two catalysts are loaded in a double-layer bed configuration, since a higher methane conversion and benzene selectivity were obtained in this configuration. This result can be ascribed to an increase in the ethylene production on the first bed loaded with oxidative coupling catalyst (La), and its subsequent conversion to benzene on the second bed loaded with aromatization catalyst (Mo).

It was observed that the use of a double layer bed has a greater influence on selectivity than on conversion, because the methane conversion for the double layer bed (Mo_La Double) is only slightly greater than the one obtained with pure aromatization catalyst (Mo), as it can be seen in Fig.2A. However, we can note that the difference in the benzene selectivity is more significant (Fig.2B).

For the run carried out with a mixture of both catalysts (La and Mo) on a single bed (Mo_La Single), a rapid decrease in the methane conversion can be observed, showing that the use of a mixture of these catalysts has a negative influence and enhances catalyst deactivation. This result shows that there is interference between the basic sites of the surface on lanthanum-based catalyst and the acid sites of the surface on molybdenum catalyst. According to previous works on oxidative coupling reaction [6]; the increase in the ethylene selectivity is due to the suppression of the acid sites and consequent inhibition of the cracking products. On the other hand, the presence of the catalyst with alkaline properties changes the aromatization properties of Mo/ZSM-5 whose acid sites, instead of promoting the aromatization, promote methane cracking, producing coke deposition and increasing the catalyst deactivation. Thus, when the reactor was loaded with a mixture of these two catalysts on a single bed, the results were poorer than the ones obtained with the use of pure Mo/ZSM-5 (Mo in Fig.2).

Considering the results for different catalyst bed arrangements and reaction atmospheres, a double-layer catalyst bed and an inert atmosphere were adopted for the study of the influence of residence time and ratio between Mo/ZSM-5 and La-Mg catalysts (Mo/La ratio).

Influence of the residence time and ratio between Mo/ZSM-5 and La-Mg catalysts.

In order to evaluate the effect of the ratio between the two catalysts on the methane conversion and selectivity to aromatization, runs # 1, 2 and 3 were carried out maintaining the residence time constant, while the Mo/La ratio was varied. Runs # 1 and 4 are related to fixed Mo/La ratio and variable residence time.

Fig.3 shows methane conversion and benzene selectivity at 700°C. The results obtained with fixed residence time (runs 1, 2 and 3) and different ratios between La and Mo catalysts show that the increasing in the Mo/La ratio increases the methane conversion (Fig. 3A). On the other hand, an opposite relation is found for the selectivity to benzene (Fig. 3B), which increases as the Mo/La ratio decreases. These results also showed that the influence was significant when the Mo/La ratio was changed from 1 to 2 and negligible when the Mo/La ratio was later increased to 4, indicating that the limit of the system is a Mo/La ratio of about 2.

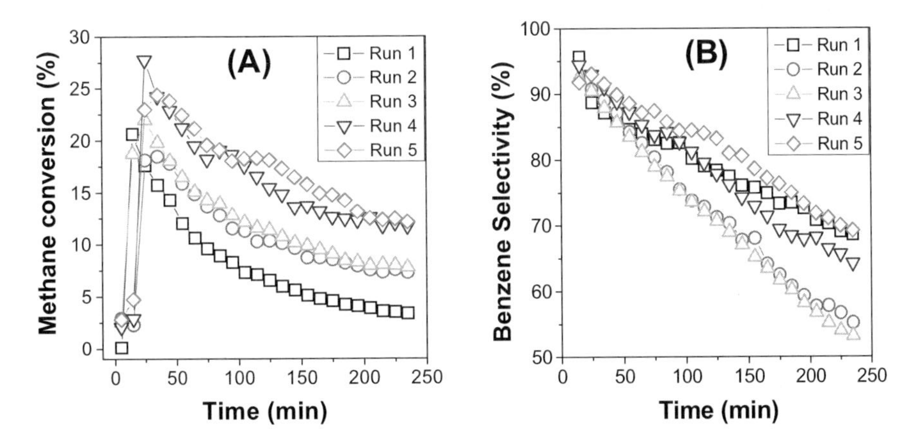

Figure 3 – Influence of the residence time and ratio Mo/La on methane conversion (A) and benzene selectivity (B).

Comparing the results with fixed Mo/La ratio and variable residence time (runs 1 and 4), we noticed that increasing the residence time increases only the methane conversion whereas the selectivity to benzene (Fig 4B) does not practically change, as expected for fixed Mo/La ratio.
An additional run increasing the residence time and considering that the best Mo/La ratio is about 2 was carried out (run 5). Comparing these results with run 2, we observed that in run 5 was obtained an increase not only on the methane conversion but also on the selectivity to benzene, showing a change in the behavior of the catalytic bed.
The results of this work show that the activity and the selectivity to aromatics depend on an adequate Mo/La ratio and residence time in order to maximize the benzene production. Another aspect is in regard to the stability of the combined double-layer bed, which exhibits better stability than the reactor loaded with pure Mo/ZSM-5 catalyst.

4. Conclusions

The combination of methane oxidative coupling and dehydroaromatization processes by La-Mg and Mo/ZSM-5 catalysts arranged in a double-layer bed leads to an increase in the benzene productivity under inert atmosphere. This effect is ascribed to an increase in the ethylene production, which plays a role of intermediary in the aromatization reaction. It was observed that the methane conversion is directly related to the Mo/ZSM-5 catalyst and that the selectivity to benzene is not only related to the La-Mg catalyst, but also to the ratio between these two catalysts in the reactor. It is also evidenced that the residence time and the ratio between the two catalysts influence benzene productivity.
In summary, the results of this work show that the activity and the selectivity to aromatics depend on an adequate Mo/La ratio and residence time in order to maximize the benzene production.

Acknowledgements

The authors acknowledge financial support from Conselho Nacional de Desenvolvimento Científico e Tecnológico - CNPq

References

[1] Lunsford, J. *Catal. Today* 63, 165–174 (2000).
[2] Liu, S.; Wang, L.; Ohnishi, R.; Ichikawa, M. *J. Catalysis* 181, 175-188 (1999).
[3] Moulijn, J.A.; Makee M., Van Diepen, A. Chemical Process Technology. John Wiley, 2001.
[4] Espindola, J.S.; Perez-Lopez, O.W. In Anais do XVII SIC - UFRGS, Porto Alegre (2005)
[5] Perez-Lopez, O.; Farias, T.; Correa, C. In 4th Mercosur Congress on Process System Engineering- (2005a).
[6] Perez-Lopez, O.; Farias, T.; Correa, C.; Marcilio, N. In Anais do XIV Congreso Argentino de Catalisis (2005b)
[7] Perez-Lopez, O.; Correa, C. In Anais do 13º Congresso Brasileiro de Catálise (2005c)

Natural Gas Conversion VIII
F.B. Noronha, M. Schmal, E.F. Sousa-Aguiar (Editors)

Bimetallic orthophosphate and pyrophosphate catalysts for direct oxidation of methane to formaldehyde

Magdaléna Štolcová [a], Christian Litterscheid [b], Milan Hronec [a], Robert Glaum [b]

[a]*Department of Organic Technology, Slovak University of Technology, Radlinského 9, 81237 Bratislava, Slovak Republic*
[b]*Institut für Anorganische Chemie; Rheinische Friedrich-Wilhelms-Universität Bonn, Gerhard-Domagk-St. 1, D-53121 Bonn, Germany*

Direct oxidation of methane to formaldehyde over bimetallic orthophosphate and pyrophosphate catalysts of the following compositions: $M^{II}_3Fe^{III}_4(PO_4)_6$, $M^{II}Fe^{III}_2(P_2O_7)_2$ and $M^{II}_5Fe^{III}_2(P_2O_7)_4$ was studied with a mixture of oxygen and nitrous oxide at atmospheric pressure. The redox behaviour of the divalent metals M^{II} = Fe, Co, Ni, Cu and the crystalline structure of the catalysts are important for the oxidation reaction. Surprising differences in the catalyst behaviour were found for the divalent metals in the compared phosphates. Amorphous orthophosphates are more active, but crystalline orthophosphates and pyrophosphates are more selective. The promising catalysts containing Fe^{II}/Fe_2^{III} and Cu^{II}/Fe_2^{III} pyrophosphates produce formaldehyde with selectivity of 45 % and almost 30 %, respectively.

1. Introduction

Direct transformation of methane to valuable products as methanol or formaldehyde is an attractive way of methane utilization. In recent decades several reviews and papers have contributed to the clarification of the reaction, however, a process available for commercial application with economically reasonable results has not been developed yet.
From the large variety of the studied catalysts reported in literature several iron

based materials exhibited activity in the conversion of methane to oxygenates. Among them promising formaldehyde production was associated with a $FePO_4$ catalyst [1-3]. High formaldehyde selectivity was observed over Fe- and Cr-promoted vanadyl pyrophosphates at very low methane conversion [4]. Several metal phosphates, such as Zn, Zr or the later one doped by Ce or La were utilized for the elucidation of methane oxidation mechanism to formaldehyde and C_2-hydrocarbons [5].
In the present contribution bimetallic orthophosphate and pyrophosphate catalysts of the following compositions: $M^{II}_3Fe^{III}_4(PO_4)_6$, $M^{II}Fe^{III}_2(P_2O_7)_2$ and $M^{II}_5Fe^{III}_2(P_2O_7)_4$ (M^{II} = Fe, Co, Ni, Cu) were studied for the partial oxidation of methane in the gas phase with a mixture of oxygen and nitrous oxide at atmospheric pressure.

2. Experimental

Catalyst preparation
The orthophosphates $M^{II}_3Fe^{III}_4(PO_4)_6$ (M^{II} = Co [6], Ni [7] and Cu [8]) were prepared by mixing of aqueous solutions of stoichiometric amounts of $Fe(NO_3)_3$, the divalent metal in half-concentrated nitric and phosphoric acid and then evaporating to dryness on a heating stirrer. The residues were calcined in air in silica-crucibles at 773 K and in a second series of syntheses at 1073 K for 72 h.
A similar procedure has been applied for the preparation of pyrophosphates $M^{II}Fe^{III}_2(P_2O_7)_2$ and $M^{II}_5Fe^{III}_2(P_2O_7)_4$ (M^{II} = Co, Ni, Cu). However, as the source for phosphate, $(NH_4)_2HPO_4$ was used and after evaporation the residues were heated at 1073 K for 72 h.
$Fe^{II}_3Fe^{III}_4(PO_4)_6$ [9] and $Fe^{II}_5Fe^{III}_2(P_2O_7)_4$ [10] were obtained as single phase products reducing $Fe^{III}PO_4$ or $Fe^{III}_4(P_2O_7)_3$ by elemental iron, $Fe^{II}Fe^{III}_2(P_2O_7)_2$ [11] by combination of $Fe^{II}_2P_2O_7$ with $Fe^{III}_4(P_2O_7)_3$ at 1073 K for 72 h, (50 - 100 mg iodine as mineralizer, sealed silica tube).
Catalyst characterization
The chemical composition and crystal structure of the catalysts was checked prior to and after the experiment by X-ray powder diffraction using $CuK_{\alpha 1}$ radiation (λ = 1.54051 Å, image-plate Guinier technique [12]; see Table 1).
Catalyst testing
The partial oxidation of methane to formaldehyde was carried out in an electrically heated flow reactor at 933 K and atmospheric pressure. The quartz reactor was filled with 0.5 g of solid catalyst with a particle diameter of 0.6 – 1 mm. The flow rate of methane, oxygen, nitrous oxide and the balance of helium, were controlled and measured by mass flow meters (total feed 3.0 $l \cdot h^{-1}$, mol. (%): CH_4 (65), O_2 (26), N_2O (1.3), He (8.7)). The reaction mixture was quenched by cold water with a flow rate of 4 $ml \cdot h^{-1}$ injected under the catalyst bed. The liquid reaction products were collected in a frozen flask containing 3

ml of distilled water. The composition of reaction mixture was analysed every hour.

Table 1 Summary of synthetic experiments.

theoretical composition	temp. [K]	product as prepared (from Guinier-photographs)	product after catalyst testing (from Guinier-photographs)
$Fe^{II}_3Fe^{III}_4(PO_4)_6$	1073	$Fe^{II}_3Fe^{III}_4(PO_4)_6$	$Fe^{II}_3Fe^{III}_4(PO_4)_6 + Fe^{III}PO_4$
$Co_3Fe_4(PO_4)_6$	773	$CoFeO(PO_4)$ [a)]	$Co_3Fe_4(PO_4)_6$
$Co_3Fe_4(PO_4)_6$	1073	$Co_3Fe_4(PO_4)_6 + FePO_4$	$Co_3Fe_4(PO_4)_6 + FePO_4$
$Ni_3Fe_4(PO_4)_6$	773	amorphous	$(Fe_{1-x}Ni_x)(PO_4)_{1-x/3}$ [b)]
$Ni_3Fe_4(PO_4)_6$	1073	$Ni_3Fe_4(PO_4)_6 + FePO_4 + NiFeOPO_4 + \alpha\text{-}Ni_2P_2O_7$	$Ni_3Fe_4(PO_4)_6 + FePO_4 + NiFeOPO_4 + \alpha\text{-}Ni_2P_2O_7$
$Cu_3Fe_4(PO_4)_6$	773	amorphous	$Cu_3Fe_4(PO_4)_6 + (Fe_{1-x}Cu_x)(PO_4)_{1-x/3}$ [b)]
$Cu_3Fe_4(PO_4)_6$	1073	$Cu_3Fe_4(PO_4)_6$	$Cu_3Fe_4(PO_4)_6$
$Fe^{II}Fe^{III}_2(P_2O_7)_2$	1073	$Fe^{II}Fe^{III}_2(P_2O_7)_2$	$Fe^{II}Fe^{III}_2(P_2O_7)_2$
$Co^{II}Fe^{III}_2(P_2O_7)_2$	1073	$Co^{II}Fe^{III}_2(P_2O_7)_2$	$Co^{II}Fe^{III}_2(P_2O_7)_2$
$Ni^{II}Fe^{III}_2(P_2O_7)_2$	1073	$\alpha\text{-}Ni_2P_2O_7 + Fe_4(P_2O_7)_3$	$\alpha\text{-}Ni_2P_2O_7 + Fe_4(P_2O_7)_3$
$Cu^{II}Fe^{III}_2(P_2O_7)_2$	1073	$Cu^{II}Fe^{III}_2(P_2O_7)_2$	$Cu^{II}Fe^{III}_2(P_2O_7)_2$
$Fe^{II}_5Fe^{III}_2(P_2O_7)_4$	1073	$Fe^{II}_5Fe^{III}_2(P_2O_7)_4$	$Fe^{II}_5Fe^{III}_2(P_2O_7)_4 + Fe^{III}PO_4$
$Co^{II}_5Fe^{III}_2(P_2O_7)_4$	1073	$\alpha\text{-}Co_2P_2O_7$, $CoFe_2(P_2O_7)_2$	$\alpha\text{-}Co_2P_2O_7$, $CoFe_2(P_2O_7)_2$
$Ni^{II}_5Fe^{III}_2(P_2O_7)_4$	1073	$\alpha\text{-}Ni_2P_2O_7$ [c)]	$\alpha\text{-}Ni_2P_2O_7$ [c)]
$Cu^{II}_5Fe^{III}_2(P_2O_7)_4$	1073	$\alpha\text{-}Cu_2P_2O_7$, $CuFe_2(P_2O_7)_2$	$\alpha\text{-}Cu_2P_2O_7$, $CuFe_2(P_2O_7)_2$

a) cfg. [13]. b) powder pattern shows strong similarity to that of $\beta\text{-}CrPO_4$. c) heating with chlorine as mineralizer led to better crystallized products showing the powder diffraction pattern of $\alpha\text{-}Ni_2P_2O_7$ and $Fe_4(P_2O_7)_3$

Analysis of the reaction mixture

The reactant and the product streams were analysed by gas chromatography (Shimadzu GC 17 and Chrompack CP9001) using thermal conductivity and flame-ionisation detectors. The columns Porapack T, Carbosphere and Porapack QS were used for the separation. The methane conversion was calculated from the mass of the determined products and methane in the feed. The selectivity was calculated from the yield of each product on a carbon base.

3. Results and discussion

Catalytic oxidation of methane with a mixture of O_2 and N_2O was investigated at the temperature of 933 K and atmospheric pressure. Orthophosphates and pyrophosphates containing Fe^{3+} together with a divalent cation (M^{2+} = Fe, Co, Ni, Cu) were tested as catalysts. The actual phase composition of the catalyst material is summarized in Table 1. Bimetallic, microcrystallines orthophosphates $M^{II}_3M^{III}_4(PO_4)_6$ (M^{II} = Fe, Cu) were obtained as single phase after heating of the starting material at 1073 K. With Co and Ni multi-phase mixtures were formed. At lower temperatures amorphous products or $CoFeO(PO_4)$ [13] instead of $Co_3Fe_4(PO_4)_6$ were obtained. $M^{II}Fe^{III}_2(P_2O_7)_2$ (M^{II} = Fe, Co, Cu) were synthesized as single phase products, instead of "$Ni^{II}Fe^{III}_2(P_2O_7)_2$" always two-phase mixtures of α-$Ni_2P_2O_7$ and $Fe^{III}_4(P_2O_7)_3$ were found. For the composition $M^{II}_5Fe^{III}_2(P_2O_7)_4$ a single phase product was synthesized for M^{II} = Fe; the other samples contained always α-$M_2P_2O_7$ (M = Co, Ni, Cu) and $M^{II}Fe^{III}_2(P_2O_7)_2$ or $Fe_4(P_2O_7)_3$, respectively.

3.1. Catalytic activity of orthophosphates "$M^{II}_3M^{III}_4(PO_4)_6$"

The calcination temperature has a strong effect on the catalytic activity and crystal structure of mixed metal orthophosphates monitored before and after oxidation of methane.

While the $Ni_3Fe_4(PO_4)_6$ and $Cu_3Fe_4(PO_4)_6$ calcined at 773 K are in fresh state amorphous, $CoFeO(PO_4)$ (cfg. [13]) plus a presumably amorphous neighbouring phase was obtained instead of $Co_3Fe_4(PO_4)_6$. After catalytic reaction at about 930 K the diffraction lines of $Fe_{1-x}Ni_x(PO_4)_{1-1/3x}$ were recognized in the "$Ni_3Fe_4(PO_4)_6$" catalyst. Two crystalline phases, $(Fe_{1-x}Cu_x)(PO_4)_{1-1/3x}$ and $Cu_3Fe_4(PO_4)_6$, were formed from $Cu_3Fe_4(PO_4)_6$. The powdered patterns of $Fe_{1-x}Ni_x(PO_4)_{1-1/3x}$ and $(Fe_{1-x}Cu_x)(PO_4)_{1-1/3x}$ show strong similarity to that of β-$CrPO_4$ [14]. Interestingly, during the oxidation reaction $CoFeO(PO_4)$ plus amorphous phase is transformed to $Co_3Fe_4(PO_4)_6$.

The metal orthophosphates calcined at 1073 K are crystalline before and after catalyses. In the diffraction pattern of the used catalyst $Fe_3Fe_4(PO_4)_6$, besides the main phase, the lines corresponding to $Fe^{III}PO_4$ are observed with low intensities. This implies a partial oxidation of the quaternary phosphate.

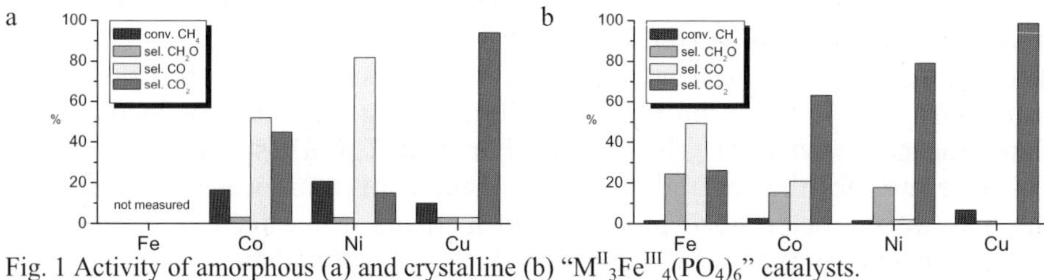

Fig. 1 Activity of amorphous (a) and crystalline (b) "$M^{II}_3Fe^{III}_4(PO_4)_6$" catalysts.

The catalytic activity of metal orthophosphates calcined at 773 K and subsequently at 1073 K were significantly different (Figs. 1a and 1b). Over amorphous catalysts about 10-19% conversion of methane was reached, however, the selectivity to CO_x prevailed. The conversion of methane over the mixed iron orthophosphate catalysts decreases in the following order: Ni > Co > Cu. Unexpected was the finding that over $Cu^{II}_3Fe^{III}_4(PO_4)_6$ and $Co^{II}_3Fe^{III}_4(PO_4)_6$ catalysts the decreased conversion leads to a rise of the CO_2 content in the reaction mixture. This indicates that CO_2 is formed directly from methane and not by the consecutive oxidation of the partial oxidation products. Since the preparation of amorphous catalysts was less reproducible and also the yield of formaldehyde is lower, the investigation of these catalysts were stopped. The crystalline bimetallic phosphates are less active but more selective. The selectivity strongly depends on the presence of the divalent metal. Interestingly, over crystalline $Cu^{II}_3Fe^{III}_4(PO_4)_6$ at 6.6 % conversion of methane the reaction product is practically only CO_2. From the tested bimetallic orthophosphates the highest selectivity to formaldehyde (about 26 %) was obtained over the single-phase $Fe^{II}_3Fe^{III}_4(PO_4)_6$ catalyst.

3.2. Catalytic activity of pyrophosphates "$M^{II}Fe^{III}_2(P_2O_7)_2$" and "$M^{II}_5Fe^{III}_2(P_2O_7)_4$"

All of the pyrophosphate catalysts were crystalline before and after the oxidation reaction. Single phase samples of the pyrophosphates were obtained for $Fe^{II}_5Fe^{III}_2(P_2O_7)_2$, $Fe^{II}Fe^{III}_2(P_2O_7)_2$, $Cu^{II}Fe^{III}_2(P_2O_7)_2$ and $Co^{II}Fe^{III}_2(P_2O_7)_2$. The latter one is isotypic to $Fe^{II}Fe^{III}_2(P_2O_7)_2$. The catalysts "$Ni^{II}Fe^{III}_2(P_2O_7)_2$" and "$Ni^{II}_5Fe^{III}_2(P_2O_7)_4$" consist of two-phase mixtures ($Fe^{III}_4(P_2O_7)_3$ + α-$Ni_2P_2O_7$). For "$M^{II}_5Fe^{III}_2(P_2O_7)_4$" with M = Co, Cu mixtures of the corresponding pyrophosphates of $M^{II}Fe_2(P_2O_7)_2$ and $M^{II}_2P_2O_7$ were observed.
Different catalytic behaviour was observed between the two types of pyrophosphate catalysts "$M^{II}Fe^{III}_2(P_2O_7)_2$" and "$M^{II}_5Fe^{III}_2(P_2O_7)_4$" ($M^{II}$ = Fe, Co, Ni, Cu) tested under similar conditions as the orthophosphates described above (Figs. 2a and 2b). Although, the conversion of methane does not vary significantly (1-3 %), the selectivity to formaldehyde is slightly higher over pyrophosphates "$M^{II}Fe^{III}_2(P_2O_7)_2$" than over the $M^{II}_5Fe^{III}_2(P_2O_7)_4$ catalysts. The latter types of catalysts produce a higher amount of CO_2. We suppose that the presence of diphosphates $M^{II}_2P_2O_7$ in the catalysts are responsible for the formation of higher amounts of CO_2 during methane oxidation.

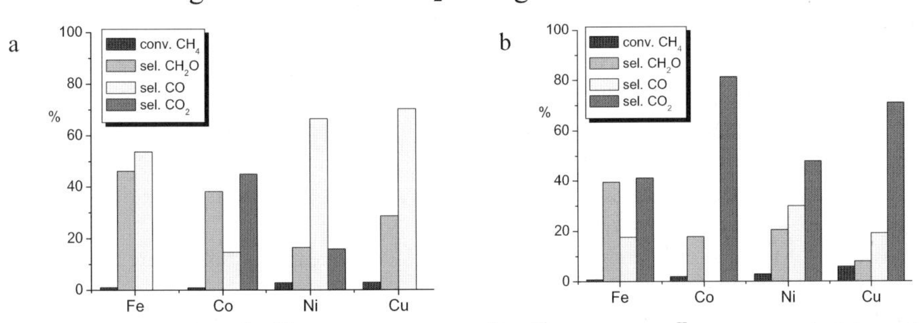

Fig. 2 Activity of (a) $M^{II}Fe^{III}_2(P_2O_7)_2$ and (b) "$M^{II}_5Fe^{III}_2(P_2O_7)_4$" ($M^{II}$ = Fe, Co, Ni, Cu) catalysts.

Promising results were observed over $Cu^{II}Fe^{III}_2(P_2O_7)_2$ and $Fe^{II}Fe^{III}_2(P_2O_7)_2$ catalysts. In their presence the only products formed during methane oxidation are formaldehyde and CO. Over $Fe^{II}Fe^{III}_2(P_2O_7)_2$ and $Cu^{II}Fe^{III}_2(P_2O_7)_2$ about 45% and almost 30% selectivity to formaldehyde was reached. However, the activity of all pyrophosphate catalysts is lower than that of phosphate catalysts.

The crystalline structures of $Fe^{II}Fe^{III}_2(P_2O_7)_2$ and $Cu^{II}Fe^{III}_2(P_2O_7)_2$ are rather different (Fig. 3). The two octahedral units $[Fe^{III}O_6]$ in $Cu^{II}Fe^{III}_2(P_2O_7)_2$ are connected by a square-planar$[Cu^{II}O_4]$ unit, but by a prismatic $[Fe^{II}O_6]$ in $Fe^{II}Fe^{III}_2(P_2O_7)_2$. It implies that both, the crystal structure and the redox couple of the building metals M^{II}/M^{III} (Fe^{2+}/Fe^{3+} or Cu^{+}/Cu^{2+}) play an important role in the partial oxidation of methane to oxygenates.

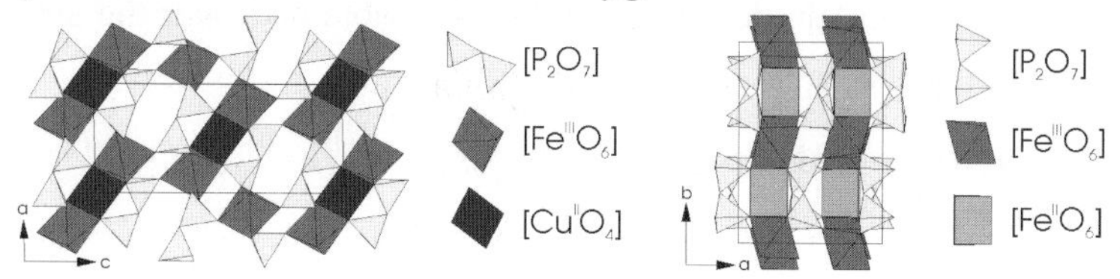

Fig. 3 The crystalline structures of $Cu^{II}Fe^{III}_2(P_2O_7)_2$ and $Fe^{II}Fe^{III}_2(P_2O_7)_2$.

Acknowledgement

The research has been financed by the Slovak VEGA project 1/2459/05 and the EU-project CONCORDE.

References

1. R.L. McCormick, G.O. Alptekin, Catal. Today, 55 (2000) 269.
2. K. Otsuka, Y. Wang, Applied Catalysis A., General 222, (2001) 145.
3. Y. Wang, X. Wang, Z. Su, Q. Guo, Q. Tang, Q. Zhang, H. Wan, Catal. Today 93-95 (2004) 155.
4. R.L. McCormick, G.O. Alptekin, A.M. Herring, T.R. Ohno, S.F. Dec, J. Catal. 172 (1997) 160.
5. M.Yu. Sinev, S. Setiadi, K. Otsuka in "New Development in Selective Oxidation II" Stud. In Surf. Sci. and Catal., V. Cortés Corberán and S. Vic Bellón (Ed) Elsevier Amsterdam, 82 (1994) 357.
6. P. Lightfoot, A.K. Cheetham, J. Chem. Soc. Dalton Trans. 89 (1989) 1765.
7. A. E. Kira, R. Gerardin, B. Malman, et al., Eur. J. Solid State Inorg. Chem. 29 (1992) 1129.
8. A.A. Belik, K.V. Pokholok, A.P. Malakho, S.S. Khasanov, B.I. Lazoryak, Zh. Neorg. Khim. 45 (2000) 1633.
9. Yu.A. Gorbunov, B.A. Maksimov, Yu.K. Kabalov, A.N. Ivashchenko, O.K. Melnikov, N.V. Belov, Dokl. Akad. Nauk SSSR 254 (1980) 873.
10. B. Malama, M. Ijjaali, R. Gerardin; G. Venturini, C. Gleitzer, Eur. J. Solid State Inorg. Chem. 29 (1992) 1269.
11. M. Ijjaali, G. Venturini, R. Gerardin, B. Malaman, C. Gleitzer, Eur. J. Solid State Inorg. Chem. 28 (1991) 983.
12. K. Maaß, R. Glaum, R. Gruehn, *Z. Anorg. Allg. Chem.* 628 (2002) 1663.
13. N.E. Khayati, N. Cherkaoui, Ch.R. El Moursli, R.J. Carvajal, G. Andre, N. Blanchard, F. Bouree, G. Collin, T. Roisnel, Eur. Phys. J. B 22 (2001) 429.
14. J.P. Attfield, P.D.Battle, A.K.Cheetham, J. Solid State Chem. 28 (1985) 983.

Natural Gas Conversion VIII
F.B. Noronha, M. Schmal, E.F. Sousa-Aguiar (Editors)

IT-SOFC operated with catalytically processed methane fuels

Baofeng Tu[a, b], Yonglai Dong[a], Mojie Cheng[a], Zhijian Tian[a] and Qin Xin[a]

[a] *Dalian Institute of Chemical Physics, Chinese Academy of Sciences, 457 Zhongshan Road, Dalian 116023, China*
[b] *Graduate School of the Chinese Academy of Sciences, Beijing 100039, China*

Abstract

An anode supported solid oxide fuel cell integrated with a catalytic reactor, which processed methane fuels by partial oxidation with air or oxygen, steam reforming and CO_2 reforming, were investigated. A supported nickel catalyst was used for methane processing, and the obtained fuels were analyzed by an online gas chromatograph. The fuels processed from partial oxidation with air and CO_2 reforming caused an apparent decrease of the cell performance, which was due to the diluting effect of nitrogen and CO_2 and the occurrence of limiting current densities. Whereas, the fuels processed from partial oxidation with oxygen and steam reforming affected very little on the cell performance.

1. Introduction

Solid oxide fuel cell is one of the most attracting energy conversion systems due to its high-energy conversion efficiency, clear and flexibility of usable fuel type. Nature gas is a cheep and abundant fuel for SOFC. Direct electrochemical oxidation of methane on the anode is an ideal way, but the serious carbon deposition cannot be avoid and deteriorate the cell structure. For internal reforming methane, the anode is used as both reforming and electrochemical catalyst. Due to the high content in the anode and high sintering temperature during cell fabrication, nickel in the anode is of micro-scale in size and usually suffers from serious coking. The developed SOFC systems are generally integrated with fuel pre-treating system. At present, SOFC systems are generally constructed with fuel processors for natural gas fuel [1-3]. There are several techniques for pre-treating methane [4-7], but rather less attention has been paid to the effect of different pre-treating techniques on cell performance. In this paper, we integrated an anode supported solid oxide fuel cell with a methane-processing reactor and investigated the effects of the various methane processing methods and fuel composition on the cell performance.

2. Experimental

NiO-YSZ anode substrate with a composition of 50 wt% NiO and 50 wt% YSZ were prepared by the tape-casting method, and then one side of the anode substrate was painted with the suspension of YSZ. Anode electrolyte assembly was obtained after sintering at 1400°C. An LSM-YSZ cathode was screen-printed on the electrolyte, and then sintered at 1200°C. The anode supported solid oxide fuel cell was operated with fuels from a catalytic reactor. A supported nickel catalyst was used for methane processing. The fuels from different processes, including POM with oxygen, POM with air, steam reforming of methane and CO_2 reforming of methane, were analyzed by using an online GC. The performances of solid oxide fuel cells with different fuels were measured.

3. Results and Discussion

3.1. Effect of POM with oxygen on cell performance

Partial oxidation of methane with oxygen produces syngas as depicted in reaction (1).

$$CH_4+0.5O_2=2H_2+CO,\ \Delta H_{298}= -\ 36KJ/mol \quad (1)$$

$$CH_4+2O_2=CO_2+2H_2O,\ \Delta H_{298}= -\ 802KJ/mol \quad (2)$$

Table 1. Fuel compositions from POM with oxygen at 1023K

Flow rates of the feed		H_2%	CO%	CH_4%	CO_2%
CH_4 (ml/min)	O_2 (ml/min)				
50	25	60.6	25.2	10.2	4.0
50	20	55.6	23.9	17.2	3.3
50	15	48.6	22.0	27.5	1.9
50	10	40.3	18.8	39.9	1.0
50	5	26.8	13.1	59.8	0.3

In the case with excess methane, mainly the reaction (1) occurs. But under the condition with excess oxygen, the complete oxidation of methane would also take place. We can obtain fuels of H_2: CO=2:1 by the POM reaction. Table 1 shows the fuel compositions from POM with oxygen at 1023K at different ratios of O_2/CH_4. With the increase of the O_2/CH_4 ratio, the conversion of methane increases, and all the concentrations of H_2, CO and CO_2 increase. At the O_2/CH_4 ratio of 1:2, the typical ratio for the conventional POM to syngas, only 10.2% CH_4 remains, and the concentration of hydrogen increases to 60.6%.

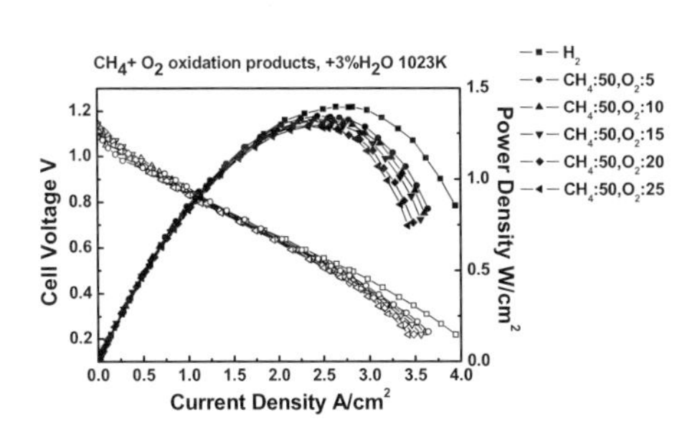

Fig. 1. Performances of the cell at 1023K with H_2 and fuels from POM with oxygen.

Fig. 1 shows the current-voltage and current-power curves of the cell with hydrogen and fuels from POM with oxygen. The maximum power density and the maximum current density of the cell varies very a little with the increase of the ratios of O_2/CH_4, When the ratio of O_2/CH_4 is increased from 5:50 to 25:50, the maximum power density decreases less than 5%. High concentration of oxygen can increase the ratio of O/C in processed methane fuels, which is better for suppressing carbon deposition but may result in the complete oxidation methane reaction [1].

3.2. Effect of POM with air on cell performance

Table 2. Fuel compositions from POM with air at 1023K

Flow rates of the feed		H_2%	N_2%	CO%	CH_4%	CO_2%
CH_4 (ml/min)	AIR (ml/min)					
40	100	34.8	44.8	14.3	3.0	3.1
40	80	33.6	42.1	13.9	7.5	2.9
40	60	32.8	37.7	13.6	13.6	2.3
40	40	33.2	30.8	14.1	21.0	0.9
50	30	27.4	24.8	12.1	35.1	0.6
60	20	20.9	17.9	9.1	52.0	0.1

Table 2 shows the fuel compositions from POM with air at 1023K at different ratios of air/CH_4. The conversion of methane increases apparently with the increase of the ratio of air/CH_4. The concentrations of H_2 and CO increase gradually when the ratio of air/CH_4 is smaller than 1. When the ratio of CH_4/air is larger than 1, the concentrations of H_2 and CO increase little for the diluting effect of nitrogen. At the CH_4/air ratio of 40:100, the conversion of methane increases to 85%, and the concentrations of nitrogen and carbon dioxide increase to 44.8% and 3.1%, respectively.

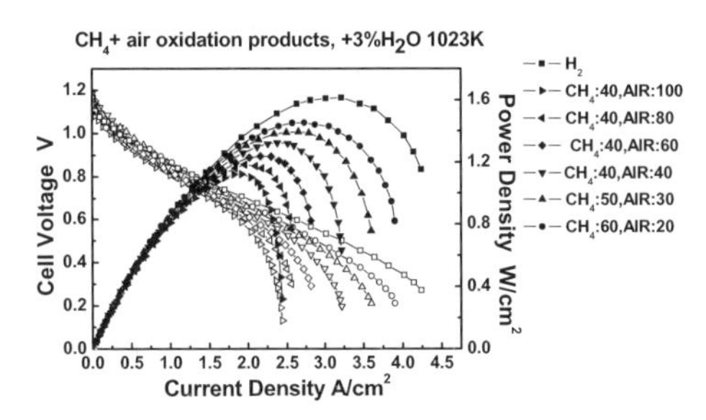

Fig. 2. Performances of the cell at 1023K with H_2 and fuels from POM with air.

Fig. 2 shows the current-voltage and current-power curves of the cell using fuels from POM with air and H_2. The cell performance decreases apparently with the increase of the air/CH_4 ratio, and the concentration polarization becomes larger, which is mainly due to the diluting effect of nitrogen. The dilution of nitrogen can not only reduce the partial pressure of hydrogen but also reduce the diffusion coefficients of hydrogen [8]. At the air/CH_4 ratio of 100:40, the maximum power density decreases to 70% of the maximum power density with hydrogen.

3.3. Effect of steam reforming of methane on cell performance

Steam reforming of methane mainly contains two reactions:

Steam reforming of methane, $CH_4+H_2O=CO+3H_2$, $\Delta H_{298}=206KJ/mol$ (3)

Water gas shift reaction, $CO+H_2O=CO_2+H_2$, $\Delta H_{298}= -41KJ/mol$ (4)

Table 3. Fuel compositions from steam reforming of methane at 1023K

Flow rates of the feed		H_2%	CO%	CH_4%	CO_2%
CH_4 (ml/min)	H_2O (g) (ml/min)				
25	75	76.9	13.4	0.4	9.3
25	50	76.5	15.9	0.7	6.9
25	25	72.9	20.0	4.3	2.8
25	13	60.6	18.6	19.5	1.3

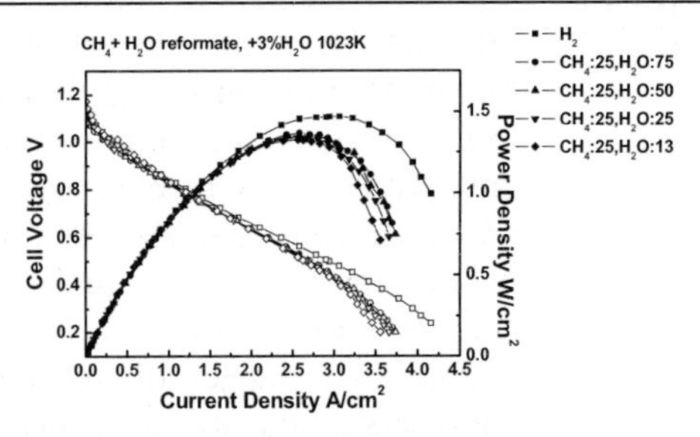

Fig. 3. Performances of the cell at 1023K with H_2 and fuels from steam reforming of methane.

High concentration of hydrogen can be obtained by steam reforming of methane, and the ratio of H_2/CO in the product is higher than 3:1.

Table 3 shows the fuel compositions from steam reforming of methane at 1023K at different ratios of H_2O/CH_4. With the increase of the ratio of H_2O/CH_4, the H_2 and CO_2 concentrations increase gradually. With the H_2O/CH_4 ratio increased from 2 to 3, the methane concentration reduces from 0.7 % to 0.4 %, but the hydrogen concentration varies little.

Fig. 3 shows the current-voltage and current-power curves of the cell with the products from steam reforming of methane and H_2. When the H_2O/CH_4 ratio is below 2, the cell performance increases slightly with the increase of the H_2O/CH_4 ratio. The reason may be that the product H_2 concentration increases when the H_2O/CH_4 ratio increases. One mole of methane reacting with steam can produce three-mole hydrogen, so the fuel flow rate can also be increased when steam-reforming reaction (3) was promoted.

3.4. Effect of CO_2 reforming of methane on cell performance

The CO_2 reforming of methane consists in the reaction of methane with carbon dioxide to produce hydrogen and carbon monoxide:

$$CH_4+CO_2=2CO+2H_2,\ \Delta H_{298}=247KJ/mol \qquad (5)$$

$$CO_2+H_2 = CO+H_2O,\ \Delta H_{298}= 41KJ/mol \qquad (6)$$

The synthesis gas generated by carbon dioxide reforming has a low H_2/CO ratio of ca.1. With the increase of the ratio of CO_2/CH_4, reaction (6) can be promoted, so the ratio of H_2/CO is less than 1:1 when the ratio of $CO_2/\ CH_4$ is lager than 1:1, which can be seen from table 4. As shown by Table 4, we can see that there are only 9.7 % CH_4 and 3.7 % CO_2 remained when the ratio of CO_2/CH_4 increases to 21:25, and the concentrations of H_2 and CO are 43.6% and 43%. When the ratio of $CO_2/\ CH_4$ increases to 63:25, there is only 0.7 % CH_4 remained, whereas the concentration of carbon dioxide increases to 22.6 % and the concentration of hydrogen decreases to 28.5%.

Table 4. Fuel compositions from CO_2 reforming of methane at 1023K.

Flow rates of the feed		H_2%	CO%	CH_4%	CO_2%
CH_4 (ml/min)	CO_2 (ml/min)				
25	63	28.5	48.2	0.7	22.6
25	42	37.9	48.3	1.6	12.2
25	21	43.6	43.0	9.7	3.7
25	13	42.1	36.5	19.9	1.5

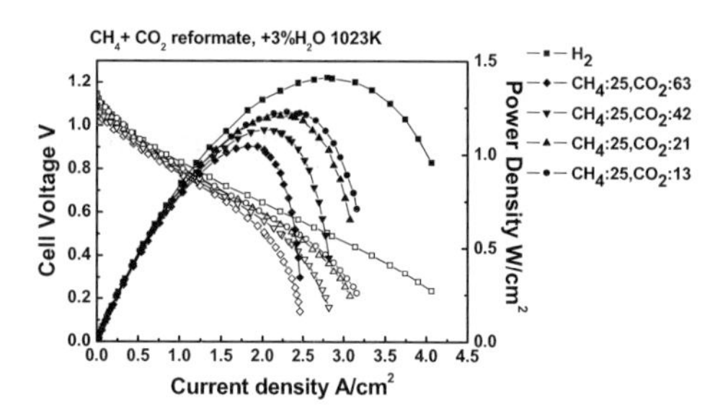

Fig. 4. Performances of the cell at 1023K with H_2 and fuels from CO_2 reforming of methane.

Fig. 4 shows the current-voltage and current-power curves of the cell with processed fuels from CO_2 reforming of methane and H_2. With the increase of the CO_2/CH_4 ratio, the maximum power density and the maximum current density of the cell decrease when the ratio of CO_2/CH_4 is lager than 1. When the ratio of CO_2/CH_4 increases to 63:25, the maximum power density decreases to 74% of that with hydrogen. Concentration polarization becomes more and more serious. The mainly reason is the diluting effect of carbon dioxide whose concentration increases with the increase of the CO_2/CH_4 ratio. On the other hand, that reaction (6) leads to the decrease of H_2 concentration and the increase of CO concentration. Yi Jiang and A V. Virkar [8] have shown that carbon dioxide could decrease the performance and large amount of CO (CO/H_2 >1) could also lead to the decrease of the performance apparently.

4. Conclusions

The effects of different processed methane fuels on cell performance were studied. The fuels from partial oxidation with air caused the occurrence of limited current densities and decreased the performance apparently, which was ascribed to the diluting effect of nitrogen. While the processed methane fuels from partial oxidation with oxygen and steam reforming affected little on cell performance in the range of our experiments. The fuels from CO_2 reforming of methane also caused the occurrence of limited current densities and decreased the performance apparently, which was ascribed to the diluting effect of CO_2 and the increase of the concentration of CO in the processed fuel.

Acknowledgments. The authors gratefully acknowledge financial supports from the Ministry of Science and Technology of China (Grant No. 2004CB719506 and 2005CB221404).

References

1. R. Peters, E. Riensche, P. Cremer, Journal of Power Sources, 86 (2000) 432-442.
2. C. Finnerty, G. A. Tompsett, K. Kendall, R. M. Ormerod, Journal of Power Source, 86 (2000) 459-463.
3. M. Sahibzada, B. C. H. Steele, K. Hellgardt, D. Barth, A. Effendi, D. Mantzavinos, I. S. Metcalfe, Chemical Engineering Science, 55 (2000) 3077-3083.
4. C. Finnerty, G. A. Tompsett, K. Kendall, R. M. Ormerod, Journal of Power Source, 86 (2000) 459-463..
5. T. Kim, S. Moon, S. -I. Hong, Applied Catalysis A: General, 224 (2002) 111-120.
6. S. Freni, G. Calogero, S. Cavallaro, Journal of Power Sources, 87 (2000) 28-38.
7. E. Riensche, J. Meusinger, U. Stimming, G. Unverzagt, Journal of Power Sources, 71 (1998) 306-314.
8. Y. Jiang, A. V. Virkar, Journal of The Electrochemical Society, 150 (7) A942-A951 (2003).

Natural Gas Conversion VIII
F.B. Noronha, M. Schmal, E.F. Sousa-Aguiar (Editors)

Activity and stability of iron based catalysts in advanced Fischer-Tropsch technology *via* CO_2-rich syngas conversion

L. Spadaro[1], F. Arena[2], G. Bonura[1], O. Di Blasi[1], F. Frusteri[1]

[1]Istituto di Tecnologie Avanzate per l'Energia (CNR-ITAE),Via Salita S. Lucia sopra Contesse 5, I-98126 S. Lucia, Messina, ITALY
[2]Dipartimento di Chimica Industriale e Ingegneria dei Materiali, Università degli Studi di Messina Salita Sperone 31, I-98166 S. Agata, Messina, ITALY

Abstract:
Several iron-based catalysts supported on different oxide carriers (CeO_2, MnO, ZnO) were synthesized by the combustion route and probed in the Fischer-Tropsch synthesis reaction in the range 250-300°C and at 10-20 bar total pressure, using a *PF*-reactor and various CO_x reaction mixtures ($H_2/CO=2$ or $H_2/CO_2=3$ or $(CO+CO_2)/H_2=2$). The prepared catalysts feature a catalytic performance superior to that of commercial FT catalysts and suitable for industrial application as a very high activity (100-60% of CO_x conversion), selectivity (low methane and high long-chain product selectivity; ASF parameter close to 0.9) and long lifetime (500 h).

1. INTRODUCTION

Fischer-Tropsch (FT) synthesis is a well know industrial process for the *syn-fuel* manufacture, in which syngas is reacted over catalysts to produce a mixture of straight and branched-chain hydrocarbons [1]. Both cobalt and iron based catalysts are adequate for the FT process coupling a low cost with a high catalytic performance [1]. In spite of the almost similar formulation of FT-fuels with respect to the refinery's products (gasoline, kerosene, diesel, etc.), these *syn-fuels* possess many valuable characteristics as the superior thermal efficiency and the complete absence of heavy metal, sulfur and aromatic compounds, implying thus many economic and environmental advantages [1-7]. Either *"Low Temperature"* (180-270°C) or *"High Temperature"* (300-350°C) FT-process are industrially performed depending on the kind of catalyst, the characteristics of available products and the typology of feedstock utilized for the catalytic generation of syngas which, in turn, may be derived from coal, natural gas, biomass or heavy oil streams [2,3,4,6]. In this field, new technologies for *syn-fuel* manufacture (*"Gas-to-Liquid",* GTL) and based on "Natural Gas processing" *via* syngas represent the strategic route engaged by the major company as BP, SASOL, SHELL, CHEVRON and EXXON-MOBIL

[2,6]. Synthetic fuels obtained *via* the Fischer-Tropsch synthesis from CO_2-rich syngas streams are of high practical interest as it allows the exploitation of remote NG sources containing high amount of CO_2, also diminishing the reforming temperature to reduce the syngas generation costs which is the heaviest voice in the economy of large-scale integrate GTL processes [2,4,5,6]. However, FTS processes *via* CO_2 hydrogenation are less effective than the conventional syngas (CO/H_2) route [4,8], because of limitations imposed by the "*reverse gas water shift reaction*" (RWGS) equilibrium [3,4,9-12]. Therefore, the present work focuses on the development of suitable effective catalytic systems to be employed in the conversion of "CO_2-rich" syngas streams, coming from NG *"LT-wet-partial-oxidation"*, into clean hydrocarbons *via* an advanced Fischer-Tropsch technology.

2. EXPERIMENTAL

Catalysts. Various Fe-based catalysts were prepared by the combustion route, according to the procedure elsewhere described [10]. The obtained catalysts were doped with potassium by a subsequent stepwise addition of K_2CO_3 aqueous solution. Before testing, all the catalysts were pre-reduced in-*situ* at 400°C in flowing H_2 (2.0 $Nlg^{-1}h^{-1}$). An industrial catalyst (SAS), developed by SASOL and considered in the literature as *state-of-the-art* catalyst [4], was used as reference system. The list of the studied catalysts is given in Table 1.

TPR and TPD measurements were performed in a flow apparatus operating in both pulse and continuos modes, according to the procedures elsewhere described [4,10].

Catalysts testing. Reaction tests were performed at 20 bar in the range 250-300°C using a PFR fed with different mixtures ($H_2/CO=2$; $H_2/CO_2=3$; $(CO+CO_2)/H_2=2$) at GHSV ranging from 1 to 5 Nl $g^{-1}\cdot h^{-1}$. The reaction stream was analysed by GC [4].

Table 1. List of catalysts.

		Composistion (wt. %)							
code	*S.A. m²/g*	*Fe₂O₃*	*SiO₂*	*CeO₂*	*ZrO₂*	*MnO*	*ZnO*	*CuO*	*K₂O*
CK	134.0	62.5	1.5	25.5	3.0	--	--	3.0	4.5
ZK	40.0	37.8	--	--	--	--	54.7	3.0	4.5
MK	36.0	71.7	--	--	--	20.0	--	3.3	5.0
SAS	210.0	79.5	13.3	--	--	--	--	2.8	3.3

3. RESULTS AND DISCUSSIONS

As the different oxide carriers (CeO_2, ZnO, MnO, SiO_2) characterisation data outline a quite different development in *Surface Area* (S.A.) ranging from ca. 30 to ca. 200 $m^2\cdot g^{-1}$ (Table 1). The TPR and TPD measurements disclose that both composition and preparation route affect the redox and chemisorption properties, highlighting a different behaviour of the prepared catalysts with respect to the commercial one. The TPR patterns of representative CK and SAS systems are shown in Figure 1. The reference and ceria-based catalysts feature a similar reduction pattern, characterized by the presence of two main hydrogen consumption peaks centred at ca. 300°C and 550°C that, according to literature, are attributable to the reduction of CuO [13] and of FeO_x to Fe^0 (Fe_2O_3 to Fe_3O_4 to Fe^0) [14], respectively. From a quantitative point of view, the reference system exhibits a higher reduction kinetics in the lower temperature zone (200-400°C) with respect to the CK one that features a concomitant reduction of Fe^{III} species. On the contrary, the CK catalyst shows a higher reduction degree in the range 400-700°C that mirrors an *over-stoichiometric* hydrogen consumption, likely pointing to the surface reduction of ceria (i.e., $Ce^{IV} \rightarrow Ce^{III}$) [10].

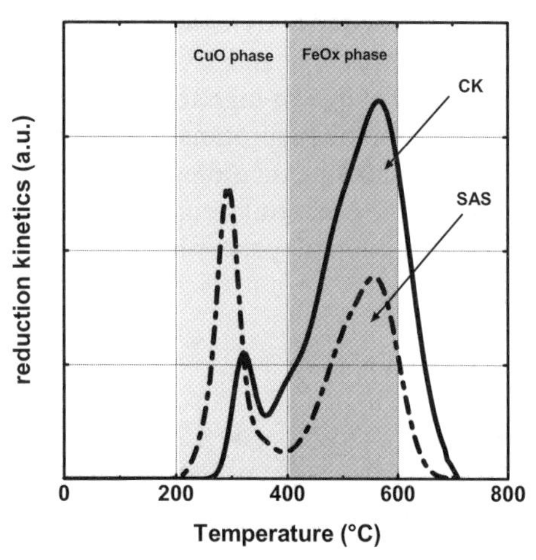

Figure 1. TPR spectra of CK and SAS system

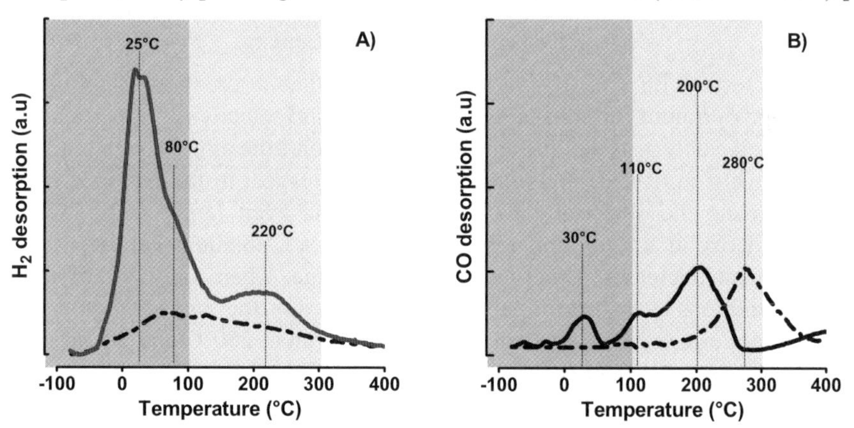

Figure 2. H_2-TPD (A) and CO-TPD (B) spectra of CK (solid line) and SAS (dotted line) catalysts.

To shed lights on the chemisorption properties of CK and SAS samples, in Figure 2 are shown the TPD patterns of H_2 and CO. In both cases the desorption process involves two main temperature zones, one of which (150-300°C) is strongly related to the FT reactions. By comparing the different profiles it immediately emerges a higher extent of H_2 and CO uptake for the catalyst prepared by the combustion route, either at low or high temperature. The lower CO desorption T range is also an index of a minor adsorption energy [4]. Crucial for the catalytic pattern is also the presence of alkali. According to literature data [4-9], the formation of hydrocarbon *via* CO_2 hydrogenation occurs through the primary conversion of CO_2 to CO. In this contest potassium dopant is helpful for improving the conversion of CO_2 to CO.

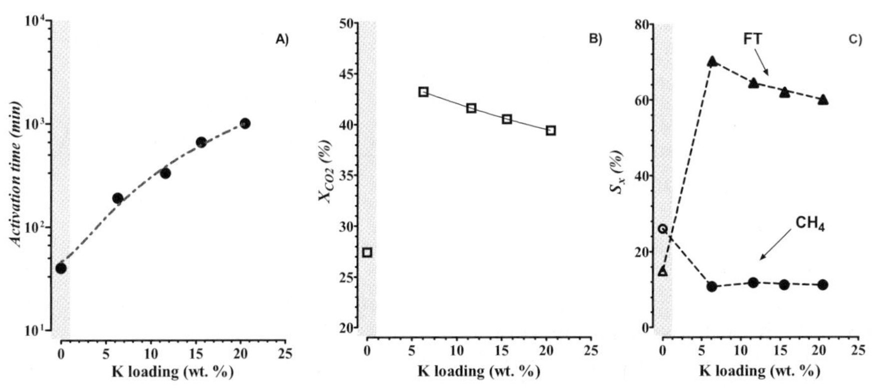

Figure 3. Effect of K loading on the performance of CK catalyst. (A) activation time; (B) CO_2 conversion; (C) product selectivity.

Preliminary tests with CO_2/H_2 stream indicated that a ca. 4 wt.% K loading represents the optimum for enhancing both activity and selectivity. In particular, for the ceria based catalysts (CK), presenting an activation time almost unchanged (Fig. 3A), the conversion of CO_2 is almost twofold with respect to the undoped catalyst (Fig. 3B), while the selectivity to hydrocarbons is even larger (Fig. 3C). The progressive decay in activity and selectivity at higher K loadings can be attributed to a consequent decrease in the surface H_2 coverage. Then, two sets of tests in different conditions were performed in order to assess the catalyst behaviour in presence or not of CO_2. Namely, the first tests were run at 250°C and 20 bar using a feed composition of H_2/CO (2:1) flowing at a GHSV of 1.4 Nl $g^{-1}\cdot h^{-1}$, while the subsequent tests were run with $(CO/CO_2=1)/H_2=2$ mixture at 300°C to promote the endothermic *"RWGS"* equilibrium. The results, collected in Tables 2 and 3, signal that the best results in terms of CO_x conversion are attained on the CK catalyst.

Table 2. Activity data at 100h, 250°C and 20 bar, using the CO/H_2 stream.

	Conversion (%)	Selectivity (%)						Productivity
code	CO	CH_4	C_2-C_5	$>C_6$	CO_2	α	P/O	(g_{CH2}·$Kgcat^{-1}h^{-1}$)
CK	80.0	6.2	25.0	30.5	38.3	0.73	1.3	133.0
ZK	73.0	3.6	18.4	41.3	36.7	0.83	0.4	129.5
MK	75.0	2.8	15.2	45.6	36.4	0.87	0.4	136.0
SAS	45.0	4.8	13.7	56.2	25.3	0.92	0.7	94.0

The CK sample attains a ca. 80% CO conversion using the CO stream while the CO is fully converted in addition to a 16% CO_2 conversion employing the CO_2-rich syngas. Similar and valuable is also the perfomance of MK and ZK catalysts. Indeed, these feature a ca. 75% CO conversion in the CO/H_2 mixture and of 95% in the second stream (Table 3), being doubled the CO_2 conversion on the ZK system (11%) with respect to the MK one (7%). Whereas, the reference catalyst exhibits a much poorer functionality toeards CO_2 hydrogenation. Comparing the data obtained using the two reaction streams, then it emerges that the catalyst selectivity depends upon type of stream and reaction conditions as well. First, notable is the rise in methane production using the second reaction stream (Table 3), though it results dramatically high for the reference catalysts (30 %). This implies a reduced amount of long-chain products (see data on C6+ and α terms in Tables 2-3) mostly for the SAS and CK systems, while the drop results lesser for the MK and ZK catalysts. In terms of olefin/paraffin distribution, the MK and ZK catalysts give mostly a more olefinic product, while on the CK and SAS catalysts the distribution depends upon reaction stream and/or reaction temperature in an opposite way. In particular, paraffins are prevailing at 250°C on the CK catalyst or, using the CO_2-rich stream, at 300°C on the SAS system. In terms of productivity, similar hydrocarbon production levels are obtained either *via* conventional or CO_2-rich syngas streams on the prepared systems (Tables 2-3). Then, long-terms stability tests of 500h were performed using the CO_2-rich composition to shed light into catalyst lifetime.

Table 3. Activity data at 100h, 300°C and 20 bar, using the CO_2-rich syngas stream.

	Conversion (%)		Selectivity (%)							Productivity
code	CO	CO_2	CH_4	C_2-C_5	$>C_6$	CO	CO_2	α	P/O	(g_{CH2}·$Kgcat^{-1}h^{-1}$)
CK	99.0	17.0	11.2	62.3	26.5	--	--	0.66	0.3	136.0
ZK	95.0	11.0	9.4	51.1	47.5	--	--	0.79	0.3	127.0
MK	94.0	7.0	9.4	53.0	37.6	--	--	0.77	0.3	121.0
SAS	74.0	0.0	33.3	35.5	15.2	--	16.0	0.54	1.2	50.5

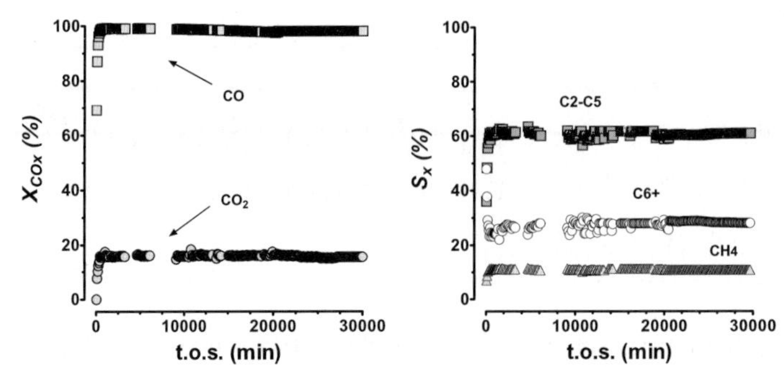

Figure 4. Long term stability test on the CK catalyst using the CO_2-rich syngas stream.

Stability tests results of the CK system, shown in Figure 4, signal a quite stable performance during all t.o.s., the C2-C5 (≈60%) and C6+ (≈29%) being the main products of the advanced FT technology on the ceria-based system (Fig. 4B).

4. CONCLUSIONS

The practical feasibility of FTS *via* hydrogenation of CO_2-rich streams using iron-based catalysts is assessed. The hydrocarbon productivity depends upon reactants chemisorption and heat of products desorption. The catalysts prepared by the combustion route drive effectively the hydrogenation of both CO and CO_2 ensuring a high-chain growth and a very high stability both in CO and CO_2 hydrogenation.

References

[1] G.P. Van der Laan, A.A.C.M. Beenackers, *Catal. Rev.-Sci. Eng.* 41 (1999) 255.

[2] J. Shen et al., Symposium on Advances in Fischer-Tropsch Chemistry. 219th National Meeting, ACS, San Francisco (2000).

[3] US-DOE research project N.: DE-AC22-94PC94055—13 on *"Technology development for iron Fischer-Tropsch catalysis"*, (1999).

[4] European research project (EU) N.: NNE5-2002-00424 on *"New GTL Based on Low Temperature Catalytic Partial Oxydation (LTCPO-GTL)"*, (2005).

[5] H. Schulz, *Top. Catal.* 26 (2003) 73.

[6] H. Fleisch, R. A. Sills, *Stud. Surf. Sci. Catal.* 147 (2004) 31.

[7] S.L. Soled et al., *Top. Catal.* 26 (2003) 101.

[8] G. D. Weatherbee, C. H. Bartholomew, *J. Catal.* 77 (1984) 352.

[9] T. Riedel et al., *Ind. Eng. Chem. Res.* 40 (2001) 1355.

[10] L. Spadaro et al., *J. Catal.* 234 (2005) 451.

[11] P. H. Choi et al., *Catal. Lett.* 40 (1996) 115.

[12] *Appl. Catal. A*, vol. 186 (1996) 1-442.

[13] F. Arena et al., *Appl. Catal.* B, 45 (2003) 51.

[14] E. E. Unmuth et al., *J. Catal.* 61 (1980) 242.

Natural Gas Conversion VIII
F.B. Noronha, M. Schmal, E.F. Sousa-Aguiar (Editors)

Highly active cobalt-on-silica catalysts for the Fischer-Tropsch synthesis obtained via a novel calcination procedure

Jelle R.A. Sietsma, Johan P. den Breejen, Petra E. de Jongh, A. Jos van Dillen
Johannes H. Bitter, and Krijn P. de Jong

Inorganic Chemistry and Catalysis, Faculty of Science, Utrecht University, Sorbonnelaan 16, NL-3508 TB Utrecht, PO Box 80003, The Netherlands

Using ordered mesoporous SBA-15 as model support and pore volume impregnation of aqueous cobalt nitrate we identified a novel calcination procedure in a diluted NO atmosphere, which enables the preparation of Co_3O_4/SiO_2 catalysts that combine highly loadings (15 -18 wt% cobalt) with high dispersions (4-5 nm Co_3O_4 particles). Experiments with silica gel demonstrated that this method is applicable to conventional supports too. After reduction the Co/SiO_2 catalyst treated via this calcination method showed to have a superior activity in FT synthesis compared to the catalyst treated via air calcination.

1. Introduction

Fischer-Tropsch (FT) synthesis is an important process step for the production of clean liquid fuels from synthesis gas. Although the FT synthesis is also catalysed by Ru, only Fe and Co are commercially applied. Despite Fe being more cost-effective, Co is generally preferred in natural gas based FT because of its superior activity, selectivity and stability. At this moment a typical industrial catalyst has relatively large Co particles of 10-20 nm in size. From an economical point of view it is important that the effectiveness of the Co is increased and an optimal Co particle size of 6-8 nm has been reported recently [1]. Next to the Co dispersion, the nature of the support has been varied with emphasis on SiO_2 [2], Al_2O_3 [3] and TiO_2 [4]. In particular with SiO_2 many fundamental studies on preparation [2], Co dispersion [5] and reducibility [6] have been carried out. In this paper we focus on preparation of highly active Co/SiO_2 catalysts.

By far the most widely used method for the preparation of heterogeneous catalysts is impregnation of a porous support with a precursor-containing solution, followed by drying. Subsequently, the precursor is converted into the desired metal via calcination and reduction treatments. This method is primarily attractive because of its simplicity, low costs and limited amount of waste produced. Use of an aqueous $Co(NO_3)_2.6H_2O$ precursor solution is attractive as it enables the preparation of highly loaded catalysts via a single-step impregnation. Moreover, the nitrate can be easily removed via combustion leaving no impurities behind. Unfortunately, its use as precursor salt typically yield catalysts that display a poor Co dispersion [7]. Few attempts have been reported to address this problem. Van der Loosdrecht *et. al.* showed that increasing the space velocity during the calcination treatment led to a higher FT activity due to an increased metal dispersion [3]. Iglesia and co-workers demonstrated that omitting the calcination step via direct reduction of the dried impregnate improved the performance significantly [8].

Here, we present a novel calcination method found using SBA-15 as a model support. Because of the ordered structure of SBA-15 detailed information could be obtained on the impact of experimental conditions on the dispersion and distribution over the support of the precursor of the active phase. This facile calcination method enabled us to obtain a Co_3O_4/SiO_2 catalyst that combined a high loading with a high dispersion. After reduction this catalyst demonstrated to be highly active in the Fischer-Tropsch synthesis.

2. Experimental

SBA-15 (pore volume= 1.00 $cm^3.g^{-1}$, average pore size= 9 nm) was synthesized according to a literature procedure [9]. Davicat 1404SI silica gel (mesopore volume= 1.25 $cm^3.g^{-1}$, average pore size= 6.5 nm) was used as received from Grace-Davison.

Prior to impregnation, the supports were dried for 2 hours in vacuum at 80 °C. Quantities of 0.25 g of SBA-15 and 1 g of silica gel were impregnated to incipient wetness with an aqueous 3 M $Co(NO_3)_2.6H_2O$ solution to provide 15 and 18 wt% Co/SiO_2, respectively. The samples were dried during 12 hours at 70 °C and labelled S/D (SBA-15) and G/D (silica gel). Subsequently, quantities of 40 to 100 mg of these samples were given a thermal treatment during 4 hours at 450 °C (heating rate= 1 $°C.min^{-1}$) in air or in 1 vol% NO/He flow using a plug-flow reactor (100 $ml.min^{-1}$). The products were denoted S/D-AC and S/D-NC, and G/D-AC and G/D-NC, respectively.

XRD patterns were obtained at room temperature from 10 to 90° 2θ with a Bruker-AXS D8 Advance X-ray Diffractometer setup using Co-$K\alpha_{12}$

radiation. N_2-physisorption measurements were performed on a Micromeritics Tristar 3000 apparatus at -196 °C. Prior to analysis the samples were dried in a He flow for 14 hours at 120 °C. The pore size distribution, micro- and meso-porosity were derived from the adsorption branch of the isotherm using NL-DFT [17]. There is no standard method for the determination of blocked mesopore volume ($V_{meso,bl}$). We used BJH theory with the Harkins and Jura thickness equation and the Kruk-Jaroniec-Sayari correction for ordered mesoporous siliceous materials to calculate from the desorption branch the cumulative pore volume distribution [18]. The total amount of $V_{meso,bl}$ was determined considering that the volume in pores with a diameter of 2 - 5 nm is (partially) blocked. Dark-field scanning transmission electron microscopy (STEM) images were obtained using a Tecnai 20 FEG microscope operating at 200 kV and equipped with a High Angle Annular Dark Field (HAADF) detector.

Fischer-Tropsch synthesis was done at 1 bar and 220 °C using a plug-flow reactor and H_2/CO volume ratio of 2. To achieve isothermal plug-flow conditions a small quantity of the sample (20 mg) was mixed with 200 mg silicon carbide particles with a diameter of 0.2 mm. The calcined catalysts were *in situ* reduced using helium containing 33 vol% of hydrogen and a total flow of 60 $ml.min^{-1}$. Reduction temperatures were varied from 350 °C to 600 °C using a heating rate of 5 $°C.min^{-1}$. The final temperature was maintained for two hours. Online product gas analysis (C_1-C_{20}) was performed using a gas chromatograph to determine the selectivity (wt%) towards C_1 and C_{5+} hydrocarbons. The activities and selectivities were determined after 16-22 hours of reaction. For this the conversion of CO was adjusted to 2% by tailoring the space velocity.

3. Results and discussion

The impact of the individual steps in preparation on the distribution of the precursor over the support was studied using SBA-15 as support. STEM results of sample S/D obtained after impregnation and drying showed that the cobalt nitrate precursor was well-dispersed over the support. Moreover, no large cobalt nitrate particles were found outside the mesopores of the SBA-15 particles. Both micro- and meso-pore volume had considerably decreased upon impregnation and drying; a decrease of 55% in meso-porosity was observed. This could be attributed to filling of the mesopores with cobalt nitrate. Low angle XRD results confirmed that the long range order of the support had been retained. XRD at higher angles demonstrated that the cobalt nitrate was amorphous. An HAADF-STEM image of sample S/D-AC obtained after calcination in air is depicted in Figure 1. The brighter areas correspond with the Co_3O_4 particles. The image shows that during this step severe sintering and redistribution of the precursor had taken place. The Co_3O_4 was found inhomogeneously distributed over the support. Moreover, precursor retained

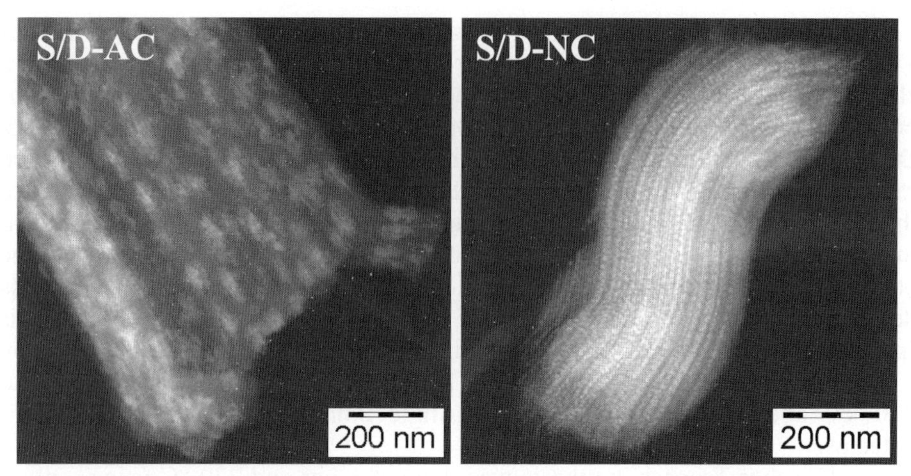

Fig. 1. Dark-field HAADF-STEM images of samples S/D-AC and S/D-NC after calcination in air or in 1 vol% NO/He, respectively.

inside the mesoporous channels seemed to have been limited in their growth by the pore walls. These particles adopted a rod-like geometry and were blocking the mesopores. The N_2-physisorption isotherm of this sample contained a forced closure of the desorption branch at a relative pressure of ~0.48, suggesting indeed Co_3O_4 plugs inside the mesoporous channels [12]. The (partially) blocked mesopore volume was found to be 82% of the total pore volume. With XRD line broadening an average Co_3O_4 crystallite size of 10 nm was found. Comparable experiments with pure helium instead of air yielded similar results.

On the right-hand side in Figure 1 an HAADF-STEM image obtained after thermal treatment under a 1 vol% NO/He flow (S/D-NC) is depicted. The electron micrograph demonstrates that with this novel calcination method sintering and redistribution of the precursor was prevented. The Co_3O_4 particles with a size of 4-5 nm were found homogeneously distributed throughout the mesoporous channels of the SBA-15. Also no Co_3O_4 appeared to have migrated to the external surface of the support particles. N_2-physisorption did not show any plugging of the mesopores, confirming that the Co_3O_4 particles were relatively small. XRD line-broadening showed an average Co_3O_4 crystallite size of 5 nm.

Experiments with silica gel were carried out to investigate whether these results could be transferred to conventional supports. Small quantities of dried impregnate G/D were calcined in air or in 1 vol% NO/He and labelled G/D-AC and G/D-NC, respectively. STEM analysis showed that the air calcination led to a catalyst with a broad particle size distribution (8-60 nm), whereas the NO/He treatment yielded small Co_3O_4 particles again with a uniform size of 4-5 nm. XRD results confirmed these observations as for sample G/D-AC and G/D-NC average crystal sizes of 11 and 5 nm were observed, respectively.

The Co_3O_4/SiO_2 catalysts were tested for their performance in the FT synthesis of hydrocarbons from synthesis gas. Moreover, the impact of the reduction temperature applied prior to catalysis was investigated. In Figure 2 the observed activities for both catalysts are plotted as a function of the reduction temperature. The activity of the catalyst treated through air calcination increased significantly by increasing the reduction temperature from 350 to 450 °C, but further increasing the temperature had little effect. On the contrary, the activity observed for catalyst G/D-NC that was treated through the novel calcination method continued to increase up to a reduction temperature of 550 °C. At higher temperatures the activity decreased presumably due to sintering of the Co particles.

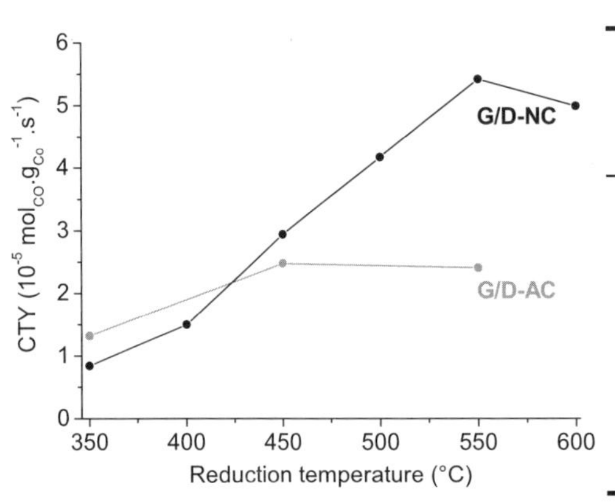

Fig. 2. Relation between FT activity and reduction for samples G/D-AC and G/D-NC

Table 1. Fischer-Tropsch selectivities

$T_{reduced}$ (°C)	C_1 selectivity (%)		C_{5+} selectivity (%)	
	AC*	NC#	AC	NC
350	20	28	51	37
400		25		41
450	17	21	57	46
500		22		44
550	17	28	57	34
600		26		37

*= sample G/D-AC, #= sample G/D-NC

The observed differences between both catalysts can be explained by the Co_3O_4 particle sizes obtained after the calcination treatment. For the air calcined sample relatively large particles of 8 to 60 nm were found. As a result reduction proceeds relatively facile at lower temperatures. Therefore, applying higher reduction temperatures had little effect on the activity. On the contrary, after NO/He calcination small Co_3O_4 particles of 4-5 nm were obtained. Since reduction tends to proceed more slowly for highly dispersed metal oxide phases, applying higher reduction temperatures improved the degree of reduction significantly. Comparison between both catalysts reduced at 550 °C showed that modification of the calcination treatment resulted in an activity increase of more than a factor of two. The CH_4 and C_{5+} selectivities of the catalysts are given in Table 1, and showed that the well-dispersed catalyst prepared via the novel calcination treatment (G/D-NC) had a lower C_{5+} selectivity than the relative poorly dispersed G/D-AC obtained via calcination in air. These results are in line with recent observations of Bezemer *et al.* that showed that the C_{5+} selectivity decreases upon reduction of the Co particle size [1].

4. Conclusions

The preparation of cobalt on silica catalysts via impregnation and drying using an aqueous nitrate precursor solution was studied in detail. During calcination in air severe sintering and redistribution was observed. The composition of the gas flow during calcination largely affects the ultimate dispersion and distribution of precursor over the support. We discovered that carrying out the calcination in a stream of diluted nitric oxide prevented mobility of the precursor. As a result a cobalt oxide on silica catalyst was obtained that combines high loadings (15-18 wt% cobalt) with high dispersions (4-5 nm cobalt oxide particles). The catalyst treated via the new calcination method demonstrated to have a superior activity in FT synthesis over the catalyst treated via air calcination.

Acknowledgements

The authors wish to thank Hans Meeldijk for his support with STEM characterisation of the samples.

References

1. L.G. Bezemer, J.H. Bitter, H. P.C.E. Kuipers, H. Oosterbeek, J.E. Holewijn, X. Xu, F. Kapteijn, A.J. van Dillen, and K.P. de Jong, J. Am. Chem. Soc. 128 (2006) 3956.
2. J-S. Girardon, A.S. Lermontov, L. Gengembre, P.A. Chernavskii, A. Griboval-Constant and A. Y. Khodakov, J. Catal. 230 (2005) 339.
3. J. van de Loosdrecht, S. Barradas, E.A. Caricato, N.G. Ngwenya, P.S. Nkwanyana, M.A.S. Rawat, B.H. Sigwebela, P.J. van Berge and J.L. Visagie, Top. Catal. 26 (2003) 121.
4. R. Riva, H. Miessner, R. Vitali and G. Del Piero, Appl. Catal., A 196 (2000) 111.
5. A. Barbier, A. Tuel, I. Arcon, A. Kodre and G.A. Martin, J. Catal 200 (2001) 106.
6. B. Ernst, S. Libs, P. Chaumette and A. Kiennemann, Appl. Catal., A 186 (1999) 145.
7. A. Martínez, C. López, F. Márquez and I. Díaz, J. Catal. 220 (2003) 486.
8. S.L. Soled, E. Iglesia, R.A. Fiato, J.E. Baumgartner, H. Vroman and S. Miseo Top. Catal. 26 (2003) 101.
9. D. Zhao, J. Feng, Q.Huo, N. Melosh, G.H. Fredrickson, B.F. Chmelka and G.D. Stucky, Science, 279 (1998) 548.
10. M. Jaroniec, M. Kruk, J.P. Olivier and S. Koch, *Proceedings of COPS-V*, Heidelberg, Germany (1999).
11. Kruk, M.; Jaroniec, M. and Sayari, A. Langmuir 13 (1997) 6267.
12. P.I. Ravikovitch and A.V. Neimark, Langmuir 18 (2002) 9830.

Natural Gas Conversion VIII
F.B. Noronha, M. Schmal, E.F. Sousa-Aguiar (Editors)

Intensification of Commercial Slurry Phase Reactors

Alex Vogel[a], Andre Steynberg[a], Berthold Breman[b]

[a]*Sasol Technology, Research and Development Division, PO Box 1, Sasolburg, 1947, South Africa*
bSasol Technology, Research and Development Division, PO Box 217, 7500 AE Enschede, The Netherlands

1. Introduction

The capital cost of building GTL plants has increased substantially in today's economic climate. Demand for engineering contractors and construction materials have increased globally, resulting in higher project costs. In addition, recent studies indicate that large scale facilities are now needed to support commercially viable projects that make a meaningful impact on the global energy scene. This adds to the incentive to save specific capital costs ($/dbbl) to lower the entrance barrier. Despite these increases in capital cost, GTL projects still remain very attractive at the current high oil prices.

Energy is in short supply placing upward pressure on natural gas prices. There is therefore now more competition for the same stranded gas resources. Large GTL facilities compete with LNG for natural gas monetization opportunities. Capital cost reductions in GTL projects will enhance the potential for GTL to secure gas reserves in the future.

While air separation and synthesis gas generation remain the main contributors to the capital cost of the GTL process units, capital savings can also be achieved in synthesis gas conversion. Improvements in Fischer-Tropsch reactor design now allows for increased production capacity from the same reactor shell size or equivalent production capacity from smaller reactors through process intensification, whilst maintaining or even improving selectivity and overall synthesis gas conversion. The manufacture of large scale reactors is costly with a limited number of suppliers who have the capacity to produce such vessels. This paper focuses on the potential of the Fischer-Tropsch slurry phase reactor to contribute to cost reductions per barrel of product through reactor intensification. The cost reduction could be realised through a combination of a lower material and construction cost per barrel of product produced, as well as having access to a larger pool of potential manufacturers. Opportunities will also be created to standardize certain components of reactor manufacture in partnership with preferred suppliers. A further benefit is the ability to match the FT reactor capacity to the increasing maximum capacity for vendor supplied air separation units and synthesis gas generation units. Single train capacities of ca. 7000 ton/day oxygen are expected to be achievable in the future [1], which corresponds to a FT reactor capacity of ca. 40,000 bpd.

The dimensions of the Oryx reactors have challenged and found the limits of scale in single unit manufacturing, logistics and lifting at site. Slurry phase reactor technology will need to process synthesis gas at higher feed gas rates to increase the single reactor capacity. In order to sustain synthesis gas conversion and selectivity into liquid products at these higher rates, the slurry phase reactor requires increased catalyst loads and/or increased catalyst activity. Recent developments in Fischer-Tropsch catalysts have also enabled the reactor to operate at an even higher reactant per pass conversion levels, which can be exploited by operating the reactor at an even higher synthesis gas due to reduced recycle needs. This has been achieved through a greater resistance to water partial pressure, a primary product of the Fischer-Tropsch reaction. This paper, however, only discussion slurry phase reactor developments.

2. Fundamental studies

Cold model work, using a non-reacting slurry phase mock-up with a mimic gas-slurry system at representative gas density, has already demonstrated stable bubble column hydrodynamic performance at gas rates giving more than double the current commercially used gas velocity and volumetric solid loads, as shown in Figure 1. It is believed that the applied gas velocities during these cold model studies are still far below the undesired transition from the churn-turbulent regime to a transported bed regime [2]. The initial small scale cold model studies were followed up with demonstrations of increased feed gas rates, 25% higher than current design, using a semi-commercial scale reactor. These tests have also shown steady reactor operation under the intensified synthesis conditions. Much of the studies on increases in catalyst concentration were also conducted in a semi-commercial scale reactor. Increasing catalyst concentrations with up to 30% above the current design basis has already been shown to be viable with relatively minor changes to the existing commercial scale reactor design.

An understanding of the interaction between gas velocity, catalyst concentration and gas void fraction is used to predict the potential performance enhancement for future slurry phase reactors. It is shown that by keeping the ratio of catalyst mass to feed gas rate, as well as the reactor volume constant, that catalysts of substantially higher activity are not necessarily required to achieve enhance volumetric production performance. The process parameters of catalyst mass and gas velocity have been shown to have significant potential for increased design values before hydrodynamic boundaries are reached.

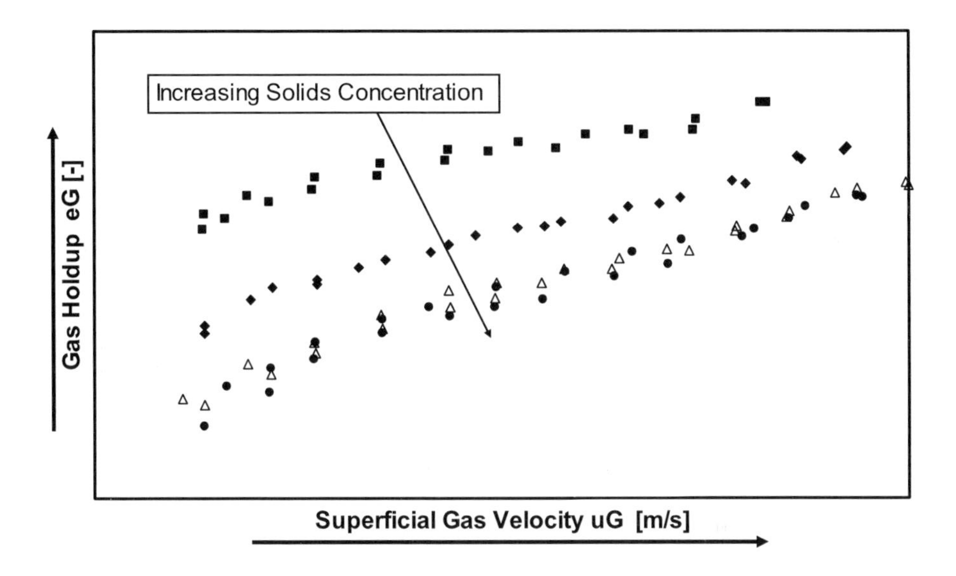

Fig 1. Gas voidage (□$_G$) for air/Al_2O_3/C_9-C_{11} paraffin system as a function of superficial gas velocity at various solid loads (vol. % solids in gas-free slurry). [Conditions: ambient temperature, pressure = 8 bar(a) (⇔ gas density = 9.6 kg/m^3).]

Successful conversion of synthesis gas at elevated velocities also requires that the Fischer-Tropsch reaction does not become mass transfer limited. Gas – liquid mass transfer studies [3] have shown that the ratio of the volumetric mass transfer coefficient to the gas hold-up does not deteriorate even up to gas velocities and catalyst concentrations as high as 0.7 m/s and 40 vol%, respectively. Correcting for the high liquid phase diffusion coefficients of hydrogen and carbon monoxide at LTFT synthesis conditions, no significant reduction of the FT rate due to mass transfer limitations is projected at increased reactor capacities for the foreseeable future.

3. Potential capacity for slurry phase reactors

The studies conducted to date indicate double the current slurry bed reactor volumetric productivity to be easily achievable for the current commercial Sasol proprietary Cobalt catalyst. The potential saving in the cost of the slurry phase reactor through such intensification is estimated to be as much as 25 to 30%. More recent studies have shown that a reactor capacity increase of 3 times the current commercial design are possible by utilizing catalysts of enhanced activity which are being developed in the pilot plant facilities in Sasolburg.

An understanding of the interaction between gas velocity, catalyst concentration and gas void fraction is used to predict the potential performance enhancement for future slurry phase reactors. It is shown that by keeping the ratio of catalyst mass to feed gas rate, as well as the reactor volume constant, that catalysts of substantially higher activity are not necessarily required to achieve enhance volumetric production performance. The process parameters of catalyst mass and gas velocity have been shown to have significant potential for increased design values before hydrodynamic boundaries are reached.

Figure 2 illustrates, with the solid lines, how the void fraction in the reactor can actually decrease slightly when moving from the current design point (circle on

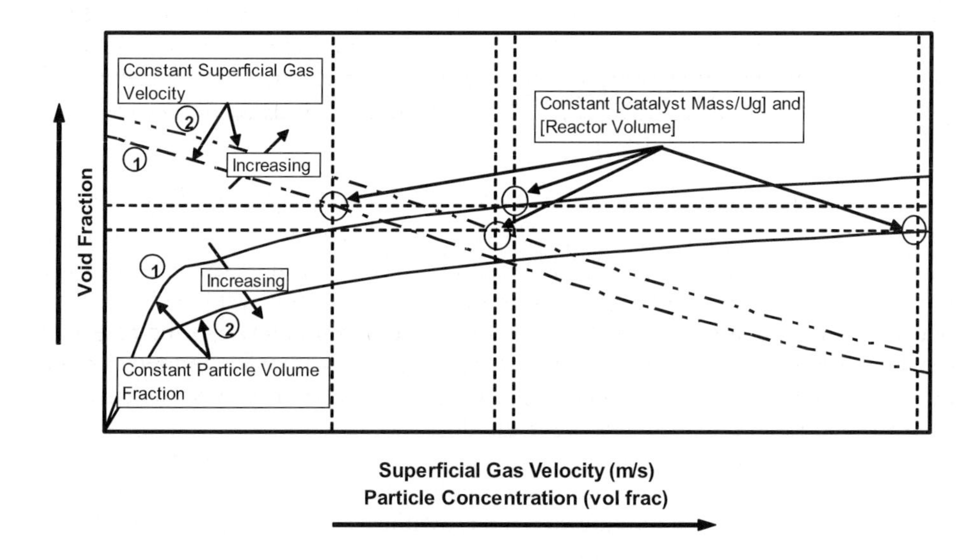

Fig 2. Slurry Bed Void Fraction vs Superficial Gas Velocity and Particle Concentration

the solid line 1) to a new proposed design point at double the superficial gas velocity (circle on the solid line 2).The circles on the dashed lines show the required increase in catalyst particle concentration to go from the current design to the double capacity design. The dashed lines also illustrate the decreasing gas voidage trend with increasing particle concentration. In effect, loading more catalyst increases the slurry volume available by decreasing the gas hold-up.

The capacity to remove heat of reaction from slurry bed reactors also needs to be increased to support reactor intensification. Significant improvements in heat removal are possible with fairly minor changes to the slurry phase reactor cooling system. The cooling tube design and layout can substantially increase the available area for heat transfer. A decrease in tube diameter and pitch can be used to increase the available cooling surface area. These improvements in cooling tube design, together with a reduction in steam generation pressure, can potentially triple the heat removal capacity of a slurry phase reactor without detracting significantly from the reactor volume available for the catalyst/wax slurry.

4. Demonstration of slurry phase reactor capacity

A nominal 1m diameter slurry phase reactor has traditionally been used in Sasolburg to develop the reactor design technology for both the Fe and Co catalyst slurry phase processes over the past 16 years. This reactor has already been used to show an increase in C_{10}+ volumetric productivity of 80% (Figure 3) over the current commercial reactor capacity installed in Qatar at superficial gas velocities and catalyst loads substantially below the upper limits tested in cold model studies as discussed above.

An additional semi-commercial scale reactor, to test and develop specific reactor design data is being constructed in Sasolburg at a cost of just over US$30 million. The new system is required because current test facilities are not able to cover the full range of potential reactor intensification studies. This reactor is termed the Fischer-Tropsch Design Reactor (FTDR). The range of process conditions available to the reactor, such as gas velocity, bed height, reactor gas composition and pressure, will allow demonstration of the capacity limits that could apply to future commercial reactors.

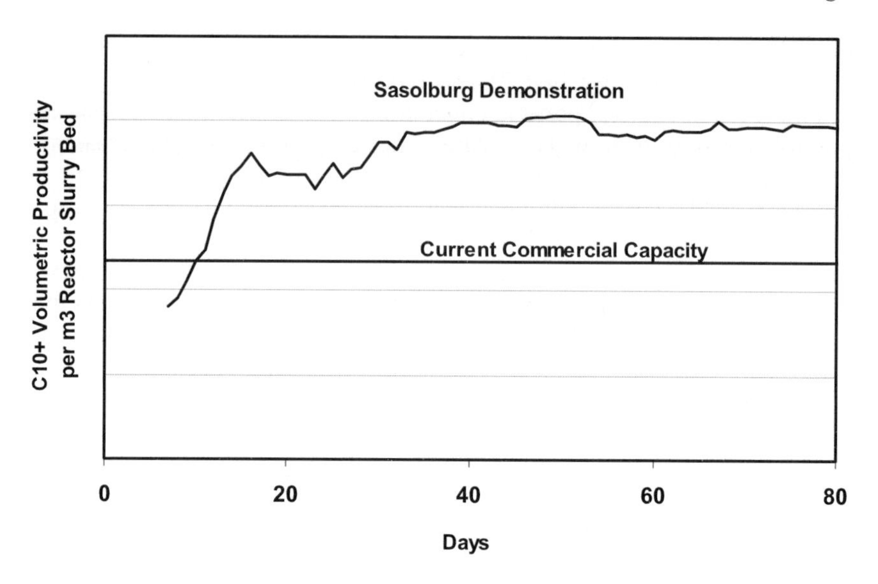

Fig 3. Relative C_{10}+ production per m^3 of reactor slurry bed

It can be concluded that the Slurry Phase Reactor for the Cobalt-based LTFT synthesis still has enormous potential for improvement relative to the economically acceptable Oryx design. The FTDR will be commissioned early in 2007. The learning from operation of the Oryx reactors and the understanding to be gained from this new test reactor will be used to substantially enhance the volumetric productivity for future slurry phase reactor designs.

5. References

[1]. Fischer-Tropsch Technology, A.P. Steynberg and M.E. Dry (editors), 2004, Studies in Surface Sciences and Catalysis, 152, page 444.

[2]. Fischer-Tropsch Technology, A.P. Steynberg and M.E. Dry (editors), 2004, Studies in Surface Sciences and Catalysis, 152, page 117.

[3]. Gas-Liquid Mass Transfer in a Slurry Bubble Column at High Slurry Concentration and High Gas velocities, C.O. Vandu, B. van den Berg and R. Krishna, 2005, Chem. Eng. Technol. 28 (9).

Natural Gas Conversion VIII
F.B. Noronha, M. Schmal, E.F. Sousa-Aguiar (Editors)

A Hydrodynamic Experimental Study of Slurry Bubble Column

Behnoosh Moshtari[a], Jafar Sadegh Moghaddas[b], Ensieh Gangi[c]

[a,b]*Department Of chemical Engineering, Shahand University Of Technology, Tabriz, Iran*
[c] *Oil Research Center, Tehran, Iran.*

1. MAIN TEXT

Introduction

A bubble column is a device in which a gas phase is injected through a column of liquid to promote a chemical or biochemical reaction in the presence or absence of a catalyst suspended in the liquid phase [1]. These reactors are typically operated at high pressure (1-80 MPa), high temperature and high superficial gas velocity (30 m/s). The design and scale up of these reactors require comprehensive knowledge about hydrodynamic behavior under these operating conditions [2].
Most studies have shown that, there are two basic flow regimes in bubble columns; homogeneous and heterogeneous [3,4]. In a system with homogeneous regime the bubble size distribution is narrow and a uniform bubble size, generally in the range 1-7 mm is found [5]. With increasing the superficial gas velocity the bubble size distributions will be changed. In heterogeneous regime small bubbles combine in cluster to form large bubbles in size ranging from 20 to 70 mm. these large bubbles travel up through the column at high velocity (in the 1-2 m/s rang) [5]. Different parameters such as: superficial gas velocity, gas distributor, liquid high, solid concentration, ... can effect on type of flow regime.
Shah et al. [6] show that gas hold up is one of the most important parameter for characterizing the hydrodynamics of bubble columns. It can be defined as the percentage by volume of the gas in the two or three phase mixture in column.

Effect of superficial gas velocity on gas hold up

The dependence of the gas hold up on superficial gas velocity is generally estimated by [6]:

$$\varepsilon_G \propto u_G{}^n$$

The value of the n depends on flow regime. For the bubbly flow regime n varies from 0.7 to 1.2, and for heterogeneous regime n takes value from 0.4 to 0.7 [7]. In fact the gas hold up seems to increase linearly with superficial gas velocity in homogeneous flow regime and the gas hold up reach a maximum where the transition from homogeneous to heterogeneous flow regime occur then more none linear increase with superficial gas velocity[8].

Effect of gas hold up

The solid suspension into a bubble column causes to formation of large bubbles. An increase of bubble size increases the bubble rise velocity, and reduces the residence time of bubble in reactors [9].

The solid particles in bubble column reactors are in micron size and suspended in liquid. When the concentration of solid particle in liquid increases, the gas hold up is reduce [5].

Effect of gas distributor

There are different types of gas distributor, with different design in size and number of orifices. Most common distributors are prose plate and perforated plat. The initial bubble size and distribution at the entire orifice could be controlled by sparger characteristics. The balance between coalescence and break up of gas bubble show that the initial bubble size created at the gas sparger would not completely describe the behavior of gas bubble size distribution in entire column [10].

A large number of correlation for gas hold up have been proposed in the literature, but the large scatter in the reported data dose not allow a single correlation. In the present work, the effects of solid concentrations, sparger type and superficial gas velocity, on gas hold up into a bubble column reactor have studied, and the best correlation for predicting the hydrodynamic behavior on slurry bubble column reactors is suggested.

2. EXPERIMENTAL SET UP

Experimental set up consist of a cylindrical glass column with 0.15 m inner diameter and 2.8 m height. The column is equipped with sparger in bottom, a perforated plate and a porous plate both with 0.1 % porosity. Designing of perforated plate is based on Weber number, this sparger consist of 21 holes with 1 mm diameter.

In all experiments the pressure at top of the column was atmospheric. Gas injected from the bottom of column. After injecting gas, the liquid bed expended and the hydrostatic pressure was changed. With measuring the differential pressure through the column, the total gas hold up can be

determined. For measuring the differential pressure through the column a manometer is used.
The liquid phase is water, gas phase is air and solid particles are silica powder. In figure 1 the set up is showed.

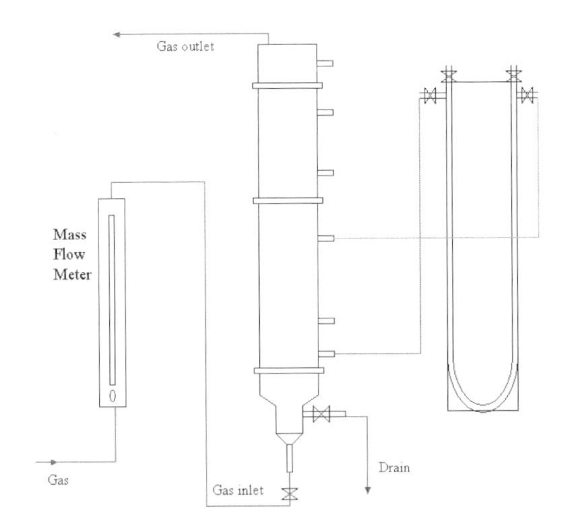

Figure 1- Experimental set up

3. RESULT AND DISCUSSION

Effect of superficial gas velocity

For studying the superficial gas velocity effect on gas hold up the two phase system is used. All the experiments show the positive effect of superficial gas velocity on gas hold up (figure 2). At low gas velocity the uniform bubble distribution is seen and all the bubbles are small, after increasing the gas velocity, gas volume through the column increase, but after finding a maximum on gas hold up the rate of increasing gas hold up decrease. This maximum is transition point and in studying the hydrodynamic properties in slurry bubble column reactors, finding the gas velocity on transition point is one of the most important parameter for predicting slurry bubble column hydrodynamic behavior.
For measuring the bubble size distribution a high speed camera was used. At low superficial gas velocity the bubble size is small (< 7 mm) and with increasing the gas velocity the bubbles are coalescences therefore at transition regime the small and large bubbles are seen through the column. At high gas velocity (more than 9 cm/s) all the bubbles will be large. This regime is heterogeneous. The large bubbles have higher rise velocity than small bubbles,

therefore residence time of large bubble decrease and cause to decrease rate of increasing gas hold up.

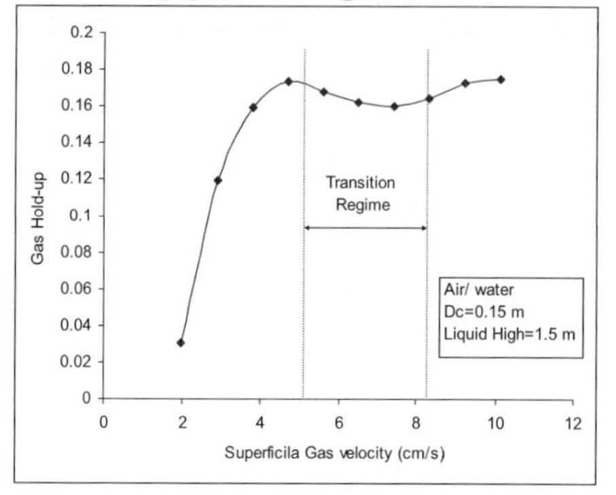

Figure 2 – Effect of superficial gas velocity on gas hold up

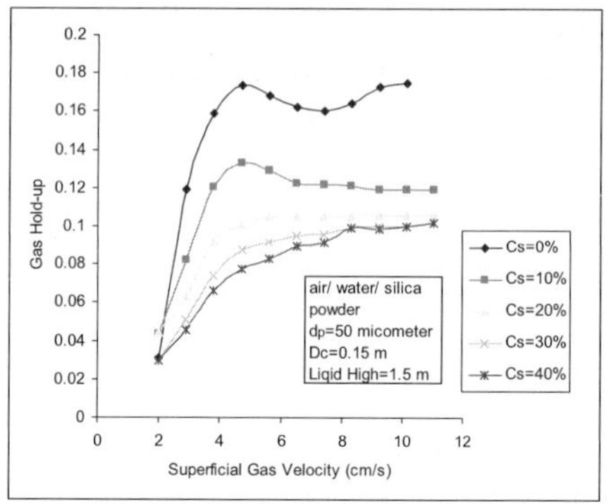

Figure 3-Effect of solid concentration on gas hold up

Effect of solid concentration

Concentrations of solid particle in the liquid phase in slurry bubble columns change the gas hold up at different superficial gas velocity. This effect is shown in figure 3. It is seen that the gas hold up decrease with increasing the solid concentration and density of powder is 2100 μm. The mean particle size of silica powder is 50 kg/m^3.

Solid particles change the physical properties of slurry, such as: density and viscosity. If the viscosity and density of liquid increase the gas hold up will be decrease. In three phase system solid particle increase the bubble, bubble coalescence, and cause to make large bubbles. An increase of bubble size increase bubble rise velocity, and reduce the resident time of bubbles in reactor, therefore the total gas hold up decrease.

Effect of sparger on gas hold up

In this work two different spargers are used: a perforated plate and a porous plate. The orifice diameter of perforated plate is 1 mm and the porous plate consists of micro size pore. The initial bubble size and distribution at the orifice could be controlled by the sparger characteristics. Several investigators have reported that gas sparger had a minimal effect on the bubble sizes and gas holdup if the orifice diameters were larger than 1-2 mm [10, 11].

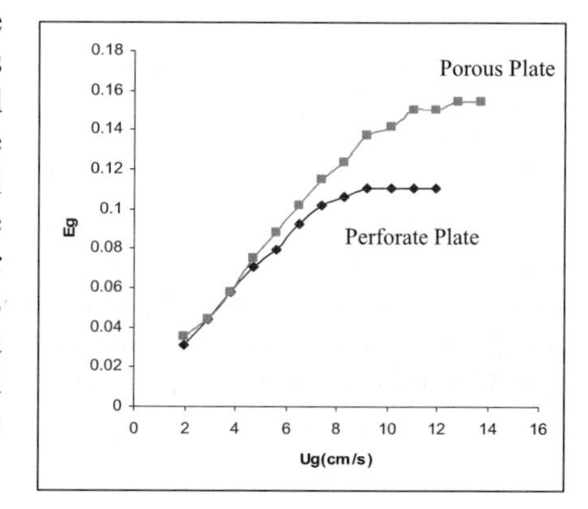

Figure 4 – Sparger type effect on gas hold up (air-water system L/d=8)

Figure 4 shows the effect of sparger type on gas hold up. When the perforated plate is used, the initial size of gas bubbles is larger, and then the gas hold up reduces. When small gas bubbles are formed, the transition from homogeneous to heterogeneous flow regime is delayed; since the rate of bubble coalescence becomes smaller. The small bubbles have larger resident time therefore cause to increase gas hold up. It is seen that the initial bubble size is not only effected reason for changing bubble size, but it is more important.
Table 1 shows some different correlations for predicting gas hold up and figure 5 shows the experimental data and predicting value for gas hold up. It can be seen that Reily et al. correlation is better than other correlations for determining gas hold up.

Table 1- Gas hold up correlations

Author	**Correlation**
Raily et al. [12]	$\varepsilon_g = 296\,U_g^{0.44}\rho_L^{-0.98}\sigma_L^{-0.16}\rho_g^{0.19} + 0.009$
Kumar et al. [13]	$\varepsilon_G = 0.728U - 0.485U^2 + 0.0975U^3$ $U = U_G[\rho_L^2\,\sigma(\rho_L - \rho_G)g]^{1/4}$
Hikita et al. [14]	$\varepsilon_g = 0.672\,(\frac{\rho_g \mu_l}{\sigma})^{0.578}(\frac{\mu_l^4 g}{\rho_l \sigma})^{-0.131}(\frac{\rho_g}{\rho_l})^{0.062}(\frac{\mu_g}{\mu_l})^{0.107}$
Hughmark [15]	$\varepsilon_g = \frac{1}{2 + (\frac{0.35}{u_g})(\frac{\rho_l \sigma}{72})^{1/3}}$

4. CONCLUSION

It can be concluded that: the total gas hold up is increased with increasing superficial gas velocity, and decrease with increasing solid concentration.
The initial bubble size which is depended on sparger type, changes the gas hold up value. Results of this research shows that the porous plate make higher gas hold up than perforated plate. Also, it is found that the Reily 's correlation predicted the gas hold up value better than other presented correlations.

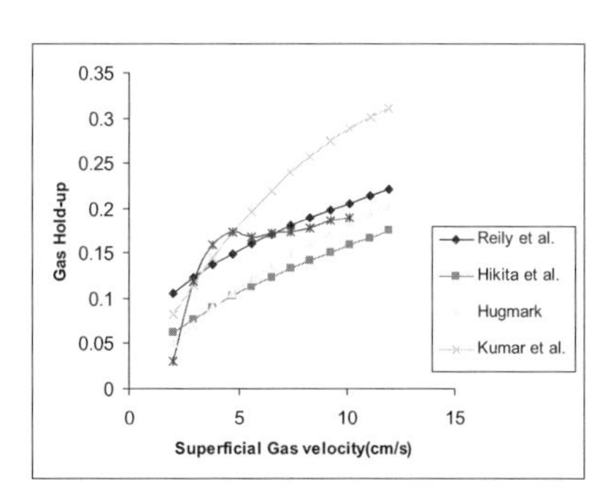

Figure 5- test of experimental data for total gas Hold-up against the correlation in table

REFERENCES

1. Amir Sarrafi, Mohammad Jamialahmadi,Hans Muller-Steinhagen, Johan M.Smith, "Gas Holdup in Homogeneous and Heterogeneous Gas-Liquid Bubble Column Reactor", The CANADIAN Journals of Chemical Engineering, vol 77 (1999).
2. Xukun Luo. DJ. Lee, Raymond Lan, Guoqiang Yang, Liang-Shih Fan, "Maximum stable Bubble size and Gas Holdup in High-Pressure Slurry Bubble Column", AICHE Journal, vol 45 No. 4, (1999).
3. R Krishna, J. Ellenberger, C. Maretto, "Flow regim Transition In Bubble Columns", Int. Comm. Heat Mass Transfer, vol. 26, No. 4, (1999)
4. Magaud, F., Souhar, M., Wild, G., Boisson, N., "Experimental Study of Bubble Column Hydrodynamics," Chemical Engineering Science, Vol. 56, (2001), pp.4597-4607.
5. R. Krishna, S. T. Sie, "Design and scale-up of the Fischer-Tropsch bubble column slurry reactor", Fuel Processing Technology 64 (2000) 73-105
6. Y. T. Shah, B. G. Kelkar, S. P. Godbole, "Design Parameter Estimation for Bubble Column Reactors", AICHE Journal, vol. 28, No. 3, (1982).
7. Deckwer, W.,-D., Bubble Column Reactor, (John Wiley and Sons, Chichester, 1992).
8. Krishna, R., Ellenberger, J., "Gas Holdup in Bubble Column Reactors Operating in the Churn-Turbulent Flow Regime," AIChE Journal, Vol. 42, (1996), pp. 2627-2634.
9. R. Krishn, "A Scale-up Strategy for a Commercial Scale Bubble Column Slurry Reactor for Fischer-Tropsch Synthesis", *Oil & Gas Science and Technology*, Vol. 55, No. 4, (2000).
10. Akita, K., Yoshida, F., "Gas Holdup and Volumetric Mass Transfer Coefficient in Bubble Columns", Industrial Engineering Chemistry Process Design and Development, Vol. 12, (1973), pp. 76-80.
11. Wilkinson, P.M., Spek, A.P., van Dierendonck, L.L., "Design Parameters Estimation for Scale-Up of High-Pressure Bubble Columns," AIChE Journal, Vol. 38, (1992), pp. 544-
12. Reilly, I.G., Scott, D.S., de Bruijn, T.J.W., Jain, A., Piskorz, J., "A Correlation for Gas Holdup in Turbulent Coalescing Bubble Columns", The Canadian Journal of Chemical Engineering, Vol. 64, (1986), pp. 705-717.
13. Kumar, A., Degaleesan, T.E., Laddha, G.S., Hoelscher, H.E., "Bubble Swarm Characteristics in Bubble Columns", The Canadian Journal of Chemical Engineering, Vol.54 (1976), pp. 503-508.
14. Hikita, H., Asai, S., Tanigawa, K., Segawa, K., Kitao, M., "Gas Hold-Up in Bubble Columns", The Chemical Engineering Journal, Vol.20 (1980), pp. 59-67.
15. Hughmark, G.A., "Holdup and Mass Transfer in Bubble Columns", Industrial Engineering Chemistry Process Design and Development, Vol. 6 (1967), pp. 218-221.

Natural Gas Conversion VIII
F.B. Noronha, M. Schmal, E.F. Sousa-Aguiar (Editors)

H-ZSM-5/cobalt/silica capsule catalysts with different crystallization time for direct synthesis of isoparaffins: simultaneous realization of space confinement effect and shape selectivity effect

Guohui Yang, Jingjiang He, Yoshiharu Yoneyama, Noritatsu Tsubaki*

Department of Applied Chemistry, School of Engineering, University of Toyama, Gofuku, Toyama 930-8555, Japan

1. Introduction

Fischer-Tropsch synthesis (FTS) as one option to produce hydrocarbons has been known since 1925 [1]. FTS products have so many advantages such as sulfur-free, aromatic free, and nitrogen-free over conventional petroleum-derived products, which decide them as ideal fuel candidates. But the FTS hydrocarbon products obtained on the traditional FTS catalysts can be only used as the substitute of diesel oil because of their high cetane number and linear structure. At present, more attention has been focused on the production of hydrocarbons rich in isoparaffin because of its wide foreground as synthetic gasoline.

Zeolites, one of the crystalline aluminosilicate, have special pores, cavities and regular channels on molecular dimensions [2]. Some zeolites are excellent hydrocracking and hydroisomerization acidic catalyst for the conversion of heavy oil to light hydrocarbons in petroleum industry. Integrating the acidic zeolite membrane with FTS catalyst, a novel zeolite capsule catalyst was successfully designed and used for the direct synthesis of light isoparaffins from syngas ($CO + H_2$). In the FTS reaction using these zeolite capsule catalysts, syngas passed through the zeolite membrane and reached Co active sites on the

* Corresponding Author: **Email:** tsubaki@eng.u-toyama.ac.jp **Phone (Fax):** 81-76-4456846.

core part of capsule catalysts. Then the formed FTS products with linear structure must enter the zeolite membrane to receive the hydrocracking and isomerization on the acidic sites of zeolite, as shown in Fig.1. As a result, the selectivity of light isoparaffins increased and all the long-chain hydrocarbons disappeared in the final products [3, 4]. In this report, three zeolite capsule catalysts constructed by growing H-ZSM-5 zeolite membrane on the external surface of Co/SiO_2 pellets with different crystallization time are tested for direct synthesis of isoparaffins via FTS reaction.

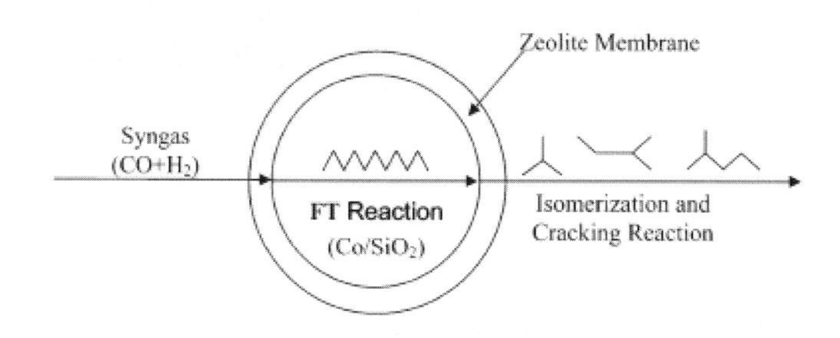

Fig.1 Schematic image of the reaction procedure on the capsule catalysts.

2. Experimental

Silica (Cariact Q-10, Fuji Silysia Chemical Ltd.) with the diameter of 0.85-1.7mm was selected as the support for preparing the conventional 10wt% FTS catalyst of Co/SiO_2 by incipient wetness impregnation of cobalt nitrate aqueous solution. In constructing the H-ZSM-5 membrane on the Co/SiO_2 pellet substrates, hydrothermal synthesis was employed with the TPAOH (tetrapropylammonium hydroxide) as the template, TEOS (tetratethylortho silicate) and $Al(NO_3)_3 \cdot 9H_2O$ as the Si and Al resources respectively. The synthesis solution was prepared based on the ratio of 0.25TPAOH: $0.025Al(NO_3)_3$: 1TEOS: 4EtOH: $60H_2O$. 10wt% Co/SiO_2 FTS catalyst was added in this synthesis sol. And then hydrothermal synthesis was performed at 453K for varied crystallization time of 1, 2 and 7d in order to get different zeolite loading amount. The final H-ZSM-5 capsule catalyst were calcined at 773K in air for 5h to remove the organic template [3].

Isoparaffin synthesis reaction via FTS using these zeolite capsule catalysts was carried out with a high-pressure flow-type fixed reactor [4]. Before the FTS reaction, zeolite capasule catalyst with 0.5g core part of Co/SiO_2 was loaded in the center of the stainless steel reactor (i.d. 8mm) and reduced in situ at 673K for 10h in H_2 flow. FTS reaction temperature and pressure were 533K and 1.0MPa respectively.

3. Results and Discussion

3.1. Characterization of capsule catalysts

The morphology of zeolite capsule catalysts was characterized by SEM-EDS. Cross sectional SEM image and EDS analysis of the capsule catalyst with 1d crystallization time (Co/SiO_2-Z-1) were showed in Fig.2. A compact zeolite membrane can be observed on the outside of Co/SiO_2 substrate, and no gaps or pinholes on the membrane suggested that it is defect-free. No cobalt signals of

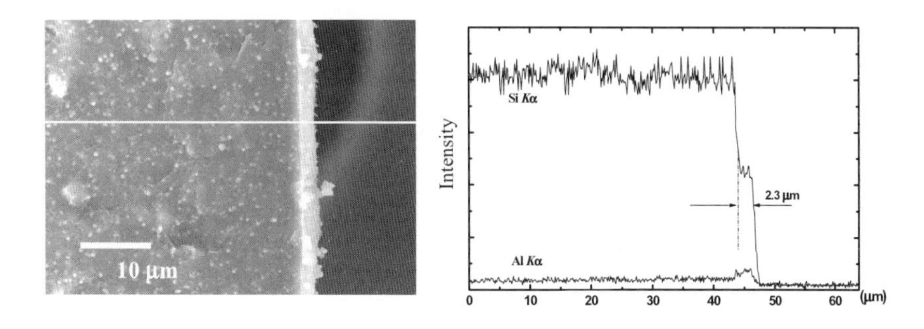

Fig.2 Cross section SEM image and its EDS line analysis of Co/SiO_2-Z-1 (one-day hydrothermal synthesis)

EDS were observed on the external surface of the capsule catalyst, proving the complete coverage of zeolite membrane onto Co/SiO_2 core. The thickness of zeolite membrane was about 2.3μm. At the interface of the Co/SiO_2 substrate and the H-ZSM-5 shell, the radial distribution of Si dropped suddenly while that of Al increased a little, indicating the phase change from SiO_2 to H-ZSM-5. Similar results were also achieved to other two zeolite capsule catalysts with crystallization time of 2d or 7d. The difference was that they had thicker zeolite membrane. For the Co/SiO_2-Z-2, the thickness of zeolite membrane was 7.6μm, and the capsule catalyst Co/SiO_2-Z-7 had 23.1μm zeolite membrane. Different zeolite membrane thickness, namely different zeolite loading amounts, also suggested that it is feasible to control the properties of this zeolite capsule using different hydrothermal synthesis conditions.

The fact that catalyst acidity changed with increasing the zeolite coating amount, had been reported in our previous paper [4]. A linear relationship between the catalyst acidity and zeolite coating amount was obtained, indicating that the property of the acidic sites was the same, despite of hydrothermal synthesis time. But the amount of the acidic sites increased with increasing crystallization time.

The XRD patterns of these samples are showed in Fig.3. Peaks of H-ZSM-5 and Co_3O_4 can be distinguished clearly. The size of Co_3O_4 measured by XRD on three zeolite capsule catalysts was similar to that of the original Co/SiO_2, indicating that hydrothermal synthesis had no obvious effect on the cobalt cluster size. The XRD peak intensity of zeolite on capsule catalysts increased with the increase of crystallization time, which suggested the increasing of zeolite loading amount on the capsule catalysts.

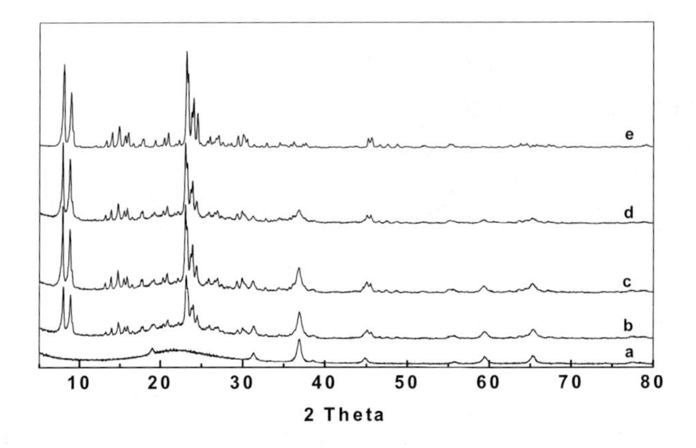

Fig.3 XRD patterns of a) Co/SiO_2, b) Co/SiO_2-Z-1, c) Co/SiO_2-Z-2, d) Co/SiO_2-Z-7 and e) H-ZSM-5

3.2. FTS performance of capsule catalyst

The catalytic experiment was performed using the catalyst of pure Co/SiO_2 and three capsule catalysts with different crystallization time, Co/SiO_2-Z-1, 2 and 7. Table.1 shows the synthesis performance properties of the catalysts. All the capsule catalysts gave similar CO conversions, which were slightly lower than that of the conventional FTS catalyst Co/SiO_2, probably because of the partly coverage of Co active sites on the core during the hydrothermal synthesis process, and the zeolite membrane may also slow the diffusion of CO and H_2. However, the activity did not decrease with the enhanced thickness of the zeolite coating, indicating that the rate-controlling step was not the diffusion of syngas through the zeolite membrane. Furthermore, the methane selectivity increased with increasing the zeolite coating amount, which should be attributed to the higher H_2/CO ratio in the core part of capsule catalyst, decided by the higher diffusion efficiency of H_2 than CO inside zeolite membrane.

From the ratios of C_{iso}/C_n in Table.1, zeolite capsule catalysts are very effective for the direct synthesis of isoparaffins via FTS. The yield of isoparaffins depended on the zeolite membrane content. With increasing zeolite membrane amounts, the ratio of isoparaffin to n-paraffin ($\geqslant C_4$) increased dramatically. Therefore, it is feasible to control the light isoparaffins selectivity of FTS reaction by controlling the crystallization time of zeolite capsule catalysts.

Table.1 The FTS properties of capsule catalysts[a]

Catalysts	Conv.%	Selectivity(%)		C_{iso}/C_n [b]
	CO	CH_4	CO_2	(n≧4)
Co/SiO_2	98.4	15.7	10.6	0
Co/SiO_2-Z-1[c]	83.6	22.7	10.0	0.37
Co/SiO_2-Z-2	85.5	31.3	10.2	0.73
Co/SiO_2-Z-7	86.1	37.4	7.0	1.88

[a] Reaction conditions: 533K, 1.0MPa, $H_2/CO=2$, $W_{Co/SiO2}/F$=10g.h/mol, 6h
[b] C_{iso}/C_n is the ratio of isoparaffin to n-paraffin of C_{4+}
[c] Numbers in catalyst's name represent crystallization time in day in zeolite capsule synthesis

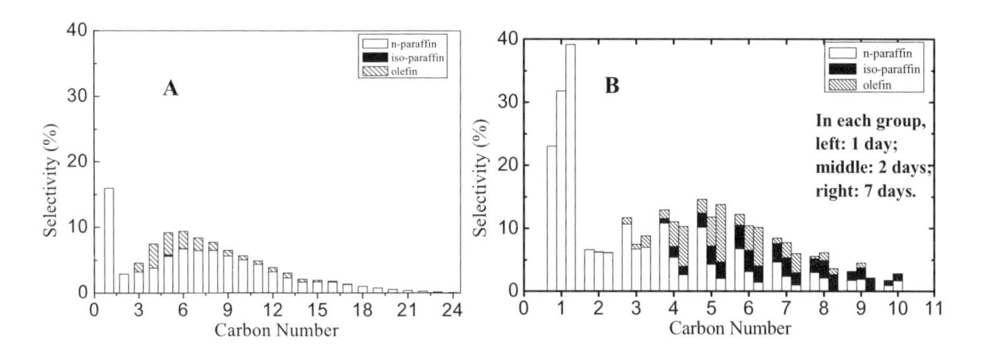

Fig.4 FTS product distributions on Co/SiO_2 (A) and Co/SiO_2-Z-1, 2, 7 (B)

By comparing the product distribution of Co/SiO_2 catalysts and capsule catalysts in Fig.4, different from the hydrocarbons distribution of conventonal Co/SiO_2 catalyst, zeolite capsule catalysts gave a very sharp product distribution. The hydrocarbons with carbon number more than 10 were suppressed totally in all capsule catalysts. In FTS reaction using the zeolite capsule catalyst, the first step was that syngas entered the capsule catalyst to undergo FTS reaction like that on the conventional Co/SiO_2 catalyst. However all the FTS products desorbed from the core part of zeolite capsule catalyst must diffuse through the H-ZSM-5 membrane. Hydrocarbons diffusion rate in zeolite membrane depends on chain length [5]. Consequently, long chain stayed in the zeolite membrane longer, resulting in that all the long-chain hydrocarbons were cracked and isomerized, contributing to suppressing the selectivity of long chain paraffins and remarkably enhancing the selectivity of light isoparaffins, due to the simultaneous effects of space confinement and shape selectivity of these capsule catalyst. This concept of capsule catalyst can be applied to a lot of consecutive reactions [6].

References

1. G. Henrici-Olive, S. Olive, The Chemistry of the Catalyzed Hydrogenation of Carbon Monoxide, Springer-Verlag, Tokyo, 1984, pp.144
2. R. M. de Vos, H. Verweij, Science. 279 (1998), 1710
3. J. He, Y. Yoneyama, B. Xu, N. Nishiyama, N. Tsubaki, Langmuir. 21 (2005), 1699.
4. J. He, Z. Liu, Y. Yoneyama, N. Nishiyama, N. Tsubaki, Chem. Eur. J. 12 (2006), 8296.
5. G. Leclercq, L. Leclercq, R. Maurel, J. Catal. 44 (1976), 68.
6. Chemical Engineering Magazine, August, 2005

Natural Gas Conversion VIII
F.B. Noronha, M. Schmal, E.F. Sousa-Aguiar (Editors)

FISCHER-TROPSCH CATALYST DEPOSITION ON METALLIC STRUCTURED SUPPORTS

L. C. Almeida[1], O. González[1], O. Sanz[1], A. Paul[2], M. A. Centeno[2], J. A. Odriozola[2], M. Montes[1]

[1]*Department of Applied Chemistry, University of the Basque Country, P. Lardizábal, 3, E-20018 San Sebastián, Spain,* mario.montes@ehu.es
[2]*Department of Inorganic Chemistry and Institute of Materials Science of Seville, University of Seville- CSIC, Av. Américo Vespucio 49, E-41092 Seville, Spain,* odrio@us.es

Abstract

Structured catalysts for Fischer-Tropsch synthesis (FTS) have been obtained by deposition of thin 20 wt. % Co/γ-Al_2O_3 catalyst layers washcoats onto the metallic surface of monolithic supports applying a preparation procedure involving: (a) preparation of metallic monoliths, (b) preparation of stable slurry of catalyst, (c) dip-coating of monolithic supports. The quality of catalyst coatings was evaluated in terms of loading, homogeneity and adhesion force. Through this procedure, Fecralloy® monoliths pretreated by thermal oxidation have been loaded with a ≈ 450 mg well-adhered catalyst coating layer and tested in FTS at 250 °C, 10 bar, under synthesis gas composition $H_2/CO = 2$. Both the coated monoliths and catalyst powder tested in FTS under similar operating conditions revealed rather comparable results in terms of activity and selectivity as a consequence of using thin layers and small particle size of powder (below 250 μm). Diffusion distances obtained for the catalytic materials resulted to be appropriate for minimizing the mass transfer effects or diffusion limitations.

Keywords:, Washcoat, Slurry, Structured support, Monolithic catalyst, Co/γ-Al_2O_3, Fischer-Tropsch synthesis.

1. Introduction

Fischer-Tropsch synthesis (FTS) is a complex series-parallel reaction including mainly hydrocarbon chain growth and termination either by hydrogenation to n-

alkane or by reductive abstraction to 1-alkene. The relative importance of these three main reactions depends on the specific activity of the active phase, on the temperature, and on the local reactants concentration on the active phase surface. Therefore, transport phenomena, both inter and intra-particle, plays a fundamental role by controlling temperature distribution and local concentration of reactants, intermediates and products due to difusional limitations.
In this sense structured catalysts and reactors offer new possibilities of temperature control in the case of microchannel reactions, different flow patterns with monoliths and foams, and reduced intraparticle limitations to diffusion and intermediates readsorption by using thin catalyst coatings.
Metals are interesting structural materials since they can be the base of microchannel reactors [1], parallel channel monoliths and foams [2]. The objective of this work is to study the controlled deposition of a Fischer-Tropsch catalyst on different structured supports, their characterization and test on the FTS.

2. Experimental

Two types of structured support are used, parallel channel monoliths and foams. Home made parallel channel monoliths uses 50 microns FeCrAlloy sheets (Goodfellow) corrugated using rollers producing 1.6 or 0.6 mm channels. Rolling up alternate flat and corrugated sheets, monoliths are produced presenting 350 and 1000cpsi. 6.8cm^3 monoliths are produced presenting 1.7mm of diameter and 3mm length. Before coating, monoliths were pretreated in air for 22h at 900°C to generate α-alumina whiskers assuring chemical compatibility and appropriate morphology to fix the catalytic coating. DUOCEL aluminium foam from ERG Materials and Aerospace of 10ppi are used. Monoliths were cut out from slabs (40mm of width) using a hollow drill with a diamond saw border of 16mm of internal diameter. Aluminium foam monoliths were pretreated by anodization in 1.6M oxalic acid at 50°C, 40min and 2A/monolith for obtaining a rough alumina surface.
Conventional 20% Cobalt on alumina FT catalyst was prepared by repeated incipient wetness impregnation of γ-alumina (SCS505) with cobalt nitrate. After impregnation the catalyst was dried and calcined at 550°C for 6h.
Structured supports were coated by washcoating in stable slurries prepared with the FT catalyst. For this purpose different parameters and properties were studied: catalyst particle size, solids content of the slurry, use of additives (colloidal alumina), rheological properties and pH of the slurry vs IEP, immersion and release velocity of the structured supports in the slurry, conditions of elimination of the slurry excess. The result of the washcoating procedure was evaluated in terms of amount of catalyst deposited, homogeneity of the coating and adhesion force.

The adherence of coating was evaluated in terms of weight loss after ultrasonic treatment. Textural properties were studied by N_2 adsorption (Micromeritics ASAP 2020) using a home made cell accepting the entire monoliths and the values obtained are the BET surface area, the total pore volume and the pore size distribution. Rheological properties of the slurry were measured in a rotational viscosimeter HAAKE, model VT 500, geometry NV. Microanalyses of the deposited films were performed using a SEM-EDS JEOL 5400. The average thickness and composition of the deposited layer was studied by means of LECO GDS-750A in the depth profile mode. In addition, the catalysts in both powdered form and monoliths were characterized by XPS.
Catalytic activity in the FTS was measured in a commercial Computerized Microactivity Reference Catalytic Reactor from PID Eng and Tech. About 0.45g of catalyst are used with a total flow adjusted to assure a space velocity of 0.15 mol kg^{-1} s^{-1} using a feed of $N_2/H_2/CO$ (9/60/30) corresponding to a H_2/CO ratio of 2. Total pressure was fixed to 10 bar, temperature at 250ºC and the reaction time to 70h. A trap at 120ºC before the pressure valve retained the high molecular products (waxes). Gaseous products were sent through a thermostatic line at 200ºC to a two ways 6890 Agilent GC, using a FID to analyse hydrocarbons from C_5 to C_{18} and a TCD to analyse H_2, CO, CO_2, N_2, H_2O, CH_4 and light hydrocarbons until C_5.

3. Results and Discussion

After studying the grinding parameters, the standard conditions chosen were: 5h at 400rpm using 50ml of catalyst suspension in water (250ml agate jar with 12 agate balls of 20 mm OD). With this grinding treatment, 90% of the solid showed particle size smaller than 5 microns [3].
The Co/Al_2O_3 catalyst showed the isoelectric point at around pH 9 with high zeta potential and then slurry stability at pH 3, that was the value used to prepare the slurries.
Viscosity is a main factor to assure homogeneous, thick and adherent coatings by washcoating [4]. Low viscosity produced homogeneous and adherent coatings but with low amount of catalyst fixed. Too high viscosity produced a high amount of catalyst deposited but in a heterogeneous form that gave low adherence. Therefore a balance between these two extreme points must be used to assure optimal results. Viscosity of the slurry was controlled by the solids content, the pH, the particle size and the use of additives like colloidal alumina. In our case the best results were obtained using 45% w/w solid content, pH 3 adjusted with HNO_3, particle size between 2 and 5 microns, that produced a stable slurry showing a viscosity of 12cp at high shear rate. The excess slurry was eliminated by centrifugation at 400rpm for 10min. After drying at 120ºC coating were repeated until reaching around 0.45g of catalyst per monolith. Three coatings were needed for the 350cpsi Fecralloy monoliths and the

aluminium foams, and two in the case of the 1000cpsi Fecralloy monoliths. After the last coating treatment, samples were calcined at 500ºC for 2h.
XPS analyses of the prepared monoliths showed mainly the existence of Co_3O_4, although the presence of a Co 2p satellite al ca. 6 eV of the main peak points also to the existence of a fraction of the deposited cobalt as Co(II) in a compound different of the Co(II,III) spinel. The Co/Al atomic ratio determined by XPS is well below the theoretical one indicating the interaction of the Co(II) species with the alumina support and low cobalt dispersion. TPR of the cobalt precursor presented a complex reduction profile due to the reduction medium-low temperature of the formed Co_3O_4 to CoO and then to Co^0, and to the high temperature reduction of the interaction product formed between cobalt and the alumina support (bidimensional aluminate like compounds).
All the coated monoliths and foams presented surface area between 104 and 110 m^2, in agreement with the amount of catalyst loaded. Mesoporosity (pore diameter of around 9 nm) correspond to the parent catalyst. XRD shows phases corresponding to the substrate metal (martensite for Fecralloy and Aluminium), the scale produced by the pre-treatment (α-alumina for Fecralloy and amorphous alumina for the anodized aluminium) and γ-alumina and cobalt oxide for the catalyst. Lateral views of the coating by SEM show local measurements of catalyst coating thickness of around of 36, 10 and 45 microns for the 350cpsi monoliths, 1000cpsi monoliths and 10ppi foams, respectively. The observed differences agree with the geometric surface of the different supports.

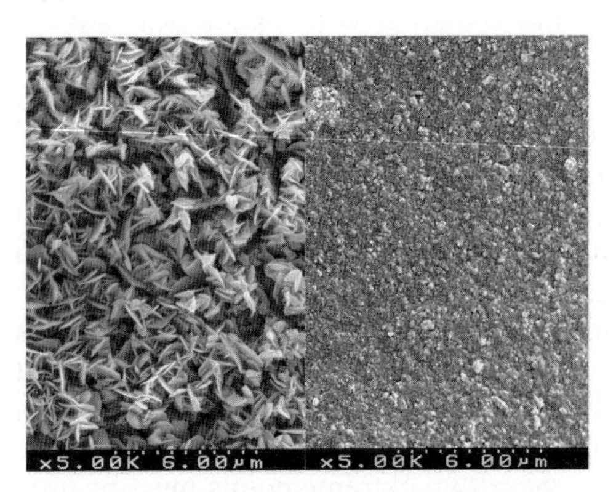

Fig. 1. SEM micrograph of alumina whiskers produced on the Fecralloy® surface by the thermal pretreatment (left side) and top view of the catalytic coating (right side).

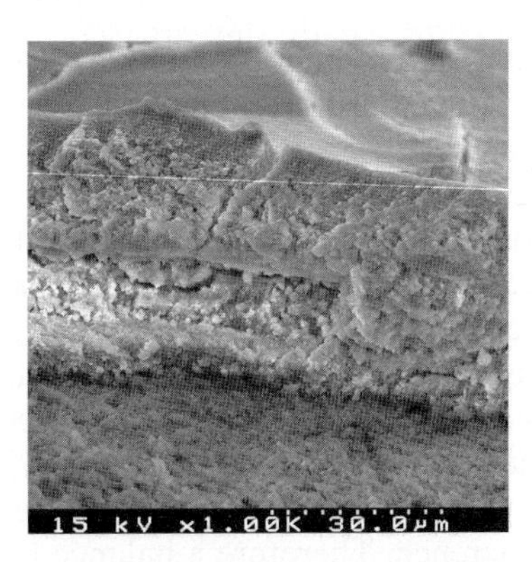

Fig. 2. SEM micrograph of Co/Al_2O_3 catalyst deposited on the surface and anchored on the alumina whiskers (lateral view).

The GDS profile of the deposited material on the FeCrAlloy support ferritic steel shows a graded composition. The outermost part of the scale is formed by a cobalt and aluminium layer with high cobalt content. Underneath this layer, an alumina rich layer is present in which cobalt is still present but with lower concentration. This bi-layer structure is the result of the roughness of the α-Al_2O_3 film produced upon oxidation of the FeCrAl support. As shown in Fig. 1 this film is formed by needle-like crystals pointing outside the support surface.

3.1. Catalytic properties

The catalytic properties in terms of activity and selectivity of the Co/γ-Al_2O_3 catalyst in form of powder and washcoated monoliths are shown in Table 1. The experimental results clearly indicate that using washcoated monolithic structures in FTS is feasible.

In general, activity and selectivity of the studied samples are rather similar. Nevertheless, the catalytic activity was slightly higher in the structured catalysts than in the powder, and the coated foam is more active than the parallel channel monoliths. From a selectivity point of view, high values of parameter α (between 0.73 and 0.82) and high selectivity toward C_{5+} (between 80% and 85%) are a clear evidence of a product distributions oriented to the fuel range (gasoline and diesel). Besides, the selectivity obtained in terms of CH_4 percentages were rather low (≈ 10%) and similar in all the samples tested.

In accordance with these results and taking in account the characteristic diffusion length calculated for the different systems (last column in Table 1) it could be concluded that all the catalytic samples tested in FTS under similar operating conditions are free of diffusion limitations, in agreement with the literature [5-8].

Table 1 Comparisons between a powder 20 wt. % Co/γ-Al_2O_3 catalyst and the corresponding structured catalysts (temperature: 250°C; pressure: 10 bar; $H_2/CO = 2:1$, W ≈ 450 mg; space velocity: 0.15 mol syngas $kg^{-1} s^{-1}$)

Catalyst	Activity [a] (x 10^{-3})	Selectivity (%)				α [b]	t [c] (μm)
		CO_2	CH_4	C_2-C_4	C_{5+}		
Powder catalyst	5.5	0.4	10.1	4.5	85.0	0.82	40
Aluminium foam	8.0	0.6	10.9	9.2	80.0	0.73	50
350 cpsi Monolith	6.0	0.5	10.4	4.8	84.3	0.80	22
1000 cpsi Monolith	8.0	0.7	11.9	4.4	83.0	0.78	11

[a] Activity in mol CO converted/kg cat*s.

[b] Chain growth parameter (ASF).

[c] Calculated diffusion distance: coating thickness for structured systems and D/6 for powder particles

4. Conclusions

Homogeneous and well-adhered Co/γ-Al_2O_3 coating layers with different thicknesses could be deposited on the pretreated surface of parallel channel Fecralloy monoliths and aluminium foams using a washcoating procedure optimized in this research. This deposition procedure consists of several steps: (a) Preparation of monolithic supports and modification of their surfaces (by thermal oxidation or anodization) (b) Preparation of stable slurry obtained by wet grinding of catalyst powder and mixture with colloidal alumina and water (c) Dip-coating of metallic monoliths by immersion and withdrawal of the structures into stable slurry at a constant speed, eliminating the excess by centrifugation and subsequently drying the sample between each step of dip-coating applied. Finally the coated samples are calcined.
The catalytic tests showed the feasibility of using metallic structured catalysts with different catalyst layer thickness in FTS. A comparison of the performances of 20wt. %Co/γ-Al_2O_3 catalyst in form of powder and washcoated monoliths reveals rather comparable results in terms of activity and selectivity as a consequence of using thin layers and small particle size of catalyst. At these dimensions the diffusion distances of the catalytic materials are appropriated to minimize the mass transfer effects or diffusion limitations.

Acknowledgments

The authors acknowledge PETROBRAS for their support of this research.

5. References

1. X. Ouyang, R.S. Besser, Catal. Today 84 (2003) 33.
2. R.M. de Deugd, R.B. Chougule, M.T. Kreutzer, F.M. Meeuse, J. Grievink, F. Kapteijn, J.A. Moujin, Chemical Engineering Science, 58 (2003) 583.
3. G. Agrafiotis and A. Tsetsekou, J. Mater. Sci., 35 (2000) 951.
4. M. Valentini, G. Groppi, C. Cristiani, M. Lrvi, E. Troconi, P. Forzatti, Catal. Today, 69 (2001) 307.
5. M.F.M. Post, A.C. van´t Hoog, J.K. Minderhoud, S.T. Sie, AIChE J. 35 (1989) 1107.
6. E. Iglesia, S. C. reyes, R. J. Madon, S.L. Soled, Adv. Catal. 152 (1993) 221.
7. A.M. Hilmen, E. bergene, O.A. Lindvag, D. Schanke, S. Eri, A. Holmen, Catal. Today 69 (2001) 227.
8. F. Kapteijn, R.M. de Deugd, J.A. Moulinjn, Catal. Today 105 (2005) 350.

Natural Gas Conversion VIII
F.B. Noronha, M. Schmal, E.F. Sousa-Aguiar (Editors)

On the origin of the cobalt particle size effect in the Fischer-Tropsch synthesis

P.B. Radstake[a,b], J.P. den Breejen[b], G.L. Bezemer[b,†], J.H. Bitter[b], K.P. de Jong[b], V. Frøseth[a,‡], A. Holmen[a]

[a]*Department of Chemical Engineering, Norwegian University of Science and Technology (NTNU), N-7491 Trondheim, Norway*
[b]*Inorganic Chemistry and Catalysis, Utrecht University, NL-3508 TB Utrecht, PO Box 80003, The Netherlands*

1. ABSTRACT

Carbon nanofiber supported Fischer-Tropsch catalysts with different cobalt particle sizes (2.6-16 nm) have been studied with steady-state isotopic transient kinetic analysis (SSITKA) at 210 °C, H_2/CO = 10. Previous studies with these catalysts for the Fischer-Tropsch synthesis had shown a strong size dependency of the surface-specific activity. From the SSITKA study it appeared that with decreasing particle size, the CH_x surface residence time increased from 10 (>6 nm) to 22 seconds (2.6 nm). For the CO residence time, an opposite trend was found. Moreover, both the CH_x and CO surface coverage decreased as function of particle size. These findings lead us a step closer to the origin of the cobalt particle size effect.

2. INTRODUCTION

In the Fischer-Tropsch (FT) reaction CO and H_2 are converted into hydrocarbons using preferably supported cobalt catalysts, which are known for their activity and selectivity.

[†]*Present address: Shell Global Solutions, PO Box 38000, 1030 BN, Amsterdam, The Netherlands*
[‡]*Present address: Statoil Mongstad, N-5954 Mongstad, Norway*

In the search to improve the catalyst performance, a previous study of our group demonstrated a clear particle size effect in the FTS performed at a total pressure of both 1 and 35 bar [1]. Catalysts with a cobalt particle size smaller than 6 nm showed a significantly lower surface-specific activity (Turnover-Frequency, TOF) compared to the larger ones (>6 nm). At 1 bar, an increase of TOF with particle sizes from 3-6 nm was found. At high pressure FT experiments (35 bar) the maximum of the cobalt weight based activity shifted from 6 nm to 8 nm. Particles larger than 6 or 8 nm, respectively, appeared to have a constant TOF.

In order to study the intrinsic cobalt particle size effect, cobalt catalysts were prepared using carbon nanofibers (CNFs) as support. In contrast to oxidic supports such as TiO_2, SiO_2 or Al_2O_3, where poorly reducible mixed compounds may form during catalyst preparation or during FT reaction, the carbon material is known for its chemical inertness. This means that the cobalt particle size effect could be studied without interferences of the support.

In the current study the focus is on the origin of the previously observed cobalt particle size effect. With the help of steady-state isotopic transient kinetic analysis (SSITKA), it aims to understand the lower TOF for catalysts with a smaller (<6 nm) cobalt particle size. SSITKA can be used to determine the coverages and surface residence times of species involved for a range of heterogeneous reactions [2], and its usefulness for FT has been proven [3]. During the steady-state FT synthesis, isotopic switches are performed and subsequently, the isotopically labeled products and reactants are monitored as a function of time. This makes it possible to distinguish different processes that add up to the total TOF [4]. In order to simplify the reaction and the number of labeled products, methanation conditions ($H_2/CO = 10$) were applied.

3. EXPERIMENTAL

4. Catalyst preparation and characterization

The carbon nanofibers (fishbone-type), with a diameter of about 30 nm, were obtained from syngas using a 5 wt% Ni/SiO_2 growth catalyst [5,6]. The support material was purified via subsequent refluxes in 1M KOH and concentrated HNO_3. The last step also implies an activation step, in which indispensible oxygen-containing surface groups, necessary to obtain high metal dispersions, are introduced [7].

Cobalt catalysts were prepared via incipient wetness impregnation. By varying cobalt loading (1 – 22 wt%), cobalt precursor (cobalt nitrate and cobalt acetate) or solvent (water and ethanol), different cobalt particle sizes over a range of 2.6 to 16 nm were obtained [1]. After impregnation, the catalyst precursors were dried overnight at 120 °C. The catalysts were reduced at 350 °C, with a flow of 10% H_2/N_2, and subsequently passivated using a diluted (0.1%) oxygen flow.

The thus obtained catalysts were analyzed with TEM, H_2-chemisorption and quantitative XPS [1] in order to determine the cobalt particle sizes. (Table 1) The catalyst were found to be metallic during FT at 1 bar pressure of H_2/CO, and deactivation due to sintering has been excluded in previous work [1].

4.1. Transient isotope experiments

SSITKA experiments were performed as described earlier [3]. Typically 100 mg catalyst, diluted with 200 mg SiC, was loaded in a plug flow microreactor. The catalyst was reduced in-situ in a flow of H_2 (10 mL/min) at 350 °C for 2 h, with a heating rate of 5 °C/min. The FT reaction was performed at 210 °C, with a flow of $^{12}CO/H_2/Ar$ = 1.5/15/33.5 mL/min and a total pressure of 1.85 bar. After 15 h FT reaction to reach steady-state conditions, a switch from $^{12}CO/H_2/Ar$ to $^{13}CO/H_2/Kr$ was performed. After reaching the (isotopic) steady-state, a backswitch was made.
The transients in ^{13}CO and ^{12}CO and of the main products $^{12}CH_4$ and $^{13}CH_4$ were followed using a mass spectrometer. From these signals the surface residence times were calculated via the area under the transient curves, where the CH_x residence time was corrected for the chromatographic effect of CO [3]. Also, the coverages of CO and CH_x species, measured via CH_4 formation, could be estimated. Besides, GC analysis was performed to determine the amount and fractions of $C_1 - C_7$ hydrocarbons.
The total activity was calculated based on the amount CO consumed, determined with GC analysis.

5. RESULTS AND DISCUSSION

All the characterization data for the as-synthesized and reduced cobalt catalysts have been published previously [1]. An overview of the different catalysts used in this work is shown in Table 1.

Table 1. Carbon nanofiber supported cobalt catalysts with their properties

	Cobalt salt	Impregnation solvent	Co loading (wt%)	Particle size (nm)[a]	Dispersion (%)[a]	$Co_{surface}$ $(mmol/g_{cat})^b$
1	$Co(NO_3)_2.6H_2O$	H_2O	22	16	6.0	0.224
2	$Co(NO_3)_2.6H_2O$	H_2O	13	8.5	11.3	0.249
3	$Co(NO_3)_2.6H_2O$	EtOH	7.5	5.9	16.3	0.207
4	$Co(NO_3)_2.6H_2O$	EtOH	5.4	5.3	18.1	0.166
5	$Co(NO_3)_2.6H_2O$	EtOH	3.7	4.7	20.4	0.128
6	$Co(NO_3)_2.6H_2O$	EtOH	1.1	4.5	21.3	0.04
7	$Co(C_2H_3O_2)_2.4H_2O$	H_2O	4.2	4.1	23.4	0.167
8	$Co(C_2H_3O_2)_2.4H_2O$	H_2O	1.0	2.6	36.9	0.063

[a] From quantitative XPS [1]
[b] Calculated from cobalt loading and dispersion

In Figure 1, the TOF for methanation conditions of the SSITKA experiments are shown as function of particle size. Literature data [1] have been plotted in this graph for comparison.

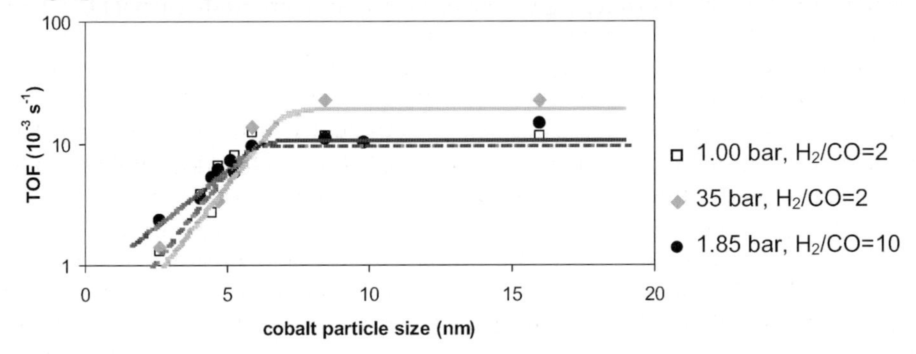

Figure 1. Turnover frequency as function of cobalt particle size

From this graph, it can be concluded that the trend found with SSITKA (H_2/CO = 10) agrees with the trends found earlier. Also the trend in weight-based activity nicely corresponds to the literature data. This indicates that results obtained with SSITKA are relevant for standard FT conditions (H_2/CO = 2) at 1 and at 35 bar. The selectivities found with SSITKA were in the range of (C_1) 79-83%, (C_2-C_4) 15-18% and (C_{5+}) 2-4%. For the selectivity data, no trend was observed with cobalt particle size, which was caused by the applied methanation conditions.

From analysis of the response curves, the surface residence time τ (s) of CH_x and CO were estimated and plotted as function of Co particle size (Figure 2).

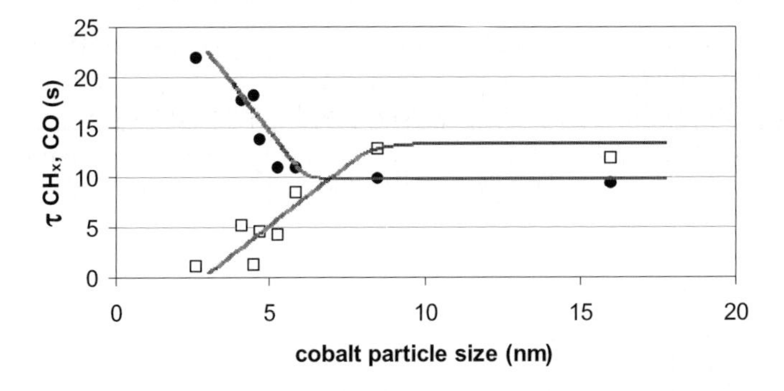

Figure 2. Corrected CH_x (●) and CO (□) surface residence times as function of cobalt particle size (210 °C, 1.85 bar, H_2/CO = 10)

Two opposite trends can be observed in this graph. First, the surface residence time of CH_x increases towards a smaller (<6 nm) cobalt particle size. This could indicate that the smaller the cobalt particle size, the stronger the CH_x species are bound to the cobalt surface. Second, the residence time of CO decreases with decreasing particle size (<6 nm). Although the total amount of metallic surface in the reactor decreases with smaller particle sizes, and the fact that the residence time of CO depends on the total amount of surface cobalt in the reactor, the trend is still the same when normalized to the number of cobalt surface atoms. Finally, both residence times appeared independent of the particle size with sizes larger than 6 nm.

From the SSITKA experiments the surface coverages were calculated as well, assuming an 1:1 stoichiometry of CO and CH_x with the Co surface atoms. As can be observed (Figure 3), the surface coverages (Θ) of CO and CH_x decreased for particle sizes below 6 nm, whereas these were more or less constant for particle sizes larger than 6 nm.

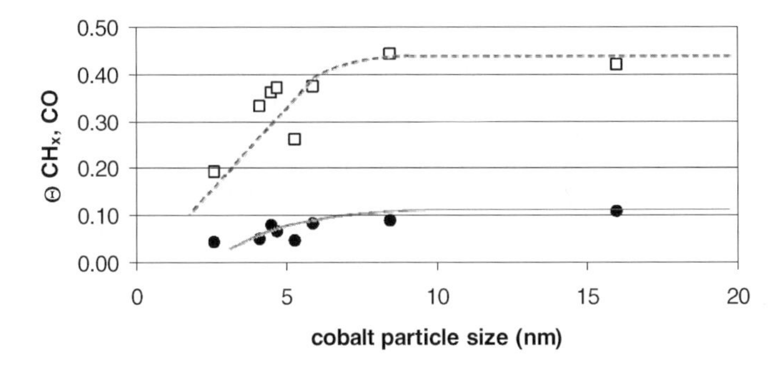

Figure 3. CH_x (●) and CO (□) surface coverage as function of cobalt particle size

Combining above trends, we may speculate that the lower intrinsic activity of the catalysts with a small cobalt particle size is caused by blocking of the reactive surface with strongly bound (or less reactive) CH_x species. These stongly bound CH_x species give rise to a lower coverage and a lower bond strength for CO, thus limiting the activation of CO on the surface of small Co particles.

5. CONCLUSIONS

Compared to 1 and 35 bar FT experiments ($H_2/CO = 2$), a similar particle size effect of the TOF is found using SSITKA experiments at $H_2/CO = 10$. This makes SSITKA a valuable tool for the investigation of the cobalt particle size effect in FT. It is shown that for particles smaller than 6 nm the surface residence times of CH_x increased, whereas the CO residence times decreased.

Larger particles (>6 nm) displayed constant residence times, comparable, at least for CH_x, to literature values [3]. Moreover, for both the CO and CH_x surface coverages a decrease for particles smaller than 6 nm is observed. Based on these results we may speculate that either the high CH_x surface residence times and/or low CO residence times contribute to a lower TOF for small particle sizes.

Although the origin for a higher CH_x and lower CO residence time can not be deduced from current measurements, these results can be regarded as a major step forward in understanding the lower activities of small cobalt particles (<6 nm) in the Fischer-Tropsch synthesis.

ACKNOWLEDGMENT

The authors acknowledge Shell Global Solutions for financial support.

REFERENCES

1. G.L. Bezemer, J.H. Bitter, H.P.C.E. Kuipers, H. Oosterbeek, J.E. Holewijn, X.D. Xu, F. Kapteijn, A.J. van Dillen, K.P. de Jong, J. Am. Chem. Soc., 128 (2006) 3956-3964
2. S.L. Shannon, J.G. Goodwin, Chem. Rev. 95 (1995) 677-695
3. V. Frøseth, S. Storsæter, Ø. Borg, E.A. Blekkan, M. Rønning, A. Holmen, Appl. Catal. A, 289 (2005) 10-15; V. Frøseth , PhD thesis, Trondheim (2006)
4. H.A.J. van Dijk, J.H.B.J. Hoebink, J.C. Schouten, Top. Catal., 26 (2003) 111-119; Top. Catal., 26 (2003) 163-171
5. M.L. Toebes, J.H. Bitter, A.J. van Dillen, K.P. de Jong, Catal. Today 76 (2002) 33-42
6. M.K. van der Lee, A.J. van Dillen, J.W. Geus, K.P. de Jong, J.H. Bitter, Carbon 44 (2006) 629-637
7. M.K. van der Lee, A.J. van Dillen, J.H. Bitter, K.P. de Jong, J. Am. Chem. Soc., 127 (2005) 13573-13582 ; G.L. Bezemer, P.B. Radstake, V. Koot, A.J. van Dillen, J.W. Geus, K.P. de Jong, J. Catal. 237 (2006) 291-302

Natural Gas Conversion VIII
F.B. Noronha, M. Schmal, E.F. Sousa-Aguiar (Editors)

Modification of cobalt catalyst selectivity according to Fischer-Tropsch process conditions

Marie-Claire Marion, François Hugues

IFP-Lyon, BP3, 69390 Vernaison, France

Abstract

The Fischer-Tropsch selectivity of a supported cobalt based catalyst has been studied at high CO conversion level. In these conditions, it is observed that a cobalt catalyst can develop a water-gas shift activity which may become significant at the expense of the Fischer-Tropsch reaction. We show that this undesired phenomena depends on a specific criteria based on the H_2O/H_2 ratio level in the Fischer-Tropsch reactor.

1. Introduction

The Fischer-Tropsch synthesis is a well known reaction which can convert synthesis gas feedstocks to produce hydrocarbons with a broad range of chain length and functionality. The Fischer-Tropsch reaction can be catalyzed by a number of metals, selected according to syngas feed characteristics and the nature of products required. As an example, iron based catalysts are usually preferred when syngas is obtained from coal gasification whereas cobalt catalyst is preferred for the conversion of syngas obtained from natural gas transformation such as in a Gas-To-Liquid (GTL) process. More generally, iron catalyst is better adapted when syngas feed is characterized by a low H_2 to CO ratio because of its activity in water-gas shift reaction. However, it is generally reported that cobalt catalysts do not show any significant water-gas shift (WGS) activity (1,2) and thus, a cobalt catalyst is generally preferred when the syngas feed presents an H_2 to CO ratio close to the stochiometric one (i.e. about 2.0).

This paper deals with the modification of Fischer-Tropsch cobalt catalyst selectivity according to process conditions. The WGS side reaction is also followed and a comprehensive study is presented too.

2. Experimental

The cobalt catalyst has been prepared, according to a state of the art recipe, by wet impregnation of aqueous cobalt nitrate on a commercial alumina from Condea, representing a total cobalt loading of 13 weight %. Then, the sample was dried, calcined in air at 400 °C for 4 hours and reduced under hydrogen at 350 °C for 8 hours. The alumina support had a BET surface area of 154 m^2/g and a pore volume of 0.5 ml/g, whereas the catalyst surface area was determined at 135 m^2/g.
The Fischer-Tropsch synthesis was carried out in a slurry pilot plant equipped with a continuous stirred tank reactor. After an initial phase where end of catalyst construction occurred under syngas (3,4), the reaction was performed at 230 °C, under pressure (15-30 bar). H_2/CO feed ratio was in the range of 2 – 2.5. After separation, gas effluent and liquid products were analyzed for conversion and selectivity determination. A wide range of conversion was experimented by changing the residence time of the reagents. Temperature was kept constant at 230 °C in order to avoid any additional parameter influence on selectivity.

3. Results and discussion

A first test was performed by using a syngas feed characterized by an H_2/CO ratio of 2.0. The initial part of the test was done at iso-conditions in order to reach stabilized performances. This period lasted for almost 835 hours at 230 °C, 30 bar during which catalyst deactivation was noticed (cf Table 1, decrease of conversion from 59 to 44% at respectively t.o.s. 283 and 835 h). Catalyst deactivation is probably due, at least in part, to high water partial pressure conditions (5). An increase of methane selectivity was also observed during this period (increase from 7 to 9.9% mol C). However, it was checked that WGS activity was limited in these conditions : CO_2 selectivity was inferior to 1%, expressed in mole of CO converted. In a second period, a high conversion range was experienced by decreasing the syngas feed flow rate. The total pressure was also decreased from 30 to 15 bar in order to soften the reaction conditions and avoid or minimize catalyst degradation which may happen under high water partial pressure (5-7). A return to initial conditions was performed at the end of the test to check catalyst change with time on stream.

Table 1 : First test results

T.O.S. (h)	Syngas flow rate (Nl/h)	P (bar)	CO conversion (%)	CH_4 selectivity (% mol C)	CO_2 selectivity (% mol C)	C_5+ selectivity (% mol C)
283	100	30	59.1	7	0.8	79.6
572	100	30	50.2	8.4	0.7	77.8
835	100	30	44.3	9.9	0.6	76.2
907	50	15	56.6	13.4	1.7	68.9
933	25	15	96.3	38.5	16.9	28.2
979	25	15	87.9	26.9	9.7	40.6
1004	30	15	72.7	16.6	4.6	60.9
1052	40	15	58.5	15.1	2.3	66.3
1099	50	15	49.2	16.2	1.7	64.0
1123	100	30	33.9	13.2	0.6	68.5
1147	100	30	33.6	14	0.6	68.0

A noticeable change of selectivity occurred when the conversion level was increased : CH_4 and CO_2 selectivity were increased whereas C_5+ selectivity was greatly decreased. The increase of CO_2 selectivity indicates the development of the water-gas shift reaction, according to reaction [2] below :

Fischer-Tropsch reaction : $CO + 2 H_2 \rightarrow -(CH_2)- + H_2O$ [1]

WGS reaction : $CO + H_2O \rightarrow CO_2 + H_2$ [2]

It seems clear that H_2O produced by the FT reaction [1] is involved in the second WGS reaction. It is also clear that the amount of water available for the WGS side-reaction increases with the CO conversion level as far as the FT reaction in concerned. However, a linear correlation between WGSR selectivity and CO conversion is not observed : it can be noticed that the proportion of CO converted into CO_2 via the WGS reaction remains quite constant in a moderate conversion range, up to 60-70 %. But, above this range, the "WGS proportion" greatly increases with conversion level, as shown in Figure 1.

In addition, it can be observed that the increase of methane formation is concomitant with the development of the WGS reaction (Figure 2). This could be explained by the extra H_2 formation obtained through WGSR and the resulting high H_2/CO conditions. In fact, a very high excess of H_2 was measured at high conversion : H_2/CO in gas effluent reached 7.3 at 96 % conversion although the inlet feed ratio was kept at 2.0. These new H_2/CO conditions also have an impact on Fischer-Tropsch selectivity, leading to an increase in the light fraction among the whole product distribution, as already reported (8-10).

At the end of the test, when getting back to initial conditions, CO_2 selectivity was greatly decreased, in relation to the conversion level, showing that the WGSR development is a reversible phenomena. Catalyst deactivation, including selectivity degradation, is nevertheless observed again : a lower conversion level, a higher methane selectivity is measured under the same initial conditions (see Table 1 data, t.o.s. > 1100 h). Again, this can be explained by the very severe water partial pressure conditions experienced (5).

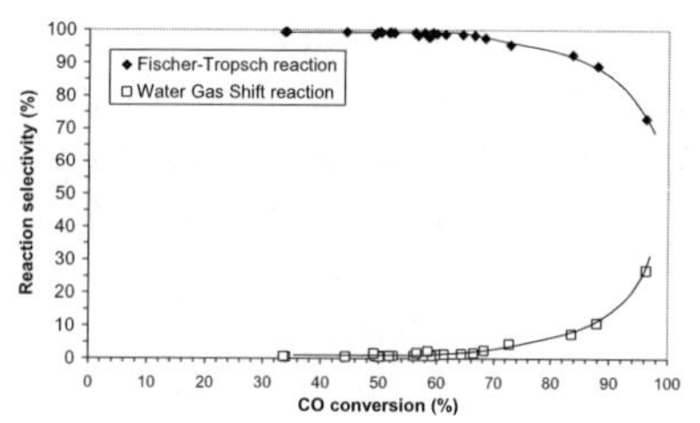

Figure 1 : Reaction selectivity

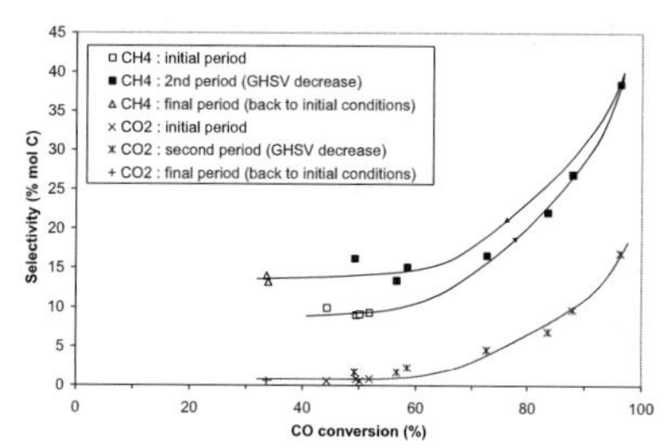

Figure 2 : CH_4 and CO_2 selectivity (feed ratio 2.0)

A new experiment was performed by using an enriched syngas feed with a 2.5 H_2/CO ratio instead of 2.0. This test was conducted in the same way as the first one : after an initial period at iso-conditions, the syngas feed flow rate was decreased to increase the conversion level up to 91%. Then, the syngas feed flow rate was increased again to return to an average conversion level.

In these conditions, the development of the WGSR is still observed at high conversion level : CO_2 selectivity increases with the conversion level. This is accompagnied by an increase of methane and a decrease of C_5+ products (Figure 3). Again, we observed that this modification of the catalyst selectivity in favor of WGSR is a reversible phenomena : the CO_2 selectivity returns to a low usual value when the CO conversion is decreased at the end of the test. However, compared with first test results, a noticeable delay can be observed in the WGSR development which occurred only above a 75-80 % conversion. Comparative results obtained in test 1 and 2, with respect to the syngas ratio,

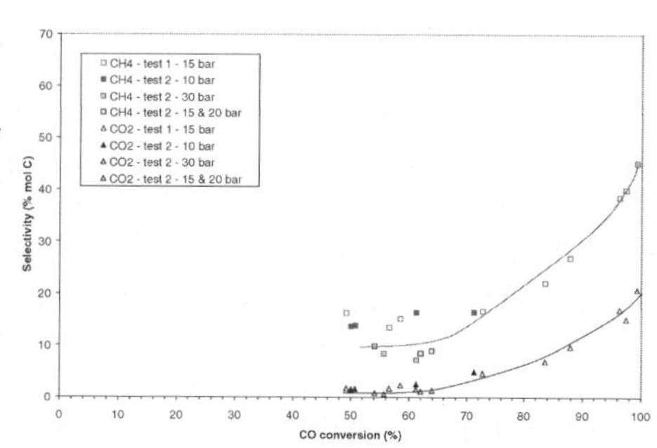

Figure 3 : Selectivity at feed ratio 2.5

are shown in Figure 4. The delay of WGS development towards a higher conversion level in case of hydrogen excess can be explained by the decrease of the H_2O/H_2 ratio at iso-conversion level. An estimation of this theoretical ratio, based on the Fischer-Tropsch reaction alone, is presented in Figure 5. For syngas ratio 2.5, case "b" takes into account an increase of light products selectivity whereas case "a" keeps the same selectivity basis as for syngas ratio 2.0. The figures obtained (case a or b) are quite similar whereas the H_2O/H_2 profile is very dependent on the syngas feed ratio. Our results show that the water-gas shift development area corresponds to conditions where the water partial pressure becomes superior to hydrogen partial pressure ($H_2O/H_2 > 1$).

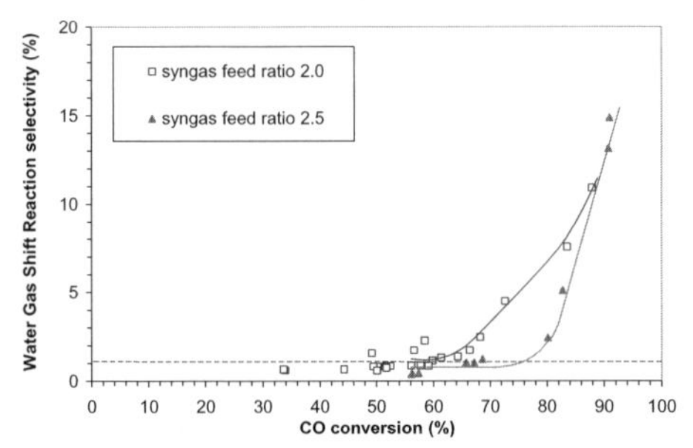

Figure 4 : Impact of syngas feed ratio on WGSR development

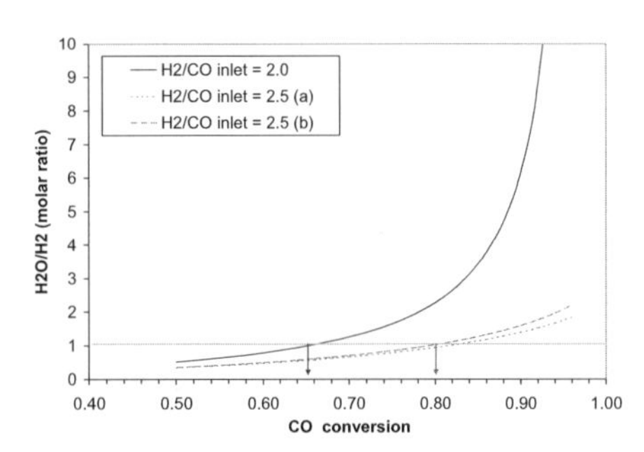

Figure 5 : H_2O/H_2 profile (calculated value)

The change of selectivity observed when H_2O partial pressure becomes superior to H_2 partial pressure could be correlated to re-oxidation of metallic cobalt in Co(II). According to this hypothesis, the oxidated cobalt would activate the WGS reaction whereas metallic cobalt would mainly catalyse the Fischer-Tropsch reaction. Actually, many authors have reported the effect of water on metallic cobalt re-oxidation (6,7,11). However, the reversibility of the process would also imply a reversible reduction of cobalt species when the theoretical H_2O/H_2 ratio get back to a low value (< 1). But, kinetic consideration of oxidation and reduction of cobalt species may not be compatible with the almost instantaneous selectivity response observed versus H_2O/H_2 ratio conditions. It is known that cobalt reduction can be completed under syngas but this usually takes some time (12-13). Another hypothesis to explain the change of selectivity versus H_2O/H_2 conditions may consider a competition between H_2 and H_2O to react with CO according to either Fischer-Tropsch reaction or water -gas shift reaction. This should imply an adsorption competition of the two reagents (H_2 and H_2O) on cobalt active sites in the close environment of CO adsorbed species. The slight CO_2 formation observed at normal FT conditions and even at low conversion level, when conditions are not favorable to cobalt oxidation and/or the presence of surface cobalt oxide species, may support this second hypothesis. In any case, our results show that it is necessary to maintain an excess of hydrogen, versus water, in order to keep the required Fischer-Trosch selectivity and avoid the development of water-gas shift reaction. This cannot be respected at high conversion level and this can be very restrictive on an industrial scale (it could mean a conversion limitation which may be prejudicial for the process economics). Our results show that adjustments in operating conditions, i.e. an increase of hydrogen excess in the syngas feed, can be used to move the conversion limitation further (Figure 4).

4. Conclusion

The Fischer-Tropsch selectivity of a supported cobalt based catalyst has been studied in a wide range of conditions. At high CO conversion level, it is observed that the cobalt catalyst can develop a water-gas shift activity which may become significant at the expense of the Fischer-Tropsch reaction.
Our study clearly shows that the development of the water-gas shift reaction is dependent on the H_2O/H_2 conditions in the reactor. Taking into account the reversibility of the process, our preferred hypothesis to explain catalyst selectivity modification is based on reactions competition consideration (i.e. competition between H_2 and H_2O to react with CO according to either Fischer-Tropsch or water-gas shift reaction). An excess of H_2 versus H_2O is necessary to avoid the development of the water-gas shift reaction. As a consequence, an excess of hydrogen in the syngas feed allows to delay the Fisher-Tropsch selectivity degradation to higher conversion level.

Reference

1. P.J. van Berge, R.C. Everson, Natural Gas Conversion IV, Stud. Surf. Sci. Catal., 107 (1997) 207.
2. B.H. Davis, Catal. Today, 84 (2003) 83.
3. H. Schulz, Z. Nie, F. Ousmanov, Catal. Today, 71 (2002) 351.
4. H. Schulz, Prepr. Pap. Am. Chem. Soc., Div. Pet. Chem., 50 (2) (2005) 155.
5. A.P. Steynberg, A.C. Vosloo, P. van Berge, US Patent N° 6 512 017 (2003)
6. D. Schanke, A.M. Hilmen, E. Bergene, K. Kinnari, E. Rytter, E. Adnanes, A. Holmen, Catal. Letters, 34 (1995) 269.
7. P.J. van Berge, J. van de Loosdrecht, S. Barradas, A.M. van des Kraan, Catal. Today, 58 (2000) 321.
8. R.L. Espinoza, J.L. Visagie, P.J. van Berge, F.H. Bolder, EP Patent N° 0 736 326 (1996).
9. D. Schanke, P. Lian, S. Eri, E. Rytter, B. Helgeland Sannæs, K.J. Kinnari, Stud. Surf. Sci. Catal., 136 (2001) 239.
10. K.J. Kinnari, D. Schanke, US Patent N° 6 022 755 (2000).
11. A.M. Hilmen, D. Schanke, K.F. Hanssen, A. Holmen, Appl. Catal., 186 (1999) 169.
12. H. Schulz, Catal. Today, 84 (2003) 67.
13. H. Schulz, Z. Nie, Stud. Surf. Sci. Catal., 136 (2001) 159.

Natural Gas Conversion VIII
F.B. Noronha, M. Schmal, E.F. Sousa-Aguiar (Editors)

Effect of water on the deactivation of coprecipitated $Co\text{-}ZrO_2$ catalyst for Fischer-Tropsch Synthesis

Lihong Shi[a,b], Jiangang Chen[a], Kegong Fang[a], Yuhan Sun[a*]

[a]State Key Laboratory of Coal Conversion, Institute of Coal Chemistry, Chinese Academy of Sciences, 27 Taoyuan Road, P. O. Box 165, Taiyuan 030001, China
[b]Graduate University of Chinese Academy of Sciences, Beijing, 100039, China

The deactivation of coprecipitated $Co\text{-}ZrO_2$ catalyst during Fischer-Tropsch synthesis was investigated in a fixed-bed reactor under dry and water addition conditions. It was found that the activity of the catalyst decreased after reaction and the catalyst deactivated more with water addition during reaction. Fresh and spent catalysts were thoroughly characterized by means of methods such as BET, X-ray diffraction (XRD), temperature-programmed reduction (TPR), Inductively Coupled Plasma (ICP), X-ray photoelectron spectroscopy (XPS), and hydrogen chemisorption to clarify the reason of the deactivation of the catalyst. Water contributed to the sintering and thus caused the deactivation of the catalyst. Still, neither cobalt loss nor the surface compound formation between cobalt and zirconia was detected.

1. Introduction

Fischer-Tropsch (F-T) synthesis is one of the major routes for converting coal-based and/or natural gas-derived syngas into chemicals and fuels. Cobalt based catalysts are the preferred catalysts for hydrocarbon synthesis because of their high F-T synthesis activity, selectivity for long-chain paraffins, and low activity for water-gas shift reaction [1].
As one of the products of CO hydrogenation, water is known to affect the kinetics and selectivity of cobalt-catalyzed F-T synthesis [2]. Recently, water

* Corresponding author. E-mail address: yhsun@sxicc.ac.cn;
Tel.: +86-351-4121-877; Fax: +86-351-4041-153.

also emerged as an important intrinsic factor to induce multiple deactivation processes.
No previous investigation, however, has been carried out on the effect of water on the deactivation of coprecipitated cobalt-zirconia catalysts. The present work is thus to study the influence of water on the catalytic performance of coprecipitated Co-ZrO_2 catalyst in Fischer-Tropsch synthesis by introducing water into the feed gas.

2. Experimental

2.1. Catalyst preparation

The coprecipitated cobalt-zirconia catalyst (ca. 20 wt.% as metal Co) was prepared in the usual batch-wise manner by adding ammonia solution to a solution containing $Co(NO_3)_2{\cdot}6H_2O$ and $ZrOCl_2{\cdot}8H_2O$ which had been heated up to 373 K. The obtained precipitate was washed by deionized water, dried at 393 K and calcined in air at 673 K for 4 h.

2.2. Catalyst test

The experiments were carried out at 2.0 MPa, 473 K, 600 h^{-1} and a H_2/CO ratio of 2.0 in a stainless-steel fixed-bed reactor of i.d. 10 mm. The catalysts were reduced in a flow of hydrogen at 673 K for 6 h (heating rate from ambient to 673 K: 1 K/min) and then cooled down to ambient before switching to syngas. Data were taken at steady state after 24 h on-stream. The gas effluent was analyzed on a GC-8A chromatographs equipped with thermal conductivity and flame ionization detectors. Liquid products and wax were collected in a cold trap and a hot trap respectively and then were off-line analyzed on a GC-920 chromatograph which was equipped with a 35 m OV-101 capillary column. We added steam by feeding water (distilled, deionized) with a liquid flow controller into a vaporizer kept at 393 K. The stream generated was mixed with synthesis gas just prior to the reactor inlet.
At the end of the run, reactor temperature was decreased to 433 K in synthesis gas. Spent catalysts were taken out and separated by magnet. To characterize spent catalysts using varied methods, two treatment ways were used. Some spent catalysts were calcined at 673 K in muffle for 4 h to get oxidation state catalysts and others were solvent-extracted using dimethylbenzene inside an Ar-filled glove box to get reduced state catalysts, which can remove enclosed wax.

2.3. Catalyst nomenclature

The fresh coprecipitated Co-ZrO_2 catalyst, spent one without water cofeeding during reaction, and that with water cofeeding during reaction were labeled as

CZ, SCZ, and WSCZ, respectively. Additionally, the dimethylbenzene-extracted and calcined catalysts were suffixed as -1 and -2, respectively.

2.4. Catalyst Characterization

BET surface area, pore volume and pore diameter were determined in a ASAP-2000 Micromeritics instrument by N_2 adsorption method. ICP-AES analysis was carried out on an instrument of Atom Scan 16 to perform elemental analysis. X-ray diffraction studies were performed in a Rigaku D/max-RA spectrometer at 40 kV and 100 mA using monochromatic Cu-Kα radiation. Temperature programmed reduction experiments were carried out in a quartz microreactor heated by an electrical furnace. The reactor was loaded with 25 mg calcined catalyst and heated at a rate of 10 K/min to 1233 K with a gas consisting of 5% H_2 in N_2. The H_2 chemisorption was performed by Micromeritics Autochem 2950 with TCD. The samples (0.2 g) were reduced under H_2 at 673 K for 6 h and then purged with Ar at 673 K for 2 h. H_2 was pulsed to the reduced catalysts at 373 K to determine the uptakes of adsorbed H_2. X-ray photoelectron spectroscopy data were obtained with an ESCALab220i-XL electron spectrometer from VG Scientific using 300W AlKα radiation. The base pressure was about 3×10^{-9} mbar. The binding energies were referenced to the C1s line at 284.8 eV from adventitious carbon.

3. Results and discussion

3.1. Catalyst deactivation

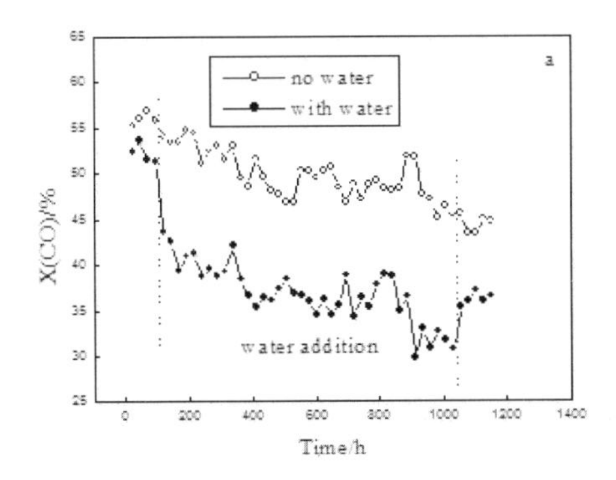

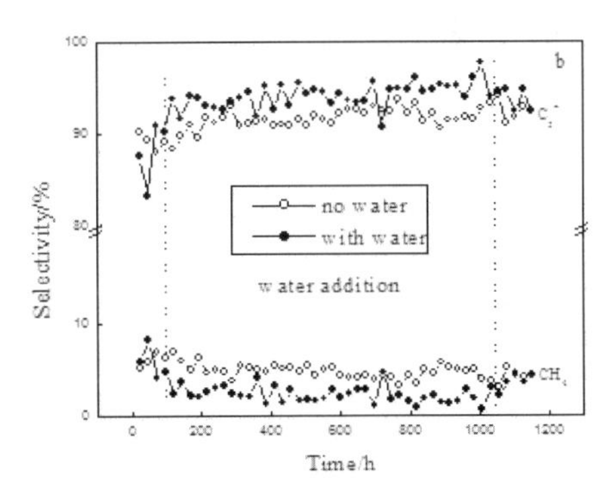

Fig.1 (a) The stability of Co-ZrO$_2$ (Reaction conditions: $H_2/CO=2.0$, GHSV=600 h^{-1}, P=2.0 MPa, T=473 K); (b) The selectivity of Co-ZrO$_2$ (Reaction conditions: $H_2/CO=2.0$, GHSV=600 h^{-1}, P=2.0 MPa, T=473 K)

Table 1 The deactivation of the catalyst under dry and water addition conditions

Reactions	Initial CO conversion (%)	Final CO conversion (%)	Decrease in CO conversion (%)
without water	55.2	44.7	10.5
witht water	52.5	36.5	16.0

The lifespan tests of $Co\text{-}ZrO_2$ catalyst are shown in Fig. 1(a). The CO conversion decreased with time on stream in two cases. Under dry and water addition conditions, the CO conversion decreased by 10.5 and 16.0% (see Table 1), respectively. It was clear that the deactivation was more serious when water was added during reaction.
The products distribution showed that the C_5^+ selectivity and CH_4 selectivity changed little with time on stream under dry condition (see Fig. 1(b)). The CH_4 selectivity decreased and the C_5^+ selectivity increased slightly when water was co-fed, which might be due to that water inhibited secondary hydrogenation of primary olefins during the reaction [3]. But the CH_4 selectivity and the C_5^+ selectivity returned initial value after water addition was terminated.

3.2. Structural change of catalysts

The textural properties of the samples are listed in Table 2. Compared with CZ, SCZ showed little difference in the surface area, but the surface area of WSCZ decreased significantly. The pore size and pore volume of three catalysts were similar. The cobalt contents of treated samples were measured with ICP (see Table 2). The close value of cobalt contents between spent catalysts and fresh catalyst suggested that the metal loss was negligible during the reaction.

Table 2 The physico-chemical properties of different catalysts

Catalysts	Surface area [m^2g^{-1}]	Average pore size [nm]	Average pore volume [cm^3g^{-1}]	Co Particle size[a] by XRD [nm]	Cobalt content[b]
CZ	101.7	12.8	0.3	<5	20.12
SCZ	104.5	11.1	0.3	9	20.07
WSCZ	82.3	13.9	0.3	11.4	20.10

[a] Calculated from XRD using Scherrer equation. [b] Obtained by ICP.

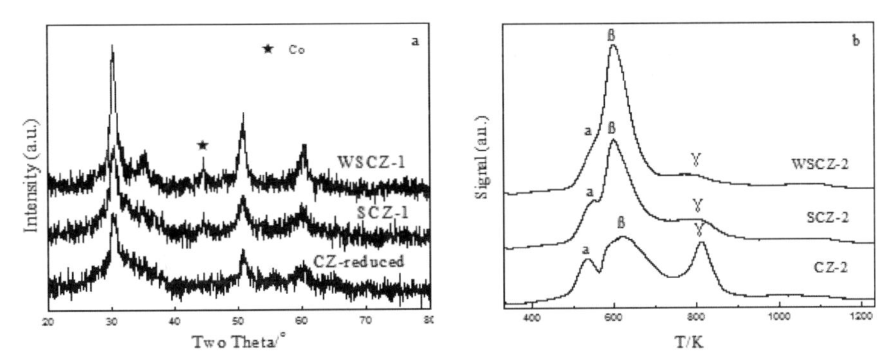

Fig. 2 (a) XRD patterns of catalysts; (b) TPR profiles of catalysts

The XRD patterns of solvent-extracted spent catalysts showed that only metallic cobalt species existed in the catalysts (see Fig. 2(a)), implying that the cobalt species were kept as metallic phase even after the extraction. It could be seen that the particle size of metallic cobalt for reduced CZ was less than 5 nm and increased to 9 and 11.4 nm for SCZ and WSCZ (see Table 2), suggesting that water promoted the growth of cobalt particles.

The H_2 chemisorption showed that the spent catalysts had lower dispersion than the fresh catalyst, especially with water addition (see Table 3), implying that cobalt particles grew after reaction and water contributed to the growth of cobalt particles. Assuming that the adsorption stoichiometry was H:Co=1, it could be concluded that the spent catalysts had less surface-exposed cobalt atoms according to the H_2 uptake data (see Table 3), which coincided with the deactivation degree of the catalyst.

Fig. 2(b) showed the TPR profiles of the catalysts. Three peaks were observed in the TPR profiles for three catalysts, which were in sequence identified to the reduction of Co_3O_4 to CoO (peak α), then CoO to Co (peak β) [4], and of cobalt species derived from a mild Co-ZrO_2 interaction to metallic Co (peak γ) [5]. However, the intensities of the corresponding peaks in the catalysts were distinctly different. Besides a slight difference for peak α, the intensity of the peak β enhanced in the order of CZ-2, SCZ-2 and WSCZ-2, whereas that of the peak γ varied inversely compared with the former, indicating that the interaction of cobalt with zirconia decreased and the Co_3O_4 particles formed. Additionally, for spent catalysts, no reduction peak appeared above 900 K, implying that zirconia was inactive towards the formation of surface compound with cobalt. However, the usual supports, such as silica, titania and alumina, were easily to form surface compound with cobalt during reaction [2, 6, 7].

Table 3 H_2-chemisorption and XPS data of catalysts

Catalysts	Dispersion (%)	H_2 uptake (μmol/g)	BE of Co2p3/2 (eV)	Surface cobalt content (%)
CZ-2	3.8	56.03	780.4	8.1
SCZ-2	2.8	50.01	780.4	5.6
WSCZ-2	2.0	41.09	780.4	4.7

The fresh and spent catalysts were characterized by binding energy of the Co2p3/2 component of 780.4 eV and a low intensity of the shake-up satellite peak at ca. 787 eV, which was typical for Co^{2+}/Co^{3+} ions in the Co_3O_4 spinel phase (see Table 3). Thus, Co_3O_4 was the predominant cobalt phase in the fresh and spent catalysts, which suggested that no surface compound between zirconia and cobalt was formed during the reaction. This was in line with TPR results. Zou et al. [7] reported that the Co2p3/2 peak was shifted toward higher energies for spent Co/SiO_2 catalyst, suggesting that cobalt silicates phase was formed on the catalyst surface. In addition, the surface cobalt content for the spent catalysts was lower than that for the fresh one, especially with water addition (see Table 3), implying that surface-exposed cobalt decreased due to the sintering of cobalt.

4. Conclusion

The activity of coprecipitated $Co-ZrO_2$ catalyst decreased after long-term run, which was more serious with water cofeeding during reaction. XRD and H_2 chemisorption of the fresh and spent catalysts indicated that metallic cobalt particles grew after reaction and water addition promoted the growth of cobalt, suggesting that water caused the sintering of cobalt and then decreased the number of surface-exposed metallic cobalt atoms.

References

1. J.L. Zhang, J.G. Chen, Y.H. Sun, J. Natur. Gas Chem., 11 (2002) 99.
2. A.M. Hilmen, O.A. Lindvag, E. Bergene, D. Schanke, S. Eri, A. Holmen, Stud. Surf. Sci. Catal., 136 (2001) 295.
3. S. Storsæter, Ø. Borg, E.A. Blekkan, A. Holmen, J. Catal., 231 (2005) 405.
4. Y. Zhang, M. Shinoda, N. Tsubaki, Catal. Today, 55 (2004) 93.
5. J.G. Chen, Y.H. Sun, Stud. Surf. Sci. Catal., 147 (2004) 277.
6. A. Barbier, A. Tuel, I. Arcon, Kodre, G.A. Martin, J. Catal., 200 (2001) 106.
7. W. Zhou, K.G. Fang, J.K. Chen, Y.H. Sun, Fuel Proce. Tech., 87 (2006) 609.

Natural Gas Conversion VIII
F.B. Noronha, M. Schmal, E.F. Sousa-Aguiar (Editors)

Effect of preparation methods on the catalytic properties of Co/SBA-15 catalysts for Fischer-Tropsch synthesis

Yuelun Wang [1,2], Jiangang Chen [1], Kegong Fang [1], Yuhan Sun[1,*]

[1] State Key Laboratory of Coal Conversion, Institute of Coal Chemistry, Chinese Academy of Sciences, No.27 Taoyuan South Road, P. O. Box165, Taiyuan, 030001, China.
[2] Graduate School of Chinese Academy of Sciences, Beijing 100039, PR China

Abstract

The influence of different preparation methods (conventional incipient wetness impregnation, vacuum impregnation, vacuum impregnation then ammonia deposition) on the physico-chemical and catalytic properties of mesoporous Co/SBA-15 catalysts for the Fischer-Tropsch synthesis has been investigated. These catalysts were characterized by N_2 adsorption-desorption, X-ray diffraction (XRD), Temperature-programme reduction (TPR), High resolution transmission electron microscope (HRTEM). Co/SBA-15 catalyst prepared by conventional incipient wetness impregnation showed the higher CO conversion, lower CH_4 selectivity and higher C_5^+ selectivity, which was attributed to the higher reduction degrees of the larger Co_3O_4 particles. Co/SBA-15 catalysts prepared by vacuum impregnation and vacuum impregnation then ammonia deposition showed smaller cobalt particles, which led to lower reducibility and lower activity, higher CH_4 selectivity and lower C_5^+ selectivity.

**Corresponding anthor, Tel: +86-351-405380; Fax: +86-351-4041153. E-mail:yhsun@sxicc.ac.cn*

1. Introduction

Fischer-Tropsch synthesis is one of the most promising ways from coal or natural gas conversion to ultra-clean fuels at economically feasible cost. In this process, supported cobalt is the preferred catalyst for the synthesis of long chain paraffins due to its high activity and low water-gas shift activity [1, 2].
It is known that preparation methods have significant impacts on the physical properties of cobalt species, which leads to different reducibility and dispersion of catalysts. For obtaining more active and high selectivity towards C_5^+ catalysts, many research groups did much work on this subject. Incipient wetness impregnation, ion adsorption, and homogeneous deposition precipitation are commonly used in order to obtain a broad range of particle size with relatively narrow particle size distribution. Ion adsorption and incipient wetness impregnation are used for the synthesis of highly dispersed catalysts at low loading, while homogeneous deposition precipitation are used for the preparation of highly loaded catalysts with intermediate dispersion [3]. In order to achieve a high density of surface active sites, incipient wetness impregnation methods are widely used for cobalt precursors dispersing on porous carriers. Mesoporous silica (SBA-15) has high surface area and pore volume, much higher hydrothermal stability [4], especialy due to its wide pores, it can diminish the diffusion resistance and provide pathways for rapid molecular transport., thus, it can be used as the preferred support to study the catalytic behaviours in FT synthesis.
The main goal of the present work is to prepare Co/SBA-15 using three different methods- vacuum impregnation then ammonia deposition, vacuum impregnation without ammonia deposition and conventional incipient wetness impregnation, respectively. The effect of different methods on catalytic performances of Co/SBA-15 was investigated in F-T synthesis.

2. Experimental

2.1. Synthesis of SBA-15 materials and catalysts preparation

SBA-15 was synthesized according to the procedure described in Ref. [5], 4.0g of P123 was stirred with 30 ml of deionized water, 30g of 2M HCl solution and 4.4g tetraethyl orthosilicate at 35℃ for 24h then transferred into a Teflon bottle heated at 100℃ for 24h. The solid was filtered, washed and calcined at 500℃ for 4h.
The catalysts were prepared as follows: 3.0 g SBA-15 was vacuumized for about 30 min using mechanical pump, added cobalt nitrate solution to achieve a loading of 15 wt, % cobalt, after the impregnation for 10h, the catalyst was added 0.5 ml ammonia solution to deposit the Cobalt precursor. This catalyst designated as Co (VA)/S (vacuum impregnation then ammonia deposition). The other catalyst prepared as the same procedure except for without adding

ammonia solution designated as Co (V)/S. The third catalyst prepared by conversional incipient wetness impregnation designated as Co(C)/S (conventional incipient wetness impregnation), Co (V)/S (vacuum impregnation), respectively. All catalysts were dried in the oven at 100℃ for 6h, and then calcined in a flow of air at 773K for 4h.

2.2. Characterization techniques and Catalytic experiments

Nitrogen isotherms were measured at 77 K using a Micromeritics Tristar 3000 sorptometer. The specific surface areas of the samples were calculated by BET method. The pore size distribution was determined using BJH method. The X-ray diffraction (XRD) analysis was carried out on a Bruker B5005 diffractometer using Cu Ka radiation. TPR was carried out in a U-tube quartz reactor at the ramp rate of 10K/min in the 5% H_2/N_2 (vol.) flow of 30 ml/min. HRTEM micrographs were obtained in a JEOL JEM-2010 electron microscope operating at 200 KV. The samples for HRTEM were prepared by directly dispersing the fine powders of the products onto holey carbon copper grids Catalysts were evaluated in a pressured fixed bed reactor at 2 MPa, $1200h^{-1}$, a temperature range from 190 ℃ to 240 ℃ and with the H_2/CO ration of 2.2 after reduction at 400 ℃ for 10 h. Wax was collected with a hot trap and the liquid products were collected in a cold trap. The gas effluents were analyzed on-line by using TCD and FID. Oil and wax were analyzed off-line in OV-101 capillary columns.

3. Results and discussion

The N_2 adsorption isotherm for SBA-15 and Co/SBA-15 catalysts presented (Fig.2a) indicative of a good quality SBA-15 material with uniform mesopores. The N_2 adsorption isotherms of Co-supported samples were similar to that of the original SBA-15, suggesting that the mesoporous structure of SBA-15 was mostly retained upon cobalt impregnation. The chemical composition and textural properties are given in Table 1. Both the BET surface area and the total pore volume significantly decreased after Co impregnation. This may be caused by a partial blockage of the SBA-15 pores by cobalt oxide clusters and/or a partial collapse of the mesoporous structure.

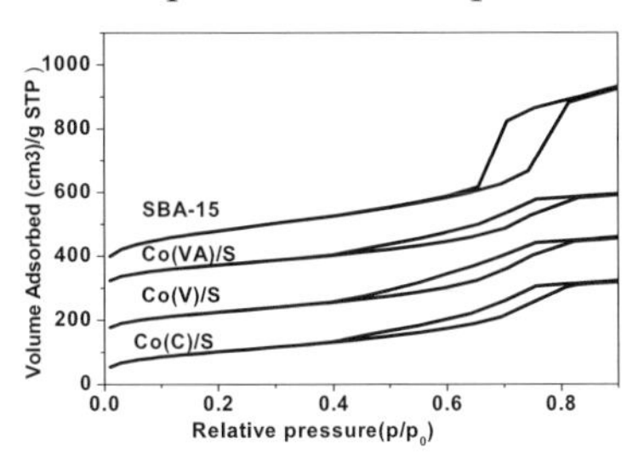

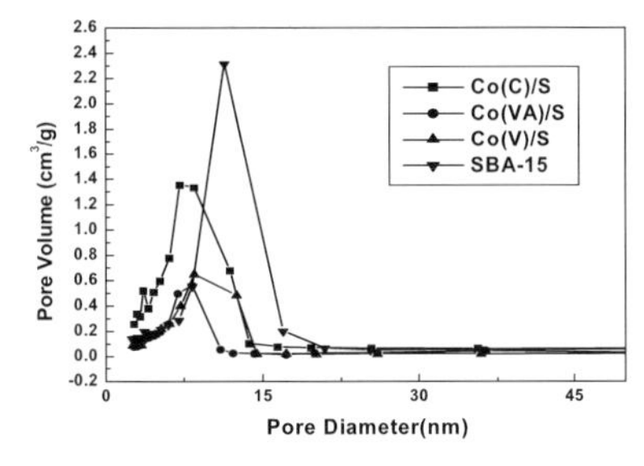

Fig1. Nitrogen adsorption isotherms and BJH pore size distribution of the samples

Table 1. Textural properties of SBA-15 supported Co catalysts

Catalysts	BET surface area (m^2/g)	Total pore volume (cm^3/g)	Pore size (nm)
SBA-15	632	1.12	9.8
Co(C)/S	414	0.52	7.4
Co(V)/S	388	0.56	7.9
Co(VA)/S	376	0.58	8.8

The XRD patterns of Co supported catalysts are presented in Fig.2. All catalysts show the characteristic peaks of Co_3O_4 at 36.7°, however, the widths of the Co_3O_4 peaks were different. The highest diffraction intensity of the main peak is observed for the catalyst prepared using normal incipient wetness impregnation, which shows the formation of the larger Co_3O_4 crystallite, while the characteristic of Co_3O_4 peaks are much broader for the catalyst by prepared ammonia deposition after vacuum impregnation, indicating the existence of smaller cobalt particles. The particle sizes calculated from XRD are given in table 2.

The corresponding reduction curves are shown in Fig. 3. Two main reduction peaks are observed. The first peak could be assigned to the subsequent reduction of Co_3O_4 to CoO, and the second one to the subsequent reduction of CoO to Co [6, 7]. Co(C)/S shows lower reduction temperature peaks, most likely corresponding to the reduction of the larger Co_3O_4 particles on the external surface. While Co (VA)/S shows broader temperature reduction peaks, implying that a lower reducibility of small cobalt particles and strong interaction between cobalt oxide species and the support. The extent of reduction of Co/SBA-15 catalysts were measured by TGA and amount of O_2 comsumed in oxygen titration oxygen titration are given in table 2.

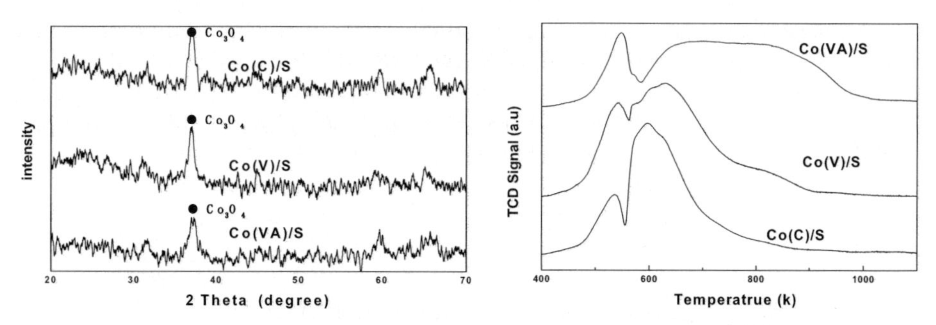

Fig.2. XRD patterns of the calcined Co/SBA-15 catalysts Fig.3. TPR profiles of catalysts

Table 3. Characterization of Co supported catalysts

catalyst	Extent of reduction measured(%TGA)	Amount of O_2 comsumed at 400 in oxygen titration, 02/Co(mole/mole)	Co_3O_4 crystallite diameter (nm)	Co dispersion (%)
Co(C)/S	93.16	0.56	10.85	11.78
Co(V)/S	82.56	0.47	9.05	14.13
Co(VA)/S	71.78	0.37	7.40	17.20

TEM shows the highly ordered hexagonal arrangement of the channels. The SBA-15-type structure was clearly maintained after cobalt impregnation and calcination. The dark areas give a contrast over the channels, which correspond to cobalt particles. It can be observed that the cobalt particles dispersed along the regular mesoporous pore stucture using vacuum impregnation. during the incipient wetness impregnation, the supports after vacuum impregnation need more the amount of water dilute nitrate solution, probably the air and water vapors inside channels were draw out, which led to the open of the microporous channels. From this point we could also conclud that cobalt specials had been inside channels. Most cobalt precursors of the catalyst prepared by conventional impregnation were located on the external surface, thus they could be grown to larger particles more easily. cobalt oxide particles appeared to be well dispersed on the channel of Co (VA)/S (see Fig.4). It is probably that higher pH value contributed to the formation of smaller cobalt particles with highly dispersion.

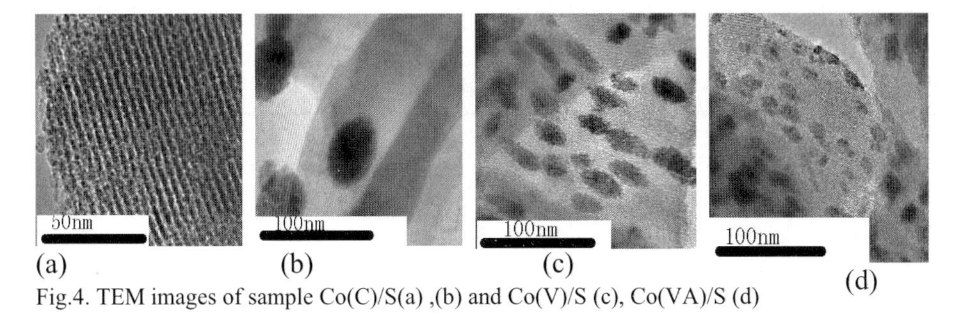

Fig.4. TEM images of sample Co(C)/S(a) ,(b) and Co(V)/S (c), Co(VA)/S (d)

The F-T reaction performances of Co/SBA-15 catalysts are summarized in Table 3. It is possible that the regular mesoporous pore stucture has a beneficial effect on mass transfer. The three catalysts show the higher selectivity to C_5^+ than reported catalysts supported conmecial silica and other mesoporous silica[8,9], the Co surface was still accessible since there was no significant loss activity conducted for 120h with increasing temperature gradually from 190 ℃ to 240 ℃, however, XRD results shown in Fig. 5 reveal that the structure of became less ordered, probably due to an effect of water vapor produced during the reaction, thus Co/SBA-15 may be potentially useful for FTS, although modified forms of SBA-15 will probably be required in order to increase its hydrothermal stablity.

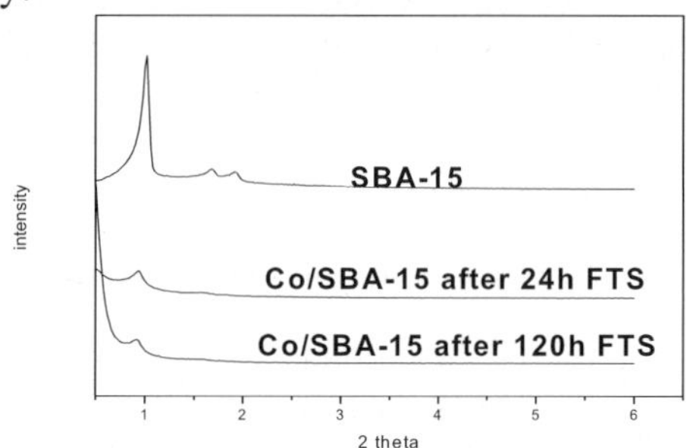

Fig.5. XRD patterns of Co/SBA-15 catalysts

Supported on SBA-15 with the same pore size, the activity and selectivity of the catalysts seemed to be sensitive to the cobalt particles which attributed to the different catalyst preparation methods. The Co(C)/S catalyst gives the highest CO conversion and lower CH_4 selectivity among the three catalysts. In principle, the activity of reduced Co catalysts should be proportional to the concentration of surface Co^0 sites. The high activity of Co(C)/S originates from its relatively large cobalt particles and high reduction degrees, thus resulting in a greater density of active surface Co^0 sites. The catalysts prepared by vacuum impregnation then ammonia deposition or without ammonia deposition show much less activity and higher CH_4 selectivity, which is related to the high cobalt dispersion. Most of the cobalt species were strongerly interacting with the support forming cobalt silicates.

Table 3 .The F-T reaction performances of the Co/SBA-15 catalysts [a]

Catalysts	CO conv.(%)	Product distribution (wt %)				
		C_1	C_{2-4}	C_{5-11}	C_{12-18}	C_{18+}
Co(C)/S	72.4	13.6	15.8	39.9	26.3	4.3
Co(V)/S	65.2	15.2	16.8	36.7	28.3	3.1
Co(VA)/S	54.5	18.9	14.8	33.2	29.6	3.6

[a]Reaction conditions: T=483K, H_2/CO=2, GHSV=1200h^{-1}, P=2MPa

4. conclusion

Co/SBA-15 catalysts with different preparation methods have been investigated. Catalysts show the higher selectivity to C_5^+ than reported catalysts supported conmecial silica and other mesoporous silica. Thus Co/SBA-15 may be

potentially useful for FTS, although modified forms of SBA-15 will probably be required in order to increase its hydrothermal stablity.
The preparation methods have significant impacts on the properties of Co/SBA-15 catalysts. Cobalt species are are in the form of aggregates both by using conversional impregnation method and vacuum impregnation method. Vacuumizing the support before incipient wetness impregnation can get smaller cobalt particles. The catalyst prepared by conversional incipient wetness impregnation shows larger cobalt particles, thus the higher CO conversion, lower CH_4 selectivity and higher C_5^+ selectivity can be observed contrasting to the catalyst prepared by vacuum impregnation then ammonia deposition. This is attributed to the higher reduction degrees of the larger Co_3O_4 particles leading to higher activity, while lower reducibility of small cobalt particles leading to lower activity. The trend of decreasing activity assigned to changes in decreasing particle size as a results of more intimate interaction of the small crystallites with the support.

Acknowledgment

This work was supported by the State Key Foundation Program for Development and Research of China (No.2005CB221402)

References

1. E. Iglesia, Appl. Catal. A 161 (1997) 59
2. C. Knottenbelt, Catal. Today 71 (2002) 437
3. G. L. Bezemer, A. J. Van Dillen, J. Am. Chem. Soc. 2006, 128, 3956-3964
4. D. Y. Zhao, Q. S. Huo, J. L. Feng, B. F.Chmelk, G.D. Stucky, J. Am. Chem. Soc. 120 (1998) 6024
5. D.Zhao, J.Feng, Q.Huo, N. Melosh, G. H. Fredrickson, B. F. Chmelka, G. D. Stucky, Science 279 (1998) 548.
6. P. Arnoldy, J. A. Moulijn, J. Catal. 93 (1985) 38
7. B. Viswanathan, R. Gopalakrishnan, J. Catal. 99 (1986) 342
8. H. Li, J. L. Li, Catal. Lett. 110(2006)71
9. A. Y. Khodakov, L. Zholobenko, J. Catal. 206 (2002) 230

Natural Gas Conversion VIII
F.B. Noronha, M. Schmal, E.F. Sousa-Aguiar (Editors)

Influence of pretreatment on the catalytic performances of Co-ZrO_2 catalysts for Fischer-Tropsch synthesis

Litao Jia [a, b], *Kegong Fang*[a], *Jiangang Chen*[a], *and Yuhan Sun* [a,*]
a State Key Laboratory of Coal Conversion, Institute of Coal Chemistry, Chinese Academy of Sciences, 27 Taoyuan south Road, P. O. Box 165, Taiyuan 030001, P.R. China
[b] *Graduate School of Chinese Academy of Sciences, Beijing, 100039, P.R. China*

The structure and performance of Co-ZrO_2 catalysts treated with H_2, syngas and CO and then the oxidation-reduction cycle was investigated by XRD, XPS, BET and TEM. It was found that such a pretreatment could improve the catalytic performance and depress the CH_4 selectivity. The presence of CO in reduction atmosphere resulted in the formation of carbonaceous deposition, which was more efficient in protecting the catalyst from oxidation.

1. Introduction

Duo to the high activity, high selectivity to linear hydrocarbons and low activity for the water gas shift (WGS) reaction, Co-based catalysts are considered to be the most important catalysts for Fischer-Tropsch synthesis (FTS) based on natural gas conversion [1]. The active species of the catalysts used in the FTS is the metallic phase, which is produced by pretreatment of the precursors in a reduction atmosphere prior to use in the reaction. However, less attention has been paid to determine how the activating agent (H_2, CO or syngas) influences on the catalyst performance. For various reasons, it is often useful to reduce the catalyst ex situ and passivation is employed to protect the surface of the reduced catalyst from rapid oxidation. But, it is less understood how the pretreatment at the different atmospheres influences on the physical properties, passivation exposing to air and the catalytic performances of

[*] Corresponding author, Tel: 86-351-4121877 Fax: 86-351-4041153 E-mail address: yhsun@sxicc.ac.cn.

catalysts. Accordingly, this work is undertaken with the aim to investigate how the different reducing agents affected the physical properties and passivation, furthermore to clarify how the re-activation of catalysts and their performances changed after second-step reduction in H_2 atmosphere.

2. Experimental

The Co-ZrO_2 catalysts with 20 wt% cobalt used in the present work were mainly prepared by co-precipitation method as reported at literature [2]. The fresh catalyst was firstly reduced under H_2, syngas and CO at 400°C, respectively. Then, it was cool down room temperature and exposed directly to air atmosphere for 24 h. The obtained samples respectively noted as C-H, C-M and C-CO.

N_2 adsorption/desorption isotherms of the samples were measured at -196 °C on a Micromeritics Tristar 3000 instrument. X-ray diffraction (XRD) was recorded in Rigaku D/max-rA with Cu target at 40 kV and 100 mA. X-ray photoelectron spectroscopy (XPS) data were obtained with an ESCALab220i-XL electron spectrometer from VG Scientific using 300W AlKα radiation. Temperature-programmed-reduction (TPR) was carried out in a home-built equipment by heating the 0.025 g sample with 10°C/min form 60 °C up to 950 °C using 60 ml (NTP)/min of 5 % (vol) H_2-Ar mixtures. The concentration of hydrogen was monitored with TCD after removal of the water product using a 3-Å molecular sieve.

The evaluation of catalysts was carried out in a pressured fixed-bed reactor with a catalyst loading of 1ml. Prior to catalytic experiments, the samples were reduced in a flow of hydrogen at 400°C for 10 h. The reaction was carried out at 2.0 MPa, 2000 h^{-1}, $H_2/CO = 2.0$ and 200 °C. The products were collected with a hot trap and a cold trap in sequence. The gaseous flow-out was analyzed with on-line chromatography. The liquid and wax were analyzed with a GC-920 chromatogram equipped with OV-101 capillary columns. Both the carbon balance and mass balance of reaction were kept among 95±5%.

3. Results and discussion

BET surface areas, pore volumes and pore sizes of the fresh catalyst and the treated samples are shown in Table 1. It could be found that there was no significant change in surface areas between the fresh catalyst and the sample C-H. However, the surface area increased for the sample C-CO and decreased for the sample C-M. These indicated that the different activating atmospheres could affect the textural properties of catalysts.

XRD diagrams of the fresh catalyst and samples after reduction-oxidation process are presented in Fig 1. For the sample C-H, the diffraction peak for Co_3O_4 phase was observed at 36.8°, which was consistent with the fresh catalyst. This suggested that the metallic cobalt produced at H_2 reduction condition was facile to oxidation upon exposure to air. Furthermore, the peak of Co_3O_4 phase became broader, indicative the presence of small cobalt particles and/or higher dispersion of cobalt species. After treated with the syngas, the sample C-M showed the diffraction lines for the cubic metallic cobalt species at 44.4^o and hexagonal metallic cobalt at both 41.7^o and 47.5^o. This demonstrated that the reduction with syngas could restrain the catalyst from the oxidation upon exposure to air. Similarly, the signal of metallic cobalt was observed in the sample C-CO. However, the strong diffraction line at 25.8^o for carbonaceous deposition was observed in TEM (no shown). Possibly, the resistance to oxidation was attributed to the presence of carbonaceous deposition (see Table 1), which originated from the Boudouard reaction ($2CO = C + CO_2$). The same results was also found by Sonia Hammache et al[3], who reported that syngas atmosphere passivation of the reduced cobalt catalyst was more efficient in protecting the catalyst from oxidation. Generally speaking, the carbon formed might be carbidic (C_α) or graphitic carbon. Because the necessary temperature for graphite formation was above 350°C [4], it was inevitable for both sample C-CO and sample C-M to produce graphitic carbon on the reduction at 400°C.

Table 1 Textural properties of the fresh catalyst and samples after reduction-oxidation process

Samples	Surface area (m^2/g)	Pore volume (ml/g)	Pore size (nm)	Carbon content (%)
Fresh	119	10.1	0.31	/
C-H	117	10.4	0.32	/
C-CO	142	11.1	0.40	72.9
C-M	62	19.1	0.31	3.0

The catalysts are characterized by XPS to clarify the transformation of the surface cobalt species during the treatment process (see Table 2). For the fresh sample, the Co $2p_{3/2}$ BE of 780.3 eV was indicative of the presence of Co_3O_4. The similar result was observed for the sample C-H. The Co $2p_{3/2}$ BE value of the sample C-M changed from 780.3 to 781.8 eV, implying the presence of Co^{2+} compounds on the surface, but it was higher than CoO (BE, 780.5 eV) and was close to the matching BE values of $CoAl_2O_4$ (BE, 781.7eV)[5]. Furthermore, the bending energy for Zr $3d_{3/5}$ shifted from 181.6 to 182.5 eV, which indicated the presence of the strong interaction between the cobalt species and ZrO_2 support on the catalyst surface. The notable decrease in the atomic ratio of cobalt implied that the surface of metallic cobalt particles was covered by partially reduced zirconia species, due to a strong cobalt species-support interaction [6]. After treated with CO, the surface atomic ratio of cobalt on the sample C-CO decreased notably and the carbon on the catalyst surface abruptly increased, which was attributed to carbonaceous deposition. In this case, the Co $2p_{3/2}$

bending energy was 779.1eV and was still higher than that of metallic cobalt, which could be due to that the carbonaceous deposition suppressed the oxidation of metallic cobalt, and, the small partial oxidation of metallic cobalt occurred possibly on catalyst surface.

Table 2 Summaries of data from XPS characterization of catalysts

Samples	$Co2p_{3/2}$ (eV)	Zr $3d_{3/5}$ (eV)	Atomic C/Zr (%)	Atomic Co/Zr (%)
Fresh	780.3	181.6	27.1	6.8
C-H_2	780.4	181.7	26.7	6.2
C-M	781.8	182.5	28.8	4.0
C-CO	779.1	185.2	96.6	0.3

TPR profiles for the fresh catalyst and samples after reduction-oxidation process are shown in Fig 2. For the fresh catalyst, there were four peaks at 224, 273, 352 and 548°C, respectively. The δ peak could be assigned to the reduction of the bigger Co_3O_4 crystal on the support surface [7]. The α peak at 273 °C could be attributed to the gradual reduction of Co_3O_4 to CoO and then β at 352°C to the subsequent reduction of CoO to metallic cobalt [8]. Their reducibility was directly related to the amount of active of Co^0 sites available for catalyzing after the standard reduction. The γ peak at 548 °C in the TPR patterns was scarcely reported in the previous literatures. Joanna £ojewska et al [9] found that the phase transition between Co_3O_4 and CoO was generated at 350°C for the surface oxides and at 550℃ for the bulk oxides. However, Dan I. Enache et al [10] pointed that the high temperature peak at 765°C was coincided with the support crystallization and a solid solution or a zirconate phase could be formed between unreduced cobalt oxides and zirconia. Thus, it was postulated that the fourth peak located at 548 °C was likely attributed to the reduction of the cobalt species interacted with zirconium in bulk but not the reduction of a solid solution or a zirconate phase.

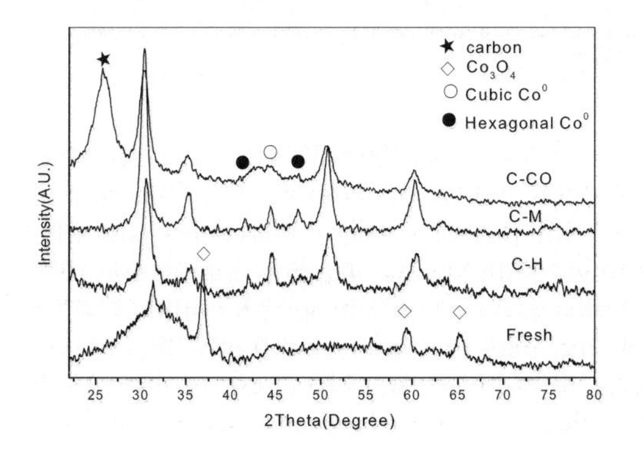

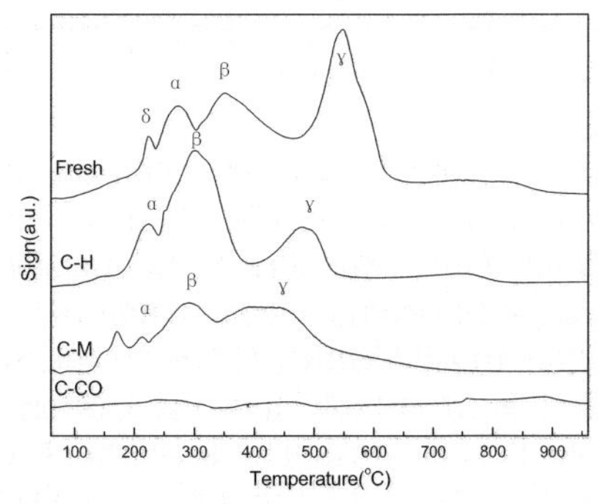

Fig 1 XRD diagrams of the fresh catalyst and samples after reduction-oxidation process

Fig 2 TPR profiles for the fresh catalyst and samples after reduction-oxidation process

After reduction-oxidation process, the sample C-CO showed the faint H_2 consumption peaks. This suggested that the main cobalt species was metallic cobalt phase for sample C-CO and it was difficult to oxidize in air, which was good agreement with the XRD results. For the samples C-M and C-H, the H_2 consumption peaks for cobalt species shifted to lower temperature, indicating that samples after treatment be facile to reduce again.

The catalytic performance of fresh catalyst and samples C-H, C-M and C-CO are shown in Table 3. It could be found that the fresh catalyst showed lower catalytic activity as compared to the samples C-H, C-M and C-CO. But the samples C-H, C-M and C-CO had lower reduction degree than that for the fresh catalyst (shown in TPR results). Thus, the high performance appeared to be related to the new Co active surface species produced by re-reduction, and hence improved the CO conversion. Exxon patent also suggested that the ROR technique is used to enhance the dispersion of the metallic cobalt [11]. V.A. de la Peña O'Shea et al found that the reduction at syngas increased the dispersion of the Co/SiO_2 catalyst [12]. However, the sample C-M showed lower CO conversion as compared to the samples C-H and C-CO. Such a decreasing activity could be related to the low reducibility of cobalt species in the sample C-M, probably in the form of strong cobalt species-support interaction as revealed by XPS measure.

Table 3 Catalytic performance of catalysts [a]

Samples	X (co) (%)	S (CH_4) (%)	S (C_2-C_4) (%)	S (C_{5+}) (%)	S (CO_2) (%)	C Balance (%)
Fresh	27.3	10.4	2.5	87.1	4.1	101.8
C-H	54.7	8.5	1.5	90.0	0.5	101.7
C-M	38.9	11.4	2.8	85.8	0.9	104.1
C-CO	50.9	46.3	7.1	46.6	2.0	99.8

[a] Reaction conditions: 2.0 MPa, H_2/CO = 2, GHSV = 2000h^{-1} T=200°C

The sample C-H showed the lowest selectivity to light hydrocarbons (C_1-C_4). But, the sample C-CO gave the highest methane selectivity although CO conversion was the same as that of the sample C-H. Generally, the high methane selectivity was reported for the catalyst having low Co reducibility. This resulted in the presence of unreduced cobalt oxides which could catalyze the WGS reaction, leading to an increasing in H_2/CO ratio on the catalyst surface. However, no significant formation of CO_2 was observed for the sample C-CO, and thus the higher methane selectivity would be ascribed to the local increase of the H_2/CO ratio near the surface Co^0 sites due to the presence of the carbon nanotube as shown in TEM result. Possibly, the other reason would be responsible for the increase in the CH_4 selectivity. As was the case for both syngas reduction and CO reduction, it was inevitable to produce carbidic (C_α) or graphitic carbon. Though the C_α might be very reactive with H_2, the complete removal of the C_α was not possible during the re-reduction condition. As a result, the residual carbidic carbon would react easily with hydrogen to form methane [13]. With the amount of carbon content changing from 3.0% for the sample C-H to 72.9% for the sample C-CO, the CH_4 selectivity significantly increased from 15.2% for the sample C-H to 50.4% for the sample C-CO at 200°C.

4. Conclusion:

This study of the influence of treatment with H_2, syngas and CO and then the oxidation-reduction cycle on the Co-ZrO2 catalyst revealed that the re-reduction could improve the catalytic activity. However, the presence of CO at initial reduction process resulted in the increasing CH_4 selectivity, which would be attributed the impact of carbonaceous deposition.

Acknowledgment

The authors acknowledge the financial support from Chinese National Natural Foundation (Contract No. 20590361 and 20303026) and State Key Foundation Program for Development and Research of China (Contract No. 2005CB221402).

References

1. E. Iglesia, Appl. Catal., 161 (1997) 59.
2. J. Chen, Y. Sun, Stud. Surf. Sci. Catal., 147 (2004) 277.
3. S. Hammache, J.G. Goodwin Jr., R.Oukaci, Catal. Today, 71 (2002) 361.
4. L.J.E. Hofer, W.C. Peebles, J. Am.Chem.Soc., 69 (1949) 893.
5. B.K. Sharma, M.P. Sharma, S. Kumar, S.K. Roy, Appl. Catal., 211 (2001) 203.
6. K.Chen, Y. Fan, Q. Yan, J. Catal., 167 (1997) 573.
7. H. Lin, C. Wang, H. Chiu, S.Chien, Catal. Lett. , 86 (2003) 63.
8. G.R. Moradi, M.M. Basir, A. Taeb, A. Kiennemann, Catal. Commun., 4 (2003) 27.
9. J. £ojewska., W. Makowski, T. Tyszewski, R. Dziembaj, Catal. Today, 69 (2001) 409.
10. Dan I. Enache, M. roy-Auberger, R. Revel, Appl. Cata. A: General 268 (2004) 51.
11. H. Beuther, T.P. Kobylinski, C.Likibby, R.B. Pannell, US Patent 4,585,798 (1986).
12. V.A. de la Peña O'Shea, J.M.Campos-Martín, J.L.G. Fierro, Catal. Commun., 5 (2004) 635.
13. I.G. Bajusz, J.G. Goodwin Jr., J. Catal., 169 (1997) 157.

Natural Gas Conversion VIII
F.B. Noronha, M. Schmal, E.F. Sousa-Aguiar (Editors)

Fischer-Tropsch synthesis. Recent studies on the relation between the properties of supported cobalt catalysts and the activity and selectivity.

Øyvind Borg[a], Vidar Frøseth[a,c], Sølvi Storsæter[a,b], Erling Rytter[a,b], Anders Holmen[a,x]

[a]Department of Chemical Engineering, Norwegian University of Science and Technology (NTNU), NO-7491 Trondheim, Norway.
[b]Statoil R&D, Postuttak, NO-7005 Trondheim, Norway.
[c]Statoil Mongstad, NO-5954 Mongstad, Norway.
[x]Corresponding author: e-mail: holmen@chemeng.ntnu.no

Abstract

The activity and selectivity of a large number of supported Co Fischer-Tropsch catalysts have been studied at 20 bar, 483 K and $H_2/CO=2/1$. The supports include different aluminas, SiO_2, TiO_2 and carbon nanofibers (CNF). Provided that the Co particle size is larger than 6-8 nm, a constant turnover frequency is observed in most cases. The deactivation behaviour and the C_5^+ selectivity, however, depend largely on the properties of the catalyst.

1. Introduction

Modern Fischer-Tropsch (FT) technology focuses on maximizing the yield of long chain paraffins (waxes), which in turn are hydrocracked to high quality diesel and other products. A key element in improved Fischer-Tropsch (FT) processes is the development of active catalysts with high selectivity to waxes. The FT synthesis is usually described by a chain growth mechanism where a C_1 unit is added to a growing chain most probably through the CO insertion mechanism [1]. α-olefins and paraffins are the primary products of the FT synthesis. α-olefins can also participate in secondary reactions adding complexity to the reaction network.

The FT synthesis is generally believed to be structure insensitive and the specific activities in terms of intrinsic turnover frequencies (TOF) at practical conditions are reported to be independent of the support and promoter [2,3]. It has, however, recently been shown by using carbon nanofibers (CNF) as supports that for Co particles smaller than 6-8 nm the turnover frequency decreases with decreasing particle size [4]. Detailed studies using steady-state isotopic transient kinetic analysis (SSITKA) have shown that the decrease in the particle size below 6-8 nm is accompanied by changes in the surface residence time and the surface coverage of CO and CH_x (at methantion conditions) [5]. The type, structure and purity of the support, the preparation variables such as cobalt precursor and solvent, cobalt loading, preparation method, pretreatment conditions (drying, calcination, reduction) and the use of promoters will influence the dispersion, particle size and reducibility and thereby the activity (pr gram catalyst) for Co supported catalysts. The above parameters have as well a large influence on the selectivity to long chain hydrocarbons (the C_5^+ selectivity). In addition, diffusion limitations will influence not only the activity, but also the selectivity in FT synthesis [6,7]. It has been claimed that the selectivity also depends on the Co site density through a re-adsorption mechanism.

Water is produced during FT-synthesis and will be present in varying quantities during the reaction, depending on the catalyst, conversion and reactor system. It is a general observation that the addition of water to the reactant mixture results in increased C_5^+ selectivity and reduced CH_4 selectivity for Co catalysts independent of the support (Al_2O_3, SiO_2, TiO_2, CNF). The effect of water on the C_5^+ selectivity was largest for CoRe supported on low surface area TiO_2 compared with Al_2O_3 and SiO_2 having larger surface area, smaller pore diameters and smaller particle sizes [8]. Water inhibits secondary hydrogenation of primary olefins [7] and from isotopic transient studies it has been proposed that the effect of water arises via an indirect effect on the active carbon inventory [8].

No unequivocal explanation has been presented for the dependence of the C_5^+ selectivity on catalyst properties. The purpose of the present contribution is to present and review some of our recent results showing the relation between catalyst properties and the catalyst performance.

2. Experimental

Catalysts containing 12 or 20 wt% Co and 0 or 0.5 wt% Re on a number of different supports (Al_2O_3, SiO_2, TiO_2, CNF) have been prepared by incipient wetness impregnation. Cobalt nitrate hexahydrate and perrhenic acid were used as cobalt and rhenium precursor, respectively. The support material has also been pretreated in different ways before impregnation. The catalysts have been characterized by standard methods such as TPR, chemisorption, XRD, pulse oxidation, BET, EXAFS, ICP, TEM and the activity and selectivity have been determined in a fixed-bed reactor (i.d. 10 mm) at 20 bar and 483 K [11]. SSITKA studies at methanation conditions have also been carried out in order to determine the intrinsic turnover frequency (TOF) [2].

3. Results and Discussion

The main focus has been on Co and CoRe on supports with large variations in pore size, pore volume etc. The supports include low surface area α-Al_2O_3 and TiO_2, γ-Al_2O_3 with surface areas between 86 and 232 m^2/g, SiO_2 and CNF. In general, the internal structure of the supports is not well-defined adding more complexity to the analysis.

Table 1. SSITKA studies at methanation conditions (483 K, 1.85 bar, H_2/CO/inert = 15/1.5/33.5 Nml/min) [2]. TOF_{FT} = turnover frequency at FT conditions (483 K, 20 bar H_2/CO = 2/1). FI1 = Fishbone CNF, incipient wetness impregnation, PW30 = Platelet CNF, wet impregnation. γ-Al_2O_3 (BET) = 160 m^2/g

Catalyst	r_{CO} $\mu mol/g_{cat}\,s$	V_{ads} H_2 $\mu mol/g_{cat}$	TOF_{FT} $10^{-3}s^{-1}$	TOF_{CH4} $10^{-3}s^{-1}$	$\tau_{CH(4)}$ s	θ_{CO}
Co/γ-Al_2O_3	1.6	67	52	7.1	10	0.24
CoRe/γ-Al_2O_3	2.7	104	52	8.6	11	0.32
CoRe/SiO_2	1.3	59	56	8.7	11	0.32
CoRe/TiO_2	1.0	24	120	15.4	11	0.41
Co/FI1	1.4	50	42	11.0	11	0.52
Co/PW3	3.5	124	38	10.5	11	0.40

For a first order irreversible reaction, the turnover frequency is given by $TOF = k \cdot \theta_i = \tau_i^{-1} \cdot \theta_i$ [2]. SSITKA makes it possible to deconvolute the turnover frequency in a rate constant and the surface coverage. SSITKA studies have shown that the intrinsic TOF seems to be independent of the support and the promoter for Co particles larger than 6-8 nm at least at the conditions used for the SSITKA studies. However, it has been observed that the apparent TOF in some cases depends on the catalyst. Reduced apparent TOF is observed for samples containing small amounts of Na. The addition of Mg or Ni to the support also shows reduced or increased apparent TOF, respectively [12]. The deviation for TiO_2 supported catalysts is a well-known phenomenon.

The effect of water on the activity depends on the support. For Co/Al_2O_3 and Co/SiO_2 the catalyst deactivates with time on stream and more so when water is added to the feed. Addition of inert to the feed instead of water confirms that the effect is due to water. For Co/TiO_2, however, catalyst deactivation is not observed during the first 100 h on stream. In fact, the activity increases upon addition of water to the feed.

It is known that improved C_5^+ selectivity can be obtained using Co or CoRe supported on low surface area α-Al_2O_3 instead of on high surface area γ-Al_2O_3. Low surface area TiO_2 as support also gives improved selectivity of C_5^+. As shown in Figure 1 the C_5^+ selectivity increases in the following order: Al_2O_3<SiO_2<TiO_2. This is in the same order as the average pore diameter: Al_2O_3 = 6.7 nm, SiO_2 = 11 nm and TiO_2 = 780 nm. The particle size as measured by H_2-chemisorption or XRD follows the same trend and the same holds for the extent of reduction of Co. However, SiO_2 has the largest pore volume and the highest surface area.

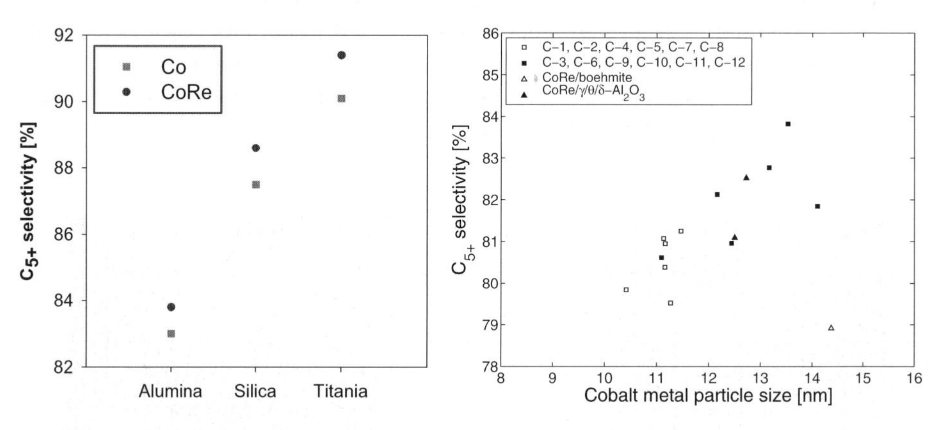

Figure 1. FT synthesis on 12wt% Co +0.5wt% Re supported on γ-Al_2O_3, SiO_2 and TiO_2. 483 K, 20 bar, H_2/CO =2. 20% H_2O added to the feed

Figure 2. FT synthesis on 20 wt% Co +0.5 Re on γ-Al_2O_3 (C-1 to C-12) with different pore sizes. 483 K, 20 bar, H_2/CO=2. No water added to the feed.

In order to study the effect of the support structure on the catalytic performance a series of γ-Al_2O_3 supported CoRe catalysts with different pore structure has been prepared and studied at FT conditions (20 bar, 483 K). The selectivity of higher hydrocarbons increases with increasing pore size of the catalyst support. However, the size of the Co particles also depends on the pore size and it is therefore not possible to conclude whether the selectivity depends on the particle size or the pore size of the support. By using incipient wetness impregnation it seems difficult to prepare catalysts with different Co particle sizes on supports with constant properties and loading. Other methods for preparation of Co catalysts supported on oxides are therefore investigated.

The effect of the particle size has been studied for Co supported on CNF. For cobalt particles larger than about 10-12 nm the selectivity did not depend on the particle size, but for smaller particle sizes it is a strong dependence [4]. However, as shown in Figure 2 for Co supported on oxides, the C_5^+ selectivity increases in the whole range of particle sizes investigated. The Co particle size for the TiO_2 supported catalyst is about 40 nm as determined by H_2 chemisorption and XRD measurements. For the CNF supported Co catalysts presented in Table 1, the C_5^+ selectivity has been found to be 82.6% (Co/FI1) and 79.0% (Co/PW3) which is in line with the much larger particle size for Co/FI1 (20nm) than for Co/PW3 (7.9nm). CoRe on boehmite given in Figure 2, however, shows a lower selectivity than expected from the particle size. The pore size (and pore volume) of boehmite is lower than for the other samples.

The catalysts have been characterized by high resolution TEM. On small pore γ-Al_2O_3 and on silica supports, Co_3O_4 appears as clusters or agglomerates of smaller particles with a broad cluster size distribution, but on the wide pore TiO_2 and α-Al_2O_3, Co_3O_4 was found as single particles, evenly distributed. Examples of Al_2O_3 and TiO_2 supported CoRe are shown on Figure 3. The Co in Figure 3 exists as Co_3O_4. It is known, however, that Co_3O_4 may crack into smaller particles during reduction [13] and the reduced catalyst in the reactor may therefore have a different appearance.

A

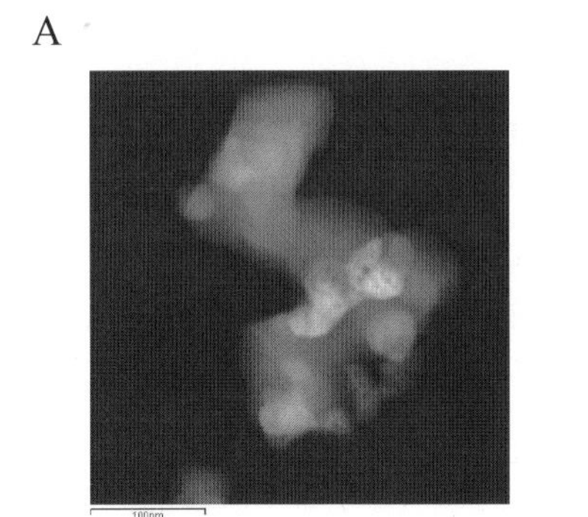

B

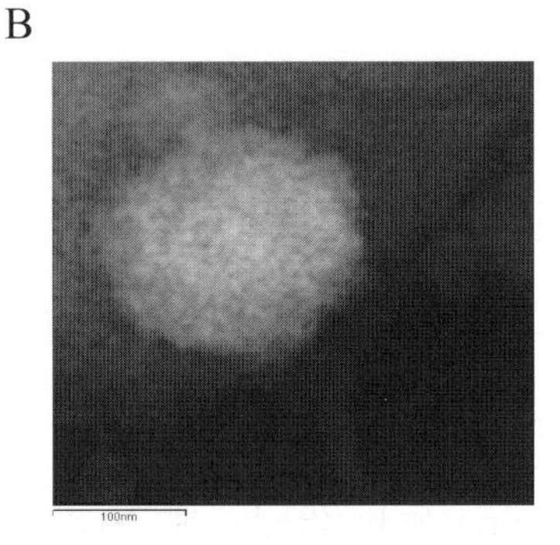

Figure 3. STEM images of CoRe/TiO_2 (A) and of CoRe/Al_2O_3 (B). The catalysts are analyzed in the oxidized state.

Re increases the activity (based on gram catalyst) and addition of Re also increases the C_5^+ selectivity for supported CoRe. TEM studies show that Re seems to be in close contact with the cobalt and more so for CoRe supported on TiO_2 and SiO_2 than on γ Al_2O_3.

Mass transfer limitations are important in FT synthesis. The C_5^+ selectivity depends on the size of the catalyst particles or the thickness of the catalyst layer in monolithic reactors. For CoRe supported on Al_2O_3, particles larger than about 300 μm result in reduced C_5^+ selectivity. The pore size of the support may also influence the C_5^+ selectivity through re-adsorption of α-olefins and it has been proposed that the difference in selectivity could partly be due to variations in the extent of re-adsortion of α-olefins [6]. The present results, however, point to the effect of the particle size, the appearance of the Co particles and the characteristics of the support material as important for determining the selectivity. Microkinetic modelling is used as a tool to understand the catalytic behaviour.

3. Conclusions

The activity (mol converted/g catalyst, time) depends on the catalytic system and the same holds for the rate of deactivation, but the intrinsic turnover frequency is constant for the systems investigated provided that the particle size is larger than 6-8 nm. The selectivity of supported cobalt catalysts depends largely on the catalyst properties and the Co particle size and appearance seem to be the most important parameter.

Acknowledgement
The Norwegian Research Council (NFR) is greatly acknowledged for financial support.

References
[1] S. Storsæter, D. Chen, A. Holmen: Surf. Sci. **600** (2006) 2051.
[2] V. Frøseth, S. Storsæter, Ø. Borg, E.A. Blekkan, M. Rønning, A. Holmen: Appl. Catal. A: Gen., **289** (2005) 10.
[3] B. Shi, B.H. Davis: Catal. Today **106 (**2005) 129.
[4]G.L. Bezemer, J.H. Bitter, H.P.C.E. Kuipers, H. Oosterbeek, J.E. Holewijn, X.D. Xu, F. Kaptejin, A.J. van Dillen, K.P. de Jong: J. Am. Chem. Soc. **128** (2006) 3956.
[5] P.B. Radstake, J.P. den Breejen, J.H. Bitter, K.P. de Jong, V. Frøseth, A. Holmen: Submitted to 8th NGCS, Brazil (2007).
[6] E. Iglesia, S.C. Reyes, R.J. Madon, S.L. Soled: Adv. Catal. **39** (1993) 221.
[7]A.M. Hilmen, E. Bergene, O.A. Lindvåg, D. Schanke, E. Eri, A. Holmen: Stud. Surf. Sci. Catal. **130^B^** (2000) 1163.
[8] S. Storæter, Ø. Borg, E.A. Blekkan, A. Holmen: J. Catal. **231** (2005) 405.
[9] Chr. Aaserud, A.M. Hilmen, E. Bergene, S. Eri, D. Schanke, A. Holmen: Catal. Lett. **94** (2004) 171.
[10] C.J. Bertole, C.A. Mims, G. Kiss: J. Catal. **221** (2004) 191.
[11] A.M. Hilmen, E. Bergene, O.A. Lindvåg, D. Schanke, S. Eri, A. Holmen: Catal. Today **105** (2005) 357.
[12] B.C. Enger, V. Frøseth, Ø. Borg, E. Rytter, A Holmen: Submitted.
[13] E.Rytter, D. Schanke, S. Eri, H. Wigum, T.H. Skagseth, N. Sincadu: ACS Petr. Chem. Div. Preprint **47** (2002).

Natural Gas Conversion VIII
F.B. Noronha, M. Schmal, E.F. Sousa-Aguiar (Editors)

Patenting Trends in Natural Gas Fischer-Tropsch Synthesis

Adelaide Antunes[a], Maria Simone Alencar[a], Fernando Tibau[a], Daniel Hoefle[a], Andressa Gusmão[a], Angela Ribeiro[a] and Rodrigo Cartaxo[a]

[a] *SIQUIM / School of Chemistry / Federal University of Rio de Janeiro / Brazil*

Abstract

The chemical transformation of Natural Gas through Fischer-Tropsch synthesis into liquid hydrocarbons which can be directly consumed as fuels has become of major interest. This is mainly due to two factors: (i) severe restrictions imposed by environmental laws promoting clean fuels; and, (ii) trends in the global energy scenario towards increased reserves of Natural Gas.

The present research consists of analysis of approximately 1200 patent documents worldwide referring to Fischer-Tropsch synthesis not using coal feedstock retrieved from the Derwent database and almost 200 relating additionally to Natural Gas. The period in question cover the last 5 years (2001-2005).

Through data/text mining techniques and technological foresight methods, identification was made of the main patent assignees, markets (with emphasis on patents in Brazil) and technological trends in the new millennium.

The results present tools for strategic decision making, thus contributing to the development of Natural Gas use in Brazil

1. Introduction

Fischer-Tropsch conversion has been known since the early decades of the 20th century, with its early development taking place in the 1920s, when the German scientists Franz Fischer and Hans Tropsch developed the process of conversion of synthesis gas, a mixture of carbon monoxide and hydrogen produced from coal, into liquid fuels. Nevertheless, it was not until the Second World War that Germany began to use this technology on an industrial scale.

Thus, it can be seen that the technology needed for this conversion already exists and is well-established. In fact, research efforts by companies in recent years have been directed at optimizing various existing processes, aiming to arrive at the best possible configuration while taking into account technical and economic aspects. The major focus of these efforts has been the development of smaller, more modern equipment, energy recovery and catalysis.

The transformation of synthesis gas into liquid hydrocarbons which can be immediately consumed as fuel has become economically attractive as result of two principal factors: (i) severe restrictions imposed by environmental laws promoting clean fuels; and, (ii) trends in the global energy scenario towards increased reserves of Natural Gas.

Given this context, the aim of this study is to identify trends in patenting with regard to Fischer-Tropsch Synthesis using Natural Gas as raw material, opposed to conventional coal fed.

2. Methodology

Research was undertaken in the DERWENT international patent database for the period 2001 to 2005, in which were identified patents related to Fischer-Tropsch synthesis not using coal feedstock (Broad View) along with the use of Natural Gas as raw material (Direct View). All patent deposits were sorted in their worldwide registers due to big companies interests in specific national markets (e.g. Brazil). The use of technological prospecting and data/text mining techniques made possible the identification of the main patent holders, the major markets and trends in patenting.

3. Results

An initial broad search was undertaken for patents related to Fischer-Tropsch synthesis not using coal as raw material. The results revealed a group of 1213 patent documents deposited in the Derwent database in the period of 2001 to 2005. In Figure 1 below may be observed the history of this patenting according to the priority date of the patent.

Figure 1 – Broad View: Patenting history – Fischer-Tropsch synthesis not using Coal (Priority Year of Patent)

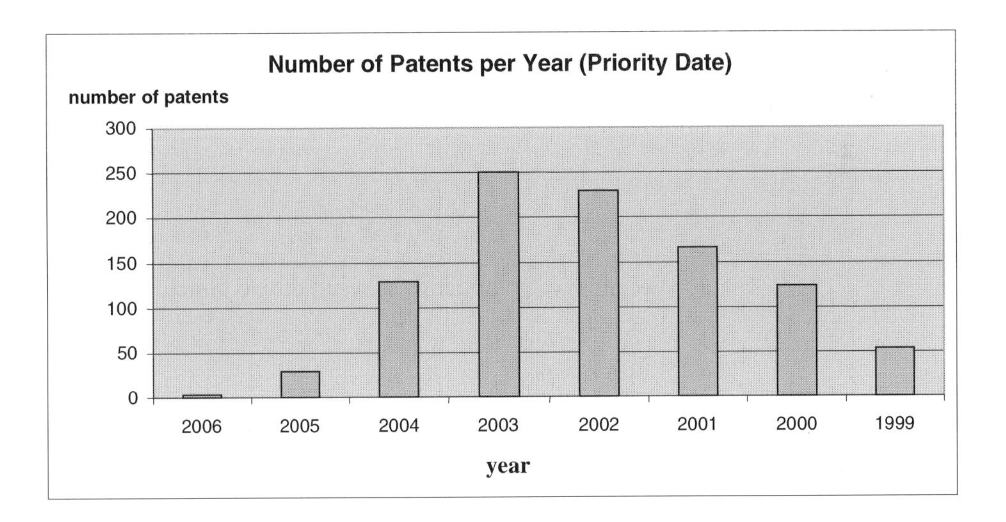

It can be seen that, from 1999 to 2003, there was a significant increase in the number of patents related to the topic under discussion. The decline presented in the last three years, accentuated in 2005 and 2006, is a result of patent secrecy, as well as the delay in the indexing of patents in the Derwent database.

Another important analytical approach in regard to this group of patents is that of markets. Of the total 1213 patents, 881 were deposited in the US Patent Office and 600 in the Australian, these being the major protected markets. 210 documents were deposited in Brazil, representing 17% of the total.

- **Fischer-Tropsch synthesis using Natural Gas:**

With the aim of obtaining results on hydrocarbon synthesis by means of the Fischer-Tropsch reaction directly focused on using Natural Gas, a new search was undertaken in the Derwent database relating both, which retrieved 174 distinct patents deposited in the period 2001 to 2005. Figure 2 below shows the history of these patents according to the priority date.

Figure 2 – Direct View: Patenting history – Fischer-Tropsch synthesis using Natural Gas (Priority Year of Patent)

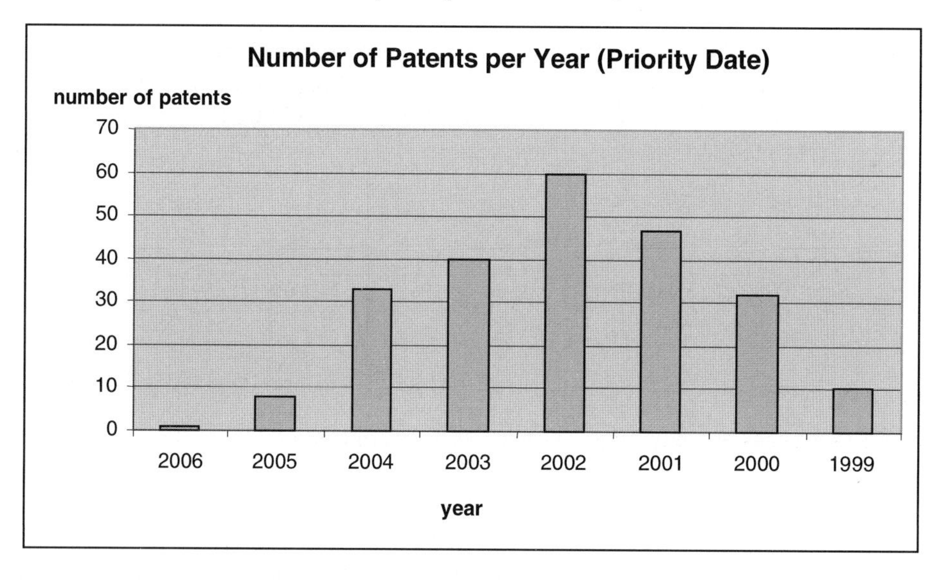

In this figure it can be observed that 2002 was the year in which there was the highest number of patents, reaching a total of 60. Once again, there is a decline in recent years, but this is attributable to patent secrecy and the lag in indexation in the database.

On the basis of *data and text mining* analyses, it was possible to identify the main global markets where these patents were deposited. Of 174 patents, 152 were deposited in US Patent Office, representing around 86% of the total. This was followed by Australia, with 83, and Japan with 54 documents, deposited in their respective patent offices. 36 patents were deposited in Brazil, representing 20% of the total. This is a significant percentage, which demonstrates the interest of foreign companies in the Brazilian market.

Following the objective of this study to identify patenting trends, identification of the major patent registrants was made. Of a total of approximately 50 distinct companies and/or universities, there exists a strong preponderance of the large US companies, as shown in Table 1, which lists the registrants with more than 10 patents, in which only Shell and Sasol are not American.

Table 1 – Main registrants

Main Registrants	Number of Patents
CHEVRON (USA)	23 (2 in partnership with SASOL)
CONOCOPHILLIPS CO (USA)	17
SHELL (UK/NL)	14
SASOL (South Africa)	12 (2 in partnership with CHEVRON)
EXXONMOBIL (USA)	11
SYNTROLEUM CORP (USA)	11

The subsequent step was an analysis of the content of the 86 patent documents listed above, as a means of identifying the trends for each company.

- **Patenting Trends among the Main Registrants:**

(i) CHEVRON: of the 23 patents analyzed, there is an observable trend towards the production of hydrocarbons and fuels, as the company owns 11 related patents.

(ii) CONOCOPHILLIPS: this company presents a diversified field of activity, with 6 patents for Fischer-Tropsch process catalysts, 5 for hydrocarbon production and 5 for production of synthesis gas.

(iii) SHELL: presents a strong trend in patenting for treatment processes, both in hydrocarbon production and in synthesis gas production, with the aim of controlling the formation of coal throughout the processes. The company owns 10 patents in this area.

(iv) SASOL: owns half of the patents (6) in hydrocarbon production, showing a clear trend in this area.

(v) EXXONMOBIL: demonstrates a significant trends in patenting catalytic systems, with 6 of the 11 patents in this area.

(vi) SYNTROLEUM: is a company showing a trend to patenting of liquid hydrocarbons via Fischer-Tropsch synthesis based on Natural Gas. Syntroleum is responsible for 9 of the 11 patents in this area.

4. Conclusion

Despite being an old technology, Fischer-Trospch synthesis is still capable of generating the interest of many globally powerful companies due to Natural Gas and other new feedstock potential. It can be observed that this is a market dominated by American companies (Chevron, Conocophillips, Exxon and Syntroleum), as these are the main global patent holders along with Shell and Sasol. This study has also demonstrated that the Brazilian market is a target for these companies, as around 20% of the relevant patents have been registered in that country.

With regard to patenting trends, we can conclude that there is a concentration of patents in the production of hydrocarbons and fuels, while there are some companies, such as Exxon, which retains the majority of patents in catalytic systems, and Conocophillips, which presents diversified activities in the production of hydrocarbons, synthesis gas and catalysts.

Fischer-Tropsch synthesis still awaits further development, as it is a clean technology whose products are not a threat to the environment.

Natural Gas Conversion VIII
F.B. Noronha, M. Schmal, E.F. Sousa-Aguiar (Editors)

Cobalt Supported on Different Zeolites for Fischer-Tropsch Synthesis

Andréia S. Zola[a] , Antônio M. F. Bidart[b], Adriano do C. Fraga[b], Carla E. Hori[c], Eduardo F. Sousa-Aguiar[b], Pedro A. Arroyo[a]

[a]*Chemical Engineering Department, State University of Maringá, Av Colombo 5790, Maringá-PR, 87020-900, Brazil.*
[b]*PETROBRAS/CENPES/PDEDS/Célula GTL, Av Jequitibá 950, Rio de Janeiro-RJ, 21941-589, Brazil.*
[c]*School of Chemical Engineering, Federal University of Uberlândia, Av. João Naves de Ávila 212 1, Bloco 1K, Campus Santa Mônica, Uberlândia-MG, 38400-901, Brazil*

ABSTRACT

In this work, the textural and chemical properties as well as the catalytic behavior of 10 wt% Cobalt supported on H-USY, H-Beta, H-Mordenite and H-ZSM-5 zeolites in the Fischer-Tropsch Synthesis (FTS) were studied. X-Ray diffraction analyses showed that the characteristic patterns for each zeolite structure were maintained after the metal addition. The average Co particle size increased in the following sequence: Co/H-Beta < Co/H-USY < Co/H-Mordenite < Co/H-ZSM-5. It was observed that the catalysts activities for FTS decreased with increasing average crystallite size values and apparently the production of longer chain hydrocarbons could be related to the presence of three-dimensional pore system and secondary porosity that can favor the chain growing in the Fischer-Tropsch Synthesis.

1. Introduction

Fischer-Tropsch Synthesis (FTS) represents a route for natural gas transformation into liquid hydrocarbon fuel via syngas. Cobalt supported catalysts have been found to be particularly convenient for higher hydrocarbons production [1], mainly in the diesel range. A few studies [2,3] have been devoted to the application of zeolites as catalyst support for the FTS with the target of controlling the product distribution. Recently, a rational strategy pointed out in the development of more active catalysts for FTS is to enhance the cobalt dispersion by decreasing the average metal particle size [4]. Therefore, this work aims to evaluate the effect of the structure and porosity on the catalytic behavior of Cobalt-supported on zeolites in the FTS for C_{10}-C_{20} hydrocarbons production, mainly liquid fraction (C_{13+}).

2. Experimental

Catalyst Preparation: The parent zeolites were commercial samples of H-Mordenite, H-USY, H-ZSM-5 and H-Beta with SAR of 13, 7, 27 and 16, respectively. A 10 wt.% Co/Zeolite were prepared by wet impregnation of zeolite with a cobalt nitrate solution. After 24h impregnation the slurry was vacuum dried at 70 °C in a rotor evaporator, dried at 90 °C overnight and then calcined at 300 °C/4 h. The samples obtained were named Co-HM 10%, Co-HU 10%, Co-HZ 10% and Co-HB 10%, where the final numbers denotes the nominal wt.% of Co.

Characterization: Cobalt content was determined by atomic absorption spectrometry using a Varian SpectrAA10-Plus spectrometer. Surface area, porosity and nitrogen adsorption/desorption isotherms were obtained by using a Quantachrome Nova 1000 Series equipment. The X-ray Diffraction (XRD) was recorded at room temperature by a Siemens Kristaloflex 4 difractometer, using KuKα radiation. The average crystallite size was estimated from Scherrer equation using the most intense reflexion of Co_3O_4 crystalline phase at ca. $2\theta = 37°$. The reduction behavior was studied by temperature-programmed reduction by flowing through the catalyst bed a 30 mL/min mixture of H_2 (1.75 vol.%) in Ar at a heating rate of 10 °C/min.

Catalytic Experiments: The carbon monoxide hydrogenation was carried out in a FTS Unit with 16 reactors in parallel. Prior the catalytic experiments, the samples were reduced in situ using a pure hydrogen flow rate of 40 mL/min at 365 °C/10 h. The experimental conditions were: T=240 °C, P=10 bar, $H_2/CO=2$ and GSVH=1287 h^{-1}. The FTS products were analysed on-line by GC, using three methylsilicone capillary columns (pre-column, C_1-C_5, C_5-C_{30}), and an IR ABB detector system. Mean CO conversion, yields and products selectivities were calculated based on the samples collected between 70-120 h on stream. A standard catalyst of 0.4%Ru-23.6%Co/Al_2O_3, optimized for C_{5+} hydrocarbons production, was also tested for comparison effect.

3. Results and Discussion

The wt.% of Co obtained by wet impregnation were 9.17, 9.81, 9.58, and 9.36 for Co-HM 10%, Co-HU 10%, Co-HZ 10%, Co-HB 10%, respectively. The N_2 adsorption/desorption isotherms of all samples were type I that are typical of microporous materials. Further, besides isotherms behavior, the surface area and micropore volume values suggest the clogging of support pores by cobalt species that makes them less accessible for nitrogen adsorption. It can also indicate that the metal is located preferentially at the external surface of the zeolite particle. This effect is more pronounced for ZSM-5 that possess the smaller pore (10-MR) and for Mordenite that possess a pseudo one-dimensional pore system. Sample USY is less sensitive to the partial pore blockage due its secondary porosity and three-dimensional pore system. The same behavior is observed for Beta zeolite since it sample also present three-dimensional pore system.

X-Ray diffraction (XRD) analyses of all catalysts are presented in Figure 1. It is possible to observe that the characteristic patterns for each zeolite structure were maintained after the metal impregnation. XRD analyses also show an intense peak at $2\theta=37^{o}$ attributed to the cobalt nanoparticles structured located at external surface of the zeolite. From XRD it was determined that zeolite ZSM-5 led to the higher average cobalt particle size, in agreement with textural analyses. Then, average Co_3O_4 crystallite sizes calculated were 11.1, 16.4, 17.6 and 24.4 nm for Co-HB 10%, Co-HU 10%, Co-HM 10% and Co-HZ 10%, respectively.

Figure 2 shows the TPR profiles of the Co-containing zeolites after calcination. It is possible to observe two reduction peaks at 300-400 °C, corresponding to the stepwise reduction of Co_3O_4 particles to CoO and then to metallic Co on the support external surface [2]. Peaks observed at temperatures higher than 500 °C

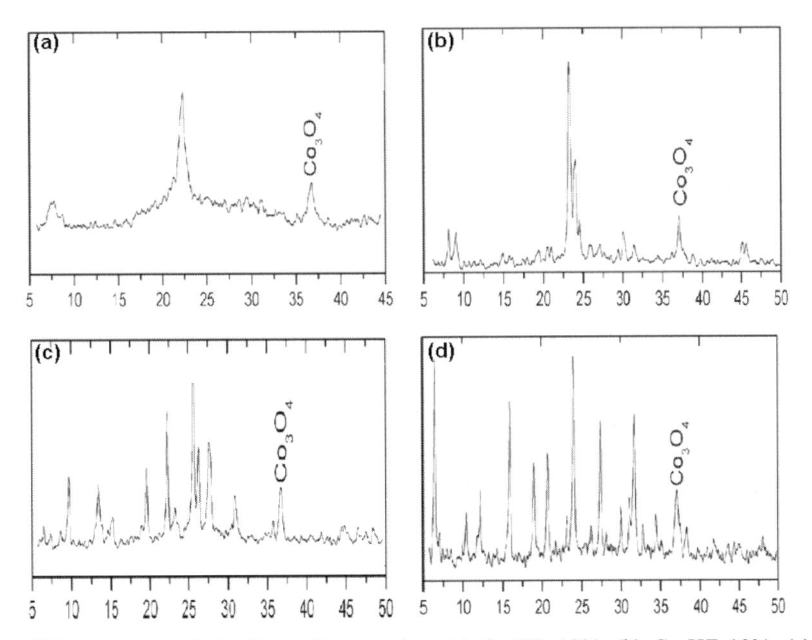

Figure 1: XRD patterns of the Co-zeolite samples: (a) Co-HB 10%, (b) Co-HZ 10%, (c) Co-HM 10% and (d) Co-HU 10%.

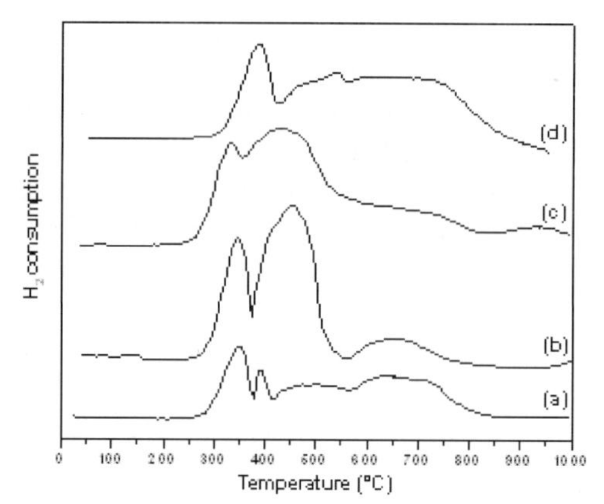

Figure 2: TPR profiles: (a) Co-HB 10%, (b) Co-HZ 10%, (c) Co-HM 10% and (d) Co-HU 10%.

suggest the reduction of smaller cobalt particles located on the external surface and inside the zeolites cages and channels. Reduction degrees obtained from H_2 uptake were 79, 77, 90 and 48% for Co-HB 10%, Co-HU 10%, Co-HM 10% and Co-HZ 10%, respectively.

Figure 3 shows mean CO global conversions and activities in the FTS for all samples tested. Most of Co-containing zeolites showed CO global conversion and activity values higher or at least as higher as a standard catalyst of 0.4%Ru-23.6%Co/Al_2O_3. Higher values of CO conversion and activity can be observed for Co-HU 10% and Co-HB 10% samples. Both structures present a three-dimensional channel system with large pores (12-MR) and secondary porosity generated during the dealumination process, in the case of USY sample, showing the influence of zeolite structure in the catalyst performance.

Figure 4 shows the influence of average Co_3O_4 crystallite size on the catalyst activity for FTS. It was observed that the catalyst activity decreases with

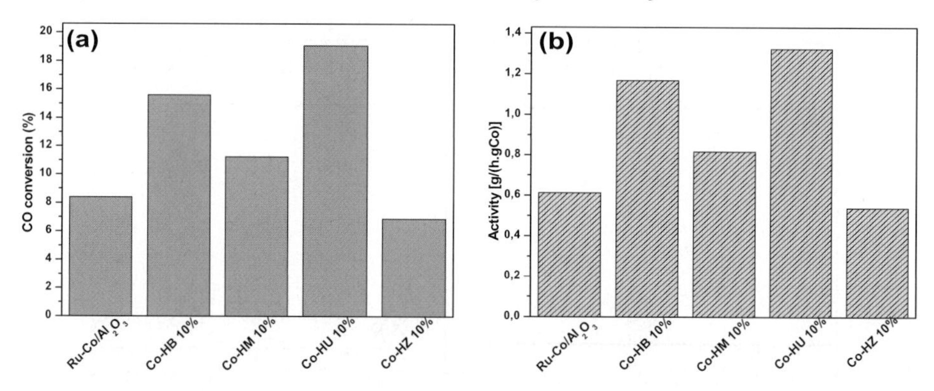

Figure 3: Catalytic behaviors of Co-containing zeolites: (a) Average CO global conversions (b) FTS Activities

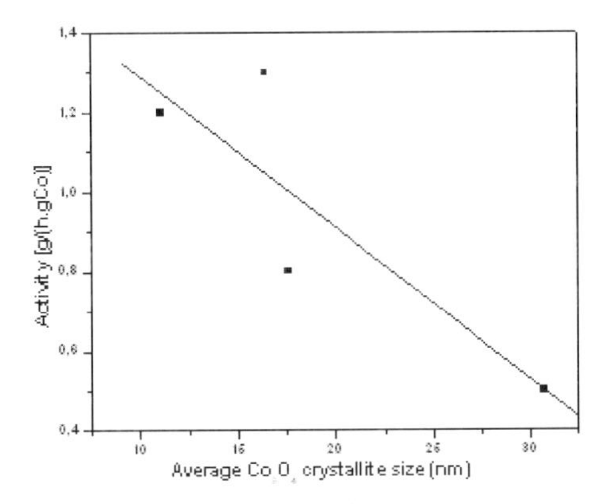

Figure 4: The influence of average Co_3O_4 crystallite size on FTS activity.

increasing average crystallite size values, in agreement with those results presented by Bezemer et al. [4].

The hydrocarbon products collected at the end of each reactor were analyzed, as previously described, and the liquid and gas yields are shown in Figure 5. All samples tested showed a higher gas yield (C_1-C_{12}) and only the sample Co-HU 10% presented measurable liquid yield (C_{13+}), closer to the 0.4%Ru-23.6%Co/Al_2O_3 standard catalyst. Apparently, production of longer chain hydrocarbons could be related to the presence of three-dimensional pore system and secondary porosity that can favor the chain growing in the FTS.

The product distribution is presented in Figure 6. It can be observed that C_5-C_9 hydrocarbons dominated the product distribution, even for optimized standard catalyst. However, the standard catalyst also shows a significant selectivity to the C_{10}-C_{20} hydrocarbons. Product distribution similar to the standard catalyst

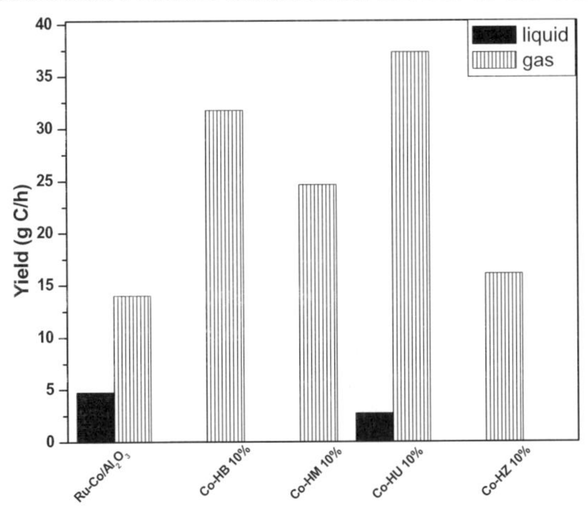

Figure 5: Liquid and gas yields for Co/zeolites and standard 0.4%Ru-23.6%Co/Al_2O_3 catalysts.

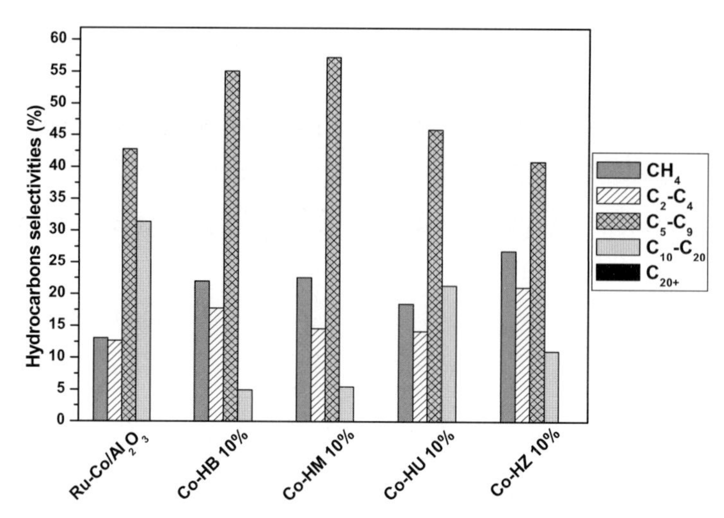

Figure 6: Selectivities toward hydrocarbons products on Co-supported zeolites catalysts.

can be observed only for Co-HU 10% sample. As discussed earlier for this sample, the higher accessibility to the catalytic sites due to the secondary porosity, big cavities and opener three-dimensional channel system favour the longer chain hydrocarbon formation. It was also noted that lower CO conversions led to the higher selectivity to methane.

4. Conclusions

Co/zeolite catalysts showed values of mean global CO conversion higher than a 0.4%Ru-23.6%Co/Al_2O_3 FTS standard catalyst. The presence of secondary porosity, cages and three-dimensional channel system with large micropores leads to a higher accessibility to the catalytic sites and favors the formation of longer hydrocarbons. An enhanced catalytic activity was also observed when the average cobalt crystallite size was decreased. Therefore, from the results presented herein it can be withdrawn that Co/USY is a potential catalyst for production higher hydrocarbons in FTS.

Acknowledgments
The authors ASZ, PAA and CEH are grateful to PROCAD/CAPES for financial support.

References

[1] Khodakov, A. Y., Bechara, R. e Griboval-Constant, A., App. Catal. A: Gen., 254, 2003, 273.
[2] Tang, Q., Wang, Y., Zhang, Q. e Wan, H., Catal. Commun. 4, 2003, p. 253
[3] Bessell, S., Appl. Catal. A: Gen., 126, 1995, p. 235
[4] Bezemer, G. L., Bitter, J. H., Kuipers, H. P. C. E., Oosterbeek, H., Holewijn, J. E., Xu, X., Kapteijn, F., Dillen, A. J., Jong, K. P, J. Am. Chem. Soc., vol. 128, nº 12, 2006, p. 3956.

Natural Gas Conversion VIII
F.B. Noronha, M. Schmal, E.F. Sousa-Aguiar (Editors)

Production of liquid hydrocarbons employing Natural Gas: a study of the technical and economical feasibility of a GTL plant in Brazil

Roberto Callari[a], Tatiana Magalhães Gerosa [b], Patrícia Helena Lara dos Santos Matai [c]

[a]*Petróleo Brasileiro S.A. – Petrobras, Avenida República do Chile65, Rio de Janeiro 20031-912*
[b]*Instituto de Eletrotécnica e Energia – Universidade de São Paulo, Avenida professor Luciano Gualberto 1289, São Paulo 05508-900*
[c]*Instituto de Eletrotécnica e Energia – Universidade de São Paulo, Avenida professor Luciano Gualberto 1289, São Paulo 05508-900*

1. Introduction

The chemical transformation of Natural Gas, mainly methane, in liquid fuels such as diesel fuel, gasoline and aviation kerosene have been until recently restricted to the African and Asian continents mainly due to the occurrence of stranded reserves (ConocoPhillips, 2004). The target of this paper is to present a technical and economical evaluation for the construction of a gas-to-liquids (GTL) plant in Brazil employing Natural Gas (NG) as raw material. The cost of capital investment as well as the production cost and diesel GTL quality concerning the world consuming market are evaluated. The main product to be considered is diesel fuel.

1.1. Investment Costs

Concerning the investment costs in GTL projects, the analysis of the projects proposed or executed by the companies that hold the technologies, allows a preliminary evaluation of the investments for the construction of a GTL plant. Table 1 shows the best investment indicators for several companies considering their production capacity.

Company	Capacity	Investment of capital	
	10^3 bbl/d	10^6 US$	US$/bbl
Sasol (South Africa)	10.0	300	30,000
	20.0	500	25,000
Shell (Malaysia)	12.5	850	68,000
Sasol/QP (Qatar)	34.0	1,000	30,000
Syntroleum (USA)	2.5 to 22.0	75 to 500	23,000 to 30,000
Texaco/Syntroleum	19.6	548	28,000

Table 1 Investment indicators in GTL projects. Source: IFP, 2001

Construction expenditures in GTL plants indicate that the investment level is between US$ 20,000 and US$ 30,000 per barrel of the plant production capacity. According to the *Energy Intelligence Group*, the estimative of investment to new projects consider the costs of US$ 28,000 per barrel for a GTL plant when compared against costs between US$ 12,000 and US$ 15,000 per barrel of the conventional refining processes. Accordingly, based on a few experiences of other projects, the operational and maintenance costs of a plant can be evaluated in 3% per year over the whole construction investment of a GTL plant (ITP Business, 2006). The GTL Shell plant located in Bintulu, Malaysia, started-up its operations in 1993, however in 1997 it occurred an accident in the oxygen production area that caused great damages to the same one, interrupting its production until middle of year 2000, when was retaken the production of 12,500 bbl/d. Including the reconstruction expenditures, the cost of capital of the plant raised to 68,000 US$/bbl, on account of a global investment of US$850 million.

1.2. Production cost

According to Arthur D. Little Global Management Consulting (2003), a possible configuration of the division of investments and costs for the implantation of a GTL plant could be presented as: cost of the raw material, NG, 22%; implantation of the syngas unit production 30% and its operational cost around 25%; the stage of FT synthesis around 18% and the stage of product finishing around 5%.

2. Diesel fuel as the main product: high quality

The quality of the GTL products, considering raw material, synthesis processes and the FT synthesis, is different when compared against the refinery diesel fuel. The diesel fuel and other liquid hydrocarbons present highly desirable characteristics considering both, energy and environmental aspects. According to the US Department of Energy GTL Diesel fuel presents high cetane index (above 70 against the refinery diesel fuel range of 48 and 50), very low sulfur emission (< 1 ppm); very low aromatics emission (< 1% in volume); low cloud point (-10°C), low emissions like 8% less NO_x, 30% less particulate matter, 38% less hydrocarbons (HC) and 46% less carbon monoxide (CO).

3. Discussion

It would be possible to expect that the economical and technical project evaluation for the construction of a GTL plant shows the feasibility of a plant in Brazil thanks to the new NG offshore reserves in Campos, Santos and Espírito Santo Basin. These three basins present a plan of exploratory development that will allow a gradual appropriation of NG reserves to Brazil until reaching the 40 million m³ per day from 2012. Although the most part of this gas is already compromised to the thermal generation program, with only 10% of this volume it would be possible to construct a GTL plant with capacity of production of more than 30,000 barrels per day. The authors start to study the possibility to construct an industrial medium-scale plant, for 15,000 barrels per day, using only one maximum of 5% of the appropriated gas. In this economic study, many premises must be assumed to allow a more intent analysis in the most important characteristics of the project, as it shows in the table 2.

Contractual period	20 years
Produced Volume	15,000 bbl/d
Conversion ratio	280 m^3 gas/bbl GTL
Daily gas consumption	4,200,000 m^3/d
Necessary reserve	1,1 TCF
Percentual of Tax	30%
Capital cost	9% pa
Transportation cost	2 US$/bbl
Contingency	20%

Table 2: Settled parameters of the project for its analysis.

The discussion and study concerning the GTL technology as presented, will show that GTL fuels will bring a quality improvement to the oil industry because the NG reserves present all over the planet can be employed. The NG industrial use should lower the emission of greenhouse gases and acid rain and also lower the emissions of contaminants into the environment. In opposition to what it is being announced by the oil refiners, that have been developing studies concerning cleaner fuels, the GTL fuels can be produced not only from stranded gas, but also they can be obtained form new reserves close to metropolitan regions. It could be the Brazilian case. Besides, this study shows that a boom in GTL technologies has been happening. Several new GTL plant projects with various installed production capacities are being located in different continents.

4. Economical analysis

In this economic evaluation, the return on the capital investment was analyzed based on the most three important variables: capital cost for barrel, feedstock cost and GTL diesel selling price per barrel. Through the calculation of the present value of the construction project and the plant operational cost, the return to the investor was calculated including all capital costs, operational costs, incomes tax and financing. The figure 1 shows the result of the return on the investment in the event which the NG was fixed in US$1/MBtu. In this situation all the projects would be viable if the GTL diesel selling price wasn't less than US$70/bbl. By the other hand the return on the investment would be around to 60% if the GTL diesel selling price is about US$100/bbl.

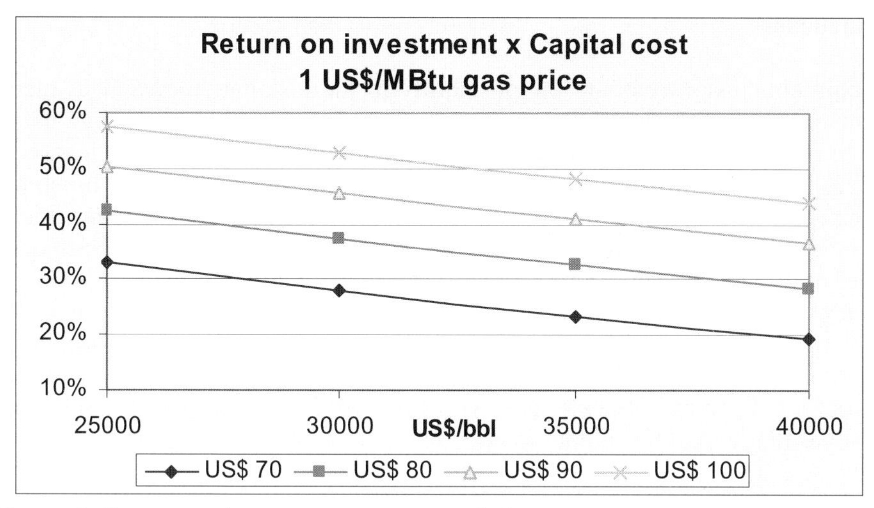

Figure 1: Return on investment versus capital cost.

The figure 2 shows that the NG price is decisive when it is related to the return on investment. To simulate a fixed capital cost of US$30,000/bbl and the GTL diesel selling price equals US$70/bbl, and a feedstock price up to US$ 1.7/MBtu the project would be unviable, and the return would be below of 10%. The GTL diesel selling price equals US$80/bbl could be viable the projects where the natural gas cost is US$2/MBtu, and where the GTL selling price is around US$90 projects which naturals gas cost would be about US$2.4/MBtu could be viable more projects. In the last simulate, where the GTL diesel selling price equals US$100/bbl, the projects could be viable even that the natural gas cost be minor or equals US$3/MBtu.

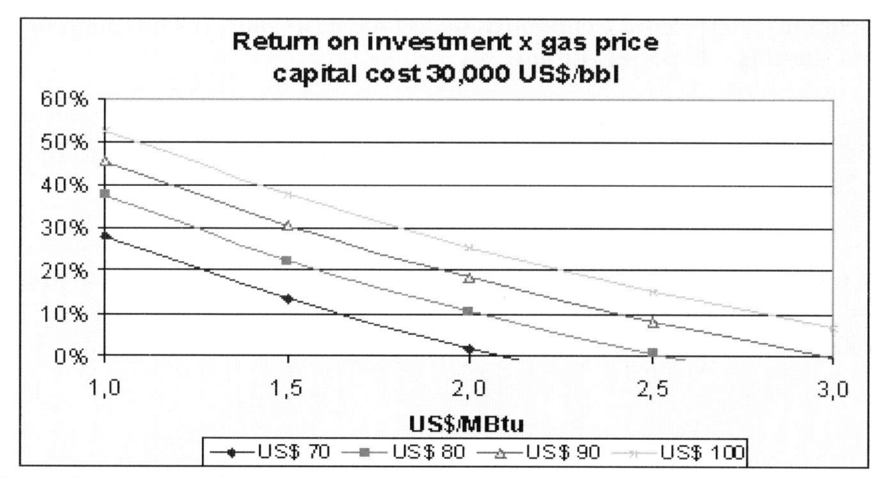

Figure 2: Return on investment versus gas price.

5. Conclusion

The main conclusion refers to the possibility of the installation of a GTL plant in Brazil close to the sea and/or close to metropolitan and industrial regions in order to manage efficiently the shipping to the international consumer markets. The feasibility construction of the GTL diesel plant becomes limited by some reasons such as gas availability, feedstock cost and GTL diesel selling price, what is led by the global oil price. The influence of these factors in the projects is more important than the technological conditions inherent to the process. It is important to emphasize that the center of Amazon region would be a good market to commercialize GTL diesel due to the lack of natural gas transportation. Therein, Urucu Basin would be an appropriated place to construct a GTL plant because there is a high volume of produced associated natural gas that would be enough to attend a medium-scale GTL plant.

6. Bibliography

BIRCH, Colin. The market for GTL diesel. Petroleum Economist, pp. 27-29, Jan, 2003. (Fundamentals of Gas to Liquids – Especial edition)

FLEISCH, T. H.; SILLS, R. A.; BRISCOE, M. D. 2002 – Emergence of the Gas-to-Liquids Industry: a review of global GTL developments. Journal of Natural Gas Chemistry, v. 11, pp. 1-14, 2002.

FLEISCH, T. H.; SILLS, R. A.; BRISCOE, M. D. GTL-FT in the emerging gas economy. Petroleum Economist, pp. 39-41, Jan, 2003. (Fundamentals of Gas to Liquids – Especial edition)

DYBKJÆR, Ib. Synthesis gas technology. Petroleum Economist, pp. 16-19, Jan, 2003. (Fundamentals of Gas to Liquids – Especial edition)

GHAEMMAGHAMI, Bahram. GTL: Progress and Prospects - Study yields generic, coastal-based GTL plant. Oil&Gas Journal, v. 99, n. 11, Mar 12, 2001.

Natural Gas Conversion VIII
F.B. Noronha, M. Schmal, E.F. Sousa-Aguiar (Editors)

New Supports for Co-based Fischer-Tropsch Catalyst

A.F. Costa,[a] H.S. Cerqueira,[a] E. Falabella S. Aguiar,[a] J. Rollán,[b] A. Martínez[b]

[a] *Petrobras, Centro de Pesquisas e Desenvolvimento Leopoldo A. Miguez de Mello (CENPES), Av. Jequitibá 950, Ilha do Fundão 21941-598 Rio de Janeiro, RJ, Brazil.*
[b] *Instituto de Tecnología Química, UPV-CSIC, Avda. de los Naranjos s/n, 46022 Valencia, Spain.*

1. INTRODUCTION

The Gas-to-Liquids (GTL) technology is an attractive route to produce high-quality liquid fuels from natural gas. This technology consists of a chemical transformation of natural gas that requires three steps: syngas generation, Fischer-Tropsch synthesis (FTS) and product upgrade. The critical step is the FT synthesis, that received significant improvements during the last decades both on the reactor [1] and catalyst technologies [2,3]. The cobalt-based catalysts are typically used at reaction temperatures in the range of 210-250°C and after the upgrading step (hydrocracking and hydrodewaxing), the main products are premium diesel (cetane number higher than 70 and virtually no S or aromatic compounds), food grade paraffin and specialty lubricants.

The desire to enhance the participation of natural gas in the energy matrix gives momentum to GTL technology, thus motivating the continuous development of new catalysts for the FT synthesis. Reduction promoters, new supports and support promoters are the focus of various research activities [4-7]. In the present paper the effect of ITQ-6 zeolite and mesoporous material SBA-15 as supports for Co-based Fischer-Tropsch catalyst was investigated, and compared with a conventional amorphous SiO_2 support.

2. EXPERIMENTAL

The all-silica SBA-15 (10 nm pore size) was synthesized in the presence of the co-polymer Pluronic P123 (Aldrich) as surfactant and tetraethyl orthosilicate (Merck-Schuchardt) as silicon source, in HCl medium [5]. The pure silica ITQ-6 support was obtained by delamination of the laminar precursor of the FER zeolite [8]. A commercial amorphous SiO_2 (Fluka, silica gel 100) was used as reference support. Unpromoted catalysts were prepared by wetness impregnation with an aqueous solution of $Co(NO_3)_2$ in order to achieve 20 wt% of cobalt in the final catalyst. Promoted Ru-Co/SiO_2, Ru-Co/ITQ-6, and Re-Co/SBA-15 samples were also prepared by co-impregnation with an aqueous solution containing $Co(NO_3)_2$ and promoters precursors, Ru(III) nitrosyl nitrate solution (Aldrich, 1.5 wt% Ru) or HO_4Re aq. solution (Across, 76.5% Re), to attain 20 wt% Co and 1 wt% promoter loadings. After impregnation, the solids were dried at 100°C and calcined at 300°C for 10 h.

After calcination the catalysts were characterized by atomic absorption spectrophotometry (AAS), X-ray diffraction (XRD), hydrogen temperature-programmed reduction (H_2-TPR) and nitrogen adsorption. XRD was performed in a Phillips X'pert diffractometer using the CuK_{α} monochromatic radiation. For H_2-TPR analysis (Autochem 2910 from Micromeritics), the temperature was increased 10°C/min from ambient to 900°C under a stream of 10%vol H_2 diluted in He. The same procedure was used to evaluate the extent of reduction of Co after the reduction at 400°C for 10 h. Textural properties were determined by the N_2 adsorption isotherms at -196°C (ASAP 2000 from Micromeritics).

After reduction, the catalysts were evaluated in a fixed bed unit under the following conditions: 2.0 MPa total pressure, temperature 220°C, H_2/CO molar ratio of 2, gas hourly space velocity (GHSV) equal to 7.2 $l.g^{-1}.h^{-1}$ referred to the syngas. The temperature in the catalyst bed was controlled to 220°C ± 1°C by means of two independent heating zones with the corresponding temperature controllers connected to thermocouples located in different positions inside the catalytic bed. During the reaction the reactor effluent passed through two traps, the first at 150°C and the second at 100°C, both kept at 2.0 MPa to condense the heaviest hydrocarbons (waxes). The stream leaving the second trap (H_2, CO, CO_2, water, alcohols and hydrocarbons from C_1 up to about C_{20}) was depressurized and analyzed *on-line*. Gas chromatography data was collected at periodic intervals in a Varian 3800 chromatograph. Details of the GC analyses can be found in previous works [5, 9]. Carbon mass balances performed at the end of the experiments including the amount of waxy products collected in the hot trap were quite satisfactory (100 ± 2%). Product distributions are given on a carbon basis.

3. RESULTS AND DISCUSSION

The amount of cobalt and promoters in the final catalysts determined by AAS matched, within experimental error, the nominal contents. XRD results showed in all cases that cobalt was in the spinel Co_3O_4 form after calcination at 300°C. The average diameter of Co_3O_4 crystallites was estimated by the

Scherrer equation [10] using the most intense reflection of Co_3O_4 at $2\theta = 36.9°$. The Co_3O_4 particle size was then converted to the cobalt metal diameter in the reduced catalyst by considering the relative molar volumes of Co_3O_4 and Co^0 as follows [11]: $d(Co^0)=0.75 * d(Co_3O_4)$. Then, Co metal dispersion was estimated from $d(Co^0)$ by assuming a spherical geometry of the metal particles with uniform site density of 14.6 atoms per nm^2 [12, 13]: $D= 96/d$, where D is the % Co^0 dispersion and d is the mean particle size of Co^0 in nm.

Table 1 summarizes the textural characterization of supports and catalysts, as well as the properties of supported cobalt particles in unpromoted and promoted catalysts comprising different supports. A decrease in surface area and pore volume was systematically observed upon incorporation of Co and precursors, which is mostly due to the dilution effect caused by the presence of the Co_3O_4 phase. Nevertheless, the relative loss of surface area was larger for the SBA-15 based catalyst, which can be attributed to partial blockage of the monodimensional pore system by the cobalt oxide clusters.

As seen in Table 1, the catalysts based on the delaminated ITQ-6 zeolite and the mesoporous SBA-15 carriers (both unpromoted and promoted) displayed better Co^0 dispersions than those based on conventional amorphous SiO_2. The use of ITQ-6 as support leads to samples with the highest dispersion, despite the larger surface area of SBA-15. This can be ascribed to the formation of large Co_3O_4 clusters on the external surface of SBA-15 as observed earlier by TEM [5] and to the very high and open external surface of the delaminated zeolite that favours the dispersion while avoiding extensive sintering of the supported cobalt particles during calcination. As also observed, the addition of Ru and Re promoters had little effect on the cobalt particle size (i.e. dispersion).

Table 1. Supports and catalysts characterization.

Sample	B.E.T.	TPV[a]	$d(Co_3O_4)$	$D(Co^0)$	ER[b]
SiO_2	387	0.81	--	--	--
ITQ-6	537	0.88	--	--	--
SBA-15	812	1.23	--	--	--
Co/SiO_2	300	0.67	13.9	9.2	95
Co/ITQ-6	350	0.61	9.3	13.8	88
Co/SBA-15	508	0.62	11.4	11.2	62
$Ru-Co/SiO_2$	301	0.58	14.4	8.9	96
Ru-Co/ITQ-6	365	0.45	9.6	13.3	96
Re-Co/SBA-15	469	0.55	11.7	10.9	96

[a] TPV: total pore volume; [b] ER: extent of cobalt reduction at 400°C for 10 h.

Figure 1 presents the reduction profiles for cobalt precursors in the unpromoted catalysts. As observed, the Co/SiO_2 sample showed two intense peaks at about 270ºC and 305ºC that correspond to the two-step reduction of Co_3O_4 to Co^0 ($Co_3O_4 \rightarrow CoO \rightarrow Co^0$). Moreover, a broad and much less intense peak centered at about 525ºC can also be observed, which is due to the reduction of some cobalt particles having a stronger interaction with the silica surface. The reduction profile of Co/ITQ-6 closely resembled that of Co/SiO_2, with two main reduction features at ca. 280ºC and 317ºC (broader than in the silica-based sample) and a less intense and broader peak at about 570ºC. A different reduction behavior was, however, found for Co/SBA-15. In this case, the two low-temperature peaks associated to the reduction of Co_3O_4 to Co^0 were not clearly distinguished and instead a single feature with a maximum at 310ºC and a shoulder at 335ºC were evidenced. Furthermore, this catalyst presented two intense and broad features in the high temperature range, with maxima at 475ºC and 650ºC, respectively. The reduction profile of Co/SBA-15 clearly indicates that part of the cobalt oxide particles, most likely those of very small diameter (< 10 nm), are interacting strongly with the walls of the mesoporous support. In fact, reduction peaks at temperatures above 600ºC have been usually associated to the formation of cobalt silicate species that are hard to be reduced [14]. In addition, the interaction of cobalt precursors with the walls of SBA-15 could also be favoured by the presence of a large surface concentration of silanols in this material. The extents of cobalt reduction in Co/SiO_2, Co/ITQ-6, and Co/SBA-15 after the H_2 treatment at 400ºC (Table 1) were 95%, 88%, and 62%, respectively, thus reflecting the reduction trends discussed above.

The reduction profiles of the promoted catalysts are shown in Figure 2. As it was seen for the unpromoted counterparts, a very similar reduction behaviour was observed for the Ru-promoted SiO_2 and ITQ-6 catalysts. In both samples, the first peak associated to the reduction of Co_3O_4 to CoO shifted to lower temperatures (213ºC in Ru-Co/SiO_2 and 228ºC in Ru-Co/ITQ-6), and the second peak related to the reduction of CoO to Co^0 became broader and shifted to higher temperatures. This, together with the absence of the reduction feature at temperatures above 500ºC clearly indicates that Ru facilitates the reduction of Co species, especially those displaying a high interaction with the support. As seen in Figure 2, the addition of Re to the Co/SBA-15 sample had a similar effect on Co reducibility than Ru in the other two catalysts. Thus, the Co_3O_4 to CoO reduction peak became clearly visible at 226ºC, and the high-temperature features disappeared in the Re-promoted catalyst. As seen in Table 1, all three promoted catalysts displayed very high and equal (96%) ER at 400ºC. It's clear that the addition of Ru and Re as promoters enhanced the reducibility of cobalt oxides without significantly affecting the dispersion. The effect of promoter on reducibility was more pronounced in Co/SBA-15 due to a larger proportion of cobalt species interacting strongly with the support.

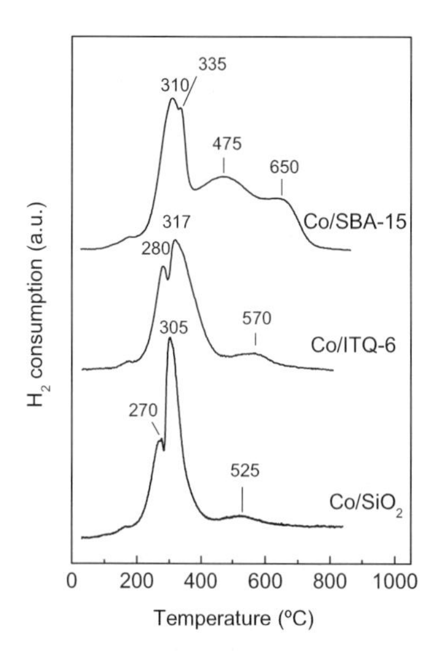

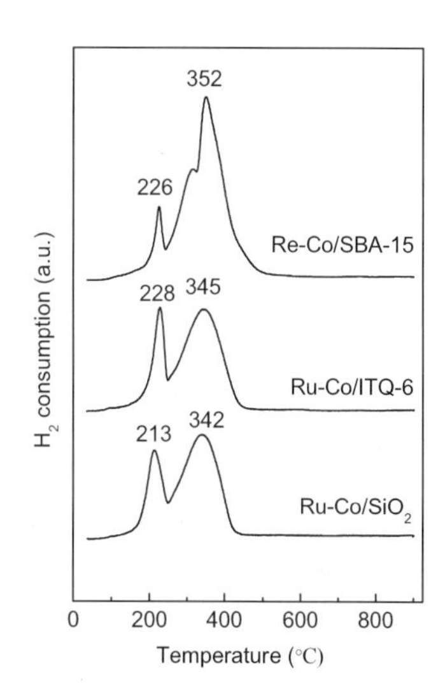

Figure 1. H_2-TPR profiles of unpromoted catalysts. Figure 2. H_2-TPR profiles of promoted catalysts.

In order to compare the catalytic performance for FTS without being disturbed by the presence of unreduced Co phases, we have evaluated the activity and selectivity of the three promoted catalysts. The results showed that the ITQ-6 and SBA-15 based catalysts were significantly more active and more selective to diesel than the reference Ru-Co/SiO_2 under the same reaction conditions (Table 2). Among all catalysts, Ru-Co/ITQ-6 displayed the highest CO conversion. This can be explained by the ability of the delaminated zeolite to efficiently disperse cobalt on its very large and accessible external surface and by the formation of highly reactive Co^0 sites, as it was previously observed by low-temperature FTIR of adsorbed CO in unpromoted Co/ITQ-6 [9].

Table 2. Catalytic test results @ GHSV=7.2 $l.g^{-1}.h^{-1}$, H_2/CO=2.0, T=220°C and P=2MPa.

Catalyst	CO conv. (%C)	Hydrocarbon distribution (%C)				
		C_1	C_2-C_4	C_5-C_{12}	C_{13}-C_{22}	C_{23+}
Ru-Co/SiO_2	18.5	9.7	11.4	39.1	25.5	14.3
Ru-Co/ITQ-6	45.4	11.4	11.9	36.2	27.3	13.2
Re-Co/SBA-15	35.6	8.7	10.1	31.3	35.9	14.0

4. CONCLUSIONS

The all-silica mesoporous SBA-15 and the delaminated ITQ-6 zeolite are promising supports for the development of improved Co-based FTS catalysts. Particularly, the presence of a very high and open external surface in ITQ-6 favours the dispersion of cobalt precursors avoiding extensive sintering during the thermal activation treatments. Unpromoted Co/ITQ-6 displays high reducibility, though still lower than Co/SiO_2. The reducibility of cobalt in unpromoted Co/SBA-15 was, however, substantially lower than in the other two supports due to a stronger Co-support interaction of the Co oxide particles confined within the mesopores in the former. Promotion with small amounts (1 wt%) of Ru and Re improved the reducibility of Co in ITQ-6 and SBA-15, respectively, to the levels obtained in Ru-Co/SiO_2 (96% extent of reduction) without significantly affecting the dispersion.

Promoted Re-Co/SBA-15 and, particularly, Ru-Co/ITQ-6 are appreciably more active for the FTS than Ru-Co/SiO_2 under the same reaction conditions. The high activity of Ru-Co/ITQ-6 can be ascribed to its better Co^0 dispersion and probably also to the formation of highly reactive sites. Moreover, Re-Co/SBA-15 displays the highest selectivity towards hydrocarbons in the diesel range, which may be due to the contribution of Co^0 particles confined within the uniform mesopores.

Acknowledgments

A. Martínez acknowledges the Comisión Interministerial de Ciencia y Tecnología of Spain (CICYT, CTQ2004-02510/PPQ) for financial support.

REFERENCES

1. B.H. Davis, Topics Catal. 32 (2005) 143.
2. E. Iglesia, Appl. Catal. A 161 (1997) 59.
3. S. Li, S. Krishnamoorthy, A. Li, G.D. Meitzner, E. Iglesia, J. Catal. 206 (2002) 202.
4. B. Viswanathan, R. Gopalakrishnan, J. Catal. 99 (1986) 342.
5. A. Martínez, C. López, F. Márquez, I. Díaz, J. Catal. 220 (2003) 486.
6. F.M.T. Mendes, F.B. Noronha, C.D.D. Souza, M.A.P. da Silva, A.B. Gaspar, M.Schmal, Stud. Surf. Sci. Catal. 147 (2004) 361.
7. G. Jacobs, J.A. Chaney, P.M. Patterson, T.K. Das, B.H. Davis, Appl. Catal. A 264 (2004) 203.
8. A. Corma, U. Diaz, M.E. Domine, V. Fornés, Angew. Chem. Int. Ed. 39(8) (2000) 1499.
9. P. Concepción, C. López, A. Martínez, V.F. Puntes, J. Catal. 228 (2004) 321.
10. B.D. Cullity, Elements of X-Ray Diffraction, Addision-Wesley, London, 1978.
11. D. Schanke, S. Vada, E.A. Blekkan, A.M. Hilmen, A. Hoff, A. Holmen, J. Catal. 156 (1995) 85.
12. N. Guan, Y. Liu, M. Zhang, Catal. Today 30 (1996) 207.
13. F.G. Botes, Appl. Catal. A 284 (2005) 21.
14. B. Sexton, A. Hughes, T. Turney, J. Catal. 97 (1986) 390.

Natural Gas Conversion VIII
F.B. Noronha, M. Schmal, E.F. Sousa-Aguiar (Editors)

FISCHER-TROPSCH SYNTHESIS ON Pd-Co/Nb_2O_5 CATALYSTS

Carlos Darlan de Souza[a], D. V. Cesar[a], S. G. Marchetti[b] and M. Schmal[a]

[a]NUCAT/PEQ/COPPE- Federal University of Rio de Janeiro, Centro de Tecnologia, Bloco G, sala 128, 21945-970, Rio de Janeiro, RJ, Brazil
[b] CINDECA-UNLP Calle 47 N° 257 1900 - La Plata, Argentina.

1. Introduction

Cobalt based catalysts have been widely used in the CO + H_2 reaction for hydrocarbons production [1-5]. They are among several others to produce hydrocarbons from CO hydrogenation due to their ability to hydrogenate dissociated carbon species and promote chain growth. Cobalt is a typical metal that may adsorb CO molecules dissociatively forming carbon atoms at the surface, which may be hydrogenated, and it is an appropriate catalyst for the formation of long chain hydrocarbons [5,6]. However, the main problem of the F-T synthesis is the wide range of product distribution when conventional F-T catalysts are being used. In order to overcome the selectivity limitations and to enhance the catalyst efficiency in CO hydrogenation, several approaches have been studied, such as the use of reducible supports and the addition of a second metal component.

In this study, we investigate the performance of the bimetallic Pd-Co/niobia systems in CO hydrogenation reaction, in an attempt to explain the effect of a second metal on the product selectivity of Co-based catalysts compared to the Co/Nb_2O_5 systems.

2. Experimental

The cobalt catalyst was prepared by incipient wetness impregnation with a $Co(NO_3)_2$. $6H_2O$ (Riedel-de Häen, 99% purity) solution. The Co content was kept constant at approximately 5% (wt.) [1, 5]. Then Pd was impregnated in sequence by wet impregnation in order to obtain 0.6 % of the metal. After impregnation, all catalysts were dried overnight at 110^0C followed by calcination at 2^0C/min up to 400^0C, kept at this temperature for 3h and stored for characterization and catalytic tests. Catalytic evaluation of the catalysts with the CO hydrogenation reaction was performed on a fixed bed. The reaction was performed at 20 atm and started after the temperature had stabilized by the CO/H_2/He mixture flow into the reactor at a rate of 30 mL/min to ensure constant space velocity for all experiments. The conditions were established for isoconversion. The total reaction time was approximately 50h. Feed gas was a mixture of 31.7% CO/64.3% H_2/4% He. Helium was used as an internal standard to calculate the total CO conversions. These gases passed through filters at room temperature to remove O_2 and water traces. Reaction products were analyzed on line by gas chromatography.

A FTIR spectrometer (Nicolet, model Nexus 470) equipped with a DRIFTS (Diffuse Reflectance Infrared Fourier Transform Spectroscopy) cell (Spectra-Tech) chamber for high temperature treatment and ZnSe mirror assembly was used to study the $CO+H_2$ reaction at different temperatures. These experiments were carried out after *in situ* reduction with pure H_2 (99.99%) at 30 mL/min at 10^0C/min up to 500^0C for 6h and then purged with He for 30 min. Subsequently the sample was cooled down to the reaction temperature. A feed mixture of $1CO:2H_2$ (32%CO: 64% H_2: 4% He) was introduced at 25 mL/min, under similar reaction conditions. Each spectrum was referenced to the spectrum of the reduced sample [5].

3. Results and Discussion

The H_2 chemisorption measurement results for Co/Nb_2O_5 and $Co\text{-}Pd/Nb_2O_5$ catalysts, reduced at 500^0C, were 14.3 and 12.3 μmols/gcat, respectively, based on the irreversible adsorption at 175^0C, suggesting the presence of big cobalt particles and very small Pd particles on top of Co particles.

Figure 1 displays the selectivity with time on stream, for a space velocity $6000h^{-1}$ and isoconversion (30%); reaction temperature 270^0C and pressure 20 bar, after reduction at 500°C. The bar diagrams of Figure 1A presents product selectivities on Co/Nb_2O_5 catalyst based on mols of carbon product in the C_i range/$\sum C_i$ formation with time on stream. Methane was very low (≈3.0%) and the diesel fraction ($C_{13\text{-}18}$) was very high (54-49%). Note that CH_4, $C_{2\text{-}4}$, $C_{5\text{-}12}$, $C_{13\text{-}18}$, and C_{19+} correspond to saturated hydrocarbons. The range $C_{2\text{-}4}$ may also includes some $C_{2\text{-}4}$ olefins (ethene and propene), which were not resolved from the more pronounced saturated hydrocarbon chromatographic peaks. Higher molecular weight olefins ($C_{5\text{-}12}$ and $C_{13\text{-}18}$) were detected on the Co/Nb_2O_5 catalyst.

The bar diagrams of Figure 1B clearly point out that the selectivity towards methane and $C_{2\text{-}4}$ hydrocarbons for the bimetallic $Co\text{-}Pd/Nb_2O_5$ changed with Pd addition and with time on stream. Note that the methane formation rate for this catalyst increased. Olefins, except $C_4^=$ were not detected before on the pure Co-containing catalysts. Light products and gasoline fractions increased, while diesel fraction decreased.

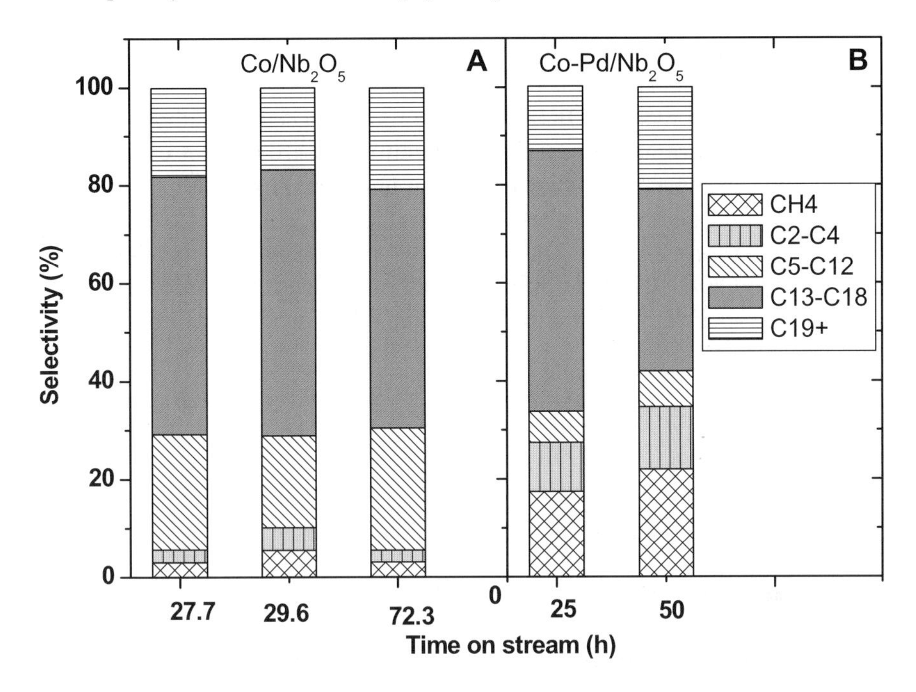

Figure 1. Catalytic Test – $H_2/CO=2$, 270°C, 20 atm, (A) Co/Nb_2O_5,, (B) Co-Pd / Nb2O5

Methane was around 17-21% and the diesel fraction decreased from 53 to 37%. The addition of the second metal indicates also higher selectivity towards alkanes. Product retention and condensation are not expected to have occurred because the overall mass balance indicates that the reactor operates steady-state. Noteworthy is that during 70 hours the catalyst was stable and deviation of mass balance was around 20%. In addition, after 50h samples were taken and coke was analysed by TG (not shown). It presented a loss of approximately 16% of carbon due to the coke deposition, which allows us to explain the deviation observed. Experiments were reproduced under similar conditions at isoconversion (around 30%).

The Anderson-Schulz-Flory equation with these experiments displayed great deviation in the C_{13}-C_{18} range. As expected, the total hydrocarbon molar compositions cannot be interpreted as an ASF distribution. Although the C_2-C_3 anomalies and the change of the chain growth probability in the range from C_3 to C_8-C_{12} range may be explained by α-olefin readsorption with secondary chain propagation and the existence of two mechanisms for chain propagation, the increasing selectivity observed between C_8-C_{12} for all experiments cannot be explained by these mechanisms [7].

Figure 2 shows the DRIFTS measurements of the CO hydrogenation at atmospheric pressure and temperature variations, searching an explanation and understanding the surface species formation under real conditions.

Typical spectra of the samples Co/Nb_2O_5 and Co-Pd/Nb_2O_5 are displayed in Figure 2A and 2B, respectively. After *in situ* reduction with H_2 during 6h, the CO/H_2/He flow mixture was introduced at a ratio of H_2/CO = 2. The reference spectrum was the reduced catalyst prior to gas admission. Spectra were obtained at different temperatures. Figure 2A presents the spectrum Co/5%Nb_2O_5, where Nb_2O_5 was used as reference spectrum. At 200^0C it displays characteristic bands of CO in the gas phase (2174 cm^{-1}) [1,3,9], a small peak for formate species at 2900 cm^{-1} (νC-H) [4].

The catalyst also shows CO bands at different temperatures 250^0C and 270^0C (b,c) indicating linear adsorption on top of metallic Co particles (2056 cm^{-1}). On the other hand, the band at 2115 cm^{-1} is attributed to the CO adsorption on Co^+ species [3]. Zhang et al [8] suggests the presence of ionic $Co^{\delta+}$ species with less electron donation capacity. At higher temperature (250 – 270^0C) this band is shifted towards 2174cm^{-1}, suggesting an increase of reduced Co species.

Formate bands were observed around 2900 cm^{-1} (νC-H) and around 1600 - 1350cm^{-1}, 1592 cm^{-1} (ν_{ass} C-O), 1393 cm^{-1} (δ_s C-H), 1371 cm^{-1} (ν_s C-O). [4,10] that increased with the temperature. Above 270^0C the band at 2350 cm^{-1}, relative to CO_2 formation was observed, which may be related to the "shift" reaction and/or the decomposition of formate species.

On the other hand, bands around 3011-2880 cm^{-1} evidence the presence of CH_x species during the reaction. According to Rygh et al [3] these bands are due to the formation of intermediate CH_X species. Besides, bands at 3017 cm^{-1}, which corresponds to the axial deformation and at 1305 cm^{-1} band of the angular deformation of C-H bond of gas methane were observed.

This catalyst displays the symmetric stretching of -CH_2 at 2880 cm^{-1}, that evidences this intermediate species. Spectra (d) and (e) confirm these bands taken in a closed chamber.

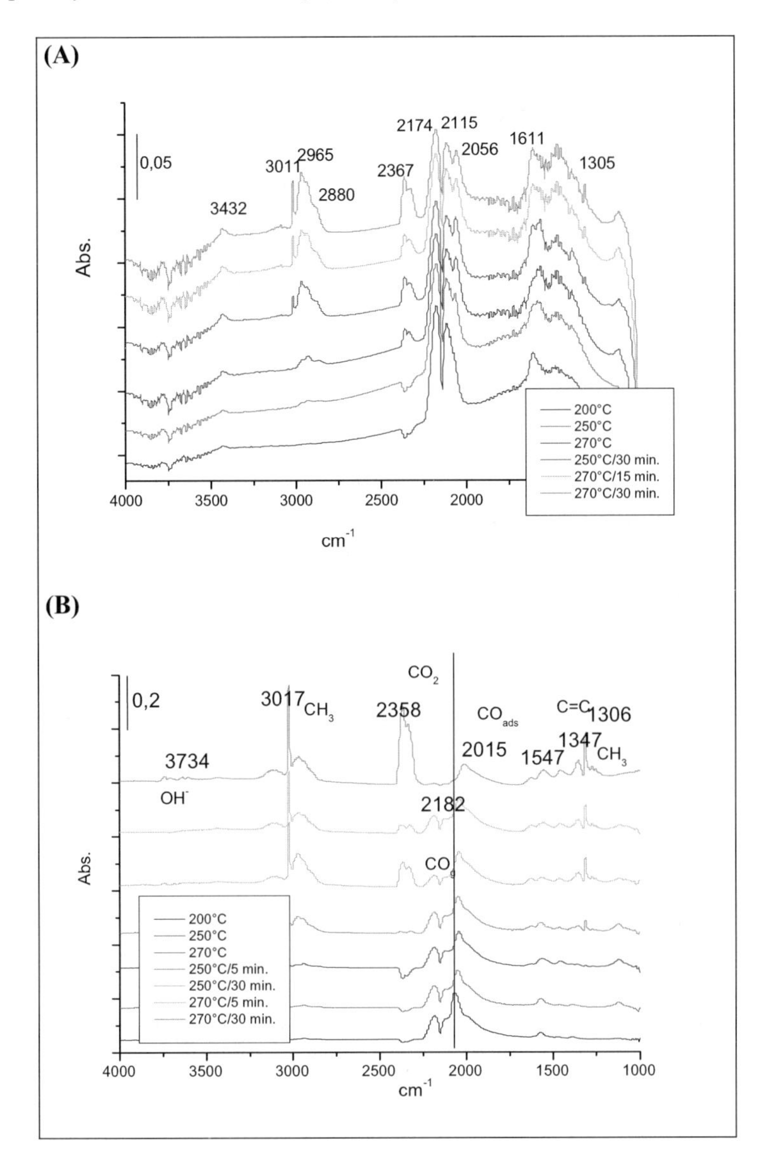

Figure 2A. Co/Nb$_2$O$_5$: (a) catalyst at 200^{0}C, (b) catalyst at 250^{0}C, (c) catalyst at 270^{0}C, (d) 250^{0}C closed chamber/30 min., (e) 270^{0}C closed chamber for 15 min.; (f) 270^{0}C closed chamber for 30 min.

Figure 2B. Co-Pd/Nb$_2$O$_5$: (g) 473K, (h) 250^{0}C, (i) 270^{0}C, (j) 250^{0}C closed chamber / 15 min., (k) 250^{0}C closed chamber / 30 min.; (l) 270^{0}C closed chamber / 5 min., (m) 270^{0}C closed chamber / 30 min.

The Co-Pd/Nb_2O_5 (Fig. 2B) displays a different behavior for CO adsorption on Co^+ species (~2115cm^{-1}), that changes with increasing temperature. However, significant modifications were observed between 3000-2800 cm^{-1} and 1600 cm^{-1}. The shoulder at 1547 cm^{-1} is assigned to the formation of -C=C-, suggesting the formation of olefins. One can see that the formation of CH_X intermediates at 250^{0}C, related to the bands at 2951cm^{-1} (ν_s CH_2), 2928 cm^{-1} (ν_{ass} CH_2), 2963cm^{-1} (ν_{ass} CH_3) and at 3017 cm^{-1} to the CH_4 gas; while at 1455cm^{-1} (δ_{ass} CH_2+CH_3) and 1375cm^{-1} (δ_s CH_3) and at 1306 cm^{-1} (δ_s C-H) to methane. The band at 1605cm^{-1} (ν C=O) is due to the adsorption of bidentade carbonate species, which are responsible for the CO_2 formation.

4. Conclusion

The Co/Nb_2O_5 showed a significant change with time on stream upon the addition of Pd. The selectivity towards heavier product range, diesel (C_{13-18}) and C_{19+}) increased with time on stream and was very stable up to 50 h.

DRIFTS results showed the formation of -C=C-, suggesting the formation of olefins upon addition of Pd. The results suggest likewise that the CH_xO species are the precursors for the formation of hydrocarbons. The Co-Pd/Nb_2O_5 shows CO adsorption on $Co^{\delta+}$ species. The CH_x^- is thought to be formed as methyl radicals on Co^0 particles.

5. References

[1] F.M.T. Mendes, C.A.C. Perez, F.B. Noronha, M.Schmal *Catal. Today* 101 p.45, 2005.

[2] E. Iglesia *Appl. Catal. A* 161, p 59, 1997.

[3] R.E.S. Rygh, C.J. Nielsen *J. Catal.* 194 p. 401, 2000.

[4] L.H. Little, *Infrared Spectroscopy of Adsorbed Species*, Academic Press, NY, 1966

[5] F. M. T. Mendes,C. A. C. Perez, F. B. Noronha, C. D. D. Souza, D. V. Cesar, H. J. Freund, and M. Schmal, *J.Phys.Chem.B* , v.110, p. 9155-9163, 2006.

[6] A. Frydman , D.G. Castner , C.T. Campbell and M. Schmal, *Journal of Catalysis*, 188, 2, p.1-13, 1999.

[7] Ahón, V.R., Lage, P.L.C., Souza, C.D.D., Mendes, F. M., Schmal, M., *Journal of Natural Gas Chemistry*, Vol.15, Issue 4, p.307-312, 2006.

[8] J. Zhangh, J. Chen, J. Ren, Y. Sun, *Appl.Catal. A*, 243, p.121, 2003.

[9] Fujimoto, K., and Oba, T., *Appl. Catal.* 51, p.289, 1985.

[10] G.Busca, J. Lamotte, J. C. Lavalley, V. Lorenzelli, *J. Am.Chem.Soc.*, 109 p. 5197, 1987.

Natural Gas Conversion VIII
F.B. Noronha, M. Schmal, E.F. Sousa-Aguiar (Editors)

Methane Conversion for Fuel Cells. The Role of Sulphur

J.R. Rostrup-Nielsen

Haldor Topsøe A/S, DK-2800 Kgs. Lyngby, Denmark

1. Reforming Technologies

Most fuel cells operate with electrochemical oxidation of hydrogen on the anode. It is possible to convert methane in high, temperature fuel cells [MCFC, SOFC], by *internal reforming* into hydrogen in the anode chamber using the electrochemical heat for the reforming reaction [1,2].

$$CH_4 + 2H_2 = CO_2 + 4H_2 \quad -Q_{ref} \quad (-\Delta H^o_{923} = -188 \text{ kJ/mol } CH_4)$$
$$4H_2 + 2O_2 = 4H_2\text{-}Q \quad +E+Q_c \quad (-\Delta H^o_{923} = 987 \text{ kJ/mol } CH_4) \qquad (1)$$

$$CH_4 + 2O_2 = 4H_2O + CO_2 \quad +E+Q_c-Q_{ref} \ (-\Delta H^o_{292} = 799 \text{ kJ/mol } CH_4)$$

The coupling of the two reactions results in higher electrical efficiency of the fuel cell. Low temperature fuel cells (PEMFC, PAFC) require external manufacture of hydrogen.

Steam reforming of natural gas is the main route for large scale production hydrogen in refineries with a typical supply pressure of 20 bar. The steam reforming process is very efficient [5]. The practical energy consumption is only 7% higher than the theoretical minimum (11.8 GJ/1000 Nm^3 H_2). However, the steam reformer furnace is expensive and since only 50% of the fired heat is absorbed by the reaction, the heat recovery easily results in export of steam. For small scale operation, this problem is solved by using a convective reformer in which the hot product gas is used as heat source for the reaction. This means that ca. 80% of the fired heat is absorbed by the process, thus eliminating the export of steam.

The reaction heat is provided by "internal" combustion in *catalytic partial oxidation* (CPO) which means that there is no need for a complex heated reactor [2]. The CPO reactor is more compact that the steam reformer as shown in Fig. 1.

Fig.1. The dark reactor shows a CPO pilot (210 Nm^3/h $CO+H_2$ (25 bar)) in front of a tubular reformer of twice the capacity.The size of the CPO catalyst bed is indicated.

CPO is not competitive with convective steam reforming for industrial hydrogen production. The typical supply pressure of ca. 20 bar requires an expensive air compressor for the CPO scheme. However, the CPO costs are reduced if oxygen is available or if the supply pressure is low. Hence, CPO becomes a real alternative to steam reforming for decentralised fuel cells.

2. The Impact of Sulphur

2.1. Sulphur Reactions

Nickel catalysts for steam reforming are sensitive to sulphur [6]. The Ni/H_2S system is well described in terms of a two-dimensional sulphide as recently demonstrated by STM measurements [7]. Nickel is more sensitive than other group VIII metals as illustrated in Fig. 2.

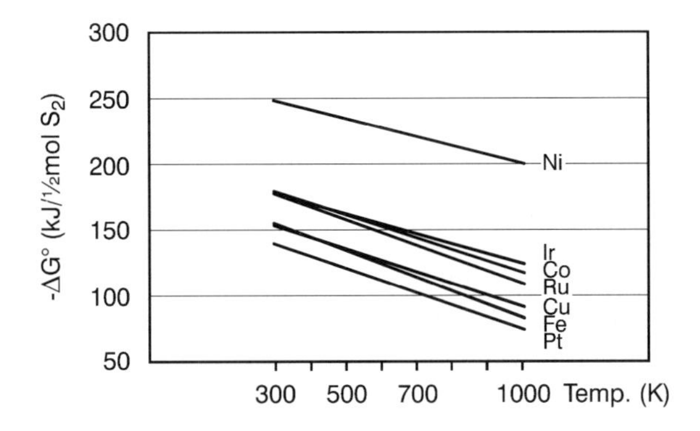

Fig. 2. Chemisorption of H_2S on metals [6]. Data from ref. [8]

As an example [2], a sulphur coverage of 0.5 at 500°C corresponds to $H_2S/H_2 = 1.6x10^{-12}$. The chemisorption process is reversible, but the driving force for desorption is low.

Table 1. Sulphur Reactions

Ni + H_2S	=	Ni-S + H_2
Ni-S + 1.5 O_2	=	NiO + SO_2
SO_2+3H_2	=	H_2S + H_2O
Rh + H_2S	=	Rh-S + H_2
Rh-S + O_2	=	Rh + SO_2

The situation for CPO is more favourable as illustrated by the reactions in Table 1. In presence of oxygen, sulphur is oxidized to SO_2, but nickel will also be oxidized which is not the case for rhodium, a typical catalyst for CPO. It means that rhodium remains active as long as oxygen is present. After depletion of oxygen, SO_2 will be reduced to H_2S which will be chemisorbed on the catalyst and down-stream catalysts and anodes. The task is then reduced to remove hydrogen sulphide from the product gas. The sulphur resistance of the CPO process is in particular an advantage when using heavy fuels (diesel) as feed. These will contain sulphur compounds which are difficult to remove at low pressure. This is crucial for the operation of low temperature fuel cells, but may not be the case for high temperature fuel cells as demonstrated by recent results on a SOFC stack [9].

2.2. Experimental

The poisoning experiments were carried out on a planar 10 cell SOFC stack (TOFC) at a test rig fed by a gas simulating the gas from catalytic partial oxidation of jet and diesel fuel from a Catator CPO unit [9]. The performance of the SOFC stack was examined at different temperatures (700°C, 800°C) load and H_2S [2,5,10,50] ppm S referred to total feedgas volume]. The sulphur balance over the cells was followed semi-quantitatively by means of photo ionisation detector.

2.3. Results

The sulphur poisoning tests showed that the electrochemical reaction is relatively insensitive to sulphur poisoning. The addition of 10 ppm H_2S had no impact on the performance at 800°C, whereas 50 ppm H_2S resulted in a slight decrease of the voltage.

At 700-720°C, the addition of 50 ppm resulted in a decrease of the voltage by ca. 10% as shown in Fig. 3. However, the tests illustrated that methane passed nearly unconverted through the stack as illustrated in Fig. 4. If extra hydrogen was added to the feed to compensate for the non-converted methane, the loss in stack performance was almost recovered. A full recovery was achieved some hours after the addition of H_2S to the feed was stopped.

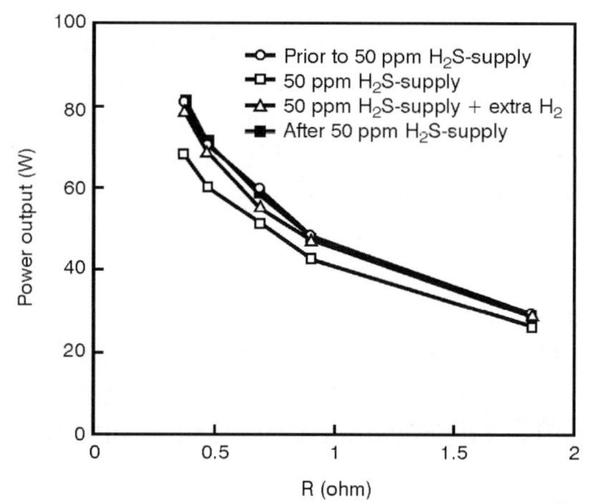

Fig. 3. CPO/SOFC performance. S-tolerance, 800°C. (1-2% CH_4, in feedgas) [10]

When increasing the current drawn from the stack, the H_2S exit the cell decreased.

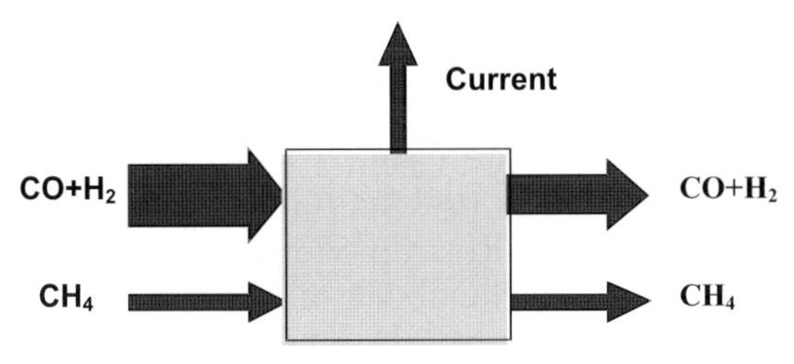

Fig. 4. SOFC performance. 50 ppm H_2S, 800°C. (1-2% CH_4, in feedgas)

The observations strongly support that whereas the methane conversion is influenced by sulphur as expected, the electrochemical conversion is non-affected. The decrease in H_2S leaving the stack when increasing the current may indicate that sulphur is converted to SO_2 by the electrochemical reaction:

$$H_2S + 3O^{--} = SO_2 + H_2O + 6e^-$$

$$H_2S + 2H_2O = SO_2 + 3\,H_2 \qquad (2)$$

thus keeping the nickel site at the interphase non-affected by sulphur poisoning.

3. Conclusions

Steam reforming is the cheapest route for large scale hydrogen production. CPO is an alternative for decentralised hydrogen production with low supply pressure.

CPO is less sensitive to sulphur than steam reforming which is an advantage for conversion of heavy fuels. The internal reforming of methane in high temperature fuel cells is sensitive to sulphur poisoning whereas the electrochemical conversion of hydrogen appears almost non-affected.

References

1. J.R. Rostrup-Nielsen, L.J. Christiansen, Appl.Catal. A, 126 (1995) 381.
2. J.R. Rostrup-Nielsen, K. Aasberg-Petersen in Fuel Cell Handbook, vol. 3, Wiley, New York 2003 p.159.
3. J.R. Rostrup-Nielsen, Phys. Chem. Chem. Phys, 3 (2001) 283
4. J.R. Rostrup-Nielsen, Cat.Rev.Sci.Eng. 46 (2004) 247.
5. J.R. Rostrup-Nielsen, T. Rostrup-Nielsen, Cattech, 6 (2002) 150.
6. J.R. Rostrup-Nielsen in Handbook of Heterogeneous Catalysis 2. Edition, Chapt. 13.11 [G. Ertl, H. Knözinger, F. Schüth and J. Weitkamp, eds], Wiley VCH [in print]
7. F. Besenbacher, J. Steensgaard, S. Lægsgaard-Jørgensen, Phys.Rev.Let. 69 (1992) 3523.
8. H. Wise, J. McCarty, and J. Oudar, in Deactivation and Poisoning of Catalysts [H. Wise and J. Oudar, eds.], Marcel Dekker, New York 1985, Chapt 1, p.1.
9. J.R. Rostrup-Nielsen, J.B. Hansen, S. Helveg, N. Christiansen, A-K. Jannasch, J.Appl.Phys. A., 85 (2006) 427.
10. F. Silversand, A-K Jannasch, J. Hansen, J. Pålsson, Abstr. Fuel Cell Seminar 2006, Honolulu.

Natural Gas Conversion VIII
F.B. Noronha, M. Schmal, E.F. Sousa-Aguiar (Editors)

Experimental Demonstration of H_2 Production by CO_2 Sorption Enhanced Steam Methane Reforming Using Ceramic Acceptors

Esther Ochoa-Fernández, Claudia Lacalle-Vilà, Tiejun Zhao, Magnus Rønning, De Chen

Department of Chemical Engineering, Norwegian University of Science and Technology, Sem Sælands vei 4, Trondheim 7491, Norway

1. Introduction

Hydrogen is considered as a new clean energy source for transport and especially, for fuel cell applications. The most common industrial process for production of hydrogen in steam methane reforming (SMR):

$$CH_4 + H_2O \rightarrow CO + 3\,H_2, \qquad \Delta H^o_{298}=206 \text{ kJ/mol} \tag{1}$$

$$CO + H_2O \rightarrow CO_2 + H_2, \qquad \Delta H^o_{298}=-41 \text{ kJ/mol} \tag{2}$$

The steam reforming process involves multiple steps and severe operating conditions. The reformer is normally operated at 1073-1123 K and 20-35 bar which needs a large energy input to maintain the reaction temperature. In addition, removal of CO_2 is necessary in order to accomplish the requirements related to green-house gas emissions. CO_2 removal results in a large penalty in the efficiency of SMR [1].
Within the last few years, the concept of multifunctional reactors combining reaction and separation has received increased attention, especially the concept of sorption enhanced steam methane reforming (SESMR) [2-6]. A CO_2 acceptor can be installed together with the catalyst in the reactor bed to remove CO_2 from the gas phase during SMR operation. Thus, the normal equilibrium limitations of the reforming and shift reactions are shifted, and the direct production of

higher H_2 yields is possible. The SESMR can be carried out at lower temperatures (673-873 K) than conventional steam reforming, which can lower investment and operation costs. Recent studies show that SESMR results in similar and even improved efficiencies than SMR depending on the selected CO_2 acceptor [1].

Previous work has identified different acceptors for CO_2 capture. For example, Lopez-Ortiz *et al.* and Johnsen *et al.* used CaO as the CO_2 acceptor for hydrogen production in a fixed-bed and fluidized-bed reactor, respectively [3,4]. A temperature swing process was used for regeneration of the acceptor. Xiu *et al.* used hydrotalcite-like compounds for the same purpose [5]. In this case, the regeneration was performed by means of pressure swing desorption. Hydrogen yields higher than 95% can be achieved by using these acceptors at different operating conditions. However, the main problem is that a large make-up flow of fresh-acceptors is necessary due to deactivation of the acceptors. Hydrotalcite-like compounds are able to accept CO_2 at a wide range of temperatures up to 673 K [7]. However, these compounds loose more than half of their capacity after several cycles. The same problem has been observed for calcium-based acceptors [2]. The use of high temperatures for the regeneration leads to sintering and subsequent pore blockage and losses in the porosity. In addition, CaO can react with steam under the reaction conditions leading to the formation of $Ca(OH)_2$ and, thus, lowering the methane conversion and hydrogen yield [8]. The development of an acceptor with relatively low temperatures for regeneration and with good multi-cycle properties is highly desirable. Recently, it has been reported that mixed oxides such as Li_2ZrO_3, Na_2ZrO_3 are promising candidates with high CO_2 capture capacity and high stability [9-13].

The objective of the present work is to demonstrate the H_2-SESMR process with the use of Li_2ZrO_3 and Na_2ZrO_3 as acceptors. For this purpose, a hydrotalcite-like Ni catalyst has been used for the reforming reaction. The reaction has been carried out in a fixed-bed reactor.

2. Experimental part

Nanocrystalline Li_2ZrO_3 and Na_2ZrO_3 have been prepared by a novel soft chemistry route in our laboratory. Details about the preparation, characterisation and properties of these materials have been reported elsewhere [10,13,14]. A Ni-Mg-Al hydrotalcite-like catalyst with 40% Ni has been prepared by a coprecipitation technique resulting in a catalyst with high activity and stability at SESMR conditions [15].

SESMR was performed in a stainless-steal fixed-bed reactor with inner diameter of approximately 16 mm. The reforming reactions were carried out at 848 K, 5 bar, steam-to-carbon ratio 5, acceptor-to-catalyst ratio 5 and total flow of 47 ml/min CH_4. A mixture of the steam reforming catalyst and the different acceptors was placed in the reactor. Prior to reaction, the samples were reduced at 923 K for 10 h in H_2/Ar=100/100 ml/min at atmospheric pressure. A heating rate of 2 K/min was used to increase the temperature from ambient to 923 K. After reduction, the temperature was decreased to 848 K and the reactive gases were introduced in the reactor. The reaction was continued until the acceptor was saturated, which corresponds to the breakthrough of the CO_2 concentration in the products stream. At this point, the CH_4/steam mixture was switched to Ar and the temperature was increased to 923 and 1023 K for Li_2ZrO_3 and Na_2ZrO_3, respectively. Multi-cycle experiments were carried out in order to study the stability of the process. The downstream gas composition was followed by an online microGC.

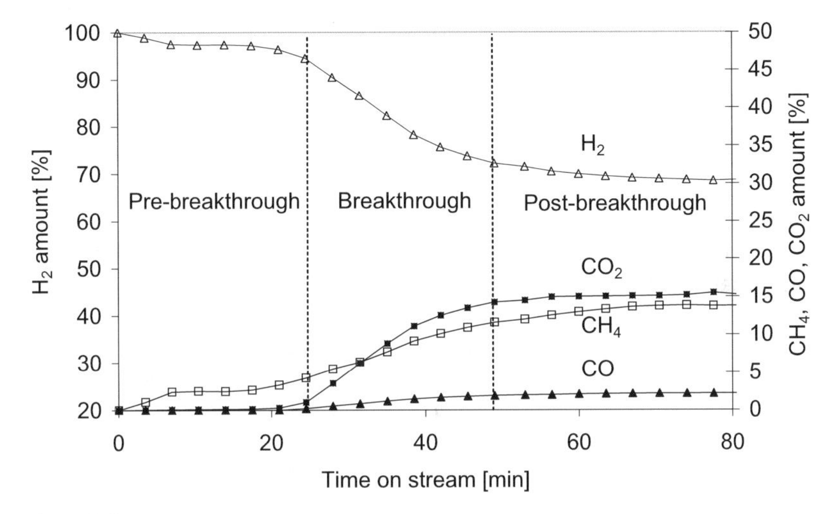

Figure 1. Effluent concentration profiles on a water free basis. 7.25 g Na_2ZrO_3, 1.45 g Ni catalyst, 848 K, S/C:5, 5 bar, 47 ml/min CH_4. Regeneration: 1023 K in Ar.

3. Results and Discussion

3.1. Equilibrium considerations

Steam methane reforming is a strongly endothermic reaction. Thus, the equilibrium concentration of H_2 will increase with increasing temperatures, reaching a value of 67.9% on dry basis at the working conditions of this study. By introducing a CO_2 acceptor the equilibrium will be shifted towards the H_2 production, as discussed above. In the case that Li_2ZrO_3 is used as acceptor, the H_2 yield could be enhanced to 91% H_2, while the thermodynamic limitation when using Na_2ZrO_3 is 92% H_2 in a single step [1]. The thermodynamic data are obtained from FactSage^{TM} 5.0 [16].

3.2. SESMR

Sorption enhanced steam reforming has been carried out by using both Li_2ZrO_3 and Na_2ZrO_3 as acceptors. Figure 1 shows a typical effluent concentration profile on a water free basis when Na_2ZrO_3 is used as acceptor. A H_2 yield above 97% is obtained for a period of time of 20 min. The main impurity was unconverted methane. The downstream composition remained unchanged until the saturation of the acceptor. At this point, breakthrough of the CO_2 took place and the conversion of methane decreased following the thermodynamics of conventional steam methane reforming. Accordingly, the H_2 yield decreased to approximately 69%.
The obtained experimental H_2 yield (97%) is higher than the H_2 yield calculated from the thermodynamics at the working conditions (92%). In fact, according to the available thermodynamic data, the equilibrium partial pressure of CO_2 at 848 K is around 0.02 bar. However, it has been tested experimentally that Na_2ZrO_3 can take CO_2 efficiently at lower concentrations. Therefore, it seems that the tabulated thermodynamic data do not completely describe the behavior of nanocrystalline Na_2ZrO_3.

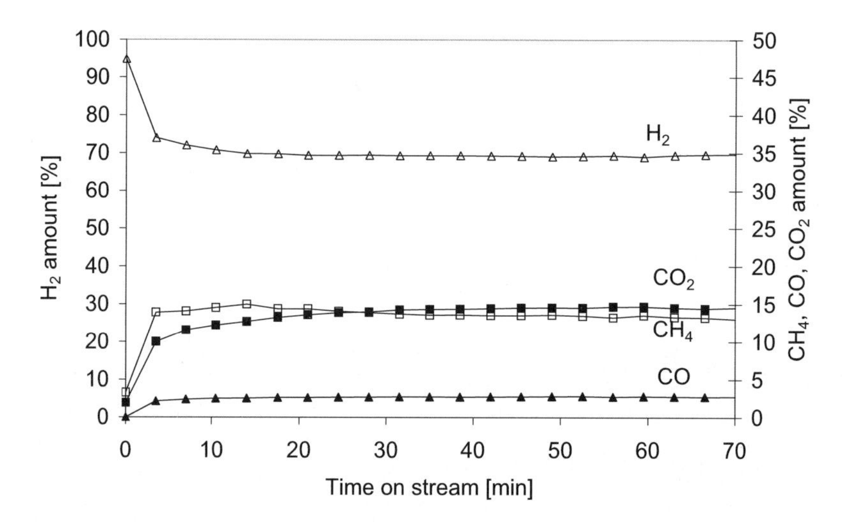

Figure 2. Effluent concentration profiles on a water free basis. 15 g Li_2ZrO_3, 3 g Ni catalyst, 848 K, S/C:5, 5 bar, 47 ml/min CH_4. Regeneration: 923 K in Ar.

The same experiment was carried out using Li_2ZrO_3 as acceptor. However, the H_2 yield was not considerably enhanced in this case as shown in Figure 2. In fact, the capture kinetics of Li_2ZrO_3 at low partial pressures of CO_2 obtained from a microbalance study are much slower than those for Na_2ZrO_3 [1]. As a result, the capture reaction cannot compete with the reforming and the CO_2 formed in the reaction zone is not completely removed. Therefore, the obtained

H_2 was very close to the thermodynamic equilibrium of conventional SMR at these conditions (67.9 % H_2) and the use of the acceptor does not enhance the reaction. Fixed-bed reactor simulations using a dynamic one-dimensional pseudo-homogeneous model have shown that higher H_2 yields can be obtained by using Li_2ZrO_3 as acceptor, but much lower space velocities than used in this work are required [17].

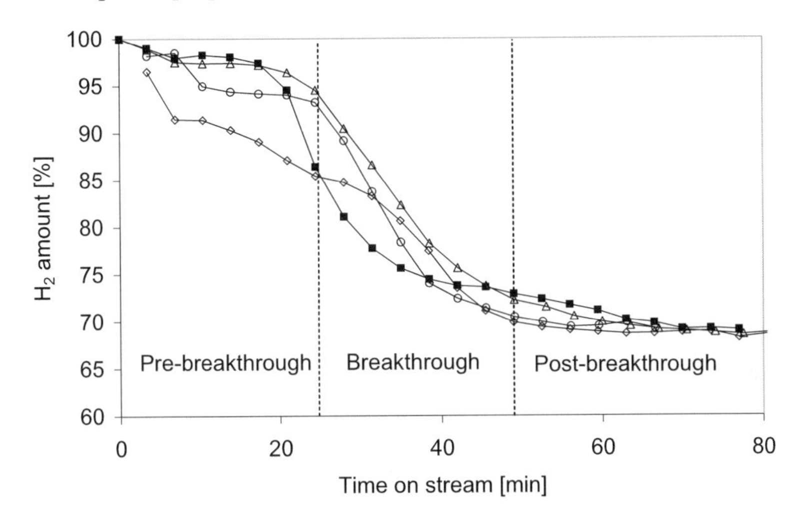

Figure 3. Effluent H_2 concentration on a water free basis. Δ: cycle1, ■: cycle2, ○: cycle3, ◊:cycle4. 7.25 g Na_2ZrO_3, 1.45 g Ni catalyst, 848 K, S/C:5, 5 bar, 47 ml/min CH_4. Regeneration: 1023 K in Ar.

3.3. Stability

After saturation of the acceptors, the samples were regenerated as described above. Later, the cycle was repeated several times to check the stability of the process. The conversion of methane and, as a consequence, the H_2 yield decreased dramatically during the second cycle. Presumably, the catalyst was poisoned by impurities from the CO_2 acceptor, since the conversion decreased although the CO and CO_2 levels were below 0.5%. This happened when using both, Li_2ZrO_3 and Na_2ZrO_3. The reason for this contamination is still not clear. However, it has been shown that the impurities could be partially removed after one operation cycle, most likely remove with steam and the acceptor could be re-used. As shown in Figure 3, several cycles were carried out after removing the poison. Similar H_2 yields as with fresh sample were obtained in a second cycle and after saturation of the acceptor the conversion followed conventional SMR. However, the breakthrough took place some minutes earlier, indicating a small loss in capacity of the acceptor during operation in wet conditions. This has been confirmed by studying the CO_2 capture properties of Na_2ZrO_3 in wet

CO_2 using a tapered element oscillating microbalance (TEOM). On the other hand, after two successive cycles the maximum H_2 yield decreased from 97% to 95% and 91%, respectively, while the CO and CO_2 levels remained under 0.5%. However, the amount of CH_4 increased considerably, indicating again catalyst deactivation. Besides, the kinetics of Na_2ZrO_3 after 5 operating SESMR cycles were checked in the TEOM and found identical to the fresh acceptor. Detailed study of catalyst deactivation mechanism is in progress.

4. Conclusion

Li_2ZrO_3 and Na_2ZrO_3 have been tested as CO_2 acceptors for SESMR. H_2 yields higher than 95% in a single step were obtained by using Na_2ZrO_3 as acceptor. No such enhancement in the H_2 yield was observed when Li_2ZrO_3 was used due to low CO_2 capture kinetics at low partial pressures of CO_2. Interaction between the catalyst and the acceptor has been observed, resulting in catalyst deactivation. The acceptor could be recovered and operated for several cycles but still some deactivation was observed. Further studies are necessary in order to understand and control the deactivation process that can make difficult the application of these materials in sorption enhanced steam methane reforming.

References

1. E. Ochoa-Fernández, G. Haugen, T. Zhao, M. Rønning, I. Aartun, B. Børresen, E. Rytter, M. Rønnekleiv, D. Chen, Prepr. Pap.-Am. Chem. Soc., Div. Fuel Chem. 51 (2006), 598.
2. B. Balasubramanian, A.L. Ortiz, S. Kaytakoglu, D.P. Harrison, Chem. Eng. Sci., 54 (1999) 3543.
3. A. Lopez-Ortiz, D.P. Harrison, Ind. Eng. Chem. Res., 40 (2001) 5102.
4. K. Johnsen, H. J. Ryu, J. R. Grace, C. J. Lim, Chem. Eng. Sci., 61 (2006) 1195.
5. G.-H. Xiu, P. Li, A.E. Rodriges, Chem. Eng. Sci., 57 (2002) 3893.
6. Y.-N. Wang, A.E. Rodrigues, Fuel, 84 (2005) 1778.
7. Z. Yong, A.E. Rodrigues, Energy Convers. Manage., 43 (2002) 1865.
8. N. Hildengrand, L. Readman, I.M. Dahl, R. Blom, Appl. Catal., A, 303 (2006) 131.
9. K. Nakagawa, T. Ohashi, J. Electrochem. Soc., 145 (1998) 1344.
10. E. Ochoa-Fernández, T. Grande, M. Rønning, D. Chen, Chem. Mater., 18 (2006) 1383.
11. K. B. Yi, D. Ø. Eriksen, Sep. Sci. Tech., 41 (2006) 283.
12. A. Lopez-Ortiz, N.G.P. Rivera, A.R. Rojas, D.L. Gutierrez, Sep. Sci. Tech., 39 (2004) 3559.
13. T. Zhao, E. Ochoa-Fernández, M. Rønning, D. Chen, In preparation (2006)
14. M. Rønning, E. Ochoa-Fernández, T. Grande, D. Chen, PCT Int. Appl., (2006)
15. E. Ochoa-Fernández, C. Lacalle-Vilá, K.O. Christensen, M. Rønning, A. Holmen, D. Chen, Top. Catal. (2006) Accepted.
16. FactSage 5. The Integrated Thermodynamic Dtabank System; www.factsage.com
17. E. Ochoa-Fernández, H. K. Rusten, M. Rønning, A. Holmen, H. A. Jakobsen, D. Chen, Catal. Today, 106 (2005) 41.

Natural Gas Conversion VIII
F.B. Noronha, M. Schmal, E.F. Sousa-Aguiar (Editors)

Effect of support on autothermal reforming of methane on nickel catalysts

Juliana da Silva Lisboa[a], Mônica Pinto Maia[a], Ana Paula Erthal Moreira[a] and Fabio Barboza Passos[a*]

[a] *Departamento de Engenharia Química e de Petróleo, Universidade Federal Fluminense, Rua Passo da Pátria, 156, Niterói, 24210-240,, Brazil.*

$Ni/\gamma\text{-}Al_2O_3$, Ni/CeO_2 and Ni/ZrO_2 catalyts were investigated in the autothermal reforming of methane and were characterized by XRD, DRS and TPR. NiO in different geometries was the main precursor for the several catalysts, but the presence of $NiAl_2O_4$ was observed for $Ni/\gamma\text{-}Al_2O_3$ after calcination at 650°C. The reaction procedeed by an indirect mechanism and activity followed the order $Ni/ZrO_2 > Ni/CeO_2 > Ni/\gamma\text{-}Al_2O_3$.

1. Introduction

Autothermal reforming is an association between steam reforming and the partial oxidation of hydrocarbons. The main advantage of this process is the fact that an exothermic reaction and an endothermic reaction are performed simultaneously, optimizing the energy costs of the industrial plant [1]. This process may be an useful alternative for generation of synthesis gas, once the H_2/CO ratio may be controlled by varying the $CH_4:O_2:H_2O$ ratio in the feed [2]. However, as in other processes for synthesis gas production, nickel catalysts are used and suffer severe deactivation during operation due to coke formation on the surface [4]. This way, there is interest in developing catalysts that are more resistant to coke formation. The support may play an important role in providing more stable catalysts for methane reforming reactions [3,4]. In the present work, nickel catalysts supported on $\gamma\text{-}Al_2O_3$, CeO_2 and ZrO_2 were tested in the autothermal reforming of methane and characterized by Temperature-Programmed Reduction (TPR), UV-Vis Diffuse Reflectance Spectroscopy (DRS) and X-Ray Diffraction (XRD). The mechanism of the autothermal reforming over these catalysts was investigated by temperature programmed surface reaction (TPSR).

2. Experimental

The ZrO_2 support was prepared by calcination of zirconium hydroxide (MEL Chemicals) at 800°C , for 2h. The CeO_2 support was obtained by calcination of $(NH_4)_2Ce(NO_3)_6$ (ALDRICH) at 800°C, for 2 h. The γ-Al_2O_3 was obtained by calcining bohemite (CATAPAL A) at 800°C. The catalysts were prepared by incipient wetness impregnation of the supports with an aqueous solution of $Ni(NO_3)_2.6H_2O$, were dried at 120°C and calcined at 650°C for 6h. TPR experiments were performed in a multipurpose unit coupled to a Balzers Omnistar quadrupole mass spectrometer . The samples (150 mg) were previously dried at 150°C for 30 min under He flow (30 mL/min). For the TPR a 5% H_2/Ar gas mixture, a heating rate of 10°C /min and a final temperature of 1000°C were used. TPSR experiments were performed in the same unit used for TPR. The samples (27mg) were previously dried at 150°C for 30 min under He flow (30mL/min), followed by reduction under H_2 flow at 800°C, for 2h and cooled in He to room temperature. The sample was then submitted to a reactant mixture containing 5%O_2/He, 20%CH_4/He, H_2O and He with flow rates of 60.2 mL/min, 60.2 mL/min, 6.0 mL/min and 73.6 mL/min, respectively. The inlet mixture ratio was CH_4:O_2:H_2O equal to 4:1:2 with a total flow rate equal to 200 mL/min. The temperature was raised to 800°C at a heating rate of 20°C/min . The samples were characterized at room temperature in a VARIAN-Cary 500 spectrophotometer equipped with a Diffuse Reflectance Accessory (Harrick). In order to separate the contribution of the support, the reflectance $R(\lambda)$ of the sample was made proportional to the reflectance of the supports, and the Kubelka-Munk function F(R) was calculated. XRD experiments of calcined catalysts and supports were performed in a Miniflex model RIGAKU spectrometer using CuKα radiation. The diffratograms were obtained between 2θ= 20 e 80° using a 0.05° step. The dispersion of the samples were estimated by measuring the activities of the catalysts in the cyclohexane dehydrogenation, a structure insensitive reaction. Cyclohexane dehydrogenation was performed under atmospheric pressure in a continuous flow micro-reactor. The reactant mixture was obtained by bubbling hydrogen through a saturator containing cyclohexane at 12°C (H_2/C_6H_{12} =13.6). The total flow rate was 100 cm^3/min and temperature was 270°C. At these conditions, the conversion was kept below 10%, and only benzene was obtained as product. The effluent gas phase was analyzed by an on-line gas chromatograph (HP 5890) equipped with a flame ionization detector and a HP-INNOWax capillary column. Autothermal reforming of methane was performed under atmospheric pressure and 800°C in a continuous microreactor. The catalytic samples (20 mg) were diluted in SiC to a total mass of 100mg. The reactor feed consisted of 2.5% O_2, 10.2%CH_4, 2.9% H_2O and 84.4%He in a total flow of 135 mL/min. Reactant and product concentrations were measured using a Varian 3800 gas cromatograph equipped with a Carboxen 1010 capillary column and a thermal conductivity detector.

3. Results and Discussion

XRD diffraction patterns for the several supports used and for the catalyts are presented in Figure 1. CeO_2 presented a cubic structure (Joint Comitee for Powder Diffraction Studies - JCPDS – 4-0593), while ZrO_2 presented a monoclinic structure (JCPDS – 13-307) , as observed by other authors [5-7]. XRD diffraction of the γ- alumina investigated confirmed its structure.

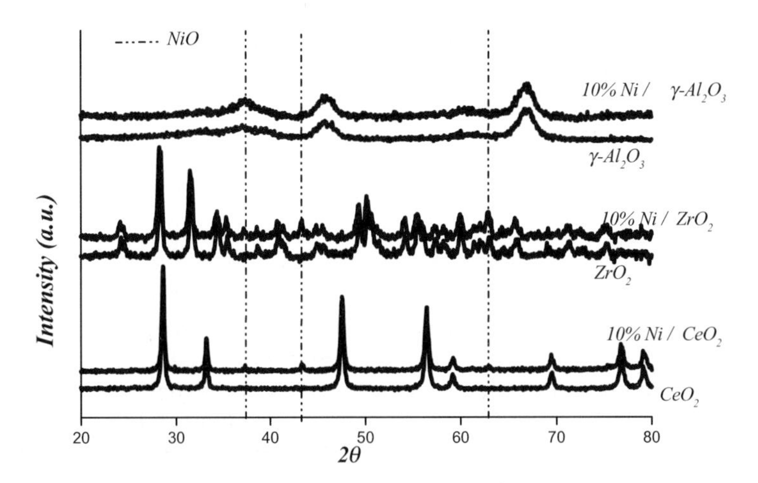

Figure 1 – XRD difraction of Ni/γ-Al_2O_3, Ni/CeO_2 and Ni/ZrO_2.

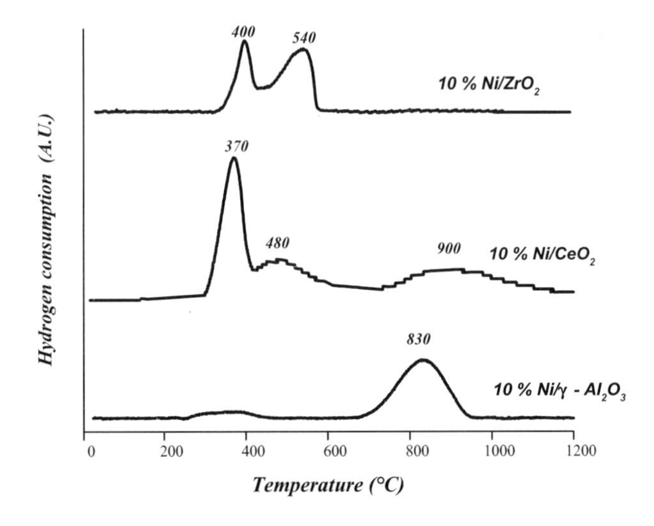

Figure 2 – Temperature Programmed Reduction of Ni/γ-Al_2O_3, Ni/CeO_2 and Ni/ZrO_2.

For the catalysts, NiO was the only nickel species that could be observed in the diffraction pattern . However, temperature programmed reduction (TPR) experiments showed also the presence of $NiAl_2O_4$ for the Ni/γ-Al_2O_3 catalysts (Figure 2) Besides NiO reduction, reduction of surface and bulk CeO_2 was observed for Ni/CeO_2 catalyst, while only reduction of NiO was observed for Ni/ZrO_2.
DRS UV-Vis results are shown in Figure 3. For the 10% Ni/ γ-Al_2O_3 and 10% Ni/ZrO_2 bands between 290 nm and 305 nm were observed, which indicate the presence of free NiO [8] . For the 10% Ni/ γ-Al_2O_3 a band around 630 nm indicated the formation of $NiAl_2O_4$ in agreement with TPR results. In the case of 10% Ni/CeO_2, a band at 430 nm was observed and it may be ascribed to the charge transfer of tetrahedral Ni^{2+} [9].

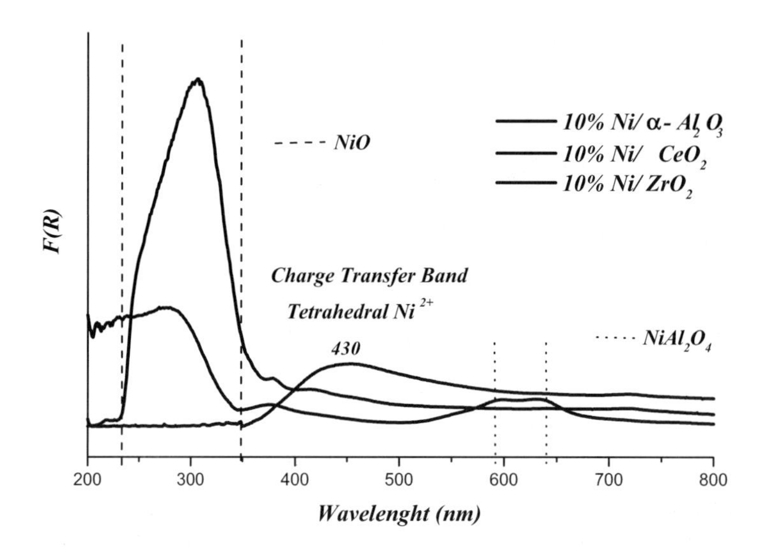

Figure 3 – DRS UV-Vis of Ni/γ-Al_2O_3, Ni/CeO_2 and Ni/ZrO_2.

Figure 4 shows the temperature programmed surface reaction profile for Ni/CeO_2. Except for Ni/γ-Al_2O_3, all other catalysts presented similar profiles. The autothermal reforming proceeded by a two-step mechanism. Methane conversion started around 450°C, with the formation of CO_2 and H_2O. Then, in the second step synthesis gas is formed via steam and CO_2 reforming, . For the γ-Al_2O_3 supported catalyst, the steam reforming was suppressed and the activity for this catalyst was lower than the observed for the other catalysts. only combustion products were observed. The conversion at the end of the TPSR was estimated and the following order was observed 10% Ni/γ-Al_2O_3 < 10%

Ni/CeO_2 <10% Ni/ZrO_2. These experiments were performed at relatively "clean" coking conditions and this activity was roughly proportional to nickel dispersion as measured by cyclohexane dehydrogenation a structure insensitive reaction (Table 1). The bad performance of the 10% $Ni/\gamma\text{-}Al_2O_3$ was ascribed to the aluminate formation which turned more difficult the reduction to metallic Ni.

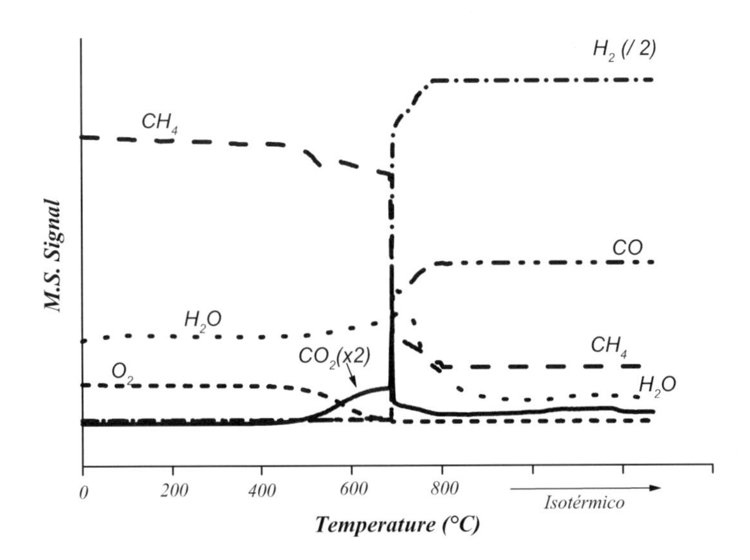

Figure 4 - Temperature Programmed Surface Reaction on Ni/CeO_2.

The same trend was observed when the catalysts were tested in the autothermal reforming of methane. Under the studied condition, Ni/ZrO_2 proved to be more active and was quite stable during 24 hours of time on stream (Table 1). The 10% $Ni/\gamma\text{-}Al_2O_3$ catalyst presented a very low activiy, probably due to the nickel aluminate formation, which is very difficult to be reduced. This behavior may be claimed to be due to interaction between Ni and ZrO_2 [10], The interfacial sites on Ni/ZrO_x would be active for CO adsorption and CO_2 dissociation helping to keep the surface free from coking [10]. However Ni/CeO_2 should have presented a higher activity, as observed for Pt/CeO_2 in the partial oxidation of methane [11] as CeO_2 is well known for its oxygen storage capacities. This may have been caused by the dilution of the feed in helium used in this study. Catalytic runs at different condition are underway to check for these incongruencies.

Table 1 – Autothermal reforming of methane at 800°C and cyclohexane dehydrogenation at 270°C on Ni catalysts

Catalyst	Reaction rate cyclohexane dehydrogenation (x 10^{-3} mol/h/gcat)	Apparent dispersion (%)	Final methane conversion at autothermal reforming [1] (%)
10 % Ni / γ-Al_2O_3	0.8	0.1	3
10 % Ni / CeO_2	3.8	0.59	18
10 % Ni / ZrO_2	4.6	0.71	47

[1] conditions described in experimental section

Conclusions

Based on Temperature-Programmed Surface Reaction (TPSR) experiments a two step mechanism for autothermal reforming was observed. First there is total combustion of methane followed by CO2 and steam reforming, except for the γ-Al_2O_3 supported catalyst, for which the steam reforming was suppressed. Under coke-free conditions, the activity is directly proportional to the number of surface sites and Ni/ZrO_2 catalyst was the most active catalyst.

References

1. M.A. Peña, J. P. Gómez, J. L. G. Fierro, Appl. Catal. A. 144, 7 (1996).
2. J. R. Rostrup-Nielsen, J. Sehested, J., J. K. Norskov, Adv.Catal. 47, 65 (2002).
3. M. Laniecki, M. Malecka-Grycz, F. Domka, Appl. Catal. A: Gen. 196 (2000) 293.
4. S. M. Stagg-Williams, F. B. Noronha, G. Fendley, D. E. Resasco J. Catal. 194 (2000) 240.
5. H. S. Roh, K. W. Jun, W. S. Dong, J.S. Chang, S. E. Park, Y. Joe J. Mol. Catal. A: Chem. 181 (2002) 137.
6. S. Pengpanich, V. Meeyoo, T. Rirksomboon, K. Bunyakiat Appl. Catal. A: Gen. 234 (2002) 221.
7. S. Irusta, L. M. Cornaglia, E. A. Lombardo, J. Catal. 210 (2002) 263.
8. J. S. Lisboa, D. C.R.M. Santos, F. B. Passos, F. B. Noronha Catal. Today 101 (2005) 15.
9. F. Delannay Characterization of Heterogeneous Catalysts, Marcel Dekker, New York, 1984.
10. M.M.V.M. Souza, D.A.G. Aranda and M. Schmal. J. Catal. 204 (2001), 498.
11. F. B. Passos, E. R. Oliveira, L. V. Mattos, F. B. Noronha, Catalysis Today 101 (2005) 23.

Selective CO removal in the H_2-rich stream through a double-bed system composed of non-noble metal catalysts

Yun Ha Kim,[a] Eun Duck Park,[a] Hyun Chul Lee,[b] Kang Hee Lee,[b] Soonho Kim[b]

[a]*Division of Energy Systems Research and Division of Chemical Engineering and Materials Engineering, Ajou University, San 5 Wonchun-dong, Yeongtong-gu, Suwon 443-749, Korea*
[b]*Energy and Materials Research Laboratory, Samsung Advanced Institute of Technology (SAIT), P.O. Box 111, Suwon 440-600, Korea*

Abstract

The CO concentration in the effluent from the water-gas shift reactor can be reduced to be less than 15 ppm from about 1 vol% through the double bed system composed of CuO-CeO_2, which can oxidize CO selectively in the presence of excess H_2, and Ni/Y-ZrO_2, which is active for CO methanation in the presence of excess CO_2, over a wide reaction temperature range.

1. Introduction

The polymer electrolyte membrane fuel cell (PEMFC) has been attracting much attention in the application to electric vehicles or residential power-generations. This PEMFC utilizes hydrogen as a fuel. Thus, the fuel processor is in need to convert various hydrocarbons into hydrogen via a couple of catalytic reactions such as a steam reforming and a water-gas shift reaction ($CO + H_2O \leftrightarrow CO_2 + H_2$) [1]. In the case of water-gas shift reaction, it is an exothermic and a thermodynamically limited reaction. Therefore, the high CO conversion can be achieved only at low temperatures. About 1 vol% of carbon monoxide usually remains in the hydrogen stream after the water-gas shift reactor. Since platinum, an anode of PEMFC, is prone to be poisoned in the presence of small amounts of CO in the hydrogen stream, carbon monoxide should be removed to a trace-level. The acceptable CO concentration is below 10 ppm at Pt anode and below 100 ppm even at CO-tolerant alloy anodes [2]. Although several different methods for the CO removal have been studied, the preferential CO oxidation

(PROX) has been considered to be most promising. In this system, the following two reactions can occur.

$$CO + (1/2)O_2 \rightarrow CO_2 \quad (1)$$
$$H_2 + (1/2)O_2 \rightarrow H_2O \quad (2)$$

The selective CO oxidation (Eq.(1)) should be occurred preferentially over the H_2 oxidation (Eq.(2)) in the presence of excess H_2. Because the hydrogen is abundant in the reactant, the following hydrogenation reactions can be carried out simultaneously.

$$CO + 3H_2 \rightarrow CH_4 + H_2O \quad (3)$$
$$CO_2 + 4H_2 \rightarrow CH_4 + 2H_2O \quad (4)$$

The CO methanation (Eq. (3)), facilitated over supported Ni and Ru catalysts, should be minimized except that CO concentration is quite low because it consumes relatively larger amounts of H_2 compared with that of the PROX [3]. In any cases, the CO_2 methanation (Eq.(4)) should be suppressed.

A number of catalysts more active for the PROX have been reported such as metal oxide catalysts, supported gold catalysts and supported noble metal (Pt, Ru, and Rh) catalysts [2]. Among them, supported Pt- or Ru-based catalysts have been accepted to be most effective. However, these noble metal catalysts have some disadvantages such as its high price and a rather low selectivity for PROX. To overcome these problems, we revisit the CuO-CeO_2 catalyst because this has a rather high selectivity for the PROX. We also introduce the double-bed system composed of the PROX catalyst and the methanation catalyst to enhance the CO removal efficiency.

2. Experimental

The CuO-CeO_2 catalyst was prepared by the co-precipitation method as described in the previous work [4]. The Cu content was intended to be 20 at.%. This CuO-CeO_2 catalyst was calcined in air at 973 K before a reaction. A Ni/Y-ZrO_2 catalyst was prepared by a conventional wet impregnation method from $Ni(NO_3)_2$ and Y-ZrO_2 (TZ-8YS, Tosoh). The Ni content was intended to be 10 wt%. This Ni/Y-ZrO_2 was calcined in air at 773 K and reduced in the hydrogen stream at 773 K before a reaction.

Catalytic activity tests for PROX were carried out in a small fixed bed reactor with catalysts that had been retained between 45 and 80 mesh sieves. In the case of the double bed system, the weight fraction of the CuO-CeO_2 catalyst was 0.2. A reactant gas flow was fed to a reactor at an atmospheric pressure. The effluent gas composition was determined through gas chromatographic analysis (HP5890A, molecular sieve 5A column) or an online gas analyzer (NGA2000,

MLT4, Rosemount Analyzer System from Emerson Process Management) of each gas component (CO at ppm level, CO_2, H_2, CH_4 at % level).

3. Result and Discussion

It is generally accepted that the effluent CO concentration can be decreased further over supported noble metal catalysts by increasing the O_2 concentration. However, this is dependent on the reaction temperature over metal oxide catalysts such as CuO-CeO_2. The PROX activity decreased over CuO-CeO_2 below 403 K but it increased above 413 K with increasing O_2 concentration as shown in Fig. 1. This decreased CO conversion at low temperatures is closely related to the formation of additional H_2O due to H_2 oxidation because the PROX activity was reported to be decreased with increasing H_2O concentration especially at low temperatures over CuO-CeO_2 [4]. Although the higher PROX activity can be achieved by increasing [O_2]/[CO] at high reaction temperatures, the temperature window showing CO concentrations below 100 ppm was rather narrow. Therefore, it can be concluded that the single CuO-CeO_2 catalyst cannot be applied to the CO removal system for PEMFC.

The effect of the inlet CO concentration on the CO methanation was examined over Ni/Y-ZrO_2 as shown in Fig. 2. As long as the CO concentration was lower than 2000 ppm, a rather broad reaction temperature region exhibiting low CO concentrations below 20 ppm, in the effluent was observed from 453 K to 495 K. However, this region became narrower and shifted to the higher temperature as the initial CO concentration increased to be above 3000 ppm. The CO_2 methanation appears to occur simultaneously with the CO methanation because the larger amounts of CH_4 was produced than that CO consumed. This is consistent with the slight decrease in CO_2 and H_2 concentration as shown in Fig. 2. Considering the extremely high concentration ratio between CO_2 and CO, this catalyst can be regarded to be active for CO methanation in the presence of excess CO_2. The increase in CO_2 concentration at about 460 K can be explained by the water gas shift reaction. From this result, the single methanation catalyst

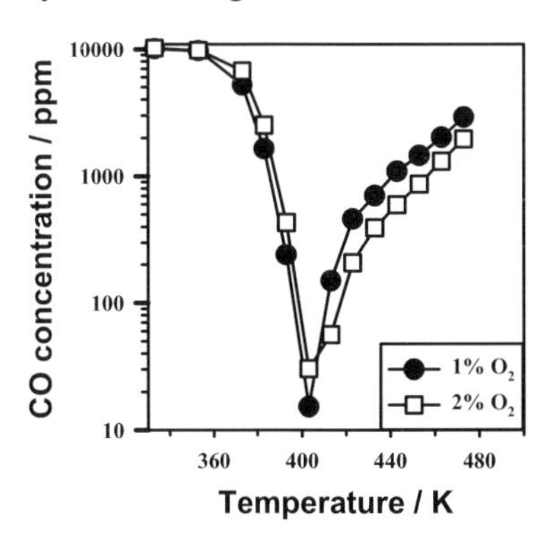

Figure 1. CO concentration in the effluent after the PROX over CuO-CeO_2 at different reaction temperatures. Reactants: 1% CO, 1%(or 2%) O_2, 2% H_2O, 50% H_2, and 20% CO_2 in He. F/W=100ml/min/g_{cat}.

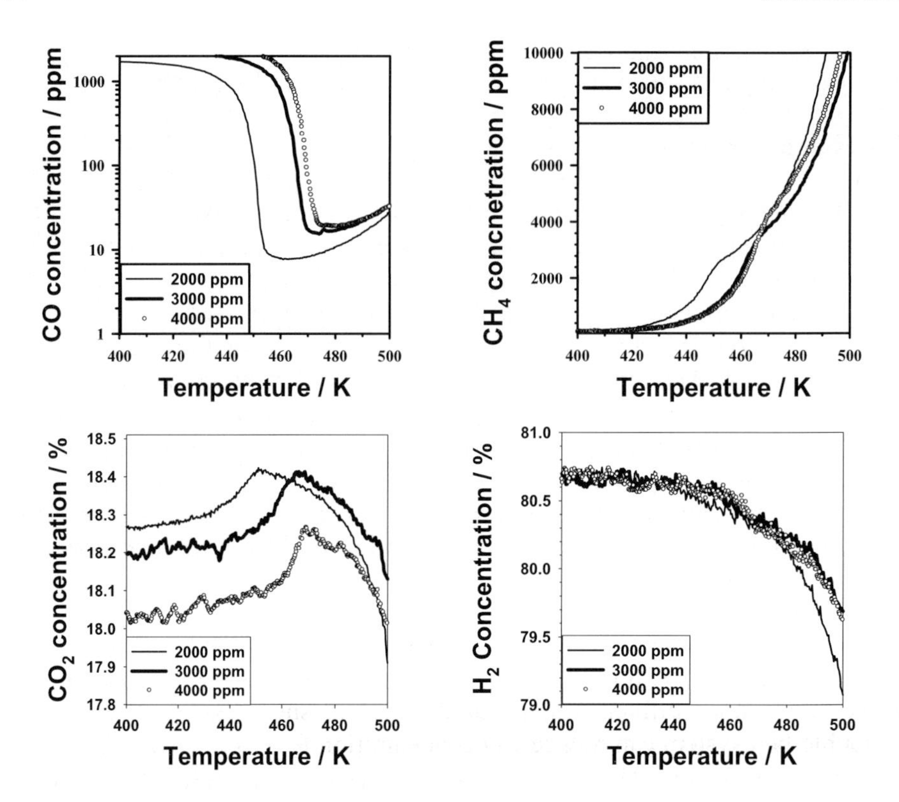

Figure 2. CO, CH_4, CO_2, and H_2 concentration in the effluent based on the dry gas composition after the CO methanation over 10 wt% Ni/Y-ZrO_2 with increasing reaction temperatures at a ramping rate of 1 K/min. Reactants: 2000 ppm (3000 or 4000 ppm) CO, 17.4% CO_2, 13.0% H_2O, balanced with H_2. F/W=100ml/min/g_{cat}.

can be applied to the CO removal system only when the inlet CO concentration is below 2000 ppm.

Figure 3 shows the double-bed system composed of CuO-CeO_2 and Ni/Y-

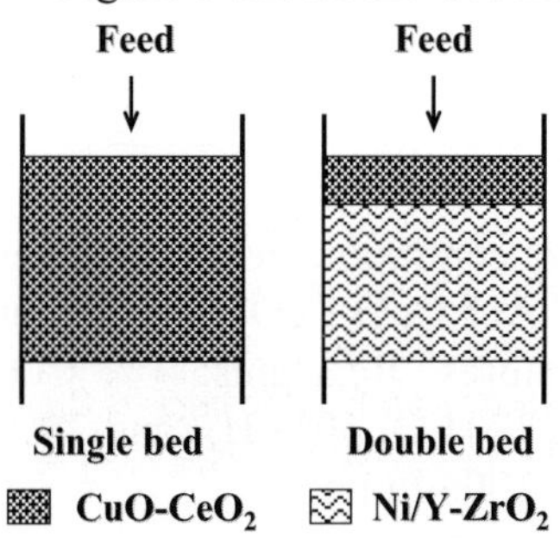

Figure 3. The schematic diagram of a single bed and a double bed system.

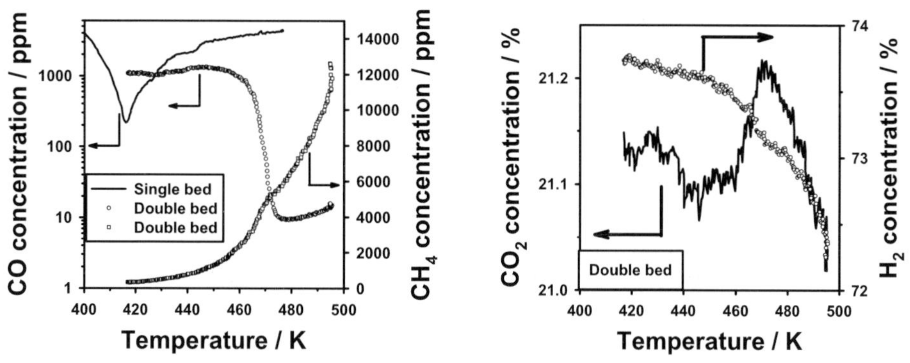

Figure 4. CO CH_4, CO_2, and H_2 concentration in the effluent based on the dry gas composition after the preferential CO oxidation over the single bed system ($CuO-CeO_2$) or the double bed system ($CuO-CeO_2$ and $Ni/Y-ZrO_2$) with increasing reaction temperatures at a ramping rate of 1 K/min. Reactants: 0.9% CO, 0.9% O_2, 17.4% CO_2, 64. 6% H_2, 13.0% H_2O, balanced with N_2. $F/W=69ml/min/g_{cat}$.

ZrO_2 to reduce the CO concentration from about 1 vol% to below 100 ppm. In the case of a single bed system composed of only $CuO-CeO_2$, the minimum CO concentration in the effluent was about 200 ppm as shown in Fig. 4. However, the double-bed system can reduce CO concentration from 0.9 vol% to below 15 ppm from 473 K to 495 K. This result implies that the high CO removal efficiency accomplished only over noble-metal catalysts can be achieved over simple transition metal-based catalyst system by introducing double-bed system as described in this report. The increasing CH_4 concentration as well as the decreasing H_2 concentration with a reaction temperature implies that the CO methanation is responsible for the further removal of CO. Similarly to the previous CO methanation result, the increase in CO_2 concentration at about 470 K and the decrease in CO_2 concentration above 480 K was observed over this double bed system, which is due to the water gas shift reaction and the CO_2 methanation, respectively.

4. Conclusion

By introducing the double bed system composed of $CuO-CeO_2$, which can selectively oxidize CO in the presence of an excess H_2, and $Ni/Y-ZrO_2$, which is active for CO methanation in the presence of an excess CO_2, the CO concentration can be reduced to be less than 15 ppm from about 1 vol% over a wide reaction temperature range. This double bed system can be applied to the CO removal system after a water-gas shift reactor for the polymer electrolyte membrane fuel cell (PEMFC).

Acknowledgement

This work was supported by Samsung Advanced Institute of Technology (SAIT) and the Brain Korea 21 program.

References

[1] C. Song, Catal. Today 77 (2002) 17.
[2] L. Shore, R.J. Farrauto, in: W. Vielstich, A. Lamm, H.A. Gasteiger(Eds.), Handbook of Fuel Cells: Fundamentals Technology and Application, Part 2, vol. 3, John Wiley & Sons Ltd., West Sussex, 2003, p.211-218.
[3] S. Takenaka, T. Shimizu and K. Otusuka, Int. J. Hydrogen Energy 29 (2004) 1065.
[4] E.-Y. Ko, E.D. Park, K.W. Seo, H.C. Lee, D. Lee and S. Kim, Catal. Today, 116 (2006) 377.

Natural Gas Conversion VIII
F.B. Noronha, M. Schmal, E.F. Sousa-Aguiar (Editors)

Hydrogen Production by Steam Reforming of Natural Gas over Highly Stable Ru Catalyst Supported on Nanostructured Alumina

Hyun Chul Lee,[a,*] Yulia Potapova,[a] Ok Young Lim,[a] Doohwan Lee,[a] Soonho Kim,[a] Jae Hyun Park,[b] Eun Duck Park[b,*]

[a]*Energy & Environmental Laboratory, Samsung Advanced Institute of Technology, Suwon, 440-600 The Republic of Korea.*
[b]*Division of Energy Systems Research and Division of Chemical Engineering and Materials Engineering, Ajou University, Wonchun-Dong, Yeongtong-Gu, Suwon, 443-749, The Republic of Korea.*

A nanostructured alumina (ANS) supported Ru catalyst was prepared and tested for the steam reforming of natural gas for hydrogen production. The Ru/ANS catalyst showed higher activity and stability through whole range of reaction conditions than catalysts consisting of commercial alumina as a support. The ANS, synthesized via a surfactant-driven pathway under a hydrothermal condition, exhibited a high thermal stability, a large pore volume and high surface area with fibrous morphologies. The results indicated that this ANS can be an efficient catalyst support for natural gas steam reforming.

1. Introduction

Recently, the polymer electrolyte membrane fuel cell (PEMFC) has been regarded as one of the promising candidates for utilizing hydrogen gas to produce heat and electricity [1]. In particular, natural gas, supplied through built-in infrastructure pipelines, is an efficient hydrogen source for residential application of PEMFC. The steam reforming of natural gas is an industrially practiced technology for H_2 production, but its application in the residential PEMFC system requires that the reactor to be small and stable under frequent on-off conditions, thus reforming catalysts to be highly active and stable. In addition, highly endothermic nature of the steam reforming reaction requires the catalysts to be stable under high reaction temperature [2]. Aluminum oxide (Al_2O_3) is widely used as adsorbents, catalyst and catalyst supports due to its thermal, chemical and mechanical stability. The synthesis of nanostructured aluminas via usual surfactant-templating methods is challenging because the fast hydrolysis of aluminum precursors in aqueous media leads to the formation of lamellar hydrated hydroxides even in the presence of surfactant molecules [3]. Nevertheless, there are some successful examples of thermally stable nanostructured alumina [3-5]. In our previous study, we reported a simple method for synthesizing nanostructured γ-alumina (ANS) of various morphologies with a high crystallinity, thermal stability and high surface area by controlling hydrothermal conditions [5]. In this study, we report that nanostructured alumina is an efficient support for the natural gas reforming in the application of residential fuel cell systems, where high catalytic activity and thermal stability is prominently important.

2. Experimental

Nanostructured alumina was synthesized as described in ref [5] by using a cationic surfactant (CTAB, $CH_3(CH_2)_{15}N(CH_3)_3Br$), an aluminum precursor (ATB, aluminum tri-sec-butoxide), and a controlled amount of water under hydrothermal condition (423 K for 72 h). The ANS was obtained after calcination of hydrothermally-synthesized alumina with air flow at 500 °C for 4 hr. In order to compare effect of support properties, commercial alumina was purchased from Aldrich Company. 2 wt% Ru catalysts supported on the nanostructured alumina (denoted as Ru/ANS) and the alumina purchased from Aldrich (denoted as Ru/γ-Al_2O_3) were prepared by an incipient wetness method with corresponding salt ($RuCl_3$), respectively. They were calcined in air at 550 °C. Alumina-supported Ru catalysts above were used without any pre-treatment such as reduction with H_2 before catalytic reaction.

Catalytic activity test for steam reforming reaction was done with pure CH_4 flow in a fixed-bed reactor with catalyst (0.5g) that had been retained between 45 and 80 mesh sieves. A reactant gas flow was fed to a reactor at an atmosphere pressure. The performance of these catalysts was compared at various gas hourly space velocities (GHSV) such as 8000, 16000, 32000 h^{-1}. Steam to carbon ratio (S/C) was fixed to be 3. The temperature of steam reforming reaction was adjusted to increase from 550 °C to 750 °C. Catalytic activity was compared at 600 °C and 700 °C over prepared alumina-supported Ru catalysts (Ru/ANS and Ru/γ-Al_2O_3). In order to investigate catalytic stability over Ru catalyst, thermal cycling test of steam reforming reaction with temperature range from 550 °C to 750 °C was done on the Ru catalysts (Ru/ANS and Ru/γ-Al_2O_3) under the reaction condition of S/C= 3, GHSV= 8000 (h^{-1}). The CH_4 conversion over Ru/ANS and Ru/γ-Al_2O_3 compared at 700 °C for thermal cyclic test

The effluent gas composition was determined with an online gas analyzer (NGA2000, MLT4, Rosemount Analyzer System from Emerson Process Management). Characteristics of the prepared catalysts were investigated with nitrogen adsorption (ASAP 2010, Micromeritics), XRD (Rigaku D/MAC-III), CO chemisorption (Autochem 2910, Micromeritics), and HRTEM (Techni F30 UT, FEI, 200 kV).

3. Results and Discussion

Table 1 shows characteristics of the supports and supported Ru catalysts investigated in this work. The synthesized ANS showed a higher surface area with a larger pore volume and average pore size (determined by the BJH model using N_2 desorption isotherm) than those of the commercial alumina. Clearly, the surfactant-driven synthesis method taken in the preparation of ANS led to these advanced physical properties. An introduction of ruthenium species on ANS supports resulted in a significant decrease in pore volume, surface area and average pore size, while a slight decrease was observed on γ-Al_2O_3 support Nevertheless, the Ru/ANS showed the higher CO uptake (13.6 μmol/gcat) than the commercial alumina (7.3 μmol/gcat) , which reflect the higher dispersion of Ru on the ANS.

Table 1. Characteristics of aluminas and corresponding Ru supported catalysts

Materials	Pore Volume, (cm^3/g)	Surface Area, (m^2/g)	BJH Pore Size, (nm)	Amount of Adsorbed CO, (μmol/gcat)
ANS	1.36	438	7.1	-
Ru/ANS	0.64	263	5.6	13.6
γ-Al_2O_3	0.33	167	4.6	-
Ru/γ-Al_2O_3	0.31	161	2.2	7.3

The low magnification TEM images of Ru/ANS (Fig. 1A.(a, b, c)) show that the samples kept their unidirectional nanostructures, while no discernable image was observed on the Ru catalyst supported on the commercial support (Fig. 1A. (d)). The Ru/ANS showed a distinct selected-area electron diffraction pattern (Fig. 1a inset) superimposed with diffused ring patterns, indicating presence of poly-crystalline alumina in forms of nanostructures. Thus, it could be confirmed that Ru/ANS was composed of the γ-alumina along with the nanostructured framework with a relatively low crystallinity. The crystallinity of this material was supported by the wide-angle XRD patterns. The γ-alumina phase was observed on calcined ANS samples in which the surfactant had been removed. As shown in Fig. 1A. (c), on Ru/ANS most Ru species were doped on the nanostructured frame and intercrystallite voids. Structural stability of support and distribution of the Ru species were also investigated by TEM images on the used catalysts. Fig. 1B shows Ru distributions on Ru/ANS (Fig. 1B (a, b)) and on Ru/γ-Al_2O_3 (Fig. 1B (c, d)) after the reaction. On Ru/ANS Ru species were well distributed into the intercrystallite voids and surface of randomly stacked nano-tubular structures, whereas on Ru/γ-Al_2O_3, Ru species aggregated forming large particles, as shown in Fig. 1B. Thus, it would be concluded that the nanostructured alumina having fibrous morphology could be an efficient catalyst support which prevent aggregation of the active species (Ru) by anchoring them in the randomly stacked intercrystallite networks and in the nano-tubular inner spaces.

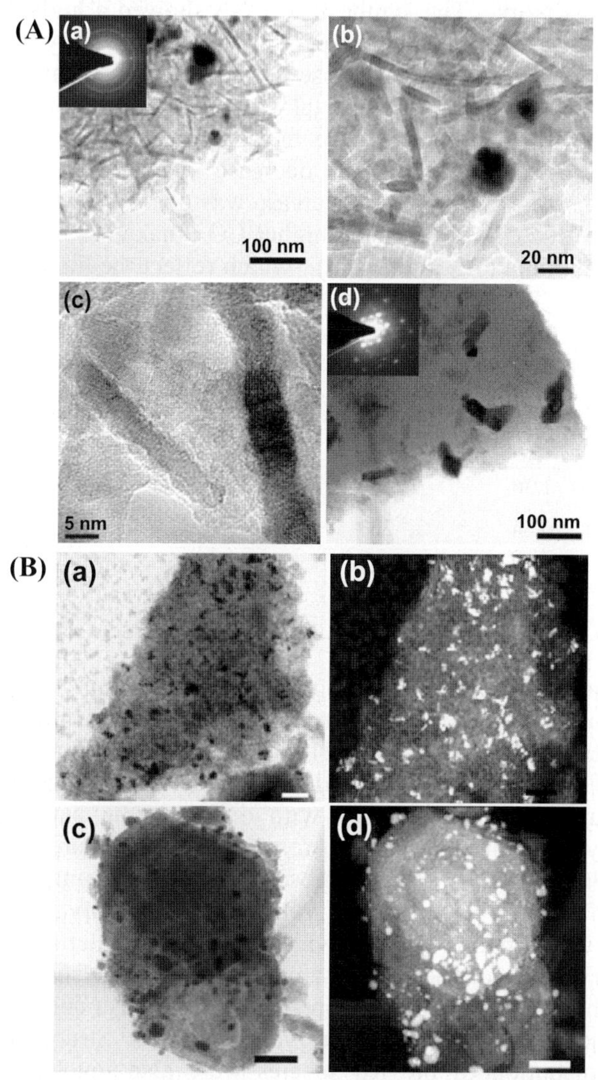

Fig. 1. TEM images and selected-area electron diffraction (SAED) patterns for the supported Ru catalysts. (A, before reaction); (a-c) Ru/ANS, (d) Ru//γ-Al_2O_3, (B, after reaction): (a, b) Ru/ANS, (c, d) Ru//γ-Al_2O_3, scale bar: 200 nm

Table 2 shows the catalytic performance of Ru/ANS on methane steam reforming in comparison with that of Ru/γ-Al_2O_3. Conversion of CH_4 was obtained under various GHSVs (8000, 16000, 32000 h^{-1}) and temperatures (600 °C, 700 °C) after steam reforming reaction for 5hr. For all catalysts, the CH_4 conversion increases as GHSVs decrease under the fixed reaction temperatures. Through all experimental condition, Ru/ANS showed higher catalytic activity than that in Ru/γ-Al_2O_3. The higher CH_4 conversion on Ru/ANS would be resulted from the enhanced Ru dispersion contributed by the excellent support characteristics such as high surface area, large pore volume, and nanostructured morphologies of ANS.

Table 2. . Conversions of CH_4 (%) on Ru/ANS and Ru/γ-Al_2O_3 under different GHSVs.

Temperature (℃)	Steam to Carbon Ratio (S/C)	GHSV (h^{-1})	Catalysts	
			CH_4 (%) (Ru/ANS)	CH_4 (%) (Ru//γ-Al_2O_3)
600	3	8000	76.0	70.3
600	3	16000	65.4	60.3
600	3	32000	51.1	46.8
700	3	8000	94.5	93.2
700	3	16000	92.1	87.9
700	3	32000	78.6	72.9

Fig. 2 shows the catalytic stability of Ru/ANS and Ru/γ-Al_2O_3 under a thermal cycling of steam reforming reaction with temperature range from 550 °C to 750 °C. The cyclic test carried out at S/C=3 and GHSV= 8000 (h^{-1}). The CH_4 conversion and product composition was obtained at 700 °C for each cycle test using on-line gas analyzer (CH_4, CO, CO_2 and H_2 in vol% unit, respectively) and indicated in Fig. 2 for the Ru/ANS catalyst. For comparison of catalytic stability, the CH_4 conversion (%) of Ru/γ-Al_2O_3 was shown with a symbol of open circle. The CH_4 conversion over Ru/ANS remained the same at 94 % through all 20 thermal cycles at 700 °C, whereas that over Ru/γ-Al_2O_3 significantly decreased during first 10 thermal cycles of steam reforming. As revealed in Fig. 1, Ru species in Ru/γ-Al_2O_3 aggregated forming large particles after reaction. Thus, Ru/γ-Al_2O_3 went to be deactivated during cyclic test, resulting in a rapid decrease in CH_4 conversion from 93.2% to 77.8%. On the other hands, Ru/ANS kept its initial activity in steam reforming of methane. Thus, ANS would be regarded as an excellent catalytic support having a high thermal stability. As a result, Ru/ANS catalyst showed high dispersion of Ru species and kept Ru dispersion after reaction by anchoring Ru species in randomly stacked intercrystallite or in the tubular spaces of ANS, as revealed in Fig. 1

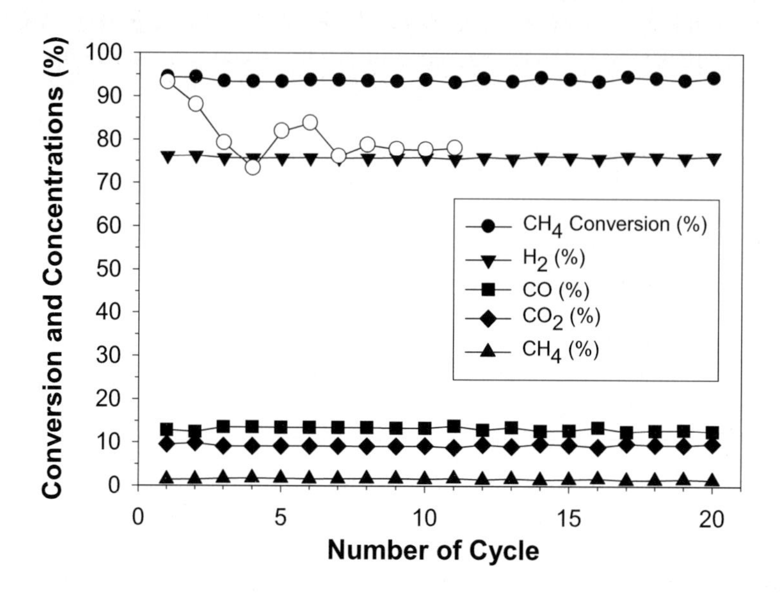

Fig. 2. Thermal cyclic test of supported Ru catalysts for CH_4 steam reforming. Filled symbols (Ru/ANS), unfilled symbol (Ru//γ-Al_2O_3), Thermal cyclic test conditions: Temperature = 700 (^oC), S/C=3, GHSV=8000 (h^{-1}).

In conclusion, nanostructured crystalline γ-alumina is an excellent catalytic support which has high surface area and pore volume, high thermal and chemical stability, and unique morphological properties. As shown, these unique properties led to an enhanced activity and stability of Ru/ANS in the steam reforming of methane.

References

1. C. Song, *Catal. Today*, **2002**, *77*, 17.
2. J. R. Rostrup-Nielsen and K. Aasberg-Petersen, in: W. Vielstich, A. Lamm, H.A. Gasteiger(Eds.), Handbook of Fuel Cells: Fundamentals Technology and Application, Part 2, vol. 3, John Wiley & Sons Ltd., West Sussex, **2003**, p.159-176.
3. S. Cabrera, J. E. Haskouri, J. Alamo, A. Beltrán, D. Beltrán, S. Mendioroz, M. D. Marcos and P. Amorós, *Adv. Mater.* **1999**, *11*, 379.
4. S. A. Bagshaw and T. J. Pinnavaia, *Angew. Chem. Int. Ed. Engl.* **1996**, *35*, 1102.
5. H. C. Lee, H. J. Kim, S. H. Chung, K. H. Lee, H. C. Lee and J. S. Lee, *J. Am. Chem. Soc.*, **2003**, *125*, 2882.

Natural Gas Conversion VIII
F.B. Noronha, M. Schmal, E.F. Sousa-Aguiar (Editors)

HEAT EFFECTS IN A MEMBRANE REACTOR FOR THE WATER GAS SHIFT REACTION

M. E. Adrover, E. López, D. O. Borio, M. N. Pedernera

Department of Chemical Engineering - PLAPIQUI
Universidad Nacional del Sur - CONICET
Camino La Carrindanga, Km 7
(8000) Bahía Blanca, ARGENTINA

1. Introduction

Most of the hydrogen is produced industrially by steam reforming of hydrocarbons or alcohols (e.g., for fuel cell applications). The process gas stream coming from the steam reformer is composed by H_2, CO, CO_2, H_2O and small amounts of unconverted reactants (CH_4). The CO concentration of the gas leaving the reformer must be reduced up to a specified level, with two main goals: 1) increase the H_2 production rate and 2) purify the process stream. To these ends, the Water Gas Shift Reaction (WGSR) is widely used:

$$CO + H_2O \Leftrightarrow CO_2 + H_2 \qquad \Delta H^{o}_{298K} = -41.09\ kJ/mol \qquad (1)$$

Reaction (1) is moderately exothermic and strongly controlled by the chemical equilibrium, which is favoured at low temperatures. In small-scale processes, such as the fuel processing for fuel cells (e.g., PEM cells) normally the WGSR is carried out in a single reactor at an intermediate temperature level [1].

An attractive alternative to increase the CO conversion is the membrane reactor (MR) (Figure 1). The main idea of this design is the selective permeation of reaction products (e.g., H_2) to

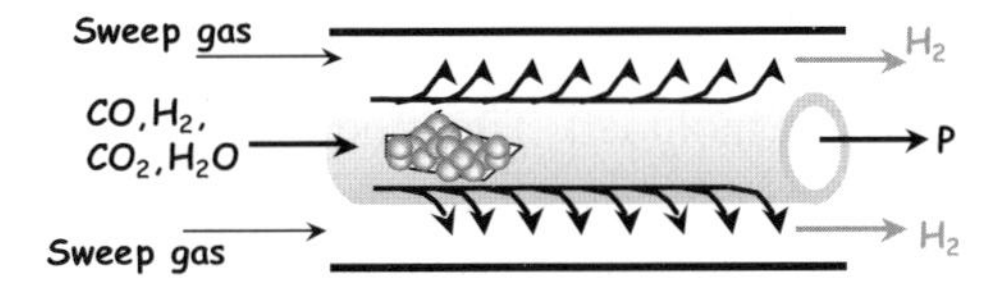

Figure 1: Scheme of the membrane reactor

shift the equilibrium towards products and consequently increase the conversion, or reduce the amount of catalyst for a desired conversion level. For this reason, MRs have deserved considerable attention in the scientific literature [2].

The H_2 removal can be carried out by means of selective dense membranes of Pd or its alloys. In order to decrease the cost and increase the permeation fluxes, composite membranes became an alternative. In these membranes a substrate of high porosity and low resistance to flow is covered by a metallic layer, which provides the selectivity [3-5].

The advantages of the MR to perform the WGSR have been demonstrated by Criscuoli et al. [6]. However, the heat effects were neglected and the reactor operation was supposed to be isothermal. This is a common assumption in the modeling of most of the MRs, which agrees with the temperature measurements inside the reactor at laboratory scale due to the high ratio between the heat transfer area and the reactor volume. This condition may not be true when higher process scales are necessary; e.g.: several membrane tubes installed in parallel within a shell where the sweep gas is circulated. Here, the usual assumption of isothermal MR should be removed and the heat effects taken into account [7-9].

In the present work the performance of the MR for the WGSR is simulated and compared with that of a fixed-bed reactor (CR), for isothermal and adiabatic operations. The analysis is further extended to a MR under non-isothermal non-adiabatic conditions.

2. Mathematical model

The simulation study is carried out by means of a 1-D pseudo-homogeneous mathematical model. The model is subject to the following hypotheses: a) axial and radial mass and heat transfer dispersions are neglected; b) isobaric conditions; c) infinite selectivity for the membrane (H_2 is the permeation gas); d) the permeation flow through the membrane follows Sievert´s law; and e) the shell is adiabatic. The kinetic model proposed for the WGSR by Podolski and Kim [10] is included. The model is represented by the following equations:

Reaction Side (catalyst tubes)

Mass Balances

$$\frac{dF_{CO}}{dz} = A_T \, r_{CO} \rho_B \quad (2)$$

$$\frac{dF_{H_2}}{dz} = A_T(-r_{CO})\rho_B - \pi \, d_{te} \, J_{H_2} \quad (3)$$

Permeation Side (shell)

Mass Balances

$$\frac{dF_{H_2,P}}{dz} = \pi \, d_{te} n_t J_{H_2} \quad (4)$$

$$\frac{dF_{SG,P}}{dz} = 0 \quad (5)$$

Energy Balance

$$\frac{dT}{dz} = \frac{A_T \rho_B (-r_{CO})(-\Delta H_r) - \pi\, d_{te} U\,(T - T_P)}{\sum_{j=1}^{N} F_j C_{pj}} \quad (6)$$

Energy Balance

$$\frac{dT_P}{dz} = \frac{\pi d_{te} n_t (T - T_P)}{\sum_{j=1}^{2} F_{j,P} C_{pj}} \left[J_{H_2} C_{pH_2} + U \right] \quad (7)$$

$$\text{where: } J_{H_2} = \frac{7.70\ 10^{-5}\ e^{(-1960.5/T)}}{\delta} \left[\sqrt{p_{H_2}} - \sqrt{p_{H_{2,P}}} \right] \quad \text{(Barbieri et al. [11])} \quad (8)$$

Boundary conditions

Cocurrent Scheme:

$$\text{At } z=0 \begin{cases} F_j = F_{j0} & for\ j = 1, 2....,N \\ T = T_0; T_P = T_{P,inlet} \\ F_{H_2,P} = 0; F_{SG,P} = F_{SG,inlet} \end{cases}$$

Countercurrent Scheme:

$$\text{At } z=0 \begin{cases} F_j = F_{j0} & for\ j = 1,2....,N \\ T = T_0 \end{cases}$$

$$\text{At } z=L \begin{cases} F_{H_2,P} = 0; F_{SG,P} = F_{SG,inlet} \\ T_P = T_{P,inlet} \end{cases}$$

The design parameters and operating conditions are given in Table 1. The CR is modeled by assuming zero for the hydrogen flow through the membrane. For both reactors, two different configurations for the gas through the shell are studied: co- and countercurrent flow.

Table 1. Geometric parameters and operating conditions used in the simulations of CR and MR (Criscuoli et al. [6]*; Giunta et al. [1]**).

L	0.21 m (CR)- 0.15 m (MR) *	U	10 W/(m^2 K)
d_{ti}	6.7 mm (CR)- 8 mm (MR) *	F_0	9.58 10^{-3} mol/s
d_{te}	13.4 mm (MR) *	Inlet CO, %	7.97 **
n_t	30	Inlet CO_2, %	10.99 **
W	9.64 g *	Inlet H_2, %	43.48 **
δ	75 μm*	Inlet H_2O, %	31.88 **
P_o	1 atm *	Inlet CH_4, %	5.68 **
$F_{SG,inlet}$	30 l/min		

3. Results and Discussion

The performance of both reactor designs is first compared for two limit situations regarding heat effects: isothermal and adiabatic operations. For the isothermal case the heat effects are neglected for both the reaction and permeation sides; whereas, for the adiabatic operation the convective heat transfer through the membrane is neglected (U=0), i.e., the only source of heat exchange between both sides is the permeating hydrogen flow.

Figure 2 shows the axial conversion profiles in the MR and CR, at a given inlet temperature, T_{oI}=360 °C. For both CR and MR, the outlet CO conversion corresponding to the adiabatic operation is lower than that of the isothermal case. These results can be attributed to equilibrium limitations. For high inlet temperatures, the assumption of isothermal operation is not conservative and the

CO outlet conversions are overestimated in both types of reactors. Conversely, for low feed temperatures, the adiabatic operation introduces conversion improvements with respect to the isothermal case, for both reactor designs, due mainly to kinetic reasons (results not shown).

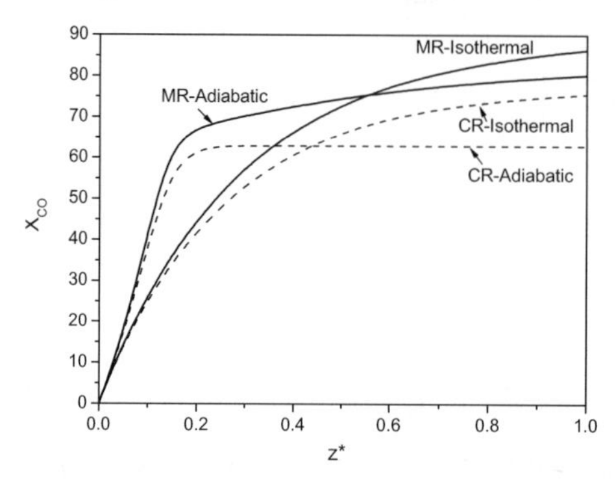

Figure 2: Axial CO conversion in MR and CR. T_{o2}=360°C. Isothermal and adiabatic conditions

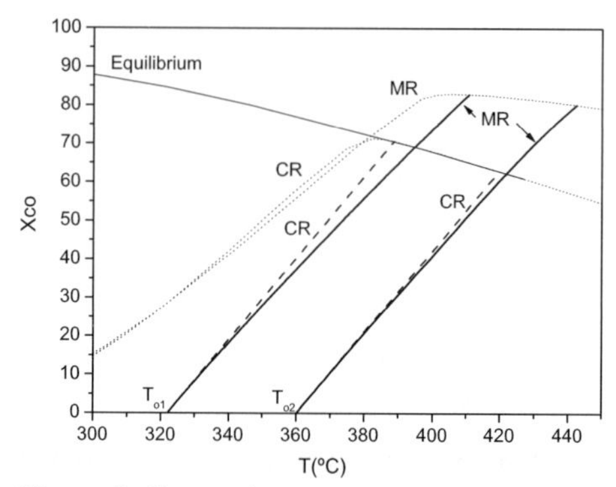

Figure 3: Conversion-temperature trajectories in MR and CR, for T_{o1}=322°C and T_{o2}=360°C.

The adiabatic operation (U=0) in the MR and the CR can also be compared in the phase-plane. Thus, Figure 3 shows the conversion-temperature trajectories inside the adiabatic reactors, for two inlet temperatures. The dotted lines join all the outlet conversion points, obtained by simulation of MR and CR at different inlet temperatures. It is clear that the total temperature rise inside the reactor is higher for the MR, for two main reasons: 1) the CO conversions are higher, i.e., larger amounts of heat are released; 2) the total gas flowrate in the reaction side diminishes along the axial position as a consequence of the H_2 permeation. Therefore, the adiabatic trajectories corresponding to the MR shift away from the classical straight trajectories corresponding to the CR (dashed lines). That is, even for the same conversion level, the temperature rise will be higher in the adiabatic MR than in the adiabatic CR.

The membrane reactor analysis is extended to non-adiabatic conditions for two different flow configurations of the gas flowing through the shell: co- and counter-current. The already reported values for the sweep gas flowrate and the overall heat-transfer coefficient (see Table 1) were kept constant for all the simulations. Figure 4 presents the axial temperature evolutions in the MR for both flow configurations. For co-current operation, the driving force for heat transfer (T-T_P) is high enough at the reactor entrance to maintain the temperature rise in the reaction side below 40 degrees. For the second half of the reactor both temperatures tends to be the same and low convective heat transfer rates are observed. Regarding the counter-current scheme, as the sweep gas temperature is higher than the reaction temperature near the entrance, the sweep gas acts as a heating medium. Consequently, the reaction temperature rises up to much higher values than in the co-current case (around 70°C). This phenomenon may cause a faster catalyst deactivation. Near the reactor outlet the cooling effect of the sweep gas causes an increase in the reaction rates.

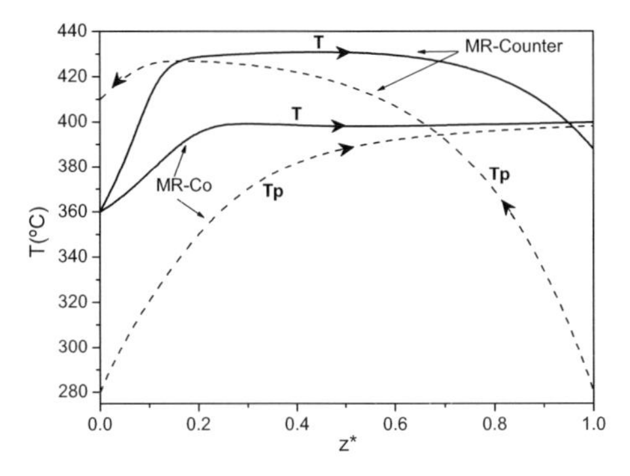

Figure 4: Axial temperature profiles of reaction (T) and permeation (T_P) sides in MR. T_o=360°C, $T_{P,inlet}$=280°C, $F_{SG,inlet}$ and U as reported in Table 1. Non-adiabatic conditions.

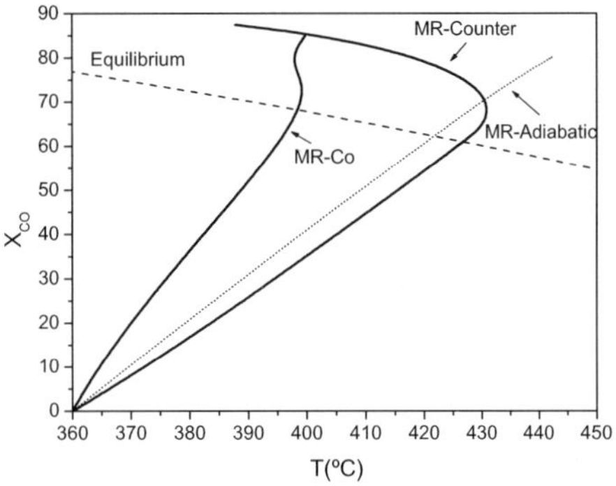

Figure 5: Conversion vs. Temperature trajectories in MR, for adiabatic and non-adiabatic operations (co- and counter-current schemes). Same conditions as in Figure 4.

Results in the phase-plane are presented in Figure 5. The CO outlet conversions under non-adiabatic conditions are around 85-90%, which determine CO levels in the order of 1-2% (dry basis), suitable to be fed to a PrOx reactor. When co-current configuration is analysed, a slight hot spot is observed. Besides, the increment in the CO conversion due to the permeation effects (shifting of the equilibrium) is the source of a slight temperature minimum (or cold spot). On the other hand, the hot spot detected in counter-current operation is considerable and the total temperature rise inside the reactor is similar to the adiabatic case. Under this flow configuration, the observed parametric sensitivity with respect to the sweep gas inlet temperature ($T_{P,inlet}$) was extremely high. This result may be a consequence of the selected sweep gas flowrate; higher sweep gas flowrates lead to lower sensitivities and flatter counter-current temperature profiles. However, higher power requirements are necessary and permeate streams with lower H_2 concentrations are obtained. For the studied operating condition, the co-current scheme appears as the most convenient configuration, as the desirable conversion level can be obtained with a temperature rise of about half of the other operations under analysis.

4. Conclusions

For adiabatic operation, the total temperature rise inside the reactor is higher for the MR due to the higher CO conversions achieved and the diminution of the total gas flowrate in the reaction side, as a consequence of the H_2 permeation. When non-adiabatic conditions are considered, the proper selection of the operating conditions for the shell side becomes a key factor to avoid undesirable temperature raises over the catalyst and high parametric sensitivity.

The previous results point out over the need to consider the heat effects in membrane reactors. Although the WGSR is moderately exothermic, the temperature changes inside the reactor are not negligible. The only reason to neglect the heat effects in MRs is the small scale of laboratory designs. When intermediate (or large) scales are under consideration, the temperature variations will affect the kinetics and the chemical equilibrium.

Regarding the selected application, the simulated non-adiabatic membrane reactors show an adequate performance in terms of reaching CO outlet concentrations of between 1 and 2%, suitable values to directly feed a final purification unit for fuel-cell applications.

5. Nomenclature

A_T = cross sectional area of tubes, m^2
C_{pj} = specific heat of component j, kJ/(mol K)
d_{te} = external tube diameter, m
F_j = molar flow of component j, mol/s
J_{H2} = permeation flow of hydrogen, mol/(s m^2)
L = tube length, m
N = number of components (reaction side)
n_t = number of tubes
p_{H2} = partial pressure of hydrogen, Pa
r_{CO} = reaction rate , mol_{CO}/(kg_{cat} s)
T = temperature, K
U = overall heat transfer coefficient, kJ/(s m^2 K)
W = catalyst mass, g
z = axial coordinate, m
δ = thickness of Pd film, m
ΔH_r = heat of reaction, kJ/mol
ρ_B = bed density, kg_{cat}/m^3

Subscripts
CO = carbon monoxide
H_2 = hydrogen
j = component j
L = at the axial coordinate $z=L$
P = at permeation side
SG = sweep gas
0 = at the axial coordinate $z=0$

6. References

[1] P. Giunta, N. Amadeo, M. Laborde, XIV Cong. Arg. Catal., Arg., (2005) 610.
[2] J. Coronas, J. Santamaría, Catal. Today, 51 (1999) 377.
[3] J. S. Oklany, K. Hou, R. Hughes, App. Catal. A, 170 (1998) 13.
[4] S. Tosti L. Bettinali, V. Violante, Int. J. Hyd. Energy, 25 (2000) 319.
[5] W.H. Lin, H.F. Chang, Surf. Coat. Tech., 194 (2005) 157.
[6] Criscuoli, A. Basile, E. Drioli, Catal. Today, 56 (2000) 53.
[7] M. Bratch, P. Alderliesten, R. Kloster, R. Pruschek, G. Haupt, E. Xue, J. Ross, M. Koukou, N. Papayannakos, En. Conv. Manag., 38 (1997) 159.
[8] M. Koukou, G. Chaloulou, N. Papayannakos, N. Markatos, Int. J. Heat and Mass Tranf., 40 (1997) 2407.
[9] N. Itoh, T. Wu, J. Memb. Sci., 124 (1997) 213.
[10] W. F. Podolski Y. G. ,Kim, Ind. Eng. Chem.,13(4) (1974) 415.
[12] G. Barbieri, A. Brunetti, T. Granato, P. Bernardo, E. Drioli, Ind. Eng. Chem. Res., 44 (2005) 7676.

Natural Gas Conversion VIII
F.B. Noronha, M. Schmal, E.F. Sousa-Aguiar (Editors)

Au/ZnO and the PROX REACTION.

Kátia R. Souza, Adriana F.F. de Lima, Fernanda F. de Sousa and Lucia Gorenstin Appel*

Instituto Nacional de Tecnologia / MCT, Av. Venezuela 82 / sala 518,CEP 20081-312, Rio de Janeiro, Brazil Fax: 55 21-21231165 E-mail: appel@uol.com.br

1. Introduction

Recently, there has been a growing interest in gold catalysts due to their potential use in many important reactions that can be employed in industrial and environmental processes. The most important application of these systems is the CO oxidation.
This occurs because preferential CO oxidation (PROX) is the cheapest way to reduce CO concentration in an H_2 stream used as a fuel cell feedstock (PEM).
Many gold catalysts have been studied for CO oxidation and some authors have suggested a very important role for the supports (1). In fact, Au/TiO_2 and Au/Fe_2O_3 are considered to be very active systems (2,3). However, it is well known that they are not stable in the presence of H_2O and CO_2.
According to some authors, Au/ZnO (4) shows good activity and stability for CO oxidation. Nevertheless, there is very little information available in relation to this catalyst in the literature (4-7). In a previous work, we discussed the synthesis of these systems. The results showed that anionic exchange provides small gold particles and also that synthesis conditions modify the gold concentration in the catalysts (8).
The present work aims at contributing to further investigation of gold based zinc oxide catalyst on the preferential CO oxidation (PROX).

2. Experimental

2.1. Preparation of Au/ZnO catalysts

An aqueous solution of $HAuCl_4$ (4.2 x 10^{-3} M) was heated up to 353 K and its pH was adjusted to 7 using one of three different types of solution: Na_2CO_3 (0.8 M) or NaOH (0.8M or 0.1M). The three samples obtained at the end of this procedure were named B1, B2, B3, respectively. The ZnO support was added to this gold solution (1 g per 100 mL of solution). The same amount of gold was used in all the preparations (Table 1). The resulting suspension was vigorously stirred and kept at 353 K for 5 hours. Finally, the precipitate was cooled, filtered, washed and dried at 393 K for 24 hours.

2.2. Characterization

The gold content of each sample was determined by gravimetric analysis. Au/ZnO was dissolved in chloridric acid, and the gold was precipitated using oxalic acid. X-ray diffraction was performed with a Rigaku Miniflex diffractometer model PW 1410, (CuKα, 0,15418nm). Two analysis conditions were used. The first applied an angular range varying from 5° to 90° with 0.02 $°s^{-1}$ and counting time of 1s per step and the second 37° to 40° with 0.02 $°s^{-1}$ and 10s per step. The last condition was used to calculate Au particle size applying Scherrer´s equation. Infrared spectra of the materials were collected with a Magna 560-Nicolet spectrometer using wafers containing 3% wt. of the sample in KBr.

2.3. Catalytic performance

Catalytic measurements of CO oxidation were carried out in a fixed-bed reactor. Before each run the catalysts were dried at 393 K during 1 h under N_2 (30 $NmLmin^{-1}$). The composition of the gas mixture used was 50% H_2, 2% CO and 2% O_2 balanced with N_2. The flow rate was 70 $NmLmin^{-1}$ and the temperature was 283K. The catalysts masses were around 44, 11 and 25 mg for B1, B2 and B3, respectively. However, the gold amount remained unchanged at $7x10^{-4}$ g_{Au} (Figure 1B). The catalytic bed was completed with 800 mg of SiC. The experimental conditions were selected aiming to test the catalysts under kinetic control, except for the 100% conversion run in which 200mg of B3 were used, the flow rate was 50 $NmLmin^{-1}$ and the temperature 313K (Figure 1A). Aiming to analyse the presence of carbonate species on B3, this sample was submitted to a thermal treatment (393 K during 1 h under N_2 at 30 $NmLmin^{-1}$) after having been previously tested during 16 h at 383K. The performance of this catalyst was observed once again after the thermal treatment (Figure 2B). In order to study the influence of ZnO in the PROX reaction, some of this oxide was used to replace SiC. Three different proportions of SiC /ZnO, i.e. 0/1, 1/1 and 1/0 were employed. For these analyses the total catalytic bed mass, the flow rate and the temperature was 800 mg, 70 $NmLmin^{-1}$ and 283K, respectively. The products and reagents were analysed by an Agilent GC-6880 gas chromatograph equipped with a FID/TCD. The selectivity and conversion values were calculated according to Mariño et al.(9)

3. Results and Discussion

Table 1 displays the main characteristics of the prepared catalysts. It can be observed that B2 shows the highest gold concentration. This probably occurred due to the dissolution of ZnO provoked by the NaOH solution which was used in the preparation. It can be inferred that, although the synthesis parameters changed the gold concentration of the Au/ZnO samples, the size of the gold particle was in most cases smaller than 3nm, as it can be seen by the XRD analysis.

Table 1 – Nominal gold (Au_{nom}, wt. %), real (Au_{real}, wt. %) concentrations values and gold size particles (d_{Au}, nm).

Sample	wt % $Au_{nom.}$	wt. % Au_{real}	d_{Au}[1]
B1	3.5	1.63	3
B2	3.5	9.59	<3
B3	3.5	2.89	<3

1 – Determined by Debye-Scherrer equation

Figure 1A displays activity and CO_2 selectivity of B3 catalyst at low space velocity. It can be inferred that this catalyst reaches high CO conversion, showing that it can clean H_2 stream up to 100 ppm. Considering that CO_2 selectivity is around 42%, one can also infer that some of the H_2 is consumed during the reaction.

Figure 1B displays CO conversion versus time for high space velocity runs. B1 and B2 show a very low CO oxidation rate, whereas a high activity and a high deactivation rate are observed for B3.

The infrared spectra of fresh Au/ZnO samples are displayed in Figure 2A. B2 and B3 show only very small bands whereas sample B1 presents absorption bands at 1513, 1401 and 1380 cm^{-1}, which can be attributed to the vibrational spectrum of carbonates (7). The high concentration of this species can be associated with the precipitant agent used during the preparation of B1 (Na_2CO_3). This suggests that the low catalytic activity observed for this sample is related to the carbonate species.

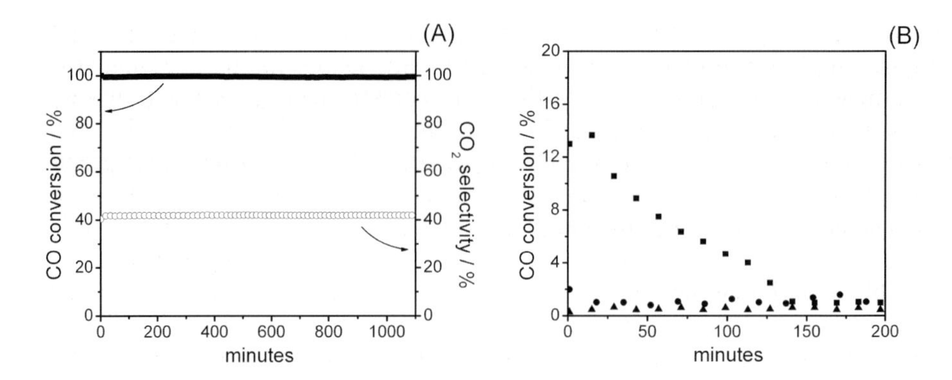

Figure 1 – (A) CO conversion and CO_2 selectivity at 313 K . Test conditions: 200 mg of B3, flow rate of 50 $NmLmin^{-1}$. (B) Catalytic results for samples B1, B2 and B3. Test conditions: $7x10^{-4}$ g_{Au}, flow rate of 70 $NmLmin^{-1}$, 283 K. The symbols ●, ▲ and ■ refers to B1, B2 and B3, respectively.

Figure 2B displays the B3 conversion for CO oxidation before and after thermal treatment, as described in the experimental section. As it can be seen, CO_2 is eliminated during thermal treatment (B region) restoring the CO conversion. This result shows that carbonate species might be the one responsible for the deactivation process. According to Manzoli et al. (7) carbonate species can be formed on gold and also on ZnO surface. Therefore, these species block the active sites of the catalytic surface.

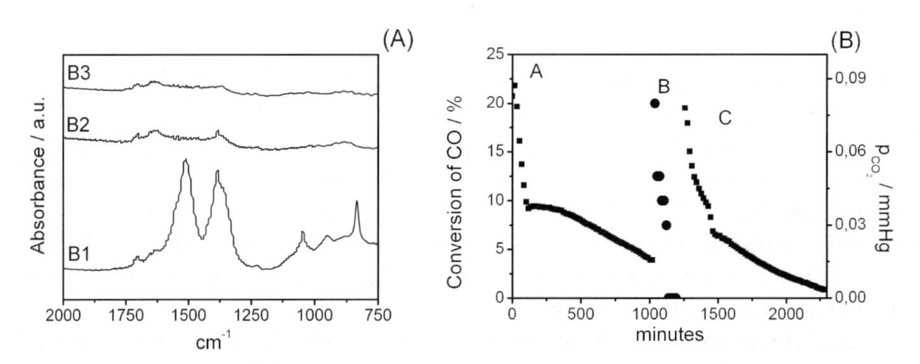

Figure 2 – (A) Infrared spectra of Au/ZnO samples. (B) Catalytic results for sample B3. Experimental conditions: 283K, 70 $NmLmin^{-1}$, catalyst mass 58 mg. Regions A, B and C before, during and after treatment (393 K during 1 h under N_2 at 30 NmLmin-1), respectively. The symbol ● refers to the CO_2 concentration analysed during the thermal treatment.

Taking into account that B3 is much more active than B2 (Figure 1B), the gold amount during each run is fixed and the gold concentration on the B2 sample is higher than on B3. One can suggest that the CO oxidation activity can be associated with the ZnO surface.
Figure 3 compares the activity of ZnO and B3 with the mixtures of B3 and ZnO. The mixtures were done replacing SiC by ZnO. It is worth stressing that all these runs employed the same amount of B3 catalyst. Although ZnO itself is not active for CO oxidation, it can be verified that the higher the ZnO concentration is in the mixture, the higher the activity, as it can also be observed in the same figure. This result is in line with the observations made in relation to catalysts B2 and B3, which suggested that ZnO in the presence of gold should show activity in this reaction. It can also be observed that the deactivation rate is the same for all the samples. Therefore, ZnO cannot be considered as a CO_2 trap.

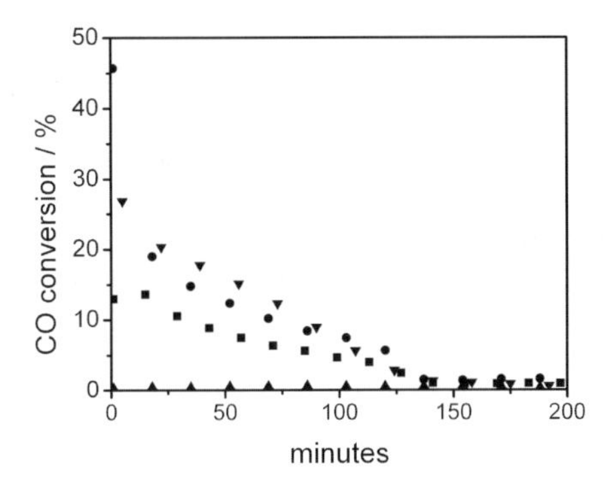

Figure 3 - Catalytic results for sample B3 and ZnO. The symbols▲, ▼, ●, ■ refer to CO conversion of ZnO and SiC; B3 and ZnO; B3, ZnO and SiC; B3 and SiC. Experimental conditions: 70 $NmLmin^{-1}$ and 283K.

Mechanisms for CO oxidation on gold catalysts have been proposed and discussed by many authors (10-12). The most accepted one suggested that CO adsorbs on gold particle and the reaction occurs between the hydroxyls on gold oxidized located in the interface between the gold particles and the support. Recently, Rossignol et al.(13) suggested a new mechanism that can be described by the following steps: (a) adsorption of CO and H_2 and dissociation of H_2 on a gold particle, (b) reaction of gas phase O_2 with adsorbed H atom and (c) reaction of the resulting oxidizing species with adsorbed CO to give CO_2. Taking into consideration that CO adsorbs on ZnO surface (14) and the results shown above, which are related to the increase of activity after adding ZnO to the catalyst bed, it can be suggested that hydrogen can spillover on the ZnO surface and steps (b) and (c) occur on this oxide.

In conclusion, it could be observed that Au/ZnO deactivation process can be related to carbonate species and ZnO in the presence of gold shows activity for preferential CO oxidation.

4. References.

1. J.M.C. Soares and M. Bowker, Appl.Catal. A: Gen., 291 (2005) 136.
2. R. Zanella, S. Giorgio, C.-H. Shin, C.R. Henry, C. Louis, J. Catal., 222 (2004) 357.
3. M.M. Schubert, A. Venugopal, M.J. Kahlich, V. Plzak, R.J. Behm, J. Catal. 222 (2004) 32.
4. G.Y. Wang, W.X. Zhang, H.L. Lian, Da Z. Jiang, T.H. Wu, Appl. Catal. A: Gen. 239 (2003) 1.
5. J. E. Bailie, G. J. Hutchings, Chem. Commun, (1999) 2151.
6. J. E. Bailie; H. A. Abdullah; J. A. Anderson; Colin H. Rochester; N. V. Richardson; N. Hodge; J-G Zhang; A. Burrows; C. J. Kielyd; G. J. Hutchings, Phys. Chem. Chem. Phys., 3 (2001) 4113.
7. M. Manzoli; A. Chiorino; F. Boccuzzi; Appl. Catal. B Environm. 52 (2004) 259.
8. K.R. Souza, A.F.F. de Lima, F.F. de Sousa, L.G. Appel, Catal. Today, *submitted.*
9. F. Mariño, C. Descorme, D. Duprez, Appl. Catal. B Environm, 54 (2004) 59.
10. S.T. Daniells, M. Makkee, J.A. Moulijn, Catal. Lett. 100 (2005) 39.
11. G.C. Bond and D.T. Thompson, Gold Bulletin, 33 (2000) 41.
12. C.K. Costello, J.H. Yang, H.Y. Lawa, Y. Wangb, J.-N. Lin, L.D. Marks, M.C. Kung, H.H. Kung, Appl. Catal. A: Gen. 243 (2003) 15.
13. C. Rossignol, S. Arrii, F. Morfin, L. Piccolo, V. Caps, J-L Rousset,J. Catal. 230 (2005) 476.
14. Th. Becker, M. Kunat, Ch. Boas, U. Burghaus,Ch. Wo, J. Chem. Phys.,113 (2000) 6334.

Natural Gas Conversion VIII
F.B. Noronha, M. Schmal, E.F. Sousa-Aguiar (Editors)

Selective Catalytic Oxidation of CO in H_2 over Copper-Exchanged Zeolites

Ivana Tachard [a], Anna C. B. da Silva [a], Fábio Argolo [a], Suzana M. O. Brito [b], Heloise O. Pastore [c], Heloysa M. C. Andrade [a], Artur J. S. Mascarenhas [a, *].

[a] *Laboratório de Catálise e Materiais, Departamento de Química Geral e Inorgânica, Instituto de Química, Universidade Federal da Bahia – UFBA , Salvador – BA, Brasil.*
[*] *Corresponding author: artur@ufba.br*
[b] *Laboratório de Catálise e Adsorção, Departamento de Ciências Exatas, Universidade Estadual de Feira de Santana – UEFS, Feira de Santana – BA, Brasil.*
[c] *Grupo de Peneiras Moleculares Micro- e Mesoporosas, Instituto de Química, Universidade Estadual de Campinas – UNICAMP, Campinas – SP, Brasil.*

Abstract

Copper-exchanged zeolites were investigated as catalysts for the selective catalytic oxidation of CO in H_2 rich feeds. Three different zeolitic topologies were chosen: MFI, MWW and FAU. Under dry conditions the following order of activity in conversion of CO to CO_2 was observed: Cu-ZSM-5 > Cu-Y > Cu-MCM-22. However, under wet conditions ($H_2O/CO = 3.0$) and stoichiometric CO/O_2 ratio of 1:0.5, Cu-ZSM-5 has presented a better performance than commercial 0.5 wt% Pt/Al_2O_3. By-products, such as methane or methanol, were not detected in the reaction effluents.

Keywords: selective oxidation, hydrogen purification, copper-exchanged zeolites

1. Introduction

The development of fuel cells has renewed the interest on hydrogen production and purification processes. Hydrogen is usually produced by methane reforming, followed by water gas shift reaction in view of purifying the feed. However, 0.5 to 1 vol. % of carbon monoxide is still present after the LTS

reactor [1]. This CO concentration is enough to poison the platinum anode of the polymer electrolyte membrane fuel cell (PEMFC) becoming necessary to reduce it to trace level [2].
Selective catalytic oxidation of CO in a hydrogen rich flow seems to be an alternative technology for purification of hydrogen for fuel cell applications. Many catalysts has been proposed to this aim, such as supported noble metals [1 – 7] and mixed metal oxides [8 – 12], but until now a catalyst that combines high activity at low temperatures, stability in long time runs, and resistance to H_2O and/or CO_2 is unknown. Recently, we have reported Cu-ZSM-5 and zinc promoted Cu-ZSM-5 catalyst as active, selective and water resistant for CO oxidation in the presence of hydrogen, but temperature of maximum activity is still high for fuel cell applications [13].
The aim of this work was investigate the influence of zeolite topology on the activity of copper exchanged zeolites in selective oxidation of CO in H_2.

2. Experimental

MCM-22 zeolite (IZA code: MWW) was chosen because in its structure there is a sinusoidal channel system, whose inner diameter is 4,0 x 5,9 Å, very similar to that found in ZSM-5 structure (IZA code: MFI). However, another independent channel system constituted by supercages of 7,1 x 7,1 x 18,2 Å is present at the MCM-22 structure [14]. Y zeolite (IZA code: FAU) was also investigated in order to compare the activity of copper sites in a more extensive environment.
Zeolite ZSM-5 (SiO_2/Al_2O_3 = 30) was synthesized by the IZA standard method [15]. The obtained gel was hydrothermally treated at 180°C for 3 days. The product was filtered, washed with distilled water, dried at 80°C for 24 h and then calcined at 500°C for 3 h under flowing synthetic air. Zeolite MCM-22 (SiO_2/Al_2O_3 = 30) was prepared by hydrothermal treatment of a gel of composition $4,5Na_2O{:}30SiO_2{:}Al_2O_3{:}18HMI{:}900H_2O$ (HMI = hexamethyleneimine) at 150°C for 10 days under non-agitated conditions [16]. The product was filtered, washed with distilled water, dried at 80°C for 24 h and then calcined at 580°C for 6 h, under flowing synthetic air. Dealuminated Y zeolite (SiO_2/Al_2O_3 = 27) was commercially obtained from Wessalith.
Catalysts were prepared by cation exchange procedures of Cu(II) solutions with the respective zeolite using molar ratio Cu/Al = 1 . The final pH was adjusted with NH_4OH to 7.5, in order to favor the formation of over-exchanged zeolites. The copper-exchanged zeolites were filtered, washed with deionized water, dried at 70°C for 8 h and calcined at 350°C for 3 h, under flowing synthetic air.
X-ray powder diffraction patterns were collected in a Shimadzu XRD 6000. Fourier transformed infrared spectra were collected in a Perkin Elmer Spectrum BX. Elemental analyses were performed by ICP-OES using a Perkin-Elmer 300-DV. Temperature programmed reduction with H_2 was performed using a

homebuilt device, with experimental parameters adjusted in order of avoiding temperature artifacts [17]. FTIR spectra of adsorbed CO were collected in situ using self-supported wafers in a vacuum cell, which were evacuated and submitted to 1.5 x 10^{-4} to 35 Torr of CO.

Activity testing of powdered catalysts was accomplished using a continuous flow microreactor coupled with a gas chromatograph (Perkin Elmer Clarus 500), operating with a Carboxen 1010 capillary column. Feed gas contained 1.0 vol.% CO, 1.0 vol.% O_2, 70 vol.% H_2 and helium balanced at GHSV = 30,000 h^{-1}. Cu-ZSM-5 catalyst was also evaluated using molar ratio $CO:O_2:H_2$ = 1:0.5:10 (helium balance) at GHSV = 15,000 h^{-1}. Activities were compared to those obtained with a commercial Pt/Al_2O_3 catalyst (Aldrich, 0.5% Pt).

3. Results and discussion

The synthesized samples have presented the X-ray diffraction patterns characteristics of MFI and MWW zeolites [18], respectively, as shown in Figure 1. In the case of MCM-22 zeolite, the diffraction pattern changes drastically under calcination, as expected on the basis of condensation of MCM-22 lamellar precursor in a three-dimensional structure [16]. Cation exchange procedures do not affect the zeolite structures, but peaks at 34.5° e 38.4° 2θ indicated the formation of segregated copper(II) oxide in the Cu-ZSM-5 and Cu-Y catalysts.

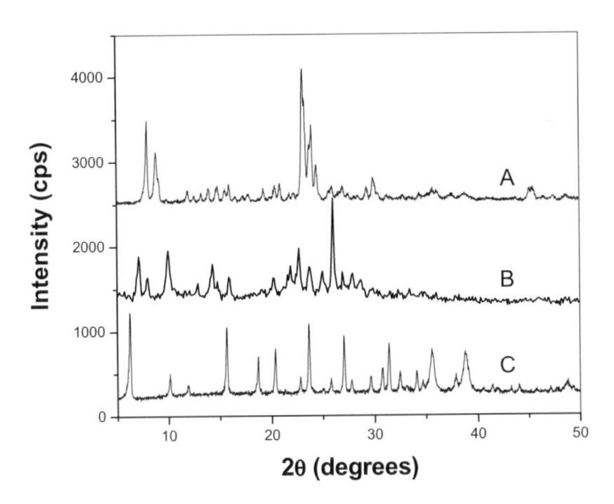

Figure 1. X-ray powder diffractograms of copper-exchanged zeolites: A. Cu-ZSM-5; B. Cu-MCM-22 and C. Cu-Y.

Sample composition and BET surface areas are shown in Table 1. Elemental analyses indicate that over-exchanged zeolites were formed (Cu/Al ≈ 1.0).

Significant decrease in the BET surface areas were observed between calcined and copper-exchanged zeolites, indicating that some pore blocking is occurring.

Table 1. Elemental analysis and BET surface areas for copper-exchanged zeolites.

Sample	SiO_2/Al_2O_3	Cu/Al	S_{BET} ($m^2.g^{-1}$)
Cu-ZSM-5	30	1.00	296
Cu-MCM-22	39	0.97	428
Cu-Y (Wessalith)	27	1.00	594

For Cu-ZSM-5 and Cu-Y catalysts, two overlapped reduction peaks were observed at 185°C e 210 – 230°C [19]. The peak at lower temperatures can be attributed to CuO deposited in the channels (MFI) or cages (FAU). External surface deposits cannot be discarded. The peak at higher temperatures can be assigned to cationic oligomeric species, such as $[Cu - O - Cu]^{2+}$, identified in the literature by other characterization techniques such as XPS, EXAFS and FTIR of adsorbed probe molecules (NO and CO) [20]. In Cu-ZSM-5 a peak at 368°C, attributed to Cu(II) isolated ions was also observed. For Cu-MCM-22 sample, three peaks at 230, 260 e 360°C were observed. These peaks were attributed as follows: i) reduction of cationic oligomeric species in MWW cages; ii) reduction of cationic oligomeric species in the sinusoidal channel; and iii) Cu(II) isolated in the cation exchanged sites in the supercage and/or sinusoidal channels. For all samples molar ratio H_2/Cu is lower than 1.0, suggesting that Cu(I) is also present in these catalysts.

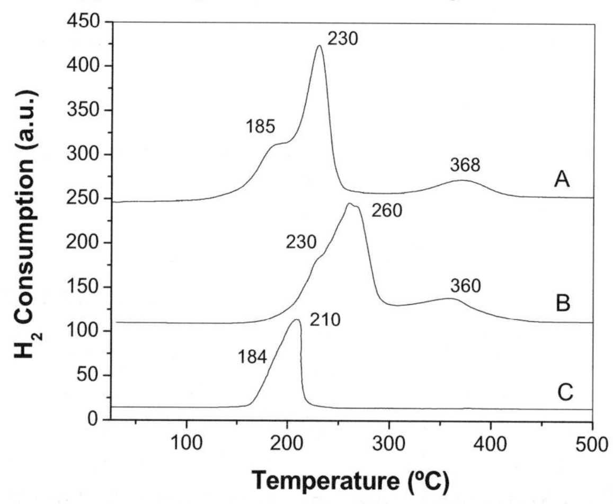

Figure 2. Temperature programmed reduction of copper-exchanged zeolites: (A) Cu-ZSM-5; (B) Cu-MCM-22 e (C) Cu-Y.

The presence of copper(I) was confirmed by FTIR spectra of adsorbed CO. For Cu-MCM-22 bands at 2178 and 2151 cm^{-1} were attributed to symmetric and asymmetric dicarbonylcopper(I) complexes, Cu(I) – $(CO)_2$, while the band at 2158 cm^{-1} was assigned to monocarbonylcopper(I) complexes, Cu(I) – (CO) [20]. At low pressures of CO, only the band at 2158 cm^{-1} was observed and presented an asymmetry at lower wavenumbers, suggesting the presence of small copper oxides aggregates, coherent with the cationic oligomeric species indicated by TPR. For Cu-ZSM-5, an additional band at 2135 cm^{-1} was also observed. This band is usually attributed to CO adsorbed on Cu(I) ions at different coordination in the MFI structure. For Cu-Y, the presence of different copper sites in FAU structure generates bands at 2135, 2145 and 2160 cm^{-1}. These results are in good agreement with literature data.

Under GHSV = 30,000 h^{-1} and absence of water vapor, all copper-exchanged zeolites have presented low CO conversions, Figure 3.a, allowing a better comparison between activities of the copper-exchanged zeolite catalysts.

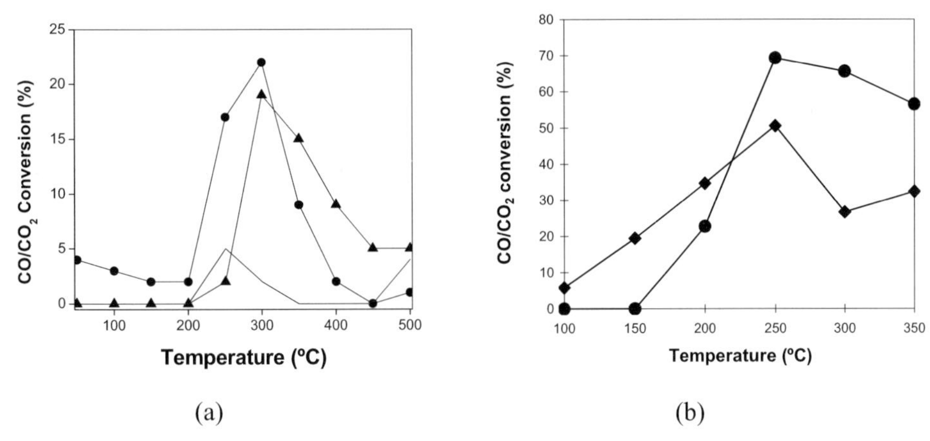

Figure 3. Conversion of CO in CO_2 over: (a) ●, Cu-ZSM-5; ■, Cu-MCM-22 and ▲, Cu-Y under GHSV = 30,000 h^{-1} and absence of water vapor; (b) ●, Cu-ZSM-5 and ♦, 0.5% Pt/Al_2O_3 under GHSV = 15,000 h^{-1} and H_2O/CO = 3.

The CO conversion increases with temperature, reaches a maximum between 200 – 300°C, and decreases at higher temperatures. The following activity order was observed: Cu-ZSM-5 (22%) > Cu-Y (19%) > Cu-MCM-22 (5%), indicating a good correlation with copper ions reducibility in these zeolite topologies. Cu-MCM-22 presented higher H_2 conversion (10%), while for Cu-Y and Cu-ZSM-5 the conversion of hydrogen do not exceed 5%. Oxygen conversion was complete over 250°C, suggesting that part of the oxygen was consumed by copper sites redox cycle. By-products, such as methane or methanol, were not detected in the reaction effluents.

Under GHSV = 15,000 h^{-1}, using H_2O/CO = 3.0 and stoichiometric ratio $CO:O_2$ = 1:0.5, Cu-ZSM-5 reached 70% of CO conversion to CO_2, while commercial

0.5 wt% Pt/Al_2O_3 resulted in 50% of conversion under the same conditions (Figure 3.b).

4. Conclusions

Copper-exchanged zeolites are active and selective for CO oxidation in the presence of H_2. Cu-ZSM-5 is more active than Cu-Y or Cu-MCM-22 catalysts, despite the similarities between structures MWW and MFI. The influence of zeolite topology on the copper sites activities needs to be deeply investigated. These results are still under the expectations for an adequate catalyst for fuel cell applications. New developments are necessary in order to obtain higher conversions at lower temperatures and long term stabilities.

5. References

1. M. L. Brown Jr., A. W. Green, G. Cohn, H. C. Andersen. Ind. Eng. Chem. 52 (1960) 841 – 844.
2. H. Igarashi, H. Uchida, M. Suzuki, Y. Sasaki, M. Watanabe. Appl. Catal. A 159 (1997) 159 – 169.
3. H. Hasegawa, K. Kusakabe, S. Morooka. J. Memb. Sci. 190 (2001) 1 – 8.
4. M. Watanabe, H. Uchida, K. Okhubo, H. Igarashi. Appl. Catl. B 46 (2003) 595 – 600.
5. I. H. Son, A. M. Lane, D. T. Johnson. J. Power Sources 124 (2003) 415 – 419.
6. P. V. Snytnikov, V. A. Sobyanin, V. D. Belyaev, P. G. Tsyrulnikov, N. B. Shitova, D. A. Shlyapin. Appl. Catal. A 239 (2003) 149 – 156.
7. A. Sirijaruphan, J. G. Goodwin Jr., R. W. Rice. J. Catal. 221 (2004) 288 – 293.
8. G. Avgouropoulos, T. Ioannides, H. K. Matralis, J. Batista, S. Hocevar. Catal. Lett. 73 (2001) 33 – 39.
9. G. Avgouropoulos, T. Ioannides, Ch. Papadopoulou, J. Batista, S. Hocevar, H. K. Matralis. Catal. Today 75 (2002) 157 – 167.
10. C. R. Jung, J. Han, S. W. Nam, T.-H. Lim, S.-A. Hong, H.-I. Lee. Catal. Today 93 – 95 (2004) 183 – 190.
11. Y. Liu, Q. Fu, M. F. Stephanopoulos. Catal. Today 93 – 95 (2004) 241 – 246.
12. J. W. Park, J. H. Jeong, W. L. Yoon, Y. W. Rhee. J. Power Sources 132 (2004) 18 – 28.
13. T. R. O. Souza, A. J. S. Mascarenhas, H. M. C. Andrade. React. Kinet. Catal. Lett. 87 (2006) 3 – 9.
14. M. E. Leonowicz, J. A. Lawton, S. L. Lawton, M. K. Rubin. Science 264 (1994) 1910.
15. H. Robson (ed.). Verified Syntheses of Zeolitic Materials. 2. ed. Amsterdam: Elsevier, 2001.
16. A. L. S. Marques, J. L. F. Monteiro, H. O. Pastore. Microp. Mesop. Mater. 32 (1999) 131.
17. M. Nele, E. L. Moreno, H. M. C. Andrade. Quim. Nova 29 (2006) 641 – 645.
18. Ch. Baerlocher, W. M. Meier, D. H. Olson. Atlas of Zeolite Framework Types. 5. ed. Amsterdam: Elsevier, 2001.
19. A. J. S. Mascarenhas, H. M. C. Andrade. React. Kinet. Catal. Lett. 64 (1998) 215.
20. A. Frache, M. Cadoni, C. Bisio, L. Marchese, A. J. S. Mascarenhas, H. O. Pastore. Langmuir 18 (2002) 6875.

Natural Gas Conversion VIII
F.B. Noronha, M. Schmal, E.F. Sousa-Aguiar (Editors)

ZrO_2-supported Pt Catalysts for Water Gas Shift Reaction and Their Non-Pyrophoric Property

Hyun Chul Lee,[a,*] Doohwan Lee,[a] Ok Young Lim,[a] Soonho Kim,[a] Yong Tae Kim,[b] Eun-Yong Ko,[b] Eun Duck Park[b,*]

[a]*Energy & Environment Laboratory, Samsung Advanced Institute of Technology, Suwon, 440-600 The Republic of Korea.*
[b]*Division of Energy Systems Research and Division of Chemical Engineering and Materials Engineering, Ajou University, Wonchun-Dong, Yeongtong-Gu, Suwon, 443-749, The Republic of Korea.*

Zirconia-supported platinum catalysts doped with cerium, nickel or cobalt (Pt-Ce/ZrO_2, Pt-Ni/ZrO_2, Pt-Co/ZrO_2) were prepared by a single step co-impregnation method and their catalytic performance on the water gas shift reaction (WGS) were investigated. A Pt/Ce/ZrO_2 catalyst was also prepared impregnating ZrO_2 sequentially with Pt and Ce in order to compare the effects of impregnation methods on the catalytic activities of the final form of the catalysts. The results showed that the co-impregnated Pt-Ce/ZrO_2 catalyst was highly active for the WGS reaction at low temperatures (250 °C, CO conversion of 95.1%, feed composition of 10 vol% CO, 10 vol% CO_2, and 80 vol% H_2 in dry base gas, H_2O/CO=6, GHSV= 6000 h^{-1}) and kept its activity even after air exposure in every cyclic test. The temperature programmed reduction (TPR) pattern of Pt-Ce/ZrO_2 indicated that the low reduction temperature of Pt induced by Ce gave rise to high activity on the WGS reaction. This Pt-Ce/ZrO_2 catalyst can be considered as an efficient catalyst for the WGS reaction, where high catalytic activity and non-pyrophoric property were required.

1. Introduction

Proton exchange membrane fuel cell (PEMFC) is efficient and environment-friendly system for generation of electricity by the electrochemical reactions of H_2 and O_2. The fuel for the PEMFC, hydrogen, can be obtained relatively easily by catalytic reforming of a variety of hydrocarbons such as natural gas, LPG, and butane. However, removal of CO, the reforming co-product which poison the fuel cell electrodes , below several to tens of ppm levels in the H_2 fuel still requires complicated procedures with the WGS reaction and subsequent preferential CO oxidation or methanation reaction. The WGS reaction (1), which is an also important process for several industrial applications [2], produces additional hydrogen from the reformed fuel.

$$CO\ (g) + H_2O\ (g) \leftrightarrow CO_2\ (g) + H_2\ (g),\ \Delta H = -41\ kJ/mol \qquad (1)$$

In general, this CO conversion is practiced in two stage using high temperature shift (HTS, ~ 400 °C) and low temperature shift (LTS, around 200 °C) catalysts in industry. In particular, the WGS reaction thermodynamics favors low temperatures for high conversion of CO to CO_2, but the limited activities of the conventional WGS catalysts led to CO concentration far above equilibrium level at low temperatures. For the fuel cell applications, the conventional WGS catalysts would not meet the requirements such as low pressure operation, small scale application, activity and stability in fuel cell duty cycles. Moreover, the pyrophoric behavior of commercial LTS catalyst (CuZn catalyst) restricts adoption of the commercial catalysts in fuel cell applications [3]. W.Ruettinger et al. [3] reported a base metal non-pyrophoric alternative to commercial CuZn catalyst. Many recent studies reported highly active Pt catalysts for the WGS reaction [4-10]. G. Grubert et al. [10] reported Cu-free non-pyrophoric catalysts discovered by a high-throughput experimentation technique. These catalysts mainly consisted of mixed oxides of Cr or Fe along with Mn and Pt supported on ZrO_2.

In this study, we prepared ZrO_2-supported Pt catalysts promoted with Ce, Ni or Co applying co-impregnation and sequential impregnation methods. The Pt-Ce/ZrO_2 catalyst prepared by co-impregnation of Pt and Ce showed a high catalytic activity for the WGS reaction in low temperatures and stable in air exposure condition which can occur during frequent on-off of fuel cell systems. This catalyst is an efficient WGS catalyst for the application of residential fuel cell systems, where high catalytic activity and non-pyrophoric behavior is particularly important.

2. Experimental

All the catalysts (Pt-Ce/ZrO_2, Pt-Ni/ZrO_2, Pt-Co/ZrO_2) were prepared by a single step co-impregnation method. An aqueous solution of $Pt(NH_3)_4(NO_3)_2$ and nitrate form of a promoter component (Ce, Ni or Co) was well mixed at ambient temperature for 10 min, then ZrO_2 support (Aldrich) was added. This mixture was stirred at 60 °C for 6 h, and the solution was evaporated. Finally, the resulting product was dried at 110 °C for 12 h and calcined at 500 °C for 4 h in dry air. Differently, a Pt/Ce/ZrO_2 catalyst was prepared impregnating the Ce nitrate and the Pt nitrate precursor subsequently. Basically, ZrO_2 was impregnated with Ce nitrate precursor at first, and dried and calcined at the same condition described above. This resulting product was then impregnated with a Pt nitrate precursor, and dried and calcined again at the same condition described above. The loading amount of Pt was adjusted to be 1.0 wt% for all the catalysts. The molar ratio of promoter (Ce, Ni or Co) to Pt was fixed at 10. The catalysts were reduced with hydrogen at 400 °C for 1 h before the reaction.

Catalytic activity tests for WGS reaction were carried out with a standard gas of 10 vol% CO, 10 vol% CO_2 and 80 vol% H_2 in dry based gas composition. The amount of H_2O added in the reactant feed was adjusted giving steam/CO molar ratio at 6.0. The gas hourly space velocity (GHSV) was fixed at 6000 (h^{-1}). The reactant gas was fed to a reactor at atmospheric pressure and the catalytic activity was measured with increasing temperature from 150 °C to 350 °C at a ramping rate of 1 °C/min. The pyrophoricity test was conducted by purging the reactor with air at 150 °C, as described in literature [3]. After completion of each cyclic reaction (150 - 350 - 150 °C), the reactor was purged with N_2 at 150 °C, then air was fed at 150 °C for 30 min with a flow rate of 100 ml/min, and the reactant was fed again to the reactor raising temperature from 150 °C to 350 °C at a ramping rate of 1 °C/min.

The effluent gas composition was analyzed using an online gas analyzer (NGA2000, MLT4, Rosemount) equipped with CO, CO_2, H_2 and CH_4 detectors. Characteristics of the prepared catalysts were investigated with CO chemisorption and Temperature Programmed Reduction (TPR) results obtained a volumetric unit (Autochem 2910, Micromeritics). The TPR was conducted over 0.2 g sample in a 10 vol% H_2/Ar stream raising temperature from 25 °C to 600 °C at a heating rate of 10 °C/min after the samples were calcined at 500 °C for 1 h.

3. Results and Discussion

Table 1 shows catalytic performance of the prepared WGS catalysts at 250 °C for the feed composition of 10 vol% CO, 10 vol% CO_2 and 80 vol% H_2 in dry based gas composition (H_2O/CO= 6, GHSV= 6000 h^{-1}). All catalyst showed no methanation reaction under the reaction condition. As concentration of CO decreased, concentration of CO_2 increased, while H_2 concentration changed a little. The thermodynamic equilibrium CO concentration was calculated for the given reactant composition and the result was also displayed in Table 1. An addition of Ni or Co with Pt on ZrO_2 support resulted in a lower conversion of CO than on the un-promoted Pt/ZrO_2. Conversely, Pt-Ce/ZrO_2 catalyst prepared by a single step co-impregnation method showed almost an order of magnitude higher CO conversion than Pt/ZrO_2 without methanation reaction ($CO + 3H_2 \rightarrow CH_4 + H_2O$, $CO_2 + 4H_2 \rightarrow CH_4 + 2H_2O$).

Table 1.CO conversion(Xco, %) for the prepared catalysts and gas composition (%) in the products (Xco* : thermodynamic equilibrium CO conversion at 250 °C with the reactant composition)

Catalyst	Steam to CO (S/CO) = 6.0						
	Temp. (°C)	Xco* (equilibrium)	Xco (catalytic)	CO	H_2	CO_2	CH_4
Pt/ZrO_2	250	96.3	11.4	9.1	79.6	11.4	0
$Pt\text{-}Ce/ZrO_2$	250	96.3	95.0	0.5	80.6	19.0	0
$Pt\text{-}Ni/ZrO_2$	250	96.3	10.1	9.2	79.6	11.2	0
$Pt\text{-}Co/ZrO_2$	250	96.3	8.7	9.4	79.4	11.2	0
$Pt/Ce/ZrO_2$	250	96.3	79.0	2.1	80.4	17.6	0

Effect of the preparation sequence on the catalytic activity was also compared in Table 1. The catalyst prepared by a single step co-impregnation ($Pt\text{-}Ce/ZrO_2$) was more active for the WGS reaction than the catalysts prepared by a sequential impregnation of Ce followed by Pt on ZrO_2 support ($Pt/Ce/ZrO_2$). Markedly, on $Pt\text{-}Ce/ZrO_2$ catalyst CO could be reduced below 0.5 % at low temperature (250 °C), which reflected much higher CO conversion activity compared to the Cu-free non-pyrophoric catalytic material (Pt-Cr-Mn-Li-Cs supported on ZrO_2) reported in ref [10] in which maximum CO conversion was about 55 % at the similar reaction condition (250 °C, 3% CO, 14% CO_2, 37% H_2, 23% H_2O, Ar balance, GHSV= 3000 h^{-1}). Notice that the experimental CO conversion on the $Pt\text{-}Ce/ZrO_2$ was close to the equilibrium CO conversion level, also reflecting an excellent catalytic activity of the catalysts obtained in this study.

TPR was conducted to find out the interaction between platinum and metal oxides over ZrO_2 supports. The results were displayed in Fig. 1. It is clear from the results shown in Fig. 1(a) that the addition of cerium in the preparation of zirconia-supported Pt catalysts ($Pt\text{-}Ce/ZrO_2$) led to much easier reduction of Pt species at low temperatures compared with those of Pt-Co and Pt-Ni catalysts. As revealed in Fig. 1(b), zirconia support did not show any reduction peak. When Pt was impregnated on ZrO_2, two broad peaks centered at 210 °C and 380 °C were observed. The peak at 210 °C could be assigned to the reduction of platinum oxide interacting weakly with the support and the peak at 380 °C could be assigned to the reduction of platinum oxide interacting strongly interaction with the support. Zirconia-supported cerium showed one reduction peak at 470 °C (Fig. 2(b)), due to the reduction of Ce^{4+} species. The $Pt\text{-}Ce/ZrO_2$ catalyst showed three reduction peaks, where the peak at 450 °C could be assigned to the reduction of CeO_2 while the peak at 290 °C and 100 °C could be assigned to the reduction of isolated Pt clusters and Pt species in strong interaction with Ce, respectively [8]. The sequence of Pt and Ce impregnation also affected on the reduction temperature as shown in Fig. 1 (b). The Pt-Ce/ZrO_2 catalyst prepared by co-impregnation of Pt and Ce precursors showed a low reduction temperature of Pt-Ce species at 100 °C, while the Pt/Ce/ZrO_2 catalyst, prepared sequential impregnation, showed a rather higher reduction temperature of Pt at 190 °C. Thus, such an early reduction of Pt and Ce interacting strongly each other seems to give rise to high activity in WGS reaction.

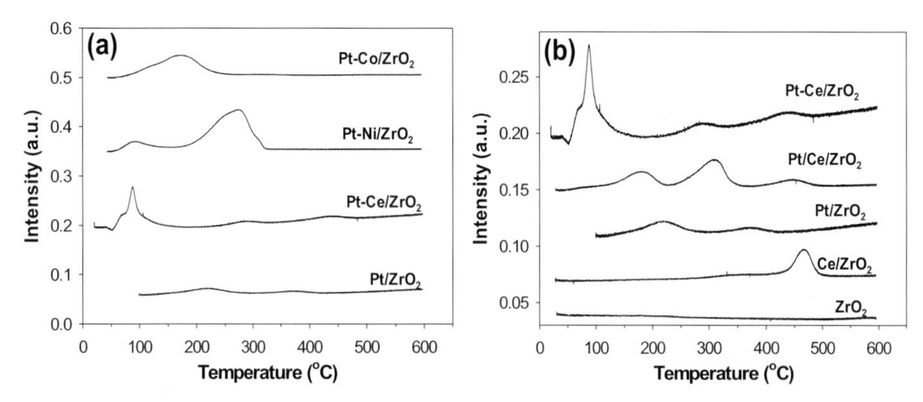

Fig. 1. TPR patterns for the prepared catalysts. (a) TPR patterns for Pt/ZrO_2 and other metal (Ce, Ni or Co) promoted zirconia-supported Pt catalysts, (b) TPR patterns for the ZrO_2 support , zirconia-supported Pt (or Ce) catalyst and (Pt, Ce)-supported on ZrO_2 with different sequence for the preparation of zirconia-supported catalysts

In order to investigate stability of Pt-Ce/ZrO_2 catalyst against exposure to air, an air flow (100 ml/min) was introduced at 150 °C for 30 min during thermal cyclic operation (from 150 °C to 350 °C). Fig. 2 shows catalytic activity after air exposure, measured at 250 °C in every cyclic test. The Pt-Ce/ZrO_2 catalyst maintained its high activity even after air exposure. Briefly, the Pt-Ce/ZrO_2 catalyst showed the high activity of WGS keeping 95 % of CO conversion and CO concentration below 0.5% through all thermal cycles including air supply step. Thus, Pt-Ce/ZrO_2 catalyst would be regarded as an excellent WGS catalyst having non-pyrophoric property.

In conclusion, Zirconia-supported platinum catalysts doped with cerium, nickel or cobalt (Pt-Ce/ZrO_2, Pt-Ni/ZrO_2, Pt-Co/ZrO_2) were prepared by a single step co-impregnation method and their catalytic performance in water gas shift reaction are compared. Pt-Ce/ZrO_2 catalyst prepared co-impregnation method is high active at low temperature (250 °C, CO conversion of 95.1%, feed composition of 10 vol% CO, 10 vol% CO_2, 80 vol% H_2 in dry base gas, H_2O/CO=6, GHSV= 6000 h^{-1}) and stable in air exposure. The low reduction temperature of Pt with Ce in Pt-Ce/ZrO_2 seems to give rise to high activity in water gas shift reaction. Thus, this catalyst would be considered as an efficient catalyst for the WGS reaction, where high catalytic activity and non-pyrophoric behavior is required.

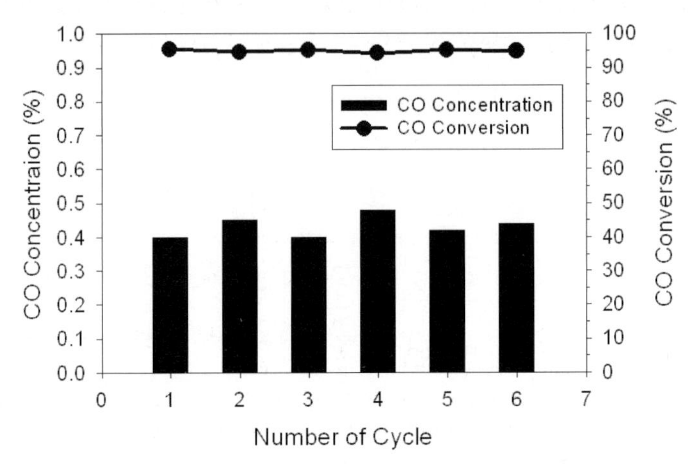

Fig. 2. Thermal cyclic test of Pt-Ce/ZrO_2 catalysts for WGS reaction. Reaction conditions: thermal cyclic temperature range from 150 °C to 350 °C, temperature for air supply = 150 °C, gas composition : 10 vol% CO, 10 vol% CO_2 and 80 vol% H_2 in dry based gas, steam to CO ratio 6.0, GHSV=6000 (h^{-1}).

References

1. C. Song, *Catal. Today*, **2002**, *77*, 17.
2. M. I. Temkin, *Adv. Catal.,* **1979**, *28*, 263
3. W. Ruettinger, O. Ilinich and R. Farrauto, *J. Power. Source*, **2003**, *118*, 61.
4. N. Schumacher, A. Boisen, S. Dahl, A. A. Gokhale, S. Kandoi, L. C. Grabow, J. A. Dumesic, M. Mavrikakis and I. Chorkendorff, *J. Catal.*, **2005**, *229*, 265.
5. S. Y. Choung, M. ferrandon and T. Krause, *Catal. Today*, **2005**, *99*, 257.
6. W. Ruettinger, X. Liu and R. Farrauto, *Appl. Catal. B*, **2006**, *65*,135.
7. A. Luengnaruemitchai, S. Osuwan and E. Gulari, *Catal Commun.*, **2003**, 4, 215.
8. P. S. Querino, J. R. C. Bispo and M. C. Rangel, *Catal. Today*, **2005**, *107-108*, 920.
9. T. Shishido, M. Yamanoto, D. Li, Y. Tian, H. Morioka, M Honda, T. Sano and K. Takehira, *Appl. Catal. A*, **2006**, *303*, 62.
10. G. Grubert, S. Kolf, M. Baerns, I. Vauthey, D. Farrusseng, A.C. Veen, C. Mirodatos, E. R. Stobbe and P. D. Cobden, *Appl. Catal. A*, **2006**, *306*, 17.

Natural Gas Conversion VIII
F.B. Noronha, M. Schmal, E.F. Sousa-Aguiar (Editors)

Steam reforming of natural gas on noble-metal based catalysts: Predictive modeling

Benjamin T. Schädel, Olaf Deutschmann

Institute for Chemical Technology and Polymer Chemistry, University of Karlsruhe (TH), Engesserstraße 20, 76131 Karlsruhe, Germany

Steam reforming of natural gas over a noble-metal based catalyst is studied experimentally and numerically at varying reactor temperature and steam-to-carbon ratio. The catalytic monolith applied in the experiment is modeled by a two-dimensional flow field coupled with detailed reaction mechanisms for both surface and gas-phase reactions and taking the complexity of natural gas compositions into account. Steam reforming of methane, ethane, propane, and butane as single components as well as of a real gas mixture is studied to develop and evaluate a multi-step surface reaction mechanism. The model developed can now be used to predict conversion and selectivity in steam reforming of natural gas of widely varying composition.

1. Introduction

Steam reforming has been described by several groups with different rate expressions; reviews are given by Rostrup-Nielsen [1,2]. In kinetic expressions, the methane chemistry is usually applied to simplify the description of natural gas conversion [3, 4]. Since natural gas contains, aside from methane, a variety of other components, such as light alkanes and CO_2, predictive modeling of chemical conversion of natural gas needs to account for the complexity of the mixture composition. In this study, we focus on the light alkanes present in natural gas. Therefore, step-by-step the single components methane, ethane, propane, and butane are studied independently, and finally, North Sea Gas H (composition given in Table 1) as example for a real natural gas is investigated. The models developed are further evaluated by comparison of numerically predicted and experimentally observed steam-reforming of a “made up” gas composition.

CO_2	1.87	vol-%
N_2	0.94	vol-%
CH_4	86.72	vol-%
C_2H_6	8.10	vol-%
C_3H_8	2.03	vol-%
C_4H_{10}	0.44	vol-%
$> C_4$	0.10	vol-%

Table 1: Composition of natural gas type "North Sea H" used in the present study

2. Experiment

The experimental set-up consists of the feed section, reactor, and analytical devices. Via liquid and mass flow controllers (Bronkhorst Hi-Tec) the gaseous components and the water are fed into a vaporizer which also serves as mixing-chamber. The small dimension of the reactor, heated by a surrounding furnace and made out of ceramic materials, and the argon dilution of the fuel guarantee isothermal conditions. Three cordierite monoliths with one centimeter in length are placed inside the flow reactor; only the central monolith is catalytically active by a rhodium/alumina coating. The monoliths in the middle and the back are honeycomb monoliths composed of approximately ninety rectangular channels. A foam monolith is placed in the front of the catalyst to ensure a nearly flat velocity profile at the catalyst inlet. Gas temperatures at the front and back of the catalytic monolith are monitored with Rh/Pt thermocouples type S. The temperature difference between those two points ranges from 0°C to 15°C. The reactor is operated at atmospheric pressure. The product composition is determined with a chemi-ionization mass spectrometer (AIRSENSE 500) and a gas chromatograph (Varian CP-3380). The experimental set-up is described in more detail in [5].

All experiments are carried out with a gas hourly space velocity (GHSV) of 40000 h^{-1}; as the flow consists of 75% argon, only 10000 h^{-1} account for the potentially reactive gases. As natural gas may contain a remarkable amount of CO_2, it is also included in the feed gas and analysis. Monolithic catalysts with two different channel densities (400 and 900 channels per square inch (cpsi)) have been examined.

3. Modeling approach

Simulations are carried out using the software DETCHEM^{CHANNEL} [6], in which the reactive flow in a single cylindrical channel is simulated in two dimensions with the axial (z) and radial (r) coordinates as independent variables. Since transport in flow direction is mainly realized by convection at the high flow rates used, axial diffusion can be neglected (radial diffusion is taken into account) leading to the well-known boundary-layer equations [7], which are solved in the computer code. The complex heat transfer between reactor and furnace overlapped by the heat consumed by the endothermic reaction is not considered due to isothermal operation. Reactions in the gas phase are modelled by a set of 757 elementary reactions among 64 chemical species, which was recently developed for combustion, partial oxidation, and reforming of light alkanes [8].

The model for the catalytic surface reactions is also based on the molecular behaviour. Here, the mean field approximation is applied, in which the adsorbed species are assumed to be randomly distributed on the surface, and the state of the catalyst is described by the locally resolved surface coverage. The temperature dependence of the rate coefficients is described by a modified Arrhenius expression:

$$k_{f_k} = A_k T^{\beta_k} \exp\left[\frac{-E_{ak}}{RT}\right] \prod_{i=1}^{N_s} \Theta_i^{\mu_{ik}} \exp\left[\frac{\varepsilon_{ik}\Theta_i}{RT}\right]$$

This expression accounts for an additional coverage dependence of the activation energy using the parameters μ_{ik} and ε_{ik} [6]. These Arrhenius coefficients were taken mainly from BOC-MP or UBI-QEB calculations following the approach of Shustorovich [9]. A surface site density (maximum number of adsorption sites) of $1.6 \cdot 10^{-9}$ mol/cm^2 is chosen for the rhodium catalyst. Chemisorption measurements resulted in a ratio of catalytic to geometric surface area of 143, which is an input parameter for the model coupling the surface chemistry and flow field [6].

4. Results and Discussion

First, steam reforming of CH_4 is studied in a temperature range of 450°C to 850°C at six different steam-to-carbon (S/C) ratios ranging from 2.2 to 4. Conversion starting at 400°C is complete at temperatures above 800°C. The 900 cpsi catalyst exhibits slightly higher conversion. CO is barely formed at low temperatures, the selectivity then increases to 50% at 800°C. A detailed multi-step surface reaction mechanism consisting of 42 reactions among 12 surface and 6 gas-phase species is derived from these experiments and the previously developed mechanism of Schwiedernoch et al. [3, 4] for modeling catalytic partial oxidation of methane over rhodium. Several additional reactions proposed by Hei et al. [10] and Lin et al. [11] are added.

Steam-reforming of ethane, propane, and butane starts between 350°C and 400°C, followed by a steep increase in conversion. Complete conversion is already achieved between 550°C and 600°C. Compared to the conversion with S/C 2.5 the conversion with S/C 4 reaches the same value at approximately 25°C lower temperature. Steam-reforming reactions on rhodium are modeled by an extension of the initial methane mechanism, in which the higher alkanes are assumed to adsorb dissociatively.

Exemplarily for a real natural gas composition, steam reforming of the natural gas type "North Sea H" was studied. The conversion of the alkanes occurs in two temperature ranges: From 350°C to 400°C ethane (60%), propane (80%), and butane (85%) are already highly converted. When conversion of methane starts, the conversion of the higher alkanes has already reached nearly 100%. The experimental results are then used to validate the multi-step mechanism.

Even though conversion in the gas phase can be neglected in the catalytic system used, the gas-phase kinetics was evaluated in additional experiments without catalytic coating. In particular for increasing C_{2+} contents in the feed and at higher temperature as well as outside the catalytic section, gas phase conversion may matter.

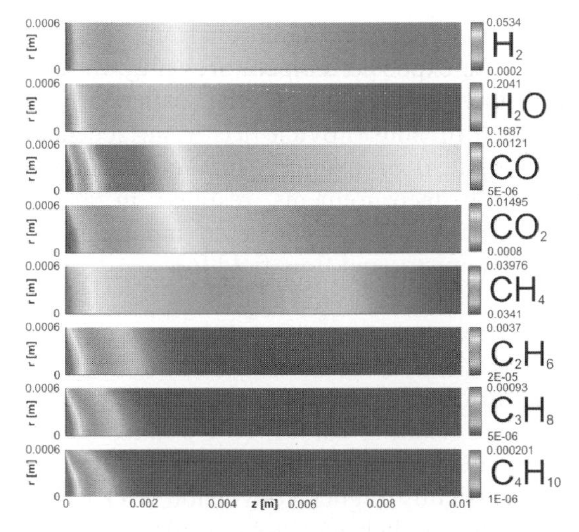

Fig. 1: Species profiles along the channel axis (flow from left to right; channel centerline at the bottom, catalytic surface at the top) for SR of gas "North Sea H" at 430°C and S/C 4.

5. Predictive Modeling

By the detailed models, the two-dimensional distributions of the gas phase species, shown in Fig. 1, and profiles of the surface coverage are calculated (not shown). The main conversion is exhibited to occur within the first millimeters of the catalytic section of the reactor.

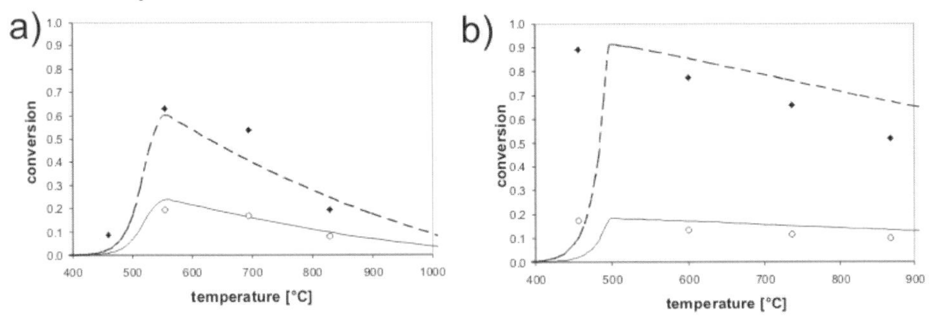

Fig. 2: Conversion for $CO/CO_2/H_2O$ mixtures a) with and b) without hydrogen – – CO simulation, ● CO experiment, — water simulation, ○ water experiment

The species profiles in the channel at lower temperatures (415 - 500 K) first show fast formation of CO which is then rapidly re-adsorbed and further reacted to CO_2. This result enforces the conclusion that water-gas shift dominates the process as also shown by additional experiments, in which the feed consisted of equal amounts of hydrogen, carbon dioxide, and carbon monoxide, and a surplus of steam. The results show a maximum in CO conversion and CO_2 formation at almost the expected temperatures (Fig. 2). Since the production of hydrogen is still low around 420°C the experiment is repeated without hydrogen. Now the temperature of the first occurrence of the water-gas-shift moves up by 50 – 100°C, which agrees with the expectation from the two-dimensional profiles (Fig. 1).

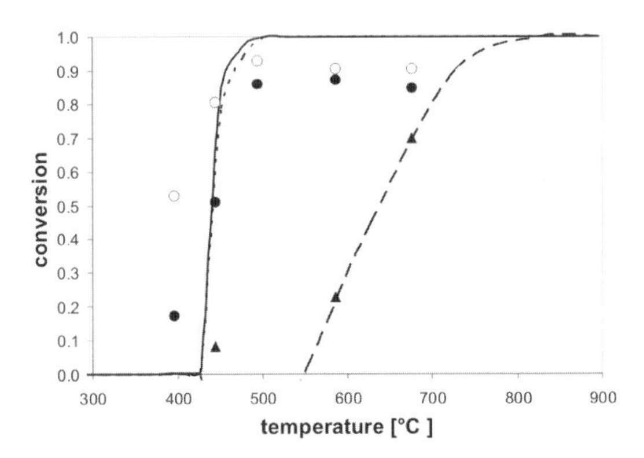

Fig. 3: Conversion in steam-reforming of the fictive "natural gas" mixture at S/C 2.5; CH_4: – – model, ▲experiment; C_2H_6: ···· model, ● experiment; C_3H_8: — model, ○ experiment.

Since all experimental observations agreed well with the predictions, eventually, an arbitrary mixed "natural gas" was studied numerically, i.e. a mixture of 70% methane, 21% ethane, and 9% propane. The reported conversion (Fig. 3) reveal that the model developed is able to predict steam-reforming of natural gas in a general way over the rhodium catalyst applied.

6. Conclusions

The study reveals that for understanding and quantitative prediction of conversion and selectivity in steam reforming of natural gas all hydrocarbon components need to be considered. All the experimentally observed conversion and selectivity data can be well-predicted by the detailed model developed for surface and gas-phase reactions. The simulation also provides a much deeper insight into the interactions of the alkanes during the reaction. The mechanisms developed with defined experiments and theoretical data can now be used to predict the conversion and selectivity of any natural gas like hydrocarbon mixture as function of temperature over the studied catalyst without any additional parameter fitting.

Acknowledgements

We would like to thank Paolo Piermartini for his assistance in the experimental study and Dr. Matthias Duisberg for many fruitful discussions.

References

1. J.R. Rostrup-Nielsen, J. Sehested, J. K. Norskov, Advances in Catalysis 47 (2002) 65
2. J.R. Rostrup-Nielsen, Catalytic Steam Reforming, Springer, 1984
3. O. Deutschmann, R. Schwiedernoch, L. Maier, D. Chatterjee, in Natural Gas Conversion VI, Studies in Surface Science and Catalysis 136, E. Iglesia, J.J. Spivey, T.H. Fleisch (eds.), p. 215-258, Elsevier, 2001
4. R. Schwiedernoch, S. Tischer, C. Correa, O. Deutschmann, Chemical Engineering Science 58 (2003) 633
5. B.T. Schädel, Untersuchung von Reformierungsprozessen von Methan an Rhodium- und Nickelkatalysatoren (Diploma thesis). Fakultät für Chemie und Biowissenschaften, Universität Karlruhe (TH), Karlsruhe, Germany, 2005
6. S. Tischer, O. Deutschmann. Catalysis Today 105 (2005) 407; www.detchem.com
7. L. L. Raja, R. J. Kee, O. Deutschmann, J. Warnatz, L.D. Schmidt, Catalysis Today 59 (2000) 47
8. R. Quinceno, J. Perez-Ramirez, J. Warnatz, O. Deutschmann, Applied Catalysis, A: General 303 (2006) 166
9. E. Shustorovich, H. Sellers, Surface Science Reports 31 (1998) 1
10. M.J. Hei M. J. Hei, H. B. Chen, J. Yi, Y.J. Lin, Y. Z. Lin, G. Wei, D. W. Liao, Surface Science 417 (1998) 82
11. Y.-Z. Lin, J. Sun, J. Yi, J.-D. Lin, H.-B. Chen, D.W. Liao, Theochem 587 (2002) 63

Natural Gas Conversion VIII
F.B. Noronha, M. Schmal, E.F. Sousa-Aguiar (Editors)

Activity of Cu/CeO_2 and Cu/CeO_2-ZrO_2 for low temperature water-gas shift reaction.

André L. M. da Silva[a] and José M. Assaf[a]

[a]*Laboratório de Catálise, Departamento de Engenharia Química, Universidade Federal de São Carlos, Rodovia Washington Luís km 235, Cx. Postal 676, CEP 13565-905, São Carlos-SP, Brasil.*

1. Abstract

The activity of Cu/CeO_2 and $Cu/Ce_{0.8}Zr_{0.2}O_2$ catalysts with 5 and 10wt% Cu was investigated in the low-temperature water-gas shift reaction (LT-WGSR). CeO_2 was synthesized by the urea gelation co-precipitation (CeO_2-UGC), hydrothermal (CeO_2-HT) and Pechini (CeO_2-Pe) methods $Ce_{0.8}Zr_{0.2}O_2$ only by the Pechini method ($Ce_{0.8}Zr_{0.2}O_2$-Pe). The catalysts were characterized by powder X-ray diffraction (XRD), hydrogen temperature-programmed reduction (H_2-TPR) and BET surface area. The catalysts showed activity in the sequence 5%Cu/CeO_2-HT > 10%Cu/CeO_2-HT ~ 5%Cu/CeO_2-UGC ~ 10%Cu/CeO_2-UGC > 10%$Cu/CeZrO_2$ > 5%$Cu/CeZrO_2$ > 5%Cu/CeO_2-Pe ~ 10%Cu/CeO_2-Pe. The same sequence was observed in the BET surface area, indicating the existence of a correlation between particle size, metal dispersion and catalytic activity.

2. Introduction

Fuel cell technology is an efficient energy process that uses hydrogen as a fuel and converts chemical energy into electric and thermal energy without combustion. The production of hydrogen from hydrocarbons involves a complex chain of reactions to obtain the high purity hydrogen required by the fuel cell. During this process, CO is also produced and needs to be removed upstream of a low-temperature fuel cell, such as the PEMFC (Proton Exchange Membrane Fuel Cell), to lower its content to acceptable limits (10-50ppm), because it poisons the anode catalyst, thus decreasing the performance and life of the fuel cell.

The reduction of carbon monoxide content may involve high temperature and low temperature water-gas shift reactions (WGSR), followed by preferential CO oxidation (PROX). The WGSR (1) is required after the reforming of hydrocarbons such as methane to decrease the CO concentration in the reformed gas [1].

$$CO + H_2O \leftrightarrow CO_2 + H_2 \qquad \Delta H = -41.2 kJ/mol \qquad (1)$$

3. Experimental

3.1. Catalyst preparation

The procedure for synthesis of CeO_2-UGC [2] consists in mixing aqueous metal nitrate solutions with urea; heating the solution to 100°C with vigorous stirring and adding deionized water; boiling the mixture for 8h at 100°C; filtering the sample and washing with deionized water at 50°C; drying overnight at 100°C; crushing the dried lump into smaller particles and calcining the powder at 400°C for 4h, with a heating rate of 2°C/min.

CeO_2-HT [4] was synthesized by the hydrothermal method as follows: $Ce(NO_3)_2.6H_2O$ was dissolved in distilled water and solution of 10% NaOH were added rapidly with stirring. After 15 minutes, all of the slurry was transferred to an sealed vessel, which was filled with deionized water up to 80% and heated at 100°C under autogeneous pressure for 10h; the system was allowed to cool to room temperature and the product was collected by filtration and washed with deionized water, then dried overnight at 60°C and calcined at 400°C for 4h, with a heating rate of 2°C/min.

The Pechini method [6] consists in mixing ethylene glycol and citric acid in the molar ratio 1:3; adding HNO_3 to zirconium carbonate and to $(NH_4)_2Ce(NO_3)_6$; adding the zirconium carbonate and $(NH_4)_2Ce(NO_3)_6$ to the mixture of ethylene glycol and citric acid; aging the whole at 120°C for 24h, drying at 120°C for 12h and calcining at 600°C for 4h, with a heating rate of 2°C/min.

The supported Cu catalysts were prepared by deposition-precipitation as follows: a 0.1M aqueous solution of $Cu(NO_3)_2$ was added to the ceria suspended in water; during this process, the suspension pH was kept constant at 9.0 by adding NaOH 0.25M solution. After an additional 60 min of continuous stirring, the precipitate was filtered and washed with deionized water, then dried overnight at 80°C and calcined at 400°C for 4h with a heating rate of 2°C/min.

3.2. Catalyst Characterization

The specific surface area (S_{BET}) of the samples was determined by nitrogen adsorption-desorption using a Quantachrome NOVA 1200. The samples were previously outgassed at 300°C for 2h under vacuum. The crystal structure of mixed oxide was analyzed by X-ray powder diffraction (Siemens D5005) employing Cu-Kα radiation in a range of 2θ between 20° and 80°. The samples were also analyzed after WGSR. Temperature-programmed reduction (TPR) analysis was performed in a flow of a 5% H_2/He mixture (30mL.min^{-1}) over 50mg of calcined samples using a heating rate of 10°C.min^{-1}.

3.3. Catalytic tests

The catalytic activity measurements were carried out under ideal conditions (without H2 and CH4) in a tubular fixed-bed flow microreactor at atmospheric pressure over a wide temperature range (150-350°C) with 180mg of catalyst, space velocity (CO) of 2000h^{-1} and H_2O:CO molar ratio of 10:1. The reactor feed gas mixture contained 5 vol.% CO in nitrogen as diluent. The catalysts employed for the measurement were previously reduced *in situ* in hydrogen at 350°C for 1h.

4. Results and Discussion

4.1. Catalyst Characterization

The XRD patterns of calcined catalysts are presented in Fig. 1(a). The presence of only the cubic phase of the CeO_2 indicates that Ce and Zr form a solid solution. The slight shift of the ceria peaks suggests the dissolution of the ZrO_2 into the cubic fluorite lattice [2].

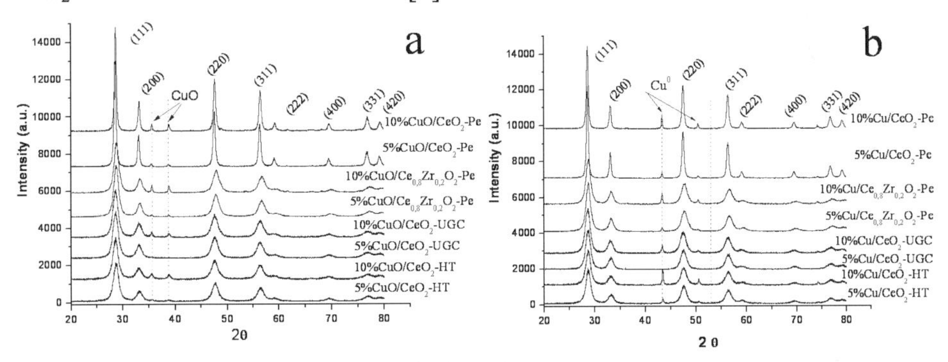

Fig.1. XRD patterns; (a) calcined catalysts (b) catalysts used in the WGSR.

All the catalysts exhibited CuO peaks, except the 5%CuO/CeO_2-HT and 5%CuO/CeO_2-UGC samples; the absence of CuO peaks indicates a fine dispersion of the CuO on the surface of the ceria [3]. Fig. 1(b) shows the XRD patterns of catalysts used in the water-gas shift reaction. All catalysts presented the Cu^0 peaks, except the 5%CuO/CeO_2-UGC samples. This indicates that after the reaction the Cu particles remain finely dispersed on the surface of the ceria.

The specific surface area (S_{BET}) of calcined samples is reported in Table 1. The CeO_2 samples prepared by UGC and the hidrothermal method had higher S_{BET} values than CeO_2 prepared by the Pechini method as a consequence of the higher calcination temperature employed in Pechini method. It can be seen that the surface area of the mixed oxide is higher than that of pure ceria, indicating that the addition of ZrO_2 to the CeO_2 lattice enhances the thermal stability regarding the growth of ceria crystallites [3]. The Table 1 also displays the average crystallite size of CuO in the fresh catalysts and Cu^0 in the catalysts after WGSR, calculated from the X-ray data by Scherrer's equation.

Table 1. Structural characteristics of samples

Samples	Temperature of Calcination (°C)	d^a (nm) fresh	d^a (nm) used	BET surface area ($m^2.g^{-1}$)
CeO_2 – UGC	400	6.7^b	-	92
CeO_2 – HT	400	6.8^b	-	131
CeO_2 – Pe	600	21.1^b	-	13
$Ce_{0,8}Zr_{0,2}O_2$ – Pe	600	7.8^b	-	81
5%CuO/ CeO_2 – UGC	400	-	-	80
10%CuO/ CeO_2 – UGC	400	13.2^c	26.5^d	71
5%CuO/ CeO_2 – HT	400	-	26.9^d	123
10%CuO/ CeO_2 – HT	400	13.3^c	31.8^d	116
5%CuO/ CeO_2 – Pe	600	21.5^c	34.8^d	14
10%CuO/ CeO_2 – Pe	600	23.0^c	38.1^d	13
5%CuO/ $Ce_{0,8}Zr_{0,2}O_2$ – Pe	600	16.2^c	28.2^d	71
10%CuO/ $Ce_{0,8}Zr_{0,2}O_2$ – Pe	600	16.3^c	29.8^d	71

[a] Measurement of average crystallite using Scherrer's equation.
[b] Measurement from line broadening of CeO_2 (111) peak
[c] Measurement from line of CuO (2 = 38.9°)
[d] Measurement from line of Cu^0 (2 = 43.3°)

The results of average crystallite size of catalysts before and after reaction show good correlation with the S_{BET} and activity in water-gas shift reaction results. The samples with lower particle size present higher S_{BET} and higher activity. Varying the copper content in the catalysts did not lead to any significant change in BET surface area.

Fig. 2(a) show the H_2-TPR profiles of catalysts with 5wt% Cu. The catalysts supported on CeO_2-HT or CeO_2-UGC show H_2 consumption peaks at lower temperatures than the catalysts supported on CeO_2-Pe or $Ce_{0.8}Zr_{0.2}O_2$-Pe. The synergistic interaction between the CuO and the CeO_2 is evident because both oxides show a shift of reduction peaks to lower temperatures [5].

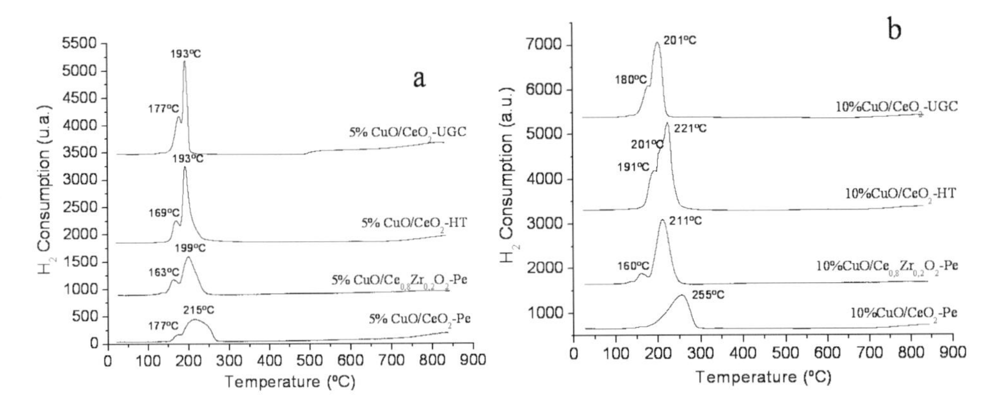

Fig. 2. H_2-TPR profiles (a) 5wt% Cu catalysts (b) 10wt% Cu catalysts.

H_2-TPR profiles of 10wt% Cu catalysts are presented in Fig. 2(b). A two-step reduction profile is observed for 5 and 10wt% Cu catalysts, in wich the reduction of CuO clusters strongly interacting with ceria occurs in the range 120-175°C and the reduction of larger CuO particles is observed near 200°C [3]. The 10wt% Cu catalysts exhibit a shift to higher temperatures of reduction indicating the presence of larger copper particles than in the 5wt% Cu catalysts.

4.2. Catalytic tests

The results of catalytic tests in WGSR, presented in Fig. 3, are consistent with S_{BET}, XRD and H_2-TPR datas. The catalytic activity was dependent on the copper particle size. The preparation of supports by hydrothermal and urea co-precipitation gelation methods resulted in ceria with a higher specific surface area, leading to higher dispersion of supported copper particles. The insertion of zirconium into the ceria lattice also gave a high surface area, but the impregnation of copper on the ceria-zirconia support resulted in larger copper particles, resulting in a low activity of Cu/Ce-Zr catalysts. Another factor was pointed out by Zhou *et al.* **[4]:** the samples prepared by the HT method may form nanorods and exhibit more effective redox properties than spherical nanoparticles; then the enhancement of catalytic activity on 5%CuO/CeO2-HT may be due to the high-energy, reactive ceria planes, which can generate strong synergetic effects between CuO and ceria. As the reaction occurs at the metal-support interface, the smaller particle, the higher the interface area and the catalytic activity in the WGSR.

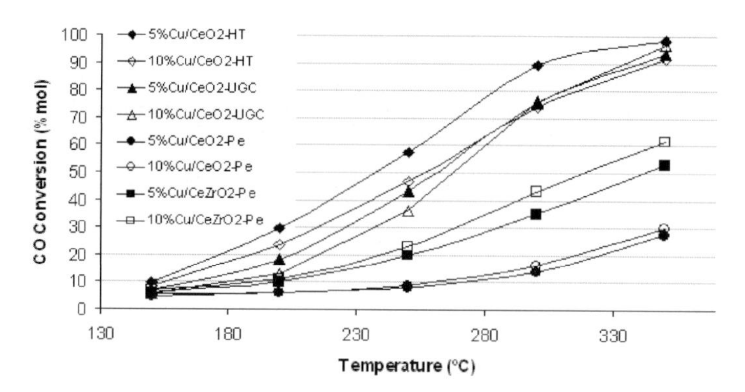

Fig. 3. CO conversion versus temperature in the WGSR on different catalysts.

5. Conclusions

The higher activity of the catalysts prepared by the HT and UGC methods has been attributed to higher surface area which leads to smaller Cu particles. These catalysts keep high dispersion of Cu species during the catalytic tests. The addition of ZrO_2 to the CeO_2 lattice enhances the thermal stability inhibiting the crystal growth of ceria. On the other hand the impregnation with Cu leads to the formation of bigger copper particles, as show the *d* values in Table 1, which proved less active to catalyze this reaction.

Acknowledgements

To CAPES - Brazil for scholarship support.

References

[1] D. L. Trimm, Applied Catalysis A: General 296 (2005) 1.
[2] Y. Li, Q. Fu e M. Flytzani-Stephanopoulos, Applied Catalysis B: Environmental 27 (2000) 179.
[3] T. Tabakova, V. Idakiev, J. Papavasiliou, G. Avgouropoulos, T. Ioannides, Catalysis Communications 8 (2007) 101-106.
[4] Kebin Zhou, Run Xu, Xiaoming Sun, Hongde Chen, Qun Tian, Dixin Shen, Yadong Li, Catalysis Letters 101 (2005) 169.
[5] Xiucheng Zheng, Xiaoli Zhang, Xiangyu Wang, Shurong Wang and Shihua Wu, Applied Catalysis A: General 295 (2005) 142.
[6] P. J. B. Marcos, D. Gouvêa, Cerâmica 50 (2004) 38.

Natural Gas Conversion VIII
F.B. Noronha, M. Schmal, E.F. Sousa-Aguiar (Editors)

Hydrogen separation from the mixtures in a thin Pd-Cu alloy membrane reactor

Xiaoliang Zhang [a,b], Guoxing Xiong [a,*], Weishen Yang [a,*]

a. State Key Laboratory of Catalysis, Dalian Institute of Chemical Physics, Chinese Academy of Sciences, Dalian 116023, China.
b. Graduate School of the Chinese Academy of Sciences, Beijing 100049, China
**Corresponding author. Tel: +86-411-84379182; Fax: +86-411-84694447.*
E-mail address: gxxiong@dicp.ac.cn (G. Xiong), yangws@dicp.ac.cn (W. Yang)

A thin dense Pd-Cu alloy membrane was deposited on porous ceramic tubular supports by vacuum electroless plating. Hydrogen permeation through this Pd-Cu alloy membrane had been investigated in the presence of CH_4, CO_2 and steam in the temperature range of 250^0C to 500^0C and the pressure range of 1.3 bar to 2.6 bar. The co-exsiting CH_4, CO_2 and steam had a negative effect on hydrogen permeation flux and hydrogen recovery. The effect increased with the amount of CH_4, CO_2 and steam in the mixtures.

1. Introduction

In recently, there is increased demand of pure hydrogen in the petroleum refining and petrochemical production processes as well as energy application related to hydrogen such as fuel cells and mobile vehicles. Methane steam reforming (MSR) is one of the most important chemical processes for the production of hydrogen, particularly in recent years owing to widespread availability of natural gas [1], which has been proven more economical than other processes such as coal gasification. About 50% of hydrogen is produced from natural gas (mainly methane). And 40% is produced by the methane steam reforming process [2]. Pd metal and its alloys (e.g. Pd-Ag, Pd-Cu) exhibit high selectivity towards hydrogen permeation and their utility for gas separation and reaction had been amply reported. In recent years, some works have published about the steam reforming of methane in Pd-based membrane reactors [3-5].

These membrane reactors can combine steam reforming of syngas, the water gas shift reaction and hydrogen purification into one operating unit. By removing the produced hydrogen with Pd-based membrane, the reaction shift in equilibrium can be obtained. However, there have been few investigations concerned with possible retardation of the hydrogen permeation through Pd-based membranes by the reactants and products of this catalytic reaction in the Pd-based membrane reactors.

The primary objective of this work is to determine whether the major constituents of MSR (CH_4, CO_2 and H_2O) have any effects on the hydrogen permeation performance in a thin Pd-Cu alloy membrane reactor. It is important to study these effects for understanding the generated hydrogen separation behavior in the MSR membrane reactors.

2. Experimental

The fabrication procedure for the vacuum electroless-plated Pd-Cu alloy membrane onto porous α-Al_2O_3 tube was described in detail previously [6]. α-Al_2O_3 porous tubular membrane was used as substrate, which had an average

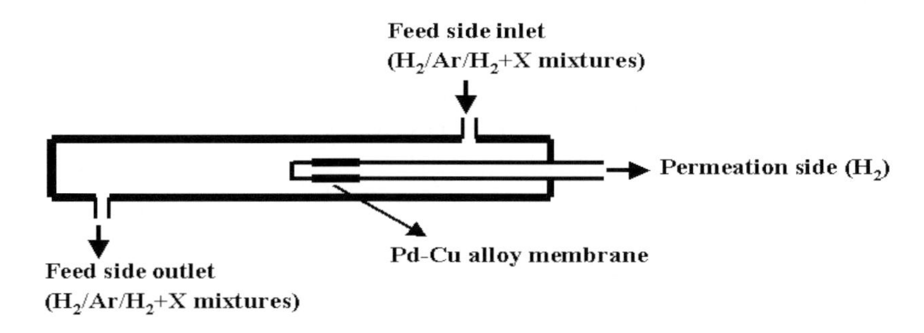

Fig. 1. Schematic of the Pd-Cu alloy membrane reactor for hydrogen separation

pore size of 0.3 μm, a porosity of 40%, an outer diameter of 11mm and a 1.5 mm thick wall. A seeding procedure was required prior to the metals plating by modified technique of sol-gel process. With seeded Pd, Pd and Cu were successively deposited on the substrate by electroless plating from Pd bath and Cu bath. In the metals deposition processes, different and certain vacuum effects were employed to both sides of the tubular membrane with two vacuum pumps respectively. After plating, the as-deposited Pd-Cu composite membrane was rinsed and dried.

As shown in Fig. 1, the as-deposited Pd-Cu membrane was positioned in a stainless steel permeator to form a membrane reactor. Hydrogen permeation through the Pd-Cu alloy membrane was measured for feeds of pure hydrogen, pure Ar and H_2-X mixtures such as H_2-CH_4, H_2-CO_2, H_2-Steam, H_2-CH_4-Steam mixtures etc. The permeation experiments were carried out in the temperature range of 250^0C to 500^0C and the pressure range of 1.3 bar to 2.6 bar. The permeation composition was measured by GC and the pressure in the permeate side was kept at ambient atmosphere. No sweep gas was used on the permeate side in all permeation experiments. After gas permeability tests, the Pd-Cu alloy membrane was slowly cooled down in the Ar atmosphere, fractured into small pieces for making characterization.

3. Results and discussion

A thin dense Pd-Cu membrane with about 6 μm thickness were prepared by the vacuum electroless plating method. After activation and annealing in hydrogen atmosphere at 500^0C, the as-deposited Pd-Cu alloy membrane formed Pd-Cu alloy membrane. Fig. 2 shows the hydrogen flux data at various temperatures from 250^0C to 500^0C as a function of transmembrane hydrogen partial pressure differences for the Pd-Cu alloy membrane. As shown in Fig. 2, the hydrogen flux is linearly proportional to the pressure difference across the membrane at each temperatures, not in the square root of hydrogen partial pressure. It illustrated that the surface process is the rate-limiting step for the hydrogen permeation process in the Pd-Cu alloy membrane.

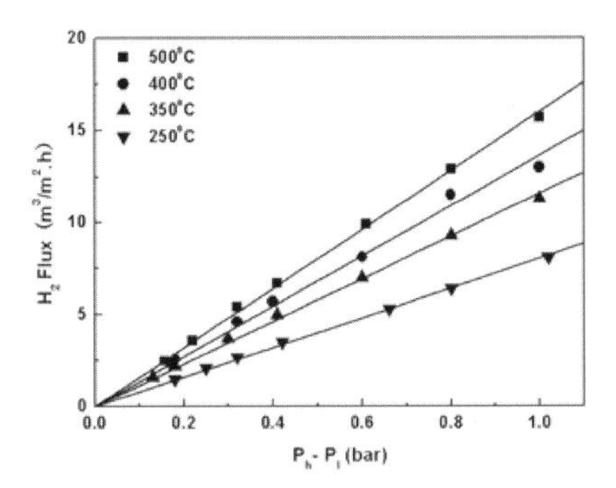

Fig. 2. Pressure dependence of the hydrogen flux for the Pd-Cu alloy membrane at different temperatures.

The hydrogen permeation flux as a function of total feed flow rate in the H_2-X (90/10, X=CH_4, CO_2, steam) binary mixtures at different temperatures is shown in Fig. 3. It can be seen that the hydrogen flux increases with feed flow rate to obtain a relative stable value when the feed flow rate is 500 ml/min. Also, from this figure, it can be seen that the hydrogen flux separating from these mixtures is obvious lower than that in pure hydrogen atmosphere as seen in Fig. 2. In Fig. 2, it displays the hydrogen flux is about 16 m^3/m^2.h, 12 m^3/m^2.h and 8 m^3/m^2.h at 500^0C, 350^0C and 250^0C, respectively, when the hydrogen partial pressure difference is 1.0 bar. It illustrates that these additives have inhibitive effects on

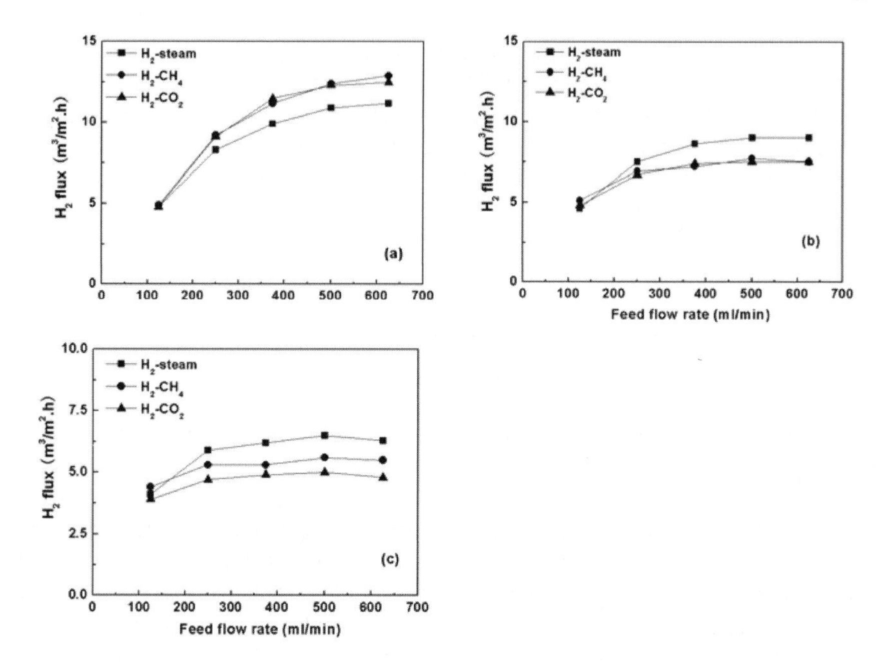

Fig. 3. Hydrogen flux of Pd-Cu alloy membrane as a function of total feed flow rate in H_2-X (90/10, X=CH_4, CO_2, Steam) mixtures at (a) 500^0C; (b) 350^0C; (c) 250^0C. The total pressure of feed mixtures is 2.0 bar.

the hydrogen pereation performance through the Pd-Cu alloy membrane as comparison to in the pure hydrogen atmosphere. Moreover, the inhibitive effect of steam on the hydrogen flux at high temperature (500^0C) is obvious higher than those of CH_4 and CO_2 at the same conditions. However, these effects of CH_4 and CO_2 on hydrogen flux at low temperatures (350^0C and 250^0C) are relatively higher than that of steam. The influence of steam, CO and CO_2 on the hydrgoen flux through Pd membrane and Pd-Ag alloy membrane is reported in the literature [7-8]. They found that the different inhibitive effects of these additives on hydrogen permeation were related to their different adsoption capability on the metallic surface at different temperatures. In this work, it could be attributed to their different competitive adsorption of CH_4, CO_2 and steam with hydrogen on the surface of this Pd-Cu alloy membrane at different temperatures.

Fig. 4 illustrates the influence of the concentrations of additives in the H_2-X (X=steam, CH_4 and CO_2) mixtures on the F/F_0 at operating conditions of 500^0C and 2.0 bar of total pressure on feed side when the feed flow rate of mixtures is 500 ml/min. The F, F_0 presents hydrogen flux for the mixtures and pure hydrogen atmosphere, respectively. F/F_0 provides a simple measure of the

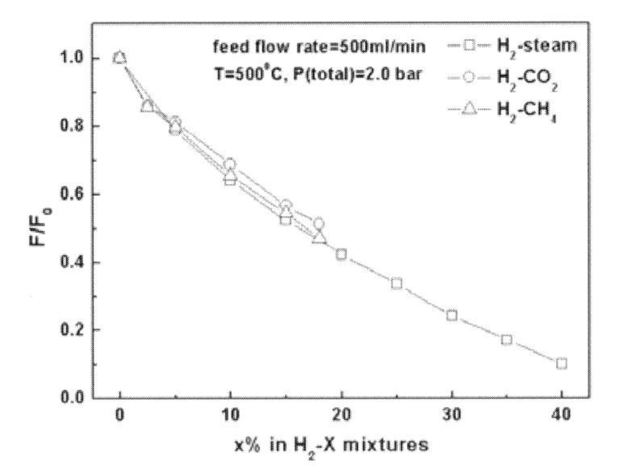

Fig. 4. The influences of composition, x% in the H_2-X mixtures (X=steam, CO_2 and CH_4), on the F/F_0 at 500^0C and 2.0 bar of total pressure on feed side.

inhibitive effects of other gas species with hydrogen on the hydrogen permeation through the Pd-Cu alloy membrane. As seen in Fig. 4, all the additives, steam, CH_4 and CO_2 displayed negative effects on hydrogen flux through the Pd-Cu alloy membrane. For instance, addition of 40% steam causes an approximately 90% decrease of hydrogen permeation flux. Such comparison of the effects among these additives at 350^0C and 250^0C were also investigated and similar trend of hydrogen permeation to be suppressed were found.

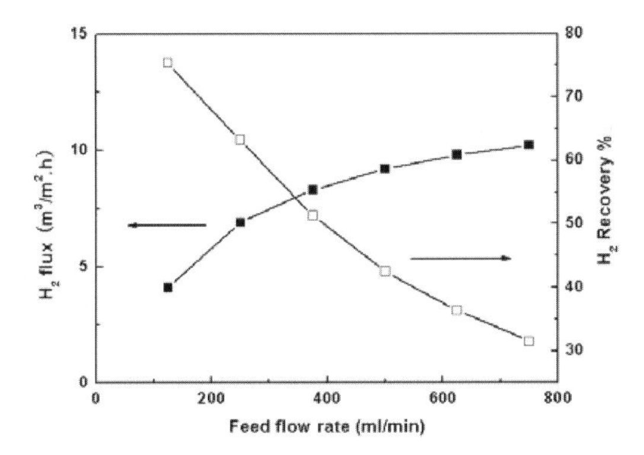

Fig. 5. The H_2 flux and H_2 recovery in the H_2-CH_4-Steam (75/5/20) mixtures at 500^0C (the total pressure of feed mixtures is 2.6 bar).

Fig.5 shows the permeation behavior of hydrogen in the H_2-CH_4-Steam (7/5/20) ternary mixtures at the operating conditions of 500^0C and a feed side total pressure of 2.6 bar. The maximum hydrogen flux is about 10 m^3/m^2.h and the maximum hydrogen recovery is over 75%. As seen in this figure, the hydrogen permeation flux increased with the feed flow rate, while the hydrogen recovery rate decreased. It maybe related to the reducing of relative effective area of Pd-Cu alloy membrane and the increasing of load-to-surface ratio of the additives on the surface of membrane with the increasing of the feed flow rate.

4. Conclusion

The effects of the major constituents of methane steam reforming such as CH_4, CO_2 and steam on the hydrogen permeation behavior through a thin Pd-Cu alloy membrane has been investigated. All these additives showed a negative effect on hydrogen permeation flux and hydrogen recovery. The effect increased with the increasing amount of CH_4, CO_2 and steam in the mixtures. Moreover, the effect of these additives showed evident difference at different temperatures due to their competitive adsorption of CH_4, CO_2 and steam with hydrogen on the surface of the Pd-Cu alloy membrane at different temperatures.

Acknowledgements

The authors are grateful to the Ministry of Science and Technology of China for financial support (Grant No. 2005CB221404).

References

1. Kikuchi E., Catal. Today, 56 (2000) 97.
2. Barbieri G., Maio F., Ind. Eng. Chem. Res., 36(1997) 2121.
3. Gallucci F., Paturzo L., Fama A., Basile A., Ind. Eng. Chem. Res., 43 (2004) 928.
4. Oklany J., Hou K., Hunges R., Appl.Catal. A: Gen., 170 (1998) 13.
5. Tong J., Matsumura Y., Suda H., Haraya K., Ind. Eng. Chem. Res. , 44 (2005) 1454.
6. Zhang X., Liu J. Sheng S., Yang W. Xiong G., in: Bredesen R. and Ræder H. (Eds.) Inorganic Membranes--Proc. of the 9th International Conference on Inorganic Membranes, Norway, 2006, PP.595.
7. Hou K., Hughes R., J.Membr.Sci:., 206(2002)119.
8. Gielens F.,Knibbeler R., Duysinx P., Tong H., Vorstman M.,Keurentjes J., J.Membr.Sci. 279(2006)176.

Natural Gas Conversion VIII
F.B. Noronha, M. Schmal, E.F. Sousa-Aguiar (Editors)

Effect of cobalt on the activity of iron-based catalysts in water gas shift reaction

Amalia L.C. Pereira[a], Nilson A. dos Santos[a], Márcio L. O. Ferreira[a], Alberto Albornoz[b] and Maria do Carmo Rangel[a]

[a]*Instituto de Química, Universidade Federal da Bahia. Campus Universitário de Ondina, Federação. 40 170-280, Salvador, Bahia, Brazil*
[b]*Instituto Venezolano de Investigaciones Científicas, Apartado 21 827, Caracas 1020-A, Venezuela*

The water gas shift reaction (WGSR) is an important step in the commercial prodution of high pure hydrogen for several applications. In order to find alternative catalysts to this reaction, the effect of cobalt on the activity of iron-based catalysts was studied in this work. It was found that cobalt changed the properties of iron oxide depending on its content. In small amounts (Co/Fe(molar)= 0.05), it did not affect significantly the specific surface area but decreased the catalytic activity, a fact which can be assigned to the production of a cobalt and iron phase, less active in WGSR than magnetite. In higher amounts (Co/Fe= 1.0), cobalt led to an increase of specific surface area and to the production of cobalt ferrite and Co_3Fe_7 alloy, which was more active and more resistant against reduction than magnetite. This catalyst can probably work under low steam conditions, decreasing the operational costs.

1. Introduction

In recent years, the water gas shift reaction (WGSR) has attracted increasing interest mainly because of its application in fuel cells [1]. By WGSR, the hydrogen production from steam reforming is increased and, most importantly, the hydrogen stream is purified by the removal of carbon monoxide which often poisons most of metallic catalysts, including the platinum electrocatalyst for fuel cells [2].
The water gas shift reaction: $CO + H_2O \rightleftarrows CO_2 + H_2$ has been widely investigated for different purposes, for instance, for ammonia synthesis and

hydrogenation reactions. Thermodynamically, the WGSR is favored by low temperatures and excess of steam, since it is exothermic (ΔH= -41 kJ.mol^{-1}) and reversible. However, high temperatures are required for industrial applications and thus the reaction is carried out in two steps in commercial processes: a high temperature shift (HTS) in the range of 370-420 °C and a low temperature shift (LTS) at around 230 °C. The HTS step is typically performed over an iron oxide-based catalyst while the LTS stage is carried out over a copper-based one. Before its use, the HTS catalyst must be converted to magnetite (Fe_3O_4), which is the active phase. During this reduction, the production of metallic iron should be avoided, since it can lead to methanation and carbon monoxide disproportionation reactions [3]. In order to ensure the magnetite stability in industrial processes, large amounts of steam are added to the feed, increasing the operational costs. There is, thus, the need for developing new catalysts which can require smaller amounts of steam and can be more resistant against the magnetite reduction. With this goal in mind, this work deals with the effect of cobalt on the activity of iron-based catalysts for the HTS stage.

2. Experimental

Samples were prepared by mixing iron nitrate and cobalt nitrate solutions with an ammonium hydroxide solution (1% v/v) through a peristaltic pump into a beaker with water, at room temperature. Samples with cobalt to iron molar ratio of 0.05 (ICA05 sample) and 1.0 (ICA10) were obtained. The final pH was adjusted to 11 using an ammonium hydroxide solution (1%). The sol produced was stirred for additional 30 min and then centrifuged (2000 rpm, 5 min). The gel obtained was washed with water and centrifuged again. These processes were repeated several times until no nitrate ions were detected in the supernatant anymore. The gel was then dried in an oven at 120 °C, for 12 h. The solid obtained was ground and sieved in 100 mesh and then calcined at 500 ºC, for 2 h. A cobalt-free sample (IA) and an iron-free sample (CA) were also prepared to be used as references.

The catalysts were characterized by chemical analysis, specific surface area measurements, X-ray diffraction (XRD), temperature programmed reduction (TPR), X-ray photoelectron spectroscopy (XPS) and Fourier transform infrared spectroscopy (FTIR).

The iron contents were determined by X-ray dispersive energy in Shimadzu EDX-700HS model equipment. The presence of nitrate groups in the samples was verified by FTIR, using a Perkin Elmer model Spectrum One equipment, in the range of 400 to 4000 cm^{-1}. The XRD powder patterns of the solids were obtained in a Shimadzu model XD3A equipment, using CuKα (λ=0.15420 nm) radiation and nickel filter, in a 2θ range between 10-80°, in a scanning speed of 2 °/min. The specific surface area measurements were carried out by the BET method in a Micromeritics model TPD/TPR 2900 equipment, using a 30% N_2/He mixture. The sample was previously heated in a rate of 10 °C.min^{-1} up to

160 °C, under nitrogen flow (60 mL. min^{-1}), for 30 min. The TPR profiles were obtained in the same equipment. The samples (0.3 g) were reduced under heating from 30 to 1000 °C, at 10 °C min^{-1}, using a 5% H_2/N_2 mixture. The XPS spectra were acquired with a VG Scientific spectrometer, Escalab model 220i-XL, with MgKα (1253 eV) anode and 400 W power, and hemispheric electron analyzer. This reference was in all cases in good agreement with the BE of the C 1s peak, at 284.6 eV.
The catalysts (0.3 g) were evaluated at 370 °C and 1 atm, in a fixed bed microreactor, with a steam to process gas (10% CO, 10% CO_2, 60% H_2 and 20% N_2) molar ratio of 0.6, providing there is no diffusion effect. During the experiments, the process gas was introduced into a saturator with water at 77 °C and then fed to the reactor. The products were analyzed by on line gas chromatography, using a CG-35 instrument, with Porapak Q and molecular sieve columns.

3. Results and Discussion

The FTIR spectra (not shown) confirmed the absence of nitrate ions in the solids. Figure 1 shows the XRD patterns of fresh and spent catalysts. Pure iron oxide (IA) displayed peaks related to hematite phase [4] while pure cobalt oxide showed the profile of the Co_3O_4 compound (JCPDF 87-1164). By adding small amounts of cobalt to hematite (ICA05 sample) no change in its pattern was noted. It means that cobalt was distributed in hematite lattice or formed a phase not detectable by XRD. However, when higher amounts of cobalt were added (ICA10 sample), cobalt ferrite ($CoFe_2O_4$) (JCPDF 74-2120) was produced.

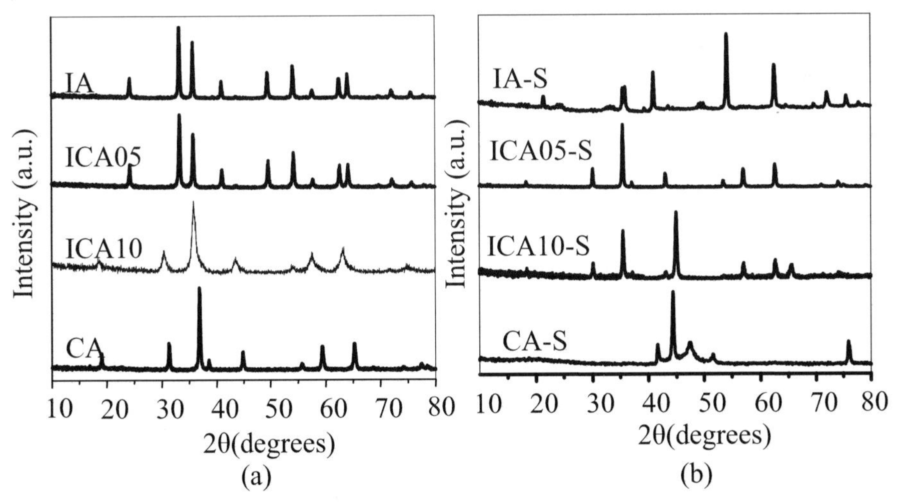

Figure 1. X-ray diffractograms of (a) fresh and of (b) spent catalysts. IA and CA samples: iron and cobalt oxide, respectively; ICA05 and ICA10 samples: with Co/Fe (molar)= 0.05 and 1.0, respectively. S represents the spent catalysts.

After WGSR, all samples showed different XRD profiles, showing that some phase transition occurred during reaction, regardless the presence of cobalt. Pure iron oxide presented a mixture of magnetite and wustite (FeO) (JCPDF 01-1121), while pure cobalt oxide showed metallic cobalt (JCPDF 75-1609 and 49-1447).On the other hand, the iron and cobalt-based samples produced different mixed phases during reaction. The sample with the smallest cobalt amount presented only the $(Co_{0,2}Fe_{0,8})Co_{0,8}Fe_{1,2}O_4$ phase (JCPDF 01-1278 and 01-1255) whereas the solid with the largest cobalt content (ICA10-S) showed cobalt ferrite ($CoFe_2O_4$) and a metallic alloy (Co_3Fe_7) (JCPDF 77-0426).
From the TPR curves (Figure 2a), it can be seen that cobalt changed the reduction profile depending on its amount. Iron oxide (IA) showed two peaks at 390 and 645 °C, assigned to the reduction of Fe^{+3} to Fe^{+2} and Fe^{+2} to Fe^0 species, respectively [4]. The additional peak at 882 °C can be related to the reduction of the residual iron oxide in the center of the particles, which was reduced at higher temperatures. It is well-known [5, 6] that the reduction of iron oxide proceeds through a surface-controlled process; once a thin layer of iron oxide with lower oxidation state (wustite, metallic) is formed on the surface, it changes to diffusional control. Therefore, this residual nucleous does not easily access the reducing gas and thus is reduced at higher temperature where the diffusional process is faster. Cobalt oxide (CA) showed a peak at 370 °C, attributed to the reduction of Co_3O_4 compound; this only peak indicates that the two reduction steps (Co_3O_4 to CoO and CoO to Co) took place at close temperatures, as found by other authors [7]. The addition of small amounts of cobalt to iron oxide (ICA05) did not change the TPR profile, but slightly shifted the first peak (390 °C) to higher temperatures (395 °C), indicating that this dopant delayed the hematite reduction. A small peak appeared at low temperatures (274 °C) that can be assigned to the reduction of the Co_3O_4 compound [7]. It was probably present as small particles or in small amounts and then was not detectable by XRD. The high temperature peaks were shifted to even higher temperatures (660, 891 °C) showing that the metallic iron production was also delayed by cobalt. On the other hand, the sample with the largest cobalt content (ICA10) presented a different profile, with three overlapped peaks (406, 534 and 648 °C), related to reduction of cobalt ferrite and a high temperature peak at 913 °C, assigned to the reduction of the residual nucleous of iron oxide. Another peak, at 800 °C, can be related to the production of the Co_3Fe_7 alloy, detected by XRD in the spent catalysts. Therefore, this alloy was also produced in the conditions of the TPR experiments. All these peaks were shifted to higher temperatures, as compared to pure iron oxide, showing that cobalt made this solid more resistant against reduction, preserving the active phase of the catalyst.
The effect of cobalt on the specific surface area depended on its content, as shown in Table 1. It can be noted that its addition increased this property and this effect increased with its amount in solids. During WGSR, the specific surface areas decreased, indicating that the phase changes were followed by the

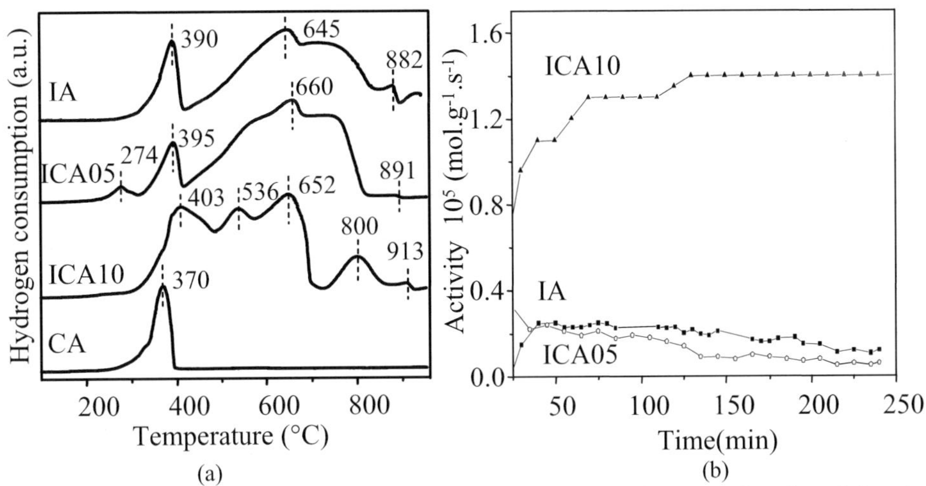

Figure 2. (a) TPR curves of the catalysts; (b) activity of the catalysts as a function of time in WGSR. IA and CA samples: iron and cobalt oxide, respectively; ICA05 and ICA10 samples: with Co/Fe (molar)= 0.05 and 1.0, respectively.

Table 1. Specific surface area of the catalysts before (Sg) and after WGSR (Sg*) and their activity (a) and the activity per area in WGSR (a/Sg*). IA and CA samples: iron and cobalt oxide, respectively; ICA05 and ICA10 samples: with Co/Fe (molar)= 0.05 and 1.0, respectively.

Sample	Sg (m^2g^{-1})	Sg* (m^2g^{-1})	a.10^5 ($mol.g^{-1}.s^{-1}$)	a/Sg*.10^7 ($mol.m^{-2}.s^{-1}$)	Surface ratio Co/Fe (atom) Fresh	Spent
IA	19	15	0.12	0.8	-	-
ICA05	23	11	0.06	0.5	0.094	0.074
ICA10	33	27	1.40	5.2	0.078	2.577
CA	15	2	-	-	-	-

coalescence of particles and pores. After reaction, the sample richest in cobalt showed a specific surface area higher than pure iron oxide indicating that cobalt acted as textural promoter decreasing sinterization during the reaction.

All solids were active in WGSR, as displayed in Table 1. The addition of small amounts of cobalt (ICA05 sample) decreased the activity, as compared to pure iron oxide, but higher amounts (ICA10) did the opposite. The values of activity per area showed that this behavior can be associated to changes in active sites. In the sample with less cobalt, this fact can probably be related to a decrease in activity of the sites, due to the production of a cobalt and iron phase, less active in WGSR than magnetite. On the other hand, the increase of activity of the sample rich in cobalt can be assigned to the production of cobalt ferrite and Co_3Fe_7 alloy which are supposed to be the active phases. The highest activity of this sample can also be related to a textural action of the dopant, avoiding

sinterization during reaction. In fact, cobalt is able to increase the specific surface area as well as to create more active sites than pure iron oxide. From these results, it can be concluded that cobalt and iron compounds are more active than magnetite in WGSR.
Figure 2 (b) shows the activity of the catalyst as a function of time in WGSR. One can see that pure oxide, as well as the sample with small amounts of cobalt, showed stable values since the beginning of reaction. However, the catalyst with high cobalt content showed an activity which increased with time, reaching stable values after 150 min of reaction. This increase can be related to the production of the Co_3Fe_7 alloy, which seems to increase the activity of the catalyst.
By comparing the surface composition (Table 1) with the bulk one, it can be noted that the sample with small cobalt content (ICA05) showed a surface richer in cobalt (Co/Fe= 0.094) than the bulk (around 0.05). On the other hand, the sample richer in cobalt (ICA10) showed a surface (Co/Fe= 0.078) with less cobalt than the bulk (around 1.0). During WGSR, cobalt migrated inside the solids in different directions in each sample, depending on the cobalt amount. If high amounts of cobalt were present, the surface became richer in cobalt after reaction; if small amounts were present, an opposite behavior was observed.

4. Conclusions

Cobalt is an efficient dopant for iron-based catalysts for WGSR at high temperatures, when added in high amounts (Co/Fe (molar)= 1.0). Cobalt ferrite and Co_3Fe_7 alloy are produced in this solid, increasing the catalytic activity, as compared to pure iron oxide. In this solid, cobalt also acts as a textural promoter, increasing the specific surface area and avoiding sintering during reaction. In addition, cobalt delays the iron reduction and thus can preserve the active phase during reaction. Therefore, this catalyst can probably work under low steam conditions, decreasing the operational costs.

References

1. G. Grubert, S. Kolf, M. Baerns, I. Vauthey, D. Farrusseng, A.C. van Veen, C. Mirodatos, E.R. Stobbe, P.D. Cobden, Appl. Catal.: A: Gen 306 (2006) 17.
2. I. Lima Júnior, J.M. Millet, M. Aouineb, M. C. Rangel, Appl Catal. A: Gen 283 (2005) 91.
3. St. G. Christoskova, M. Stoyanova, M. Georgieva, Appl. Catal. A: Gen 208 (2001) 235.
4. J. C. González, M. G. González, M. A. Laborde and N. Moreno, Appl. Catal., 20 (1986) 3.
5. T. Wiltowski, K. Piotrowski, H. Lorethova, L. Stonawski, K. Mondal, S.B. Lalvani, Chem. Eng. Proc. 44 (2005) 775.
6. K.Piotrowski, K. Mondal, H. Lorethova, L. Stonawski, T. Szymański, T. Wiltowski, Int. J. Hydrogen Energy 30 (2005) 1543.
7. W.J. Wang, Y.W. Chen, Appl. Catal. 77 (1991) 223.

The role of carbon nanospecies in deactivation of cobalt based catalysts in CH_4 and CO transformation

L. Borkó[a], Z. E. Horváth[b], Z.Schay[a], L. Guczi[a]

[a]*Department of Surface Chemistry and Catalysis, Institute of Isotopes, Hungarian Academy of Sciences, P. O. Box 77, , Budapest, H-1525, Hungary*
[b]*Research Institute for Technical Physics and Materials Science - MFA, Hungarian Academy of Sciences, P. O. Box 49, H-1525, Budapest, Hungary*

ABSTRACT

Deactivation of $Pt_{10}Co_{90}$ catalyst supported on Al_2O_3 and NaY in non-oxidative methane transformation and CO disproportionation that may have importance in H_2-deficient FT process was investigated in a constant flow system at 1023 K, 10 bar. Link can be found between deactivation of catalysts, especially of bimetallic Pt-Co nanoparticles and the formation, decomposition of meta-stabile CoC_x (XRD) and formation of carbon nanotubes (TEM). The appearance of encapsulated Co particles inside the nanotubes (EDS) proves the restructuring of the Pt-Co bimetallic particles resulting in irreversible deactivation.

1. INTRODUCTION

CO hydrogenation [1] and non-oxidative methane conversion [2] to higher molecular weight hydrocarbons are important reactions in C_1 chemistry. In both processes the formation of inactive carbonaceous surface species deactivates the catalysts [3,4]. In case of CO hydrogenation this is only one of the various effects resulting in deactivation [5], while in CH_4 transformation this has a key importance and should be hampered to obtain a working catalyst. Among the Co-based mono- and bimetallic (Co, Ru-Co, Re-Co, Pt-Co, Pd-Co) samples we studied earlier the NaY and Al_2O_3 supported Pt-Co catalysts are the best in both catalytic process [4.6,7]. The subject of our present study is to reveal the character and role of carbon nanospecies in deactivation of Al_2O_3 and NaY supported Pt-Co bimetallic catalysts in CO and non-oxidative methane transformations, respectively.

2. EXPERIMENTAL

2.1. Sample preparation and chemical characterization

Al_2O_3 and NaY supported Pt-Co catalysts were synthesized by the procedure described in ref. [4]. The samples were analyzed with X-ray fluorescence technique, X-ray diffraction (XRD), X-ray photoelectron spectroscopy, temperature programmed reduction and CO chemisorption. The physical-chemical properties of the samples are listed in Table 1.

Table 1. Physical-chemical properties of the supported $Pt_{10}Co_{90}$ samples [8].

Support	Composition wt.% (atomic ratio) (XRF) Pt	Co	CO_{ads} μmol/g	Extent of reduction TPR*	XPS Co 2p peak	Kind and size of metallic particles (EXAFS)
Al_2O_3	1.9 (8)	6.4 (92)	54	0.46	0.55	Co, Pt-Co(1nm)
NaY	1.8 (8)	5.9 (92)	42	0.98	0.83	Co, Pt-Co(1nm)

*Reduction of PtO and Co_2O_3 into metallic Pt and Co is assumed

By XRD no crystalline metal or metal oxide phases could be detected after reduction, thus Co and Pt must be mainly in form of nanoparticles (~99%) in zeolite and alumina supported $Pt_{10}Co_{90}$ catalysts. According to EXAFS studies two kinds of nanometer scale metallic clusters, i. e. monometallic Co nanoparticles and bimetallic Pt-Co nanoparticles, with almost only Pt-Co bonds co-exist. Size of Co and Pt-Co particles inside the zeolite is about 1 nm as estimated from EXAFS data.

2. 2. Catalytic reaction

Methane conversion and CO transformation reactions were investigated in a flow system detailed elsewhere [4] using 100 mg catalyst at 1023K. The products were analyzed by a gas chromatograph type CHROMPACK CP 9002 using a 50m long plot fused silica column (0.53mm ID) with stationary phase of

CP-Al_2O_3/KCl for hydrocarbons and FID detector, and 30m long Carboxen 1006 PLOT column (0.53 ID) for the permanent gases and TC detector in a temperature programmed mode. The catalysts and carbon formations after reactions were analyzed with XRD, transmission electron microscopy (TEM), electron diffraction (ED) and energy dispersive X-ray spectroscopy (EDS) methods.

3. RESULTS AND DISCUSSION

It has been calculated on the basis of ref. [9] that the ΔG^o value of carbon formation in non-oxidative methane conversion decreases with increasing temperature and reaches a negative value at 1000 K (-19 kJ/mol^{-1}), while in CO disproportionation it is -102 kJ/mol at 400 K and increase from 400K to 1000K (Table 2). At increased pressure the carbon formation is favored in case of CO and hindered in case of CH_4.

Table 2. Gibbs free energy changes (ΔG^o) of methane and CO transformations to hydrocarbons and carbon formations

No	Reaction	$*\Delta\nu$	$**\Delta G^o$, kJ mol^{-1}CH_4			
			400 K	600 K	800 K	1000 K
1	$CH_4 \rightarrow C + 2H_2$	1	42	23	2.3	-19
2	$2CO = C + CO_2$	-1	-102	-66	-30	5.4

$*\Delta\nu = \Sigma\,\Delta\nu_{gas\ phase\ products} - \Sigma\,\Delta\nu_{gas\ phase\ reactants}$; ν-number of moles;
$**\Delta G^o = \Sigma\,\Delta G^o_{products} - \Sigma\,\Delta G^o_{reactants}$

Table 3 shows the amount of carbon formed by catalytic process from methane and CO during 1 h time on stream expressed as number of carbon atoms per active sites (calculated from dispersion). TEM revealed the presence of carbon nanotubes (CNT) (Fig. 1a, a survey picture), similar to those observed by J. B. Nagy and his group [10] using catalytic process for CNT production. The Fig. 1b shows, that the corresponding EDS spectra contain peaks of both Co and Pt. Additionally, Co particles encapsulated in the carbon nanotubes could be observed on most of the micrographs taken from our samples, while no Pt could be detected in the nanotubes (as e.g. Fig. 1c,d and Fig. 1e,f). The deposited carbon is located partly in the carbon nanotubes, partly on the catalyst surface in different forms {as e.g. CoC_x (Co_3C, Co_2C), amorphous carbon}.

Table 3. Methane and CO transformation on supported $Pt_{10}Co_{90}$ catalysts to carbon nano-formations.

Reactant	Support	Yield[a] (%)	Number of carbon atoms		Content of catalyst after reaction		
			N_{tot}[b]	N_{act}[c]	XRD, ED	TEM CNT (nm)	EDS Inside CNT
CH_4*	Al_2O_3	88	93	1352	CoC_x, Co, Pt	10	Co
CH_4*	NaY	101	68.2	2523	CoC_x, Co, Pt	20-40	Co
CO**	NaY	17	11.5	425	CoC_x, Co, Pt	20	Co

Reaction conditions during 1h time on stream:
*8% CH_4/He, 25 cm^3 min^{-1}, 10 bar, 1023K,
**5.1% CO/He, 25 $cm^3 min^{-1}$, 10 bar, 1023K
[a]Yield=$m_C/m_{cat.}$(%); [b]$N_{tot} = N_C/N_{Mtotal}$ -the number of carbon atoms deposited divided by the total number of Pt and Co atoms estimated by TPR; [c] $N_{act} = N_C/N_{Msurface}$ -the number of carbon atoms deposited divided by the number of surface Pt and Co atoms measured by CO chemisorption.

Font Freide et.al [3] mentioned that in FT process even the extremely low level of deposited carbonaceous species on the cobalt crystallite sites led to deactivation of Co/ZnO catalyst. At low hydrogen content carbon was formed from CO by Boudouard reaction to produce carbon on the surface and CO_2. In our experiments full poisoning was reached by the extension of reaction time to 1 h both in CH_4 transformation and CO disproportionation.

The mechanism of the two processes is as follows. From the interaction of methane and the surface cobalt atoms $CoCH_x$ species are formed, which either are converted into C_{2+} hydrocarbons on the surface, or if it is grafted hard to the cobalt surface, they loose all hydrogen atoms and meta-stabile CoC_x is created. At decomposition of the metastable CoC_x carbon nanotubes and other carbon forms are generated on the catalyst surface. In the course of the nanotube growing cobalt particles are disrupted off the surface and encapsulated inside the tubes becoming inaccessible for the reactants. Since Pt could not be found in the carbon nanotubes the decomposition of Pt-Co bimetallic particles [4] should be assumed. Due to this scenario the catalyst becomes irreversibly deactivated. The mechanism may be applied also for CO disproportionation that may have importance in H_2-deficient FT process.

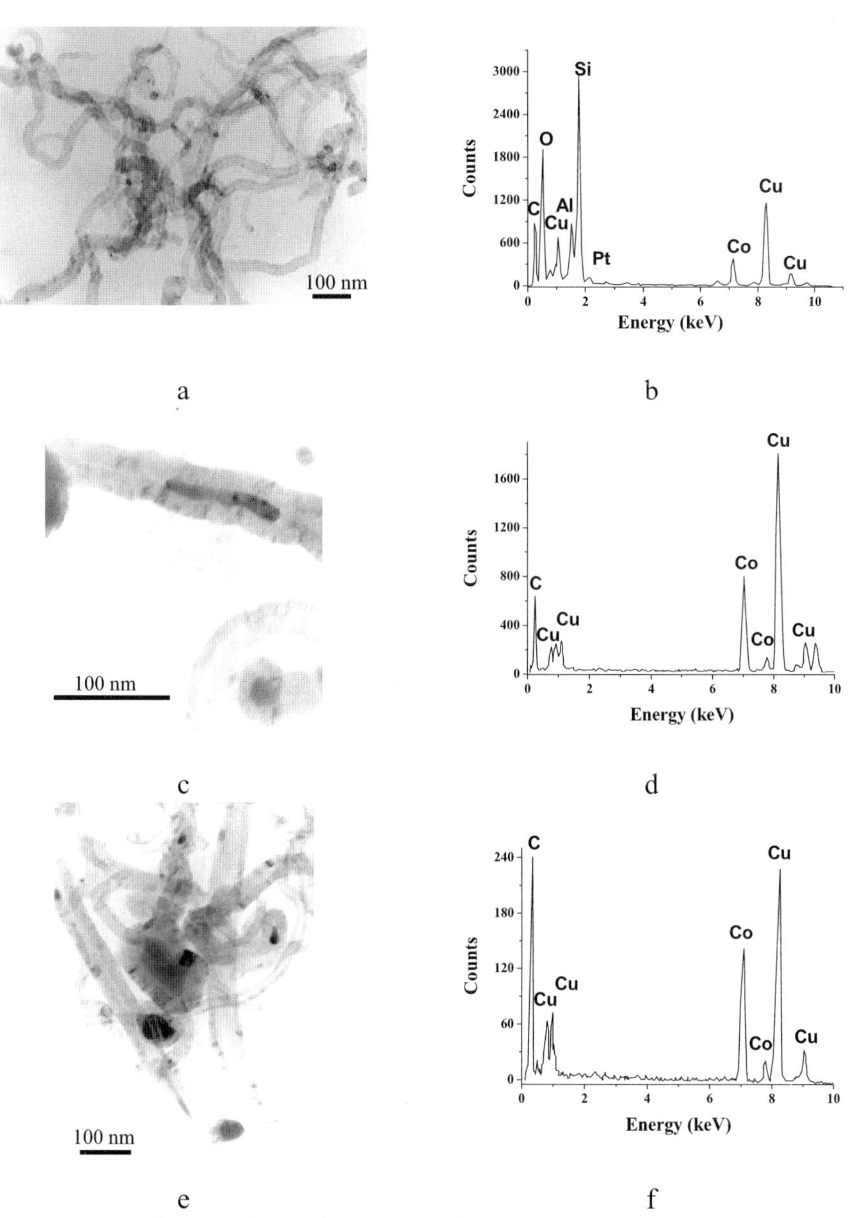

Fig. 1. TEM micrographs and corresponding EDS spectra of CNT formed in non-oxidative methane transformation on $Pt_{10}Co_{90}/NaY$ catalyst: a,b – typical survey picture of the sample; c,d–enlarged picture of a CNT with encapsulated metal rod; e,f – enlarged picture of CNT-s with encapsulated metal spheres. (Cu signal originates from the grid.)

4. CONCLUSIONS

Link can be found between deactivation of supported PtCo catalysts in the CH_4 and CO transformation and the formation and decomposition of meta-stabile CoC_x and formation of carbon nanotubes. The appearance of encapsulated Co particles inside the nanotubes proves the restructuring of the Pt-Co bimetallic particles resulting in irreversible deactivation.

ACKNOWLEDGEMENTS

The authors are indebted to the Hungarian Science and Research Found (grant T-043521) for financial support of the research. The authors thank O. Geszti for the TEM measurements.

REFERENCES

[1] M.E. Dry, Catal. Today 71 (2002) 227.
[2] T.V. Choudhary, E. Aksoylu and D.W. Goodman, Catal. Rev.-Sci. Eng., 45 (2003) 151.
[3] J. J. H. M. Font Freide, T. D. Gamlin, J. R. Hensman et al., J. Nat. Gas. Chem., 13 (2004) 1.
[4] L. Borkó and L. Guczi, Top. Catal., 39 (2006) 35.
[5] W. Zhou, J.G. Chen, K.G. Fang, Y.H. Sun, Fuel Processing Technology, 87 (2006) 609.
[6] L. Guczi, K.V. Sarma and L. Borkó, Catal. Lett., 39 (1996) 43.
[7] L. Borkó and L. Guczi, Stud. Surf. Sci. Catal., 147 (2004) 601.
[8] L. Guczi, D. Bazin, I. Kovács, L. Borkó, Z. Schay, J. Lynch, P. Parent, C. Lafon, G. Stefler, Zs. Koppány, I. Sajó, Top. Catal., 20 (2001) 129.
[9] D.R. Stull, E.F. Westrum Jr., G.C. Sinke, The Chemical Thermodinamics of Organic Compounds, John Wiley and Sons, Inc., New York-London-Sydney-Toronto, 1969.
[10] A. Fonseca, K. Hernádi, P. Piedigrosso, J-F. Colomer, K. Mukhopadhyay, R. Doome, S. Lazarescu, L. P. Biró, Ph. Lambin, P. A. Thiry, D. Bernaerts, J. B. Nagy, Appl. Phys. A, 67 (1998) 11.

Natural Gas Conversion VIII
F.B. Noronha, M. Schmal, E.F. Sousa-Aguiar (Editors)

Effective and stable CeO_2-W-Mn/SiO_2 catalyst for methane oxidation to ethylene and ethane

Bing Zhang, Jiaxin Wang, Lingjun Chou, Huanling Song, Jun Zhao, Jian Yang, Shuben Li *

State Key Laboratory for Oxo Synthesis and Selective Oxidation, Lanzhou Institute of Chemical Physics, Chinese Academy of Sciences and Graduate School of Chinese Academy of Sciences, Lanzhou 730000, PR China

ABSTRACT

CeO_2-W-Mn/SiO_2 catalyst is found to be effective for the oxidative coupling of methane (OCM); ca. 30 % of C_2+ and 22 % of C_2H_4 yields have been obtained in single pass at 800 °C. During the 500 h continuous reaction, no significant decrease of activity and selectivity is observed.

Keywords: OCM, CeO_2-W-Mn/SiO_2, stability

1. Introduction

A number of strategies have been explored and developed for the conversion of methane to more useful chemicals and fuels.[1-6] But to date, no direct processes have progressed to a commercial stage due to generally small product yields operating in a single-pass mode. For the oxidative coupling of methane (OCM), as one of the more promising direct methods for the chemical utilization of methane, the conversions and selectivities obtainable are such that C_2+ (ethane + ethylene) yields are generally< 25%.[7] Nevertheless, the strategic implications of being able to use natural gas as a feedstock continue to provide an incentive to develop the OCM process. The essential way is to explore effective and stable catalysts to improve C_2+ yields or, as an alternative, to enhance the ultimate C-efficiencies of methane conversion.

Here, CeO_2-Na_2WO_4-Mn/SiO_2 catalyst has been investigated for the OCM reaction and the results show that CeO_2-Na_2WO_4-Mn/SiO_2 has good stability and can obtain a high C_2 yield of OCM reaction.

2.Experimental

The CeO_2 modified 5% Na_2WO_4-2% Mn/SiO_2 catalysts were prepared by wet impregnation method. The catalysts were prepared on silica particles (40-55 Mesh) provided by Qindao Ocean Chemical Plant with a surface area of 156 m^2/g. The catalysts were prepared by incipient wetness impregnation method with $Mn(NO_3)_2$, Na_2WO_4•$2H_2O$ and $Ce(NO)_3$•$6H_2O$ respectively. The impregnated silica particles were then evaporated to dryness overnight at 120 °C and calcinated in air at 850 °C for 8 hrs (the surface area of the resulting catalyst = 6.8 m^2/g). Herein, the amounts of the various components were expressed in wt.%.

The catalytic runs were carried out in the integral mode using a fixed-bed vertical-flow reactor constructed from a quartz tube at ambient pressure. The fresh catalyst of 1.0 g was loaded in the reactor and the length of the catalyst bed was approximately 12 mm. Quartz chips were placed above and below the catalyst bed to preheat the reagents and also to decrease the free volume. In experiments, the reactant gases were co-fed into the reactor. At the reactor outlet a cold trap was used to remove water from the exit gas stream. Blank runs with quartz chips showed negligible conversion at the reaction conditions. The reaction products were then analyzed with an on-line gas chromatograph equipped with a TCD, using a Poropak Q column for the separation of CH_4, CO_2, C_2H_4, and C_2H_6, and a molecular-sieve 5A column for the separation of H_2, O_2, CH_4, and CO.

3.Results and discussion

The performance of CeO_2-Na_2WO_4-Mn/SiO_2 catalyst for the OCM reaction was examined by presence different diluents in reactant gases and the results were presented in Table 1. With addition of steam as diluent, obvious effect to increase CH_4 conversion and C_2+ selectivity for CeO_2-W-Mn/SiO_2 catalysts was observed. CH_4 conversion and C_2+ selectivity increased from 43.4 to 47.8 % and 47.5 to 61.8 %. As the results, at 800 °C, methane GHSV = 6667 $ml \cdot g^{-1} \cdot h^{-1}$, $CH_4/O_2/H_2O$ = 2/1/5, C_2+ and C_2H_4 yields reached at 29.6 % and 21.6 % respectively. Because the deep oxidation of hydrocarbon was usually attribute to gas oxygen in OCM reaction, CO_x selectivity decreased with decreasing of O_2 conversion after the steam addition. A possible explanation is that steam can act as a gas phase radical trap, and can interrupt the chain reaction of methane combustion to CO_x, just as the case of a CaNiK catalyst.[8]

He and N_2 as diluents were also investigated to compare with steam for the OCM reaction over CeO_2-W-Mn/SiO_2 catalyst. The comparative results of different diluents at the same reaction condition were listed in Table 1. The addition of He and N_2, to some extent, resulted in higher C_2+ selectivity and lower CO_x selectivity. This can be explained by the fact that the formation of ethylene and ethane are of a lower reaction order than the production of carbon oxides.[9] However, generally speaking, the positive functions of He and N_2 were inferior to that of steam. Because of He and N_2 were much inerter than steam under OCM reaction conditions, it was expectable that the addition of steam to the gas feed had taken out thermal by reaction because H_2O has higher thermal capacity. Moreover, an important incentive for the use of steam as diluent is that almost all of the steam will condense in the quench step following the catalytic reactor and will not have to be handled in the downstream compression and separation systems.

Table 1. Comparison of catalytic performances of CeO_2-W-Mn/SiO_2 catalyst for OCM reaction with addition of different diluents.[1]

Diluents	Conversion. %		Selectivity %				Yield %	
	O_2	CH_4	C_2+	C_2H_4	CO_2	CO	C_2	C_2H_4
No diluent	99.3	43.4	47.5	33.2	34.9	17.5	20.6	14.4
Steam	96.6	47.8	61.8	45.1	19.4	18.8	29.6	21.6
He	99.9	46.1	54.0	37.5	32.5	13.5	24.9	17.3
N_2	99.6	43.2	53.3	36.6	29.5	17.2	23.0	15.8

[1]Reaction conditions: methane GHSV=6667 ml •g-1•h-1, CH_4 /O_2/diluent =2 /1/5, T=800 °C

Table 2 shows the influence of varying ratio of CH_4/O_2 for OCM reaction over CeO_2-W-Mn/SiO_2 catalysts at 800 °C, methane GHSV = 6667 $ml \cdot g^{-1} \cdot h^{-1}$, steam/$CH_4$= 2.5/1. With the increase of CH_4/O_2 ratio, C_2 selectivity increased and CO_x selectivity decreased, accompanying that CH_4 and O_2 conversions decreased sharply. It suggested that gas phase oxygen played a key role in the oxidation of methane and C_2 hydrocarbons to CO_x. At a CH_4/O_2 of 4, the C_2+ selectivity increased to 74.4% with 30.0% CH_4 conversion, which is desirable from the practical viewpoint. Although a higher CH_4/O_2 was beneficial to high C_2+ selectivity, C_2+ yield was dropped by the decrease of CH_4 conversion. Achieved the highest yield was at the point of nearly total exhaustion of the oxygen supply. This observation agrees with experimental observations reported by a number of researchers.[10,11]

With steam as diluent for OCM reaction, one could easily consider that the steam would accelerate the loss of active phases during long reaction period and hence be deleterious to the stability of catalyst. So a 500h stability test over CeO_2-W-Mn/SiO_2 catalyst has been carried out in a 10ml fixed-bed quartz reactor. As described above, lower CH_4/O_2 can obtained high yield of C_2+ hydrocarbon but also decrease the C_2 selectivity. So, in long-term stability test, the reaction conditions were chosen with 800℃, methane GHSV = 6667 $ml \cdot g^{-1} \cdot h^{-1}$ and $CH_4/O_2/H_2O$ = 4/1/4.

Table 2. The effect of CH_4/O_2 ratio for OCM reaction over Ce-W-Mn/SiO_2 catalyst[1]

CH_4/O_2	Conversion. %		Selectivity %				Yield %	
	O_2	CH_4	C_2+	C_2H_4	CO_2	CO	C_2	C_2H_4
2	96.6	47.8	61.8	45.1	19.4	18.8	29.6	21.6
4	83.2	30.0	74.4	47.6	12.4	13.2	22.3	14.3
6	51.5	22.2	75.6	43.6	11.4	13.0	16.8	9.7

[1]Reaction conditions: methane GHSV=6667 $ml \bullet g^{-1} \bullet h^{-1}$, steam/$O_2$=2.5/1, T=800 ºC

Figure. 1 shows that, during 500h period continuous reaction without recharging the catalyst, CeO_2-W-Mn/SiO_2 catalyst remained fairly stable for OCM reaction. The C_2+ selectivity was stabilized at about 70% and the best CH_4 conversion reached 30-31% during the initial several days. The decrease of C_2 and C_2H_4 yield was mainly due to the methane conversion decreased slightly. It was interesting that the CO selectivity increased with time on stream. The sum of C_2+ and CO selectivity remained at 95%, and the CO_2 selectivity decreased from 10% to 5% at 500 h on stream. Unexpectedly H_2 began to form after 116h and the yield of H_2 maintained in the range of 3.2-4.8% on stream. H_2 can be formed via many of possible pathways. H_2 produced with the increase of CO selectivity and the decrease of CO_2 selectivity, which indicated that water-gas shift reaction was not a major pathway for the production of H_2. By the analysis of product distribution, the H_2: CO ratios lay in between about 0.4 to 1, which suggested that hydrogen was probably produced mainly via CO_2 reforming reactions or partial oxidation reactions, but steam reforming reactions was not important. Among methane, ethane and ethylene, it is generally known that methane is much less reactive than the others and the reactivity of ethane is comparable to that of ethylene [12,13]. Therefore, the major reactions for the production of H_2 over Ce-W-Mn/SiO_2 catalyst can be said to be the CO_2 reforming and partial oxidation of ethane and ethylene. This explanation can be exactly consistent with the results shown in Fig. 1 that the production of H_2 was accompanied with the decreases of C_2 and CO_2 selectivity and the increase of CO selectivity. Although the addition of steam improved the performance of Ce-W-Mn/SiO_2 for OCM reaction, one could easily consider that the steam would accelerate the loss of active phases during long reaction period and hence be deleterious to the stability of catalyst.

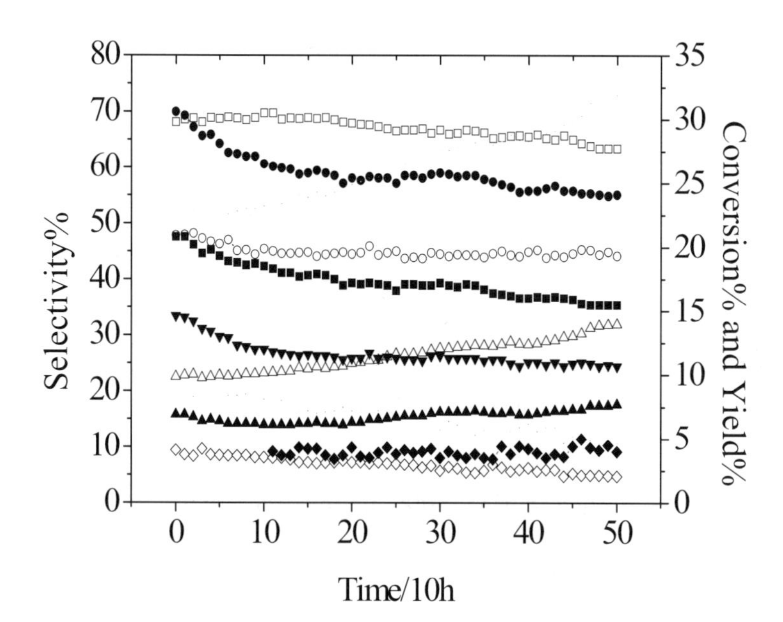

Figure. 1. 500h stability test over 10 ml CeO_2-W-Mn/SiO_2 catalyst at methane GHSV=6667 ml •g^{-1}•h^{-1}, CH_4/O_2/steam=4/1/4, T=800 °C. Conversion: (●) CH_4; Selectivity: (□) C_2+, (○) C_2H_4, (◊) CO_2, (Δ) CO; Yield: (●) C_2, (■) C_2H_4; (▼) CO_2, (▲) CO, , (♦) H_2.

4. Conclusion

In summary, CeO_2-W-Mn/SiO_2 catalyst has been found to exhibit not only excellent catalytic performance for the oxidative coupling of methane but also excellent stability in long-term reaction. The best C_2 and C_2H_4 yields reached at nearly 30 % and 22 % in single pass respectively. Nevertheless, further work is in progress to elucidate the nature of CeO_2-W-Mn/SiO_2 catalyst through catalyst characterization and carry out reaction on lager experimental scales.

5. Acknowledgments

Financial supports from the 973 Project of the Ministry of Science and Technology of China (Grant 2005CB221405) and Nature Science Foundation of China (Grant 20378081) are gratefully acknowledged.

References

1. G.E. Keller and M.M. Bhasin, *J. Catal.* **73**, 9 (1982).
2. Y. Xu, L. Lin, *Appl. Catal. A* **188,** 53 (1999).
3. J.H. Lunsford, *Catal. Today* **63,** 165 (1990).
4. L. Yu, W. Li, V. Ducarme, C. Mirodatos and G. A. Martin, *Appl. Catal. A* **175,** 173 (1998).
5. A. Malekzadeh, A. Khodadadi,M. Abedini, M. Amini, A Bahramian, A.K. Dalai, *Catal. Commun.* **2,** 241(2002).
6. Y. Cai, L. Chou, S. Li, B. Zhang, J Zhao, *Chem. Lett.* **8,** 828 (2002).
7. J.H. Lunsford, *Angew. Chem. Int. Ed. Engl.* **34** (1995) 970.
8. D.M. Ginter, E. Magni, G.A. Somorjai, H. Heinemann, *Catal. Lett.* **16,** 197 (1992).
9. K. Wiele, J.W.M.H. Geerts, J.M.N. Kasteren, in "Methane Conversion by Oxidative Processes," ed. by E.E. Wolf, Van Nostrand Reinhold, New York (1992), Chap. 8, p 259.
10. T. Ito and J.H. Lunsford, Nature, **314,** 721 (1985).
11. I. Matsuura, Y. Utsumi and T. Doi, Appl .Catal. **47,** 299 (1989).
12. J.R. Rostrup-Nielson, Catal.-Sci. Technol. **5** (1984) 1.
13. J.H. Hong, K.J. Yoon, Appl. Catal. A **205** (2001) 253.

Natural Gas Conversion VIII
F.B. Noronha, M. Schmal, E.F. Sousa-Aguiar (Editors)

Very Large Scale Synthesis Gas Production and Conversion to Methanol or Multiple Products

Kim Aasberg-Petersen[a], Charlotte Stub Nielsen[a], Ib Dybkjær[a]

[a]*Haldor Topsøe A/S, Nymøllevej 55, 2800 Lyngby, Denmark*

Abstract

Very long scale production of methanol or products requires high single line capacity to utilise the economy of scale. Autothermal Reforming (ATR) at low steam-to-carbon ratio is the preferred and most cost effective technology for the cost intensive synthesis gas production in those plants. This paper describes technology for very large scale chemical plants and options for reducing the investment including reduction of the steam-to-carbon ratio and combinations with heat exchange reforming. It is illustrated that a common synthesis gas production unit for synthesis of several products (methanol, ammonia, ...) is more economical than several smaller parallel plants.

1. Introduction

The increasing level of oil prices makes schemes for production of synthetic fuel by the GTL (Gas-to-Liquids) route attractive. The size of plants for production of chemicals is increasing and a world scale plant for production of methanol is 5000 MTPD. Even much larger plants are considered to provide the feedstock for the Methanol-to-Olefins process. "Gas hubs" for combined production of several products using a common synthesis gas production unit are also pursued.

The capital cost of the large scale plants is substantial. Synthesis gas production may account for 40-50% of the investment. There is a considerable incentive to reduce the capital cost by technology development and/or by increasing the capacity to utilise the economy of scale. This paper describes preferred technologies for large scale production of synthesis gas and methanol with emphasis on maximum single line capacity. Optimal schemes for production of multiple products are outlined and illustrated by process economical calculations.

2. Synthesis Gas and Methanol Technologies

The ideal synthesis gas composition depends on the desired end product (see Table 1). The optimal technology for synthesis gas production from natural gas for a 5000 MTPD methanol plant is tubular steam reforming followed by oxygen blown secondary reforming [1] (2-step reforming). This technology allows production of gas with the desired composition by adjustment of the operating conditions.

End Product	Ammonia	Methanol	Low Temp. FT
Synthesis gas composition	$H_2/N_2 = 3$ No CO, CO_2	$M = \frac{H_2 - CO_2}{CO + CO_2} \approx 2$	$H_2/CO \approx 2$ Low CO_2

Table 1: Desired synthesis gas compositions, FT = Fischer-Tropsch.

In plants for production of synthetic fuels by the FT synthesis (GTL Plant) and in other high capacity plants, the preferred and most cost effective technology is Autothermal Reforming (ATR) at low steam-to-carbon (S/C) ratio. The process scheme and the ATR reactor are illustrated in figures 1 and 2. A plant based on this process scheme for production of synthesis gas in a GTL plant with a capacity of 2 x 17,000 barrels per day has been started in Qatar [2].

The compact nature of the ATR allows production of synthesis gas for more than 10,000 MTPD of methanol with a single reactor. Plant design at this scale is normally optimised for minimum investment considering various options for maximum single line capacity, as is discussed below.

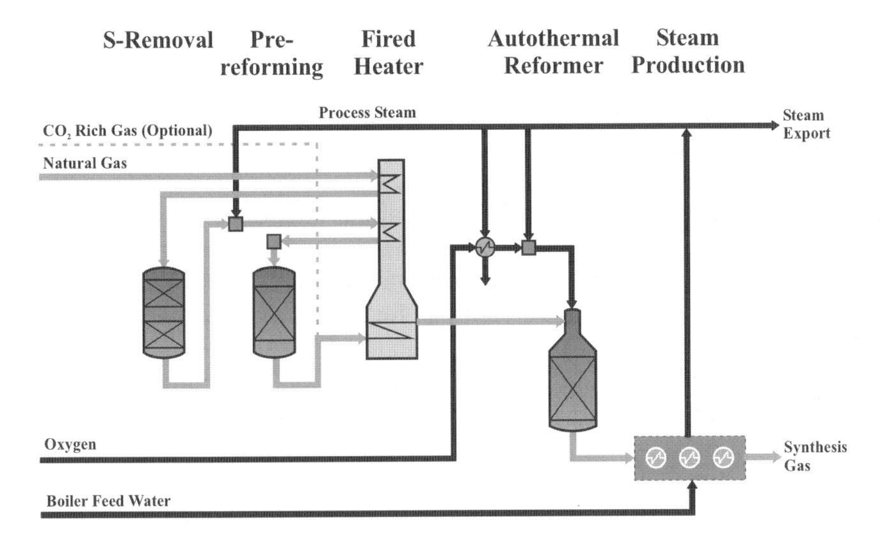

Figure 1: ATR based process scheme for production of synthesis gas.

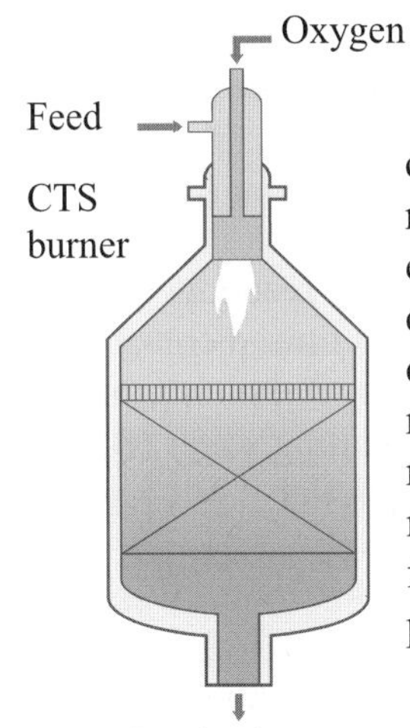

Figure 2: Autothermal Reformer.

The ATR produces a synthesis gas being too rich in carbon as compared to the ideal composition for methanol given in Table 1. M (see Table 1) in the ATR exit gas is typically 1.75-1.85 depending upon feedstock composition and operating conditions. The synthesis gas composition may be adjusted to the ideal value either by removal of carbon dioxide or by extraction and recirculation of hydrogen from the purge gas in the methanol synthesis loop [1]. The single line capacities for these options are illustrated in Table 2 for a specific process study.

Module Adjustment	H_2-recirculation	CO_2-removal
Synthesis Loop Carbon Efficiency (%)	90	94
Plant Single Line Capacity (Index)	100	96

Table 2: Effect of methanol module adjustment technology on maximum single line capacity.

The synthesis loop carbon efficiency in the case with recirculation of hydrogen is limited by the substoichiometric hydrogen content in the synthesis gas. The conversion of hydrogen into methanol or by-products (including water) is above 95%. The carbon efficiency is higher in the case with removal of carbon dioxide. However, the more carbon efficient synthesis loop only partly compensates for the loss of carbon in the carbon dioxide removal step resulting in a lower single line capacity in this specific case.

The single line capacity may be increased by reducing the S/C-ratio allowing a higher throughput of feed gas for the same equipment size. This is illustrated for methanol and GTL plants in Table 3. In the latter case tail gas simulated as carbon dioxide has been added to the ATR inlet to achieve an H_2/CO-ratio of 2 in the synthesis gas.

S/C-ratio	Synthesis Gas Capacity Index (methanol)	Synthesis Gas Capacity Index (GTL)
0.8	96	93
0.6	100	100
0.4	104	108

Table 3: Relative ATR synthesis gas production capacities as a function of the S/C-ratio.

Methanol is produced from synthesis gas by exothermic reaction between hydrogen and carbon oxides. The synthesis loop design is optimised considering investment, energy efficiency and by-product content. The synthesis is typically carried out in reactors with catalyst filled tubes cooled by boiling water (BWR). For large capacities a large number of tubes in multiple parallel BWR-reactors are required.

The BWR reactors comprise a large part of the synthesis loop investment and the economy of scale advantage is not large. As an alternative a process scheme with a series of adiabatic reactors for production of 10,000 MTPD of fuel grade methanol has been evaluated. The results are given in Table 4 along with the effect of reducing the S/C-ratio from 0.6 to 0.4.

S/C-ratio	0.6	0.6	0.4
Methanol synthesis reactor type	BWR	Adiabatic	BWR
Investment index	100	97	96

Table 4: Relative investment for production of 10,000 MTPD of fuel grade methanol.

Table 4 illustrates that the plant capital cost is reduced by using adiabatic reactors and by reducing the S/C ratio.

3. Heat Exchange Reforming for Increased Single Line Capacity

Heat exchange reforming (HTER) may be combined with ATR either in series or parallel [1]. In both cases, the plant efficiency is increased and the oxygen consumption reduced. The latter may be important to avoid additional Air Separation Units (ASUs) in case the capacity limit has been reached.

Installation of the HTER in parallel (HTER-p) increases the plant capacity without introducing additional ATRs. This introduces an additional natural gas conversion reactor, but most other parts of the plant may often be maintained as common equipment. In a methanol plant the inclusion of a heat exchange reformer furthermore increases the module, M (Table 1) in the synthesis gas. Some effects of combining a heat exchange reformer and an ATR are illustrated in Table 5.

Process	ATR	ATR and HTER-p	ATR and HTER-p
Fraction of feed to HTER-p (%)	0	11	16
M (see Table 1)	1.85	1.95	2.00
Oxygen consumption (index)	100	90	86
Synthesis gas produced (index)	100	110	116

Table 5: Effect of using parallel heat exchange reforming (HTER-p) in production of synthesis gas for methanol synthesis.

A heat exchange reformer has been in operation in parallel to an ATR in South Africa for more than 3 years [3]. The synthesis gas produced in this tandem unit has been increased by more than 30% compared to ATR along with increased energy efficiency and a substantial reduction of the oxygen consumption.

4. Synthesis Gas Production for Multiple Products

It may be desirable to combine production of multiple products in one location in so-called “gas hubs”. In most cases it may be possible to take advantage of the large single line capacities of the synthesis gas production unit to decrease investment by economy of scale. An example where multiple products are produced is Sasolburg in South Africa [4]. Two ATR plants started in 2004 operating at an S/C-ratio of 0.6 each with a capacity of more than 200,000 Nm^3/hr (H_2+CO) provide feed gas for the various synthesis end products.

An optimal choice between smaller and separate plants and a plant with common synthesis gas production may not be evident if products with very different requirements to the synthesis gas composition are desired. One example is the combined production of methanol and ammonia. A process economic study has been made to compare two stand-alone plants and a plant with common synthesis gas production and parallel synthesis. The capacity of the study was 5000 MTPD of methanol and 2000 MTPD of ammonia.

The most economical scheme for the integrated plant was to design a 2-step reforming synthesis gas unit to give a composition close to ideal for methanol. Part of the synthesis gas leaving the oxygen blown secondary reformer was split and converted to ammonia synthesis gas by water gas shift, carbon dioxide removal, and nitrogen addition. Key figures for this scheme are compared with figures for two separate plants in Table 6. The investment is significantly lower for the integrated concept mainly because of the economy of scale of the synthesis gas generation unit.

Technology	Separate Plants	Common Synthesis Gas Unit
Methanol synthesis	BWR-reactor	BWR-reactor
Ammonia synthesis	S-300 converter	S-300 converter
M (see Table 1)	-	2.10
Energy consump. (index)	100	101
Investment (index)	100	87

Table 6: Production of 5000 MTPD of methanol and 2000 MTPD of ammonia.

References

1. K. Aasberg-Petersen et al: “Synthesis Gas Production for FT Synthesis”, Stud. Surf. Sci. Cat., 152, p. 258, 2004
2. Bakkerud, Per: “Oryx GTL, Initial Operating Experience”, 8th Natural Gas Conversion Symposium, Brazil, 2007
3. Thomsen, S. G, Han, Pat, Loock, Suzelle, Ernst, Werner:”The first Industrial Experience with A Hador Topsøe Exchange Reformer”, The 51st Annual Safety in Ammonia Plants and Related Facilities Symposium, Vancouver, BC, Canada, 2006
4. A.P. Steynberg “Introduction to Fischer-Tropsch Technology”, Stud. Surf. Sci. Cat., 152, p. 1, 2004

Natural Gas Conversion VIII
F.B. Noronha, M. Schmal, E.F. Sousa-Aguiar (Editors)

Autothermal Reforming of Methane under low Steam/Carbon ratio on supported Pt Catalysts

Juan A. C. Ruiz[a], Fabio B. Passos[b], José M. C. Bueno[c], E. F. Souza-Aguiar[d], Lisiane V. Mattos[a], Fabio B. Noronha[a*]

[a]Instituto Nacional de Tecnologia, Av. Venezuela 82, CEP 20081-312 Rio de Janeiro – RJ, Brazil
[b]Universidade Federal Fluminense, Rua Passo da Pátria 156, CEP 24210-240 Niterói– RJ, Brazil
[c]Universidade Federal de São Carlos-UFSCar, São Carlos, SP, Brazil
d Universidade Federal do Rio de Janeiro, CENPES/PETROBRAS, Rio de Janeiro-RJ-Brazil

Abstract

$Pt/Ce_{0.75}Zr_{0.25}O_2$, $Pt/Ce_{0.14}Zr_{0.86}O_2$ and Pt/CeO_2 catalysts proved to be more stable than $Pt/Ce_{0.50}Zr_{0.50}O_2$, Pt/CeO_2 (prec), Pt/ZrO_2 and Pt/Al_2O_3 catalysts on the autothermal reforming of methane under low S/C ratio (0.2). The results showed that the good stability of the $Pt/Ce_{0.75}Zr_{0.25}O_2$, $Pt/Ce_{0.14}Zr_{0.86}O_2$ and Pt/CeO_2 catalysts is related to a proper amount of oxygen available per metal-support interfacial area on these materials. The high reducibility and oxygen storage/release capacity of support promotes the mechanism of continuous removal of carbonaceous deposits from the active sites, which takes place at the metal-support interfacial perimeter.

1. Introduction

Autothermal reforming (ATR) technology is the most suitable technology for natural gas conversion into synthesis gas to GTL plants. ATR produces a synthesis gas with a H_2/CO ratio equal to 2, which is suitable for the subsequent use in the Fischer-Tropsch synthesis [1]. This H_2/CO ratio can be achieved through recirculation of CO_2 or a CO_2 rich off gas as well as reducing the amount of steam in the feed. Operation at low steam to carbon (S/C) ratio not only improves the syngas composition but also reduces the CO_2 recycle, which decreases the investment and energy consumption. Commercial plant operation at S/C ratio of 0.6 is reported in the literature. However, the reduction of S/C ratio favors soot formation in the ATR reactor and carbon formation in the pre-reformer [2]. Then, there is a great interest on designing catalysts that are able to operate under deactivating conditions. The presence of oxygen vacancies in the support proved to be fundamental in keeping metal particles free from coke deposition on dry reforming and partial oxidation of methane [3,4]. The aim of this work was to study the effect of the support on the performance of supported Pt catalysts on the autothermal reforming of methane. The effect of low S/C ratio (0.2) on the catalyst stability was evaluated.

2. Experimental

Al_2O_3, CeO_2, ZrO_2 and $Ce_{0.14}Zr_{0.86}O_2$ were prepared by calcination of γ-Al_2O_3, $(NH_4)_2Ce(NO_3)_6$, $Zr(OH)_4$ and $Ce_{0.14}Zr_{0.86}O_2$ (Magnesium Elektron, Inc.), respectively at 800°C for 1h. CeO_2 (prec) and $Ce_xZr_{1-x}O_2$ (x=0.50, 0.75) were prepared by a precipitation method as described by Hori et al. [5]. The catalysts were prepared by an incipient wetness impregnation technique using an $H_2PtCl_6.6H_2O$ aqueous solution and were dried at 120 °C. After impregnation, the samples were calcined under air at 400°C for 2 h. All samples contained 1.5 wt% of platinum.
The BET surface area was measured by nitrogen adsorption at -196°C, by using an ASAP 2010 (Micromeritics) equipment. The catalysts were analyzed by powder X-ray diffraction (Philips PW3710), using Cu Kα radiation (λ = 1.5406 Å). The XRD data were collected in the range of 2θ from 20 to 80° (0.04°/step; 1 second/step). Oxygen storage capacity (OSC) measurements were carried out in a multipurpose unit connected to a quadrupole mass spectrometer (Balzers Omnistar). The samples were reduced under H_2 at 500°C for 1h and heated to 800°C in flowing He. Then, the samples were cooled to 450°C and oxygen uptake was measured by passing a 5%O_2/He mixture through the catalyst. Temperature Programmed Surface Reaction (TPSR) experiments were performed in the same apparatus used for OSC measurements. After the reduction, the sample (300 mg) was purged in He at 800°C for 30 min, and cooled to room temperature. Then the sample was submitted to a flow of CH_4/O_2/He (2:1:27) at 30 cm^3/min while the temperature was raised up to 800°C at heating rate of 20°C /min. The platinum dispersion was determined through cyclohexane dehydrogenation reaction, since more traditional techniques such as H_2 or CO chemisorption are not recommended for these catalysts due to the possibility of adsorption of both gases on ceria [6]. Cyclohexane dehydrogenation was performed in a fixed-bed reactor at atmospheric pressure. The catalysts were reduced at 500°C for 1 h and the reaction was carried out at 270°C and WHSV = 170 h^{-1} [7]. The ATR of CH_4 was carried out in a quartz reactor at atmospheric pressure. The reaction was performed at 800°C and WHSV = 260 h^{-1}. The reactant mixture contained

CH_4/O_2 and H_2O/CH_4 ratio of 2.0 and 0.2, respectively. The reaction products were analyzed using a gas chromatograph (Agilent 6890) equipped with a TCD and Carboxen 1010 column.

3. Results and Discussion

The results of BET, OSC and metallic dispersion are shown in the Table 1. The addition of ZrO_2 to CeO_2 increased the surface area from 9 to 43 m^2/g as well as the oxygen storage capacity from 0 to 696 $\mu mols/g_{cat}$, depending on the support composition. On the $Pt/Ce_{0.50}Zr_{0.50}O_2$ and $Pt/Ce_{0.75}Zr_{0.25}O_2$ catalysts, the incorporation of ZrO_2 into CeO_2 lattice promoted the CeO_2 redox properties while the $Pt/Ce_{0.14}Zr_{0.86}O_2$ catalyst exhibited OSC values slightly higher than Pt/CeO_2 catalyst. The presence of ZrO_2 strongly increases the oxygen vacancies of the support, increasing its reducibility. The enhancement of oxygen vacancies is due to the high oxygen mobility of the solid solution formed, which was identified by our XRD data. The X-ray diffraction patterns of CeO_2 and ZrO_2 exhibited the lines characteristic of a cubic and a monoclinic phase, respectively. An homogenous ceria-zirconia solid solution was detected on $Pt/Ce_{0.50}Zr_{0.50}O_2$ and $Pt/Ce_{0.75}Zr_{0.25}O_2$ catalysts while two phases were identified on $Pt/Ce_{0.14}Zr_{0.86}O_2$ catalyst: (i) isolated ZrO_2; (ii) a ceria enriched solid solution with tetragonal structure. The metallic dispersion increased in the order: Pt/CeO_2 (prec) ~ $Pt/Ce_{0.50}Zr_{0.50}O_2$ < $Pt/Ce_{0.75}Zr_{0.25}O_2$ < Pt/ZrO_2 ~ $Pt/Ce_{0.14}Zr_{0.86}O_2$ < Pt/CeO_2 ~ Pt/Al_2O_3.

Table 1: Results of BET, OSC and metallic dispersioncalculate from cyclohexane dehydrogenation reaction.

Catalysts	BET (m^2/g)	O_2 uptake ($\mu mols/g_{cat}$)	Pt Dispersion (%)
Pt/Al_2O_3	180	0	48
Pt/CeO_2 (prec)	9	18	15
Pt/CeO_2	14	194	48
Pt/ZrO_2	20	9	29
$Pt/Ce_{0.75}Zr_{0.25}O_2$	34	626	22
$Pt/Ce_{0.50}Zr_{0.50}O_2$	43	696	14
$Pt/Ce_{0.14}Zr_{0.86}O_2$	43	220	34

Figure 1 shows the TPSR profile obtained for $Pt/Ce_{0.75}Zr_{0.25}O_2$ catalyst. All catalysts showed similar profiles. At 370°C, it is observed the consumption of CH_4 and O_2 simultaneously to the formation of H_2O (not shown) and CO_2. Above 450°C, large CH_4 consumption is detected but now, H_2 and CO are also formed. At temperatures above 600°C (not shown), the H_2 and CO formation keeps growing while CO_2 and H_2O production decreases. These results are consistent to the authotermal reforming of methane proceeding through a two-step mechanism (indirect mechanism). According to this mechanism, the first step comprehends the CH_4 combustion, producing CO_2 and H_2O. In the second step, syngas is produced through CO_2 and steam reforming of unreacted CH_4.

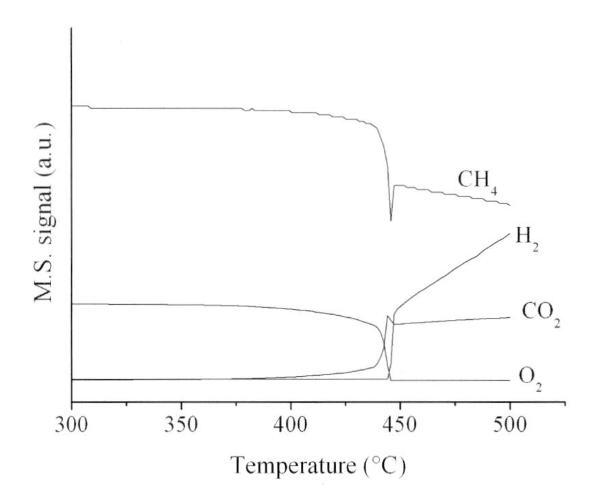

Figure 1: Temperature Programmed Surface Reaction for $Pt/Ce_{0.75}Zr_{0.25}O_2$ catalyst.

The Pt/CeO_2 and Pt/CeO_2(prec) catalysts showed the higher and lower initial activity on ATR, respectively (Figure 2). Furthermore, the Pt/Al_2O_3, Pt/ZrO_2, Pt/CeO_2(prec) and $Pt/Ce_{0.50}Zr_{0.50}O_2$ catalysts deactivated during the reaction. A slightly deactivation is observed on Pt/CeO_2 and $Pt/Ce_{0.14}Zr_{0.86}O_2$ catalysts. The $Pt/Ce_{0.75}Zr_{0.25}O_2$ catalyst practically did not loose their activity after 24 h time on stream (TOS). For Pt/Al_2O_3, Pt/CeO_2(prec) and $Pt/Ce_{0.50}Zr_{0.50}O_2$ catalysts, the H_2 selectivity (not shown) decreased during the reaction, whereas it remained practically unchanged for the other samples. The CO and CO_2 production (not shown) pratically did not change during the reaction on $Pt/Ce_{0.75}Zr_{0.25}O_2$, $Pt/Ce_{0.14}Zr_{0.86}O_2$, Pt/CeO_2 and Pt/ZrO_2 catalysts. For the other catalysts, the CO formation decreased, while the production of CO_2 increased during TOS.

We have recently studied the performance of Pt/Al_2O_3, Pt/ZrO_2 and $Pt/Ce_{0.75}Zr_{0.25}O_2$ catalysts on CO_2 reforming and partial oxidation of CH_4 [8]. $Pt/Ce_{0.75}Zr_{0.25}O_2$ catalyst proved to be more stable than Pt/Al_2O_3 and Pt/ZrO_2 catalysts. The results showed that the stability of this material on these reactions is due to the high oxygen storage capacity of the support. The higher amount of lattice oxygen near the metal particles promotes the mechanism of carbon removal from the metallic surface. According to this mechanism, the first step comprehends the decomposition of CH_4 on the metal particle, resulting in the formation of carbon, which can partially reduce the oxide support, generating CO_x species and oxygen vacancies. In the absence of a reducible oxide, carbon will deposit around or near the metal particle. The second path is the dissociation of CO_2 on the support followed by the formation of CO and O, which can reoxidize the support.

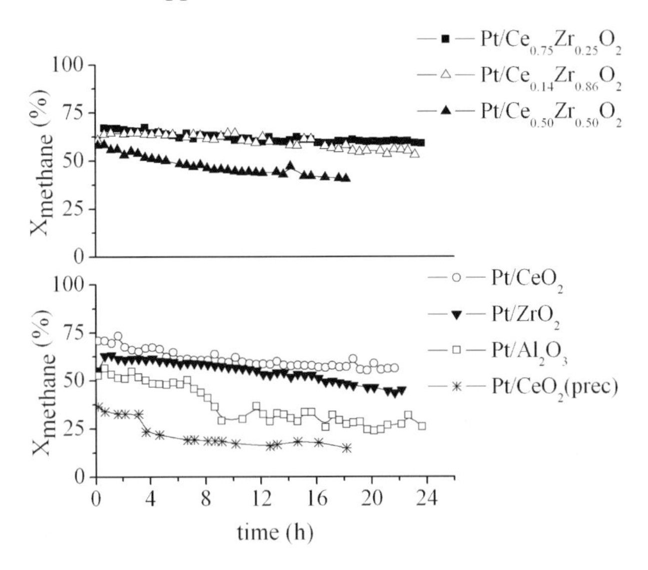

Figure 2: Methane conversion on the autothermal reforming as a function of time on stream (CH_4/O_2 and H_2O/CH_4 ratio of 2.0 and 0.2, respectively; $T_{reaction}$= 800°C and WHSV = 260 h^{-1}).

Since Al_2O_3 support is not a reducible oxide, carbon deposits around the metal particle. The metal dispersion also affected the performance of $Pt/Ce_{0.75}Zr_{0.25}O_2$, $Pt/Ce_{0.50}Zr_{0.50}O_2$ e $Pt/Ce_{0.25}Zr_{0.75}O_2$ on partial oxidation of methane [3]. $Pt/Ce_{0.50}Zr_{0.50}O_2$ catalyst presented the lower stability during this reaction, in spite of its higher amount of oxygen vacancies. The behavior of this material was assigned to its small metal dispersion. The increase of metal particle size decreases the metal-support interfacial area, reducing the effectiveness of the cleaning mechanism of metal particle. These results suggest that the stability of the catalysts on partial oxidation of methane is associated to a proper balance between oxygen transfer ability of the support and metal dispersion. In this work, the deactivation observed for Pt/Al_2O_3 catalyst on autothermal reforming of methane is due to the carbon deposition on metal particle, since Al_2O_3 support does not exhibit OSC. In the case of Pt/ZrO_2 and Pt/CeO_2(prec) catalysts, the deactivation can be related to the low OSC values obtained for these materials (Table 1), which led to the increase of carbon formation on the metal particle. The coke formation affects the CO_2 dissociation. Since the second step of the autothermal reforming reaction is CO_2 reforming of methane, as revealed by TPSR experiments (Figure 1), this step is inhibited. In the case of

$Pt/Ce_{0.50}Zr_{0.50}O_2$ catalyst, the deactivation detected was attributed to its small dispersion, which decreases metal-support interfacial area. On the other hand, the good stability of the $Pt/Ce_{0.75}Zr_{0.25}O_2$, Pt/CeO_2 and $Pt/Ce_{0.14}Zr_{0.86}O_2$ catalysts can be attributed to a proper balance between oxygen transfer ability of the support and metal dispersion. In order to evaluate only the effect of the reducibility of the support on the stability of the catalysts, the performance of $Pt/Ce_{0.14}Zr_{0.86}O_2$ and Pt/ZrO_2 catalysts should be compared, since these samples exhibit the same dispersion (Table 1). $Pt/Ce_{0.14}Zr_{0.86}O_2$ catalyst was more stable than Pt/ZrO_2 catalyst and had a higher OSC value. These results show clearly the role of the reducibility of the support on the cleaning mechanism. Then, all the results presented above showed that the stability of the supported Pt catalysts on autothermal reforming of methane depend on the amount of oxygen vacancies of the support and the metal particle size.

4. Conclusion

The support reducibility and metal dispersion affected the stability of the supported Pt catalysts on autothermal reforming of methane. $Pt/Ce_{0.75}Zr_{0.25}O_2$, $Pt/Ce_{0.14}Zr_{0.86}O_2$ and Pt/CeO_2 catalysts were quite stable, which was attributed to a proper balance between oxygen storage capacity of the support and metal dispersion. The high rate of oxygen transfer keeps the metal surface free of carbon. This mechanism of carbon removal takes place at the metal-support interfacial area. Since the increase of metal particle size decreases the metal-support interfacial area, the effectiveness of the cleaning mechanism is reduced for low metal dispersion.

Acknowledgements
The authors wish to acknowledge the financial support of the PETROBRAS (0050.0007696.04.2).

References

1. K. Aasberg-Petersen, T. S. Christensen, C. S. Nielsen, I. Dybkjaer, Fuel Process. Techn. 83 (2003) 253.
2. Per K. Bakkerud, Catalysis Today 106 (2005) 30.
3. S.M. Stagg, F.B. Noronha, G. Fendley, D.E. Resasco J.Catal. 194 (2000) 240.
4. F.B. Passos, E.R. de Oliveira, L.V. Mattos, F.B. Noronha, Catal. Today 101 (2005) 23.
5. C.E. Hori, H. Permana, K.Y. Ng Simon, A. Brenner, K. More, K.M. Rahmoeller, D. Belton, Appl. Catal. B 16 (1998) 105.
6. P. Pantu, G. Gavalas, Appl. Catal. A: Gen. 223 (2002) 253.
7. Silva, P.P., Silva, F. A., Souza, H. P., Lobo, A.G., Mattos, L.V., Noronha, F.B., Hori,C.E., Catal. Today 101 (2005) 31.
8. L.V. Mattos, E.R. de Oliveira, D.E. Resasco, F. B. Passos, F.B. Noronha, Fuel Process. Tech., 83 (2003) 147.

Conversion of natural gas to higher valued products: light olefins production from methanol over ZSM-5 zeolites

Zilacleide da Silva Barros, Fátima Maria Zanon Zotin, Cristiane Assumpção Henriques*

*Programa de Pós-Graduação em Engenharia Química – Instituto de Química – UERJ, Rua São Francisco Xavier, 524, Rio de Janeiro, CEP: 20559-900, Brazil. *cah@uerj.br*

Abstract

The conversion of methanol to light olefins catalyzed by ZSM-5 zeolites with different SAR or impregnated with phosphorus was investigated. The increase in SAR reduced the density and the strength of the acid sites, favoring both the catalytic stability and the production of light olefins, particularly propene. The incorporation of phosphorous also reduced the density and the strength of the acid sites of HZSM-5 zeolite; besides, BET specific area and microporous volume linearly decreased. The increase of P content decreased the activity and improved the stability. The highest propene/ethene molar ratio was observed for the sample with 4 wt.% of P.

1. Introduction

The increase in worldwide natural gas reserves over the last decade has led to the development of processes for the conversion of natural gas (particularly CH_4) to higher valued products. The conversion of natural gas to light olefins via methanol is an example of a viable alternative route for obtaining petrochemical inputs [1]. It involves the conversion of methane to synthesis gas (steam reform) which in turn produces methanol. The latter is then transformed to light olefins through the so-called MTO process. In this process, methanol is converted to an equilibrium mixture of methanol, dimethylether and water which can be catalytically processed to olefins. The most effective catalysts for MTO process are ZSM-5, a medium-pore size zeolite with a three dimensional pore system, and the small-pore silicoaluminophosphates SAPO molecular sieves. Over medium-pore zeolites, propene is the major olefin produced but C_5^+ hydrocarbons also appear among the reaction products. However, ZSM-5 selectivity toward light olefins can be improved either by optimizing reaction conditions or by adjusting zeolite acid properties [2]. In this work, the conversion of methanol to light olefins catalyzed by ZSM-5 zeolites was studied. The influence of both SiO_2/Al_2O_3 molar ratio (SAR) and phosphorus impregnation was investigated aiming to associate the acid properties of the modified zeolites with their catalytic performance.

2. Experimental

The studied catalysts were three commercial HZSM-5 zeolites with different SiO_2/Al_2O_3 molar ratios (30, 80 and 280), named HZ(30) (from CENPES/PETROBRAS), HZ(80) and HZ(280) (purchased from Zeolyst Int.), respectively, and four P-containing ZSM-5, prepared by wet impregnation of 10g of HZ(30) with 10 mL of H_3PO_4 solution at the appropriated concentration in order to obtain the desired P content. Water was evaporated under reduced pressure at 80°C for 24 h. These samples were named 1PHZ(30), 2PHZ(30), 4PHZ(30), and 6PHZ(30) and contained 1 wt.%, 2 wt.%, 4 wt.%, and 6 wt.% of phosphorus, respectively. The chemical composition of the studied zeolites was determined by X ray fluorescence using a Rigaku spectrometer Rix 2100. X-ray powder diffractograms were recorded in a Rigaku Miniflex X-Ray Diffractometer and textural properties such as specific surface area (BET), micropore volume (t-plot), and mesopore volume (BJH) were determined by N_2 adsorption-desorption at – 196°C in a Micromeritics ASAP 2000. The density and the strength distribution of the acid sites were determined by temperature-programmed desorption (TPD) of NH_3. The catalytic tests were carried out at 500°C with a methanol partial pressure of 0.08 atm (methanol/N_2 molar ratio equal to 0.09) in a fixed bed glass micro-reactor with an "on line" gas chromatograph (CP-PoraPlot Q-HT 25m capillary column, FID). Catalytic activity was compared at a similar WHSV (31 $g_{MeOH}/h.g_{cat}$) and at isoconversion conditions (92 ± 5 %). Before the catalytic tests, the zeolites were pre-treated under N_2 (50mL/min) from room temperature to 500°C, at a heating rate of 2°C/min, and kept at this temperature for 1 h.

3. Results and Discussion

The chemical composition of the studied zeolites, expressed by their chemical SAR, Na_2O wt.% and P wt.%, is shown in Table 1.
As can be observed, P was incorporated at the desired levels and its introduction by wet impregnation did not influence the chemical SAR of the parent zeolite.
The X-ray diffractograms of samples HZ(30), HZ(80) and HZ(280) indicated the characteristic pattern of pure and high crystalline HZSM-5 zeolites. Concerning the phosphorous-containing zeolites, no significant changes were observed in X-ray diffractograms of samples 1PHZ(30), 2PHZ(30), and 4PHZ(30), evidencing that the impregnation of phosphorous species neither affects the crystallinity nor produces additional phases. It suggests that phosphorous species are well dispersed on the zeolite structure. On the other

hand, for the sample with the highest phosphorous content (6PHZ(30)), a decrease in crystallinity and formation of an amorphous phase were observed.

Table 1. Chemical composition, textural characteristics and acid sites density of the studied samples.

	HZ(30)	HZ(80)	HZ(280)	1PHZ(30)	2PHZ(30)	4PHZ(30)	6PHZ(30)
Na_2O (%)	0.36	< 0.005	< 0.05	0.57	0.43	0.43	0.40
P (%)	-	-	-	0.97	1.8	3.7	6.8
SAR**	27.7	81.5	280	26	26.9	26.8	26.7
S_{BET} (m^2/g)	360	422	367	329	314	257	149
S_{ext} (m^2/g)	9.5	53	19	7.1	11.2	7.5	3.6
V_{mic} (cm^3/g)	0.165	0.173	0.166	0.153	0.144	0.119	0.068
V_{mes} (cm^3/g)	0.021	0.098	0.060	0.020	0.025	0.006	0.005
Acidity* $\mu mol_{NH3}/g$	1628	542	216	1211	920	590	350

* Total acid sites density **SAR = SiO_2/Al_2O_3 molar ratio

Textural characteristics of the P-containing samples, shown in Table 1, indicate that BET specific surface area and micropore volume linearly decrease with the increase in phosphorus content. These effects are more significant for sample 6PHZ(30) and can be associated to the formation of amorphous species containing phosphorus that would block the porous structure of the zeolite. For the samples with different SiO_2/Al_2O_3 ratio, the textural characteristics are quite similar, except for HZ(80), which presented a slightly higher BET specific area. NH_3-TPD results, shown in Table 1 and Figures 1 and 2, indicate a reduction in the total acid site density with the increase in phosphorous content or with the decrease in aluminum content. For all samples, the TPD profiles present two desorption peaks; the first peak was associated to weak acidity (maximum between 250 and 270°C) and the second with strong acidity (maximum between 370 and 500°C). Their comparison suggests that the increase in phosphorous content did not influence the strength of the weak acid sites while the strength and the relative contribution of the strong acid sites decrease significantly. On the other hand, for the samples with different SAR (HZ (30), HZ (80) and HZ (280)), the acid strength of weak sites slightly decreases with the decrease in the aluminum content whereas the sample with the smallest Al content seems to possess the strongest acid sites.

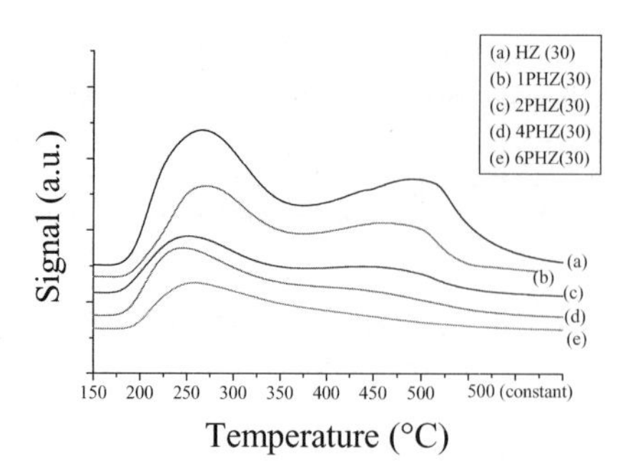

Fig. 1. NH_3-TPD profiles for P-containing samples.

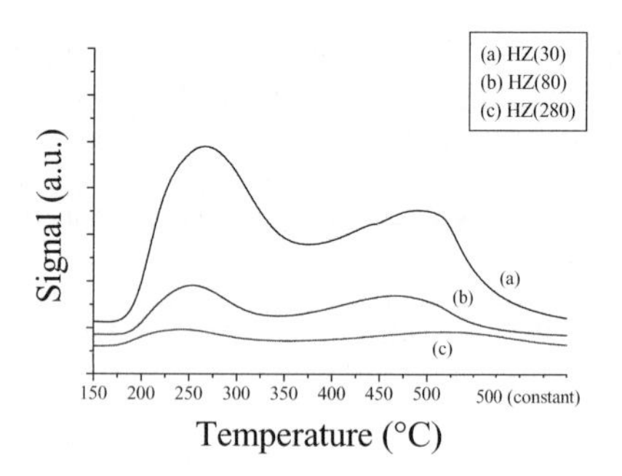

Fig. 2. NH_3-TPD profiles for samples HZ(30), HZ(80) and HZ(280).

In Figures 3 and 4, catalytic performance of the studied samples was compared at similar reaction conditions: 500°C, 31g_{MeOH}/h.g_{cat}, p_{MeOH}=0.08 atm for 4.5 h. The results showed that propene and ethene were the main reaction products along with C4 olefins and $C5^+$ (compounds with 5 or more carbon atoms). Dimethylether (DME) is an intermediary reaction product (Scheme 1) and was detected only for lower conversions or after catalyst deactivation.

$$2\ CH_3OH \underset{1}{\overset{-H_2O}{\rightleftharpoons}} CH_3-O-CH_3 \xrightarrow[2]{-H_2O} C_2^{=} - C_5^{=} \xrightarrow[3]{} \text{alkanes, cycloalkanes, aromatics}$$

Scheme 1. Reaction scheme of methanol conversion into hydrocarbons.

As can be seen in Figure 3, for samples HZ(30), HZ(80) and HZ(280), catalytic activity (MeOH conversion) decreases with the increase in SiO_2/Al_2O_3 molar ratio but the resistance to deactivation by coke increases. It is important to notice that sample HZ(280) did not deactivate after 4.5 h under the studied conditions. Considering that the differences in textural characteristics of samples HZ(30), HZ(80) and HZ(280) are not significant (Table 1), the observed results can be associated to their acid properties (acid sites density and strength distribution). Then, the higher the density and the strength of the acid sites the higher the catalytic activity, not only for methanol conversion into olefins but also for the transformation of the latter into coke molecules responsible for catalyst deactivation. The formation of DME increased upon deactivation, mainly over HZ(30), suggesting that the active sites that catalyze steps 2 and 3 (Scheme 1) are strongly deactivated by coke.
By comparing the P-containing zeolites, Figure 4 shows that the catalytic activity follows the order: HZ(30) ≈ 1PHZ(30) > 2PHZ(30) ≈ 4PHZ(30) > 6PHZ(30) and the stability of the catalysts increases with P-content.

These results indicate that the reduction in acid sites density, specific surface area and micropore volume with the amount of phosphorus influences the catalytic performance of the P-containing samples, although these properties could not explain the similar initial activities of samples HZ(30) and 1PHZ(30) as well as 2PHZ(30) and 4PHZ(30) (Figure 4)

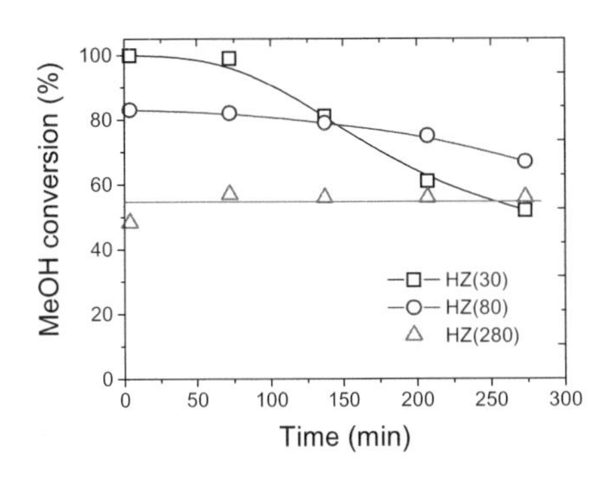

Fig. 3. Evolution of MeOH conversion for zeolites with different SAR with time-on-stream.

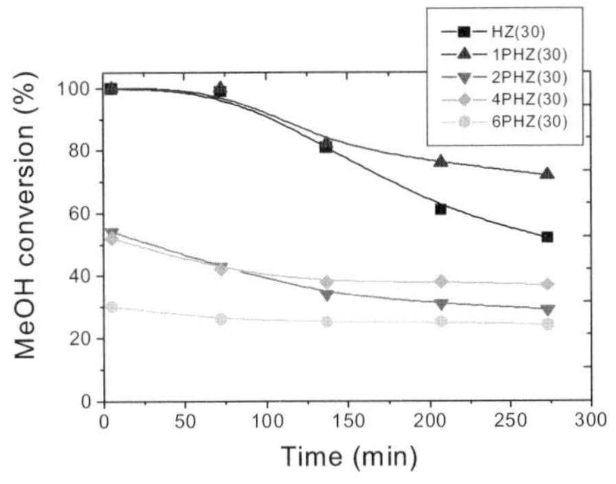

Fig. 4. Evolution of MeOH conversion for P-containing zeolites with time-on-stream.

Aiming to evaluate the effect of both phosphorus content and SiO_2/Al_2O_3 molar ratio on catalyst selectivity and light olefins yields, catalytic performance was evaluated at initial isoconversion (92 ± 5%) and in the absence of coke effects (time-on-stream (t.o.s) = 5 min).

For all samples, Table 2 shows that for high initial conversions ($X_{MeOH} \approx 90\%$) propene is the main reaction product. As expected for these high levels of MeOH conversion, DME was not identified among the reaction products. By increasing the SiO_2/Al_2O_3 molar ratio, ethene production decreases and the formation of $C4^=$ and C5 is favored. On the other hand, the selective production of propene does not seem to depend on acid properties change (SAR increase). As a consequence, propene/ethene molar ratio increases with SAR. These results are in accordance with those reported by Dehertog *et.al* [3] and Chang *et.al* [4]. Both authors reported that, for methanol conversion into light olefins catalyzed by HZSM-5 with different SAR, the acid sites density has a strong influence in the activity, the stability and product selectivities.

The incorporation of low levels of phosphorous (sample 1PHZ(30)) seems to slightly decrease the catalytic activity without significant influence on stability. For higher P-contents, the catalyst activity decreases but the catalyst stability is improved. Concerning P-containing zeolites, there is a clear tendency of increasing propene/ethene molar ratio with P content (Table 2). For the highest P content (6PHZ(30)), a damage in zeolite structure could explain the decrease in propene/ethene molar ratio.

Table 2. Influence of SiO_2/Al_2O_3 ratio and of P-incorporation in the product distribution (molar basis) of the transformation of MeOH over ZSM-5 catalysts (T = 500°C, t.o.s.= 5 min., $X_{MeOH} = 92 \pm 5$ %)

	HZ(30)	HZ(80)	HZ(280)	1PHZ(30)	2PHZ(30)	4PHZ(30)	6PHZ(30)
WHSV (h^{-1})	33.1	21.0	12.4	22.8	19.2	19.6	3.3
X_{MeOH} (%)	97	88	90	87	95	90	97
CH_4	1.1	1.3	0	2.89	1.5	0	2.1
C_2H_4	24.1	14.1	8.4	21.9	23.8	15.4	34.4
C_3H_6	52.5	39.9	50.4	31.9	39.9	46.1	38.9
DME	0.7	1.9	0	3.2	0	2.3	0
$C_4^=$	14.4	13.4	16.6	11.19	12.7	13.7	9.7
C_5	1.2	6.1	8.8	4.9	5.2	5.8	3.7
C_6	1.2	9.7	5.7	8.1	10.1	5.7	7.1
C_3H_6/C_2H_4	2.2	2.8	6.0	1.5	1.7	3.0	1.1

4. Conclusions

The conversion of methanol to light olefins over HZSM-5 was strongly influenced by the SiO_2/Al_2O_3 ratio of the zeolite. The increase in this parameter reduced the density and the strength of the acid sites, favoring both the catalytic stability and the production of light olefins, particularly propene.
The incorporation of phosphorous also reduced the density and the strength of the acid sites of HZSM-5 zeolite; besides, BET specific area and microporous volume linearly decreased. The increase of P content decreased the activity and improved the stability. The highest propene/ethene molar ratio was observed for the sample with 4 wt.% of P.

Acknowledgements

The authors would like to thank CENPES/PETROBRAS – Brazilian Petroleum Company for the technical and financial supports and NUCAT/COPPE/UFRJ for XRD and XRF analysis.

References

1. P. Barger, Methanol to Olefins (MTO) and Beyond, in Zeolites for Cleaner Technologies, G. Guisnet and J. P. Gilson, eds, Imperial College Press, 2002, p.239.
2. M. Stöcker; Microp. Mesop. Mat. 29 (1999) 3.
3. W.J.H.Dehertog and G.F.Froment, Appl. Catal. 71 (1991) 153.
4. C.D.Chang, C.T.Chu and R.F.Socha, J.Catal. 86 (1984) 289.

Natural Gas Conversion VIII
F.B. Noronha, M. Schmal, E.F. Sousa-Aguiar (Editors)

Lurgi's Methanol To Propylene (MTP®) Report on a successful commercialisation

Harald Koempel, Waldemar Liebner

Lurgi AG, Lurgiallee 5, D-60295 Frankfurt am Main, Germany

The idea and the first steps

About ten years ago researchers and technologists at Lurgi pondered on the potential of the gas market and on the necessary answers in development. This considered among others the already broad experience in gas technology like synthesis gas preparation and cleaning as well as methanol synthesis. The following points were conceived:

- A zeolite catalyst of type ZSM5 was at hand which was known to support the oligomerisation of olefins as well as formation of olefins from methanol and dimethylether
- A midterm shortage in propylene was developing due to rising demand especially for polypropylene. This demand seemed to be uncovered by the conventional production methods like steam crackers and FCCs. Figure 1 shows this "propylene gap".
- At that time more than 100 billion cubic metres of natural gas and

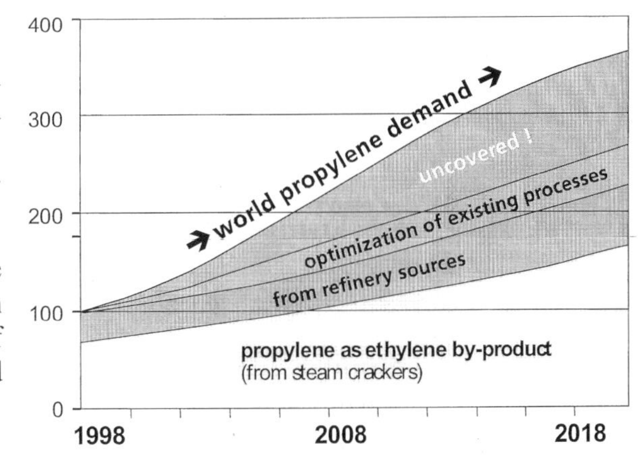

Figure 1 The Propylene Gap

oil associated gas were flared for technical reasons or for lack of markets. A consciousness was developing that this was an intolerable waste of resources in addition to greenhouse gas emission. This explains the incentive for engineers and environmentalists as well to come up with novel ideas for the utilisation of this gas.

- Lurgi at that time started to introduce its new groundbreaking MegaMethanol® process for plants with a production of 5,000 tons of methanol per day and more. [Streb, Göhna, 2000] Methanol would be available at a constant low price in the foreseeable future. This called for the development of downstream technologies for the conversion of methanol to more valuable products.

The team of researchers and technology-strategists then made some basic decisions: A process should be developed for the production of olefins from methanol – and not just any olefins but predominantly propylene which was seen as the highest value product.
The process had to utilize the catalyst owned by the catalyst partner, Süd-Chemie and it had to be based on fixed bed reactor technology. Fixed-bed belonged to the core competences of Lurgi's engineers and it is the simplest to scale up from pilot plant to commercial size – with scaling factors in the thousands and even in the ten thousand range not uncommon.

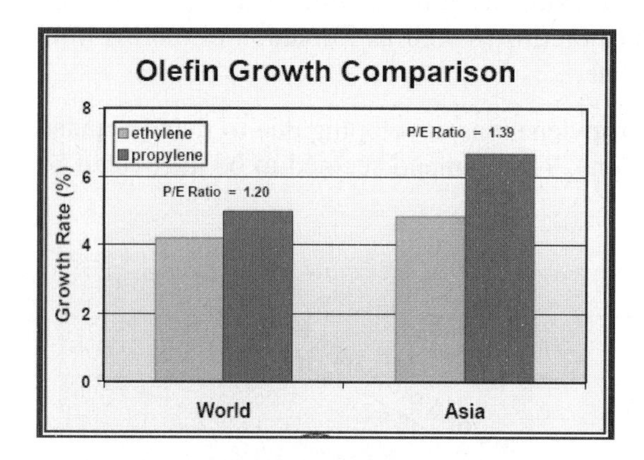

Figure 2 Propylene over Ethylene Demand

After that a R&D program was started, first using bench-scale "process demonstration units" in the lab near the head office in Frankfurt/Main and after about another five years a pilot plant at Statoil's methanol plant in Tjeldbergodden, Norway. These developments have been reported in detail earlier. [Rothaemel, Holtmann 2001] [Koempel, Liebner, 2004]

Propylene – Still an attractive product with high value

Demand growth of propylene is projected at higher than 4% worldwide with marked regional spikes as e.g. for Iran, India, PR China. Specifically, the growth in Asia is projected at more than one percent higher than overall. Also, the propylene to ethylene ratio is significantly higher in Asia as seen in Figure

2. Polypropylene is by far the largest and fastest growing of the propylene derivatives, and requires the major fraction of about 60 % of the total propylene as shown in Figure 3. The increasing substitution of other basic materials such as paper, steel and wood by PP will induce a further growth in the demand for PP and hence propylene. Other important propylene derivatives are acrylonitrile, oxo-alcohols, propylene oxide and cumene. The average growth rate for propylene itself is estimated very conservatively to be 4.5 % per year for the next two decades. How to satisfy this demand for propylene?

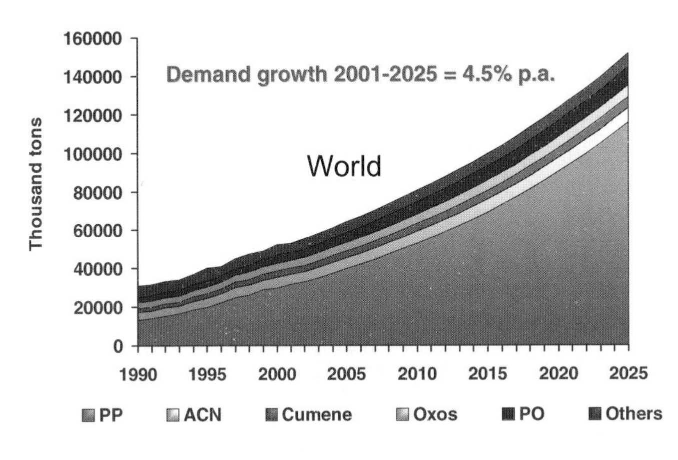

Figure 3 Propylene Demand Growth

Most current forecasts indicate a lasting gap of propylene production that has to be filled by other sources. Lurgi's new MTP process has already started to fill that gap.

Lurgi's Gas to Propylene (GTP® / MTP®) Technology

GTP® and MTP® are registered terms for the same process – they are used synonymously, the first "G" in the context of a complete process chain starting with natural gas or any synthesis gas from other sources, the second "M" when looking at the core process.

Lurgi's new MTP® process is based on an efficient combination of the most suitable reactor system and a very selective and stable zeolite-based catalyst. Lurgi has selected a fixed-bed reactor system because of its many advantages over a fluidised-bed. The main points are the ease of scale-up of the fixed-bed reactor and the significantly lower investment cost.

Furthermore, Süd-Chemie AG manufactures a very selective fixed-bed catalyst commercially which provides maximum propylene selectivity, has a low coking tendency, a very low propane yield and also limited by-product formation. This in turn leads to a simplified purification scheme that requires only a reduced cold box system as compared to on-spec ethylene/propylene separation.

With Figure 4 a brief process description reads: Methanol feed from the MegaMethanol® plant is sent to an adiabatic DME pre-reactor where methanol is converted to DME and water. The high-activity, high-selectivity catalyst used nearly achieves thermodynamic equilibrium. The methanol/water/DME stream is routed to the MTP® reactor together with steam and recycled olefins. Methanol/DME are converted by more than 99%, with propylene as the predominant hydrocarbon product. Process conditions in the five or six catalyst beds per reactor are chosen to guarantee similar reaction conditions and maximum overall propylene yield. Conditions are controlled by feeding small streams of fresh feed between the beds.

Two reactors are operating in parallel while the third one is in regeneration or stand-by mode. Regeneration is necessary after about 500-600 hours of cycle time when the active catalyst centres become blocked by coke formed in side-reactions. By using diluted air, the regeneration is performed at mildest possible conditions, nearly at operating temperature, thus avoiding thermal stress on the catalyst.

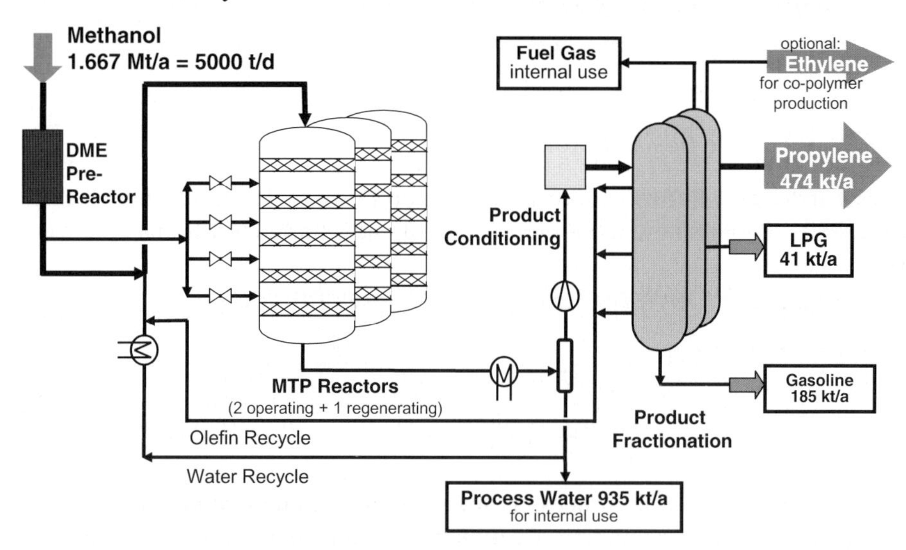

Figure 4 Simplified Process Flow Diagram

The product mixture from the reactors is cooled and the product gas, organic liquid and water are separated. The product gas is compressed and traces of water, CO_2 and DME are removed by standard techniques. The cleaned gas is then further processed yielding chemical-grade propylene with a typical purity of more then 97%. Several olefin-containing streams are sent back to the main synthesis loop as an additional propylene source. To avoid accumulation of inert materials in the loop, a small purge is required for light-ends and the C_4/C_5 cut. Gasoline is produced as a by-product. Water is recycled to steam generation for the process; the excess water resulting from the methanol conversion is purged. This process water can be used as supplemental raw water to the complex or for irrigation after appropriate and inexpensive treatment.

An overall mass balance is included in the figure based on a combined MegaMethanol® / MTP® plant. For a feed rate of 5,000 tons of methanol per day (1.67 million tons annually), approx. 471,000 tons of propylene (as 100%) are produced per year. If needed for co-polymer production, a small amount of ethylene product is available. By-products include fuel gas (used internally) and LPG as well as liquid gasoline and process water.

The technological status of MTP® in the areas of process and catalyst can be summarised as follows: The basic process design data were derived from more than 9,000 operating hours of a pilot plant at Lurgi's Research and Development Centre. Besides the optimisation of reaction conditions also several simulated recycles have been analysed. Parallel to that Lurgi decided to build a larger-scale demonstration unit to test the new process in the framework of a world-scale methanol plant with continuous 24/7 operation using real methanol feedstock. After a co-operation agreement with Statoil ASA was signed in January 2001 the Demo Unit was assembled in Germany and then transported to the Statoil methanol plant at Tjeldbergodden (Norway) in November 2001. Later in 2002 Borealis joined the co-operation.

The Demo Unit was started up in January 2002, and the plant was operating almost continuously since then. As of September 8th 2003, the Demo Unit completed the scheduled 8000 hours life-cycle test. With that the main purpose of the test was achieved: to demonstrate that the catalyst lifetime meets the commercial target of 8000 hours on stream. Cycle lengths between regenerations have been longer than expected. Deactivation rates of the methanol conversion reaction decreased with operation time. Propylene selectivity and yields were in the expected range for this unit with only a partial recycle. Also, the high quality of the by-product gasoline and the polymerisation grade quality of the propylene were proven.

After the life-cycle test the demo-unit was operated for another 3000 hours with a second batch of catalyst to obtain verification and new results from variation of operating conditions. Thus, after successfully logging 11,000 hours in an industrial environment the demo unit was brought home to Lurgi's R&D centre. There, for further investigations and fine-tuning a new process demonstration unit (PDU) was erected which is a full representation of a commercial unit in all relevant process and recycle streams. This PDU continuously delivers additional data for corroboration of the simulation model as well as for product composition and quality studies - especially of the gasoline by-product. It should be mentioned in this context that the gasoline is an excellent blending stock for refineries due to its total lack of sulphur and benzene and hid high octane number.

The catalyst development is completed and the supplier Süd-Chemie commercially manufactures the catalyst. Nevertheless, since there is always room for improvement, new studies have been initiated to possibly raise propylene yield and/or lifetime. Due to the sales success described below Süd-Chemie is expanding current manufacturing facilities.

Commercialization has started

Today, Lurgi offers the process on fully commercial terms. A contract has been signed on the very first plant with a capacity of 100 kt/a of propylene in the Middle East. Basic engineering work is complete and the bidding procedure for the next phases is prepared by the client.

By end of 2005 another contract was signed in China, for a "full size" plant of 471kt/a capacity. Here the basic engineering work is completed and procurement of the long-lead items is well under way. Contractually, the complex which comprises coal-gasification, Rectisol® syngas-cleaning, Lurgi MegaMethanol®, MTP® and a polypropylene plant will start up by end of 2008.

For another similar project in China the contract was signed by end of June 2006. Basic Engineering is near completion. Also here procurement of long-lead items has started.

Several projects around the world are in discussion/study/project development phases. They are based on natural gas or coal and also on methanol feed directly.

Conclusions

There are abundant natural gas reserves providing low cost feedstock for methanol production and aiming at better use of natural resources especially in the case of associated gases being flared. Especially propylene produced from methanol will increase the value of natural gas considerably and offers an exciting potential of growth and a high earnings level. Excitingly enough a similar reasoning holds true for regions where low-grade coal is found in abundance. The higher investment necessary for coal gasification is compensated by the extremely low feedstock cost afterwards.

Lurgi's MegaMethanol® technology can bring down the net methanol production cost below US$ 50 per ton, wherever low cost natural gas or coal are available. This opens up a completely new field for downstream products like DME and propylene.

Based on simple fixed-bed reactor systems, conventional processing elements and operating conditions including commercially manufactured catalysts, Lurgi's MegaDME and GTP/MTP® provide attractive ways to "monetise" natural gas and abundant low-grade coal as well.

Commercialisation of GTP/MTP® has started in earnest and in fullest scale: Aside from a smaller project in Iran, two world-scale plants of 471 kt/a propylene capacity are scheduled for erection in China – with procurement of long lead items already under way.

REFERENCES

S. Streb and H. Göhna: "MegaMethanol® - paving the way for new downstream industries", World Methanol Conference, Copenhagen (Denmark), November 8 – 10, 2000

M. Rothaemel and H-D. Holtmann: "**MTP, Methanol To Propylene - Lurgi's Way",**
DGMK-Conference "Creating Value from Light Olefins – Production and Conversion", Hamburg, October 10 – 12, 2001

H. Koempel, W. Liebner and M. Rothaemel: "Progress Report on MTP with focus on DME", AIChE Spring National Annual Meeting, New Orleans, April 25-29, 2004, Session: Olefins Production

Natural Gas Conversion VIII
F.B. Noronha, M. Schmal, E.F. Sousa-Aguiar (Editors)

Partial oxidation of hydrocarbon gases as a base for new technological processes in gas and power production

Vladimir S. Arutyunov

Semenov Institute of Chemical Physics, Russian Academy of Sciences, Kosygina 4, Moscow 119991, Russia

Abstract

Simple low-scale technologies based on the direct oxidation of hydrocarbon gases can significantly facilitate the development of low-resource and remote gas fields and also contribute to the solution of critical environmental problems of oil and gas industry and power generation.

1. Introduction

It is obvious that oil and gas will remain the main sources of energy at least up to the end of this century. As the most accessible resources deplete, the efficient exploitation of less accessible, remote, low-capacity, low-pressure and non-traditional resources of hydrocarbons becomes more and more important. Especially severe is the problem of natural and associated gas production, processing and transportation. As the industry demands simple and low-scale technologies, processes of light alkane partial oxidation to oxygenates, including the direct methane oxidation to methanol (DMTM) [1-5], becomes more atractive. Although it is a complex chain-branched process, the yield of main products can be roughly expressed by the following gross-equation

$$2CH_4 + 2O_2 \begin{cases} \nearrow 0.9CH_3OH + 0.1(CH_2O + H_2) \\ \searrow 0.9(CO + 2H_2O) + 0.1(CO_2 + H_2O + H_2) \end{cases} \quad (1)$$

Several recently suggested and developed prospective applications of DMTM for natural gas processing and power production are discussed below.

2. Low-scale methanol for preventing formation of gas hydrates

The most obvious and practically advanced application of DMTM in gas industry is low-scale production of methanol for preventing a gas hydrate formation in production, processing and transportation of natural gas from remote gas fields. Annual consumption of methanol for these purposes in Russian gas industry values at 300 000 ton. There are dozens gas production sites with critical need of uninterrupted methanol delivery, but without convenient transport infrastructure, spaced at huge distance from one another. Each of them requires 3-5 thousand ton of methanol per annum. A more than two-fold excess of the methanol cost at a gas well over the market price is usually taking place due to the transportation problems. Technical and economic estimations show that very simple DMTM installations can cover the demand at several times lower methanol self-cost. The estimated term of capital costs return, is less than 3-5 years. Such unit with a capacity of about 5,000 t/a designed for one of Russian gas fields consists of the simplest set of equipment and uses air as an oxidant. Thus produced rough methanol can be also used locally as a liquid fuel or its component.

Based on our experimental studies and kinetic modelling of the DMTM process different types of reactors have been designed to restrict the temperature jump during the oxidation and thus insure a high yield of methanol. Several versions of the process design for different specific cases were evaluated, including those with a direct injection of the methanol-containing reaction mixture without a preliminary separation into pipe-line to prevent the down-stream hydrate formation.

3. On-site production of methanol for sulphur removal

We believe that it is prospective to use such on-site produced low-scale methanol also as an absorbent for sulphur and some other admixtures removal from raw natural gas as in the well-know Rectizol process, developed by Lurgi. Despite its relatively lower absorption efficiency as compared with ethanolamine, on-site produced methanol is significantly cheaper. In addition, one and the same installation helps to remove sulphur compounds and CO_2, and to prevent hydrate formation in gas wells and in downstream pipe-lines. Such integrated technology is under development now.

4. Processing of ethane fraction from associated and natural gas separation plants

Another group of processes that now is under consideration is aimed at solving a very important economic and environmental problem of remote oil and gas producing regions, namely, preventing flaring or venting of associated gases. There are four main groups of products that arise at associated and natural gas separation. Whereas gas condensate and LPG fractions can be transported as normal or pressured liquids and dry natural gas can be transported via existing pipe-lines, it is very difficult, and thereby usually unprofitable, to transport the ethane fraction from remote gas processing plants to potential consumers. That is why in many cases it is simply burnt or vented. For example, only few percent of ethane produced with natural gas in Russia is collected and utilised in one way or another. More than ninety percent of this fraction is lost. But this fraction is an ideal raw material for the direct oxidation into oxygenates. According to our recent results [6], it can be processed at a relatively low pressure, less than 3 MPa, providing a high yield of such valuable products as methanol, formaldehyde, ethanol, etc. Inclusion of this process in a gas separation plant allows to combine units and heat flows and make the overall process more flexible and profitable.

5. Production of chemicals. Oxidative reforming of fat gases

Two main objections against DMTM process are usually considered. First, a low conversion per pas that leads to the need in intensive re-circulation of process gases. Another one is a relatively low selectivity to methanol (~45%) mainly due to the formation of CO according to equation (1). It leads to a substantial decrease in carbon efficiency. Nevertheless, in some cases it is possible to overcome these drawbacks and to make simple low-scale DMTM-based technologies competitive not only at remote sites. If carbon monoxide, the main gas product of the process, is used as a chemical feedstock with or without separation from reaction gases, the complete carbon efficiency can be significantly increased. For example, it was suggested to combine DMTM-based methanol production with subsequent methanol carbonylation to produce acetic acid [7]. Some other carbonylation products, e.g. methyl formate [8] and dimethyl carbonate [9] can be considered to utilize CO in reaction gases.

Another obvious route is the use of outgoing process gases as raw material or energy source. For example, one can use such property of direct oxidation as much easier oxidation of heavier hydrocarbons. Preliminary oxidative reforming can be used, e.g. as the first stage of catalytic steam reforming of fat associated gases to decrease the concentration of heavier components and thus to decrease the possibility of soot formation.

6. Methanol production on power plants and decreasing NOx emission

The most promising, to our point of view, is the possibility to produce low cost methanol on numerous power plants fed by natural gas. The integrated DMTM production of methanol with the use of outgoing gases for power generation is the best way to avoid the above-mentioned drawbacks of the process (its relatively low selectivity to methanol and low conversion per pass), because all produced heat and unconverted gases can be utilized for power generation. In this case the specific consumption of methane per ton of produced methanol can be as low as in the traditional process via syngas. At the same time, it retains all principal advantages of DMTM, such as the simplest one-stage process with very simple equipment without recycling and with air as an oxidant instead of costly oxygen (fig. 1). Such integration is especially profitable when power plant is fed directly from a high pressure pipe-line which makes the additional gas compression unnecessary. The optimized combination of heat fluxes allows increasing the methanol recovery by approximately 15%. Rough methanol from a number of such integrated plants can be further purified or processed at a separate chemical plant.

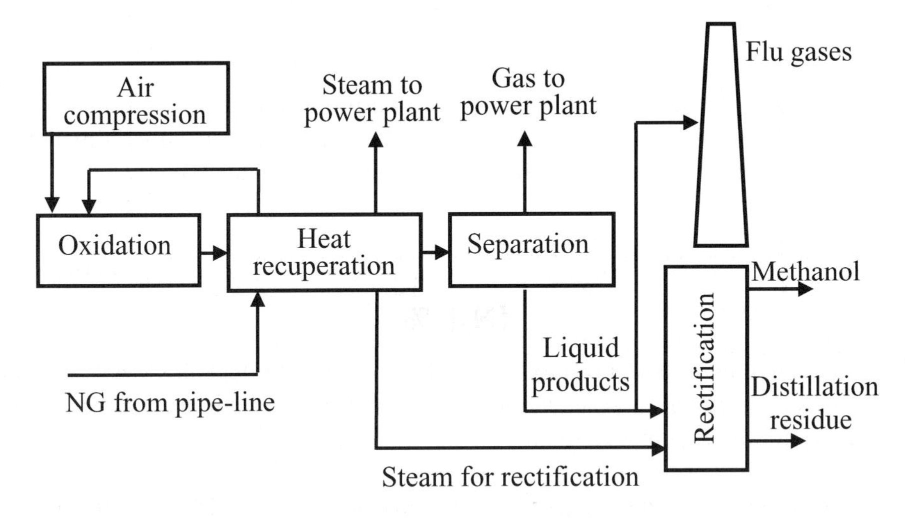

Fig. 1. A schematic sketch of DMTM application for power plant.

In addition, the integrated process can noticeably contribute to solve one of the main environmental problems of natural gas combustion for power generation. It can help to reduce the level of nitrogen oxides emission. High equilibrium concentration of NO in exhaust gases (up to 2000 ppm) is due to the high combustion temperature of natural gas. Existing ways for preventing NOx formation or their abatement are very costly and can increase the power plant capital cost and produced energy price by a factor of 1.5. The combination of a power plant with a methanol unit would lead to the production of the cheapest methanol accompanied with a substantial decrease of NOx level in power generation unit exhaust. The latter is due to a decrease of caloric value of DMTM outgoing gases diluted with nitrogen [10].

The possibility of decreasing NOx production by dilution of combustible gases with up to 30% of neutral components, e.g. N_2 or re-circulating flue gas itself, is well-known in power production. It leads only to a minor decrease in efficiency, about 0.02-0.03% per 1% of added neutral component, but too costly due to the lack of cheap sources of such neutral components. The proposed integration with DMTM methanol production solves this problem (fig. 2).

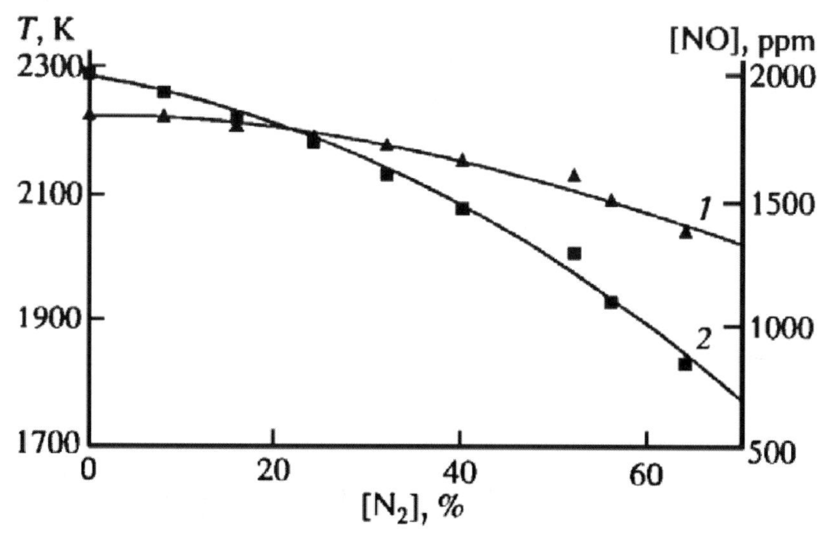

Fig. 2. Calculated temperature (1) and NO concentration (2) at adiabatic combustion of stoichiometric methane-air mixture as function of the nitrogen concentration in methane [10].

Moreover, the on-site produced rough methanol can be used to promote the NO oxidation to NO_2 in flue gases [11] (fig. 1). The subsequent capture of NO_2 in scrubbers leads to a virtually total abatement of NOx emission.

7. Conclusions

The direct partial oxidation of hydrocarbon gases is not so developed and elaborated as widely used route of hydrocarbons processing via syngas production. Newertheless, it is very flexible both from capacity and feeding gas composition points of view and has numerous and prospective areas for application (fig. 3). It can provide with an effective use of different hydrocarbon feedstocks alongside with significant environmental benefits and production of valuable chemicals and fuels.

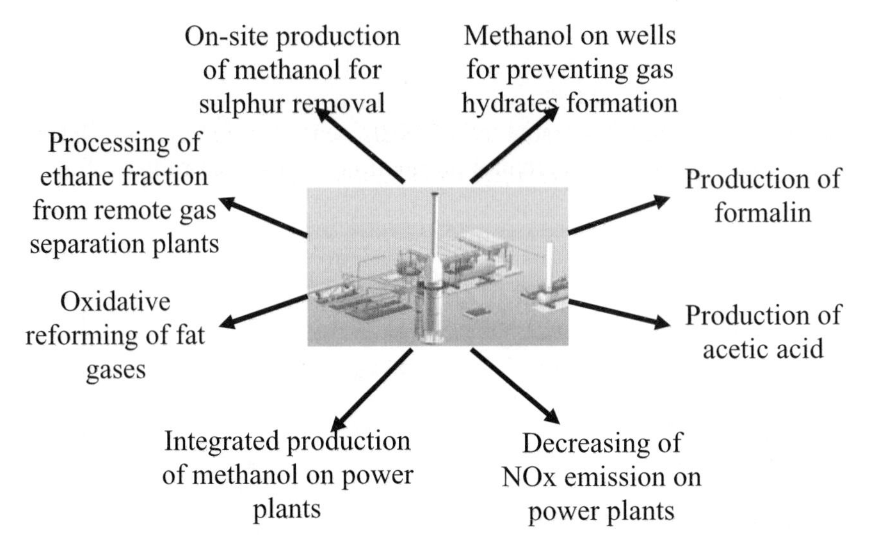

Fig.3. Prospective areas for the application of the direct partial oxidation of hydrocarbon gases.

References

[1] V.S. Arutyunov, V.Ya. Basevich, V.I. Vedeneev, Ind. Eng. Chem. Res., 34 (1995) 4238.
[2] V.S. Arutyunov, V.Ya. Basevich, V.I. Vedeneev, Russian Chemical Reviews, 65 (1996) 197.
[3] V.S. Arutyunov, O.V. Krylov, Oxidative Conversion of Methane. Moscow, Nauka, 1998 (in Russian).
[4] V.S. Arutyunov, J. Nat. Gas Chem., 13 (2004) 10.
[5] V.S. Arutyunov, O.V. Krylov, Russian Chemical Reviews, 74 (2005) 1111.
[6] E.V. Sheverdenkin, V.S. Arutyunov, V.M. Rudakov, V.I. Savchenko, and O.V. Sokolov, Theor. Foundations of Chem. Eng., 38 (2004) 311.
[7] A. Mac Farlan et al. Natural Gas Conversion VI. Studies in Surface Science and Catalysis vol. 136. Eds. E. Iglesia, T.H. Fleish. Elsevier Science B.V. Amsterdam, 2001.
[8] L. Chen, J. Zhang, P. Ning, Y. Chen, W. Wu, J. Nat. Gas Chem., 13 (2004) 225.
[9] X. Ma, Z. Li, B. Wang, G. Xu, ACS Fuel Chemistry Division Preprints, 2003, vol. 48(1), p. 426.
[10] V.S. Arutyunov et al., Eurasian Chemico-Technological Journal, 3 (2001) 107.
[11] V.M. Zamansky et al., Combust. Sci. Technol., 120 (1996) 255.

Natural Gas Conversion VIII
F.B. Noronha, M. Schmal, E.F. Sousa-Aguiar (Editors)

Effect of chemical treatment on Ni/fly-ash catalysts in methane reforming with carbon dioxide

Shaobin Wang[a], G.Q. Lu[b]

[a]*Department of Chemical Engineering, Curtin University of Technology, GPO Box U1987, Perth WA 6845, Australia*
[b]*Nanomaterials Centre, The University of Queensland, St Lucia, Brisbane QLD 4072, Australia*

1. Introduction

Fly ash is the major solid waste produced from coal-fired power stations. More than 150 millions tons of fly ash are generated each year in the world. Currently, fly ash is mainly used as civil construction materials, and there is a limit to the demand for coal ash by construction industries. Different applications should be thus considered. Any successful utilisation of the waste is of both great economical and environmental interests. The major chemical compounds of fly ash are SiO_2 and Al_2O_3, which are highly thermostable and thus offering the property as catalyst supports. However, few investigations have been reported using fly ash as catalyst support [1, 2].
Carbon dioxide reforming of methane producing low H_2/CO ratio has several advantages over steam reforming and partial oxidation of methane. In recent years, extensive researches have been done on the catalytic reforming of methane with carbon dioxide to produce synthesis gas [3, 4]. Ni, Pd, Pt, Ru, Rh, and Ir have been proved to be effective catalytic components for this reaction. Noble metals are found to be much less sensitive to carbon formation than nickel does. Economical considerations for industrial application still require for development of non-noble metal catalysts with high activity and long-term stability. It has been reported that some Ni-based catalysts exhibit higher activity and stability depending on supports. Several researchers have also reported that zeolite, which is alluminosilicate, supported metal catalysts could be effective for CO_2 reforming of methane [5-7].
In this paper, we report an investigation of exploring fly ash as support for Ni-based catalysts. The fly ash was firstly treated with different chemical methods to change the physicochemical properties as supports for Ni catalysts and then tested in CO_2 reforming of methane.

2. Experimental

Raw fly ash (FA) was obtained from Tarong Power Station of Queensland, Australia. The raw ash was washed with distilled water to remove residual carbon and iron oxides. The washed ash was further treated by different chemical methods before catalyst preparation. Two portions of the ash were dispersed into 0.5 N NH_3 and 2 N HNO_3 solutions, respectively, at ambient temperature for 24 h under stirring. They are labelled as FA-NH_3 and FA-HNO_3. One portion of the ash was dispersed into a saturated $Ca(OH)_2$ solution and another portion was mixed with CaO at a ratio of ash:CaO=95:5 and the mixture was dispersed into water, too. The last two dispersions were kept at 150 °C for 24 h. The solids were filtrated, washed and dried, named as FA-$Ca(OH)_2$ and FA-CaO, respectively.

Nickel supported on the washed fly ash, the four chemically treated ash samples, and two commercial metal oxide samples, SiO_2 and γ-Al_2O_3, were prepared by wetness impregnation method. $Ni(NO_3)_2{\cdot}6H_2O$ was used as the precursor compound of Ni and the loading of Ni was 5% in weight of the catalysts. The catalysts were calcined in air for 4 h at 500 °C and reduced *in situ* at 500 °C for 3 h in H_2/N_2 flow prior to catalytic tests. La_2O_3 promoted FA-$Ca(OH)_2$ and FA-CaO supported Ni catalysts were also prepared by simultaneous impregnation of nitrate salts, drying, and calcination at the same conditions as described above.

The surface areas of the catalysts were determined by N_2 adsorption at –196 °C on a NOVA 1200 adsorption system (Quantachrom, USA). Before adsorption, all samples were degassed at 300 °C in high vacuum for 2 h. A Philips PW 1840 powder X-ray diffractometer was employed to analyse the fresh and used catalyst phases. The amount of Ni on catalyst was determined by XRF.

Catalytic dry reforming of methane with CO_2 was conducted under atmospheric pressure with a flow system including a vertical quartz reactor and an on-line gas chromatograph (GC-17A). The reaction gases were of CO_2 and CH_4 with molar ratio of 1:1 with a GHSV of 18000 $cm^3\ h^{-1}g^{-1}$. The catalyst was reduced *in situ* at 500 °C for 3 h in a flow of 10% H_2/N_2, prior to reaction. The catalytic activity was studied in the temperature range of 500-800 °C. All the products were analysed by the GC equipped with a TCD.

3. Results and discussion

The XRF measurements indicate that FA is mainly consisted of aluminosilicate (92.2%) with minor other compounds (TiO_2, Fe_2O_3, K_2O, MgO and CaO). After chemical treatment, no significant changes in chemical compositions occur on HNO_3 and NH_3 treated samples. However, $Ca(OH)_2$ and CaO treatments increase CaO content and CaO treatment sample shows much reduced contents of silica and alumina. N_2 adsorption also shows that the chemical treatment results in an increase in BET area for all samples except the one treated by NH_3.

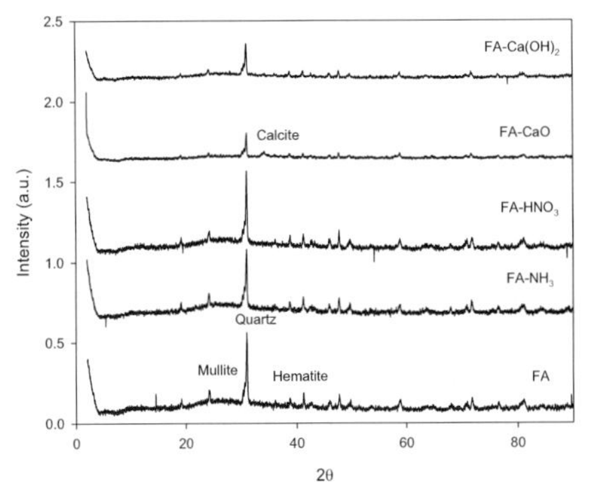

Fig.1 XRD patterns of fly ash supports.

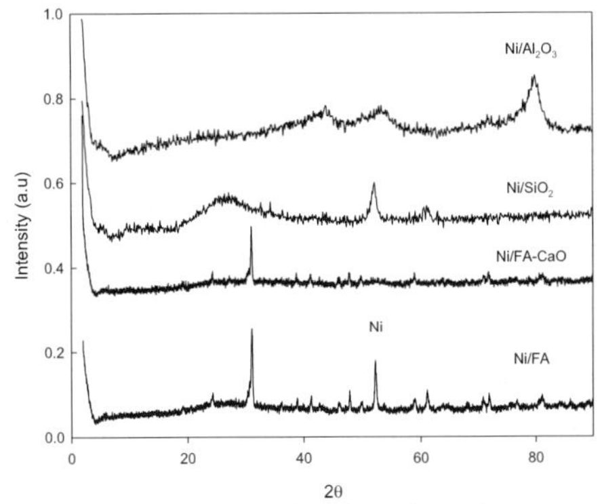

Fig.2 XRD patterns of Ni-based catalysts.

Fig.1 shows the XRD patterns of all fly ash samples. As shown that raw fly ash, NH_3 and HNO_3 treated fly ash display similar XRD patterns. The major phases of the three samples are quartz, mullite and hematite. After treated by $Ca(OH)_2$ and CaO, fly ash samples display different XRD patterns. The intensities for quartz and mullite diffraction peaks are greatly reduced and a new phase, calcite, appears, suggesting the phase transformation.

Table 1. Physico-chemical properties and methane conversion (%) of various Ni/fly-ash.

	S_{BET}	Ni_w	Ni_p	Temperature (°C)						
	(m^2/g)	(%)	nm	500	550	600	650	700	750	800
Ni/FA	2.1	5.89	18.5	7.6	13.4	18.5	22.1	29.1	35.8	42.7
Ni/FA-HNO_3	3.5	6.52	20.1	1.7	2.0	2.3	2.7	2.3	3.1	1.2
Ni/FA-NH_3	1.2	6.44	17.5	9.0	23.8	38.1	47.6	50.7	51.2	61.9
Ni/FA-$Ca(OH)_2$	4.8	6.68	-	12.1	23.8	37.8	49.2	54.7	62.9	70.7
Ni/FA-CaO	30	6.05	-	15.1	29.3	46.4	62.2	74.6	82.8	90.3
Ni/γ-Al_2O_3	157	5.5	-	14.0	31.2	45.1	61.3	78.1	87.0	94.1
Ni/SiO_2	238	4.8	12.0	21.3	38.8	54.4	69.7	80.9	89.4	96.2
CH_4 Equilibrium Conversion				19.2	32.3	48.8	64.0	78.0	92.0	93.9

Table 1 presents some properties of Ni-based fly ash catalysts and their methane conversion as well as thermodynamic conversion at different temperatures. As can be seen that the Ni loading is much similar for all catalysts. However, these catalysts exhibit varying activities depending on chemical treatment. Ni/FA shows lower activity with methane conversion of 43% at 800 °C. HNO_3 treatment results in a negative effect on the catalytic activity, making Ni/FA-HNO_3 showing little activity for this reaction even at 800 °C. The treatments with bases (NH_3, $Ca(OH)_2$, and CaO) produced active catalysts. The stronger and larger amount of base (CaO) was used, the better the activity of the resultant catalysts. Ni/FA-CaO catalyst demonstrates higher activity in terms of CH_4 conversion, being much closer to that of Ni/SiO_2 catalyst. CH_4 conversions over Ni/SiO_2 and Ni/FA-CaO catalysts are approaching those expected at thermodynamic equilibrium. For all active catalysts, the CH_4 conversion increases with increasing temperature and CH_4 conversion can reach over 90% at 800 °C.

There is no new crystal phase observed from the X-ray diffraction patterns after the treatments for FA-HNO_3 and FA-NH_3, suggesting that a little phase transformation occurred. However, the catalytic activity of the final Ni/FA, Ni/FA-NH_3 and Ni/FA-HNO_3 catalysts is quite different indicating that the reaction is very sensitive to the change on the surface property of the fly ash. For FA-$Ca(OH)_2$ and FA-CaO, XRD has shown that a new phase has occurred. The surface becomes certainly more basic by a treatment with a base (NH_3 or CaO) and the adsorption of the acidic reactant CO_2 could be enhanced. All catalysts based on the supports treated with a base show higher CO_2 conversion than that of Ni on the washed ash. Several researches have shown that the catalytic activity is strongly influenced by the nature of support. It has been found that acid treatment of supported Ni catalysts decreased the catalytic activity [8] while addition of basic promoter (CaO) [3, 9-11] generally enhanced the conversion of CO_2 in reforming of methane with CO_2. BET surface areas for ash supports show that BET surface areas of base-treated ash supports are generally increased. XRD measurements show that the nickel crystallite size on Ni/FA-CaO are less than those on Ni/FA (Fig.2), which suggests that base treatment increases the nickel dispersion on support and thus increasing the catalytic activity.

It has been reported that La_2O_3 is a good promoters for CO_2 reforming of methane over several Ni systems [9]. Fig.3 presents the comparison of activity over La_2O_3 promoted Ni/FA-$Ca(OH)_2$ and Ni/FA-CaO catalysts. As seen that La_2O_3 exhibits a promoting effect on the catalytic activity and the loading affects the activity for two catalyst systems. Higher La_2O_3 loading on catalysts produces higher methane conversion. For Ni/FA-$(CaOH)_2$, the promoting effect is more significant than Ni/FA-CaO. At 800 °C, methane conversion will be 70%, 88% and 90% for Ni/FA-$(CaOH)_2$, Ni/FA-(CaOH)-1%La_2O_3, and Ni/FA-(CaOH)-5%La_2O_3, respectively.

Fig.4 shows the stability performance of Ni/FA-CaO, Ni/SiO_2 and Ni/Al_2O_3 catalysts at 700 °C. The conversions over Ni/FA-CaO and Ni/Al_2O_3 catalysts decrease slightly faster at the first 3 h, probably due to accumulation of carbon deposits, and then the deactivation rates are much lower whereas Ni/SiO_2

catalyst shows a continuous deactivation. After 24 h performance, CH_4 and CO_2 conversions over Ni/FA-CaO catalyst decrease from the initial 79% and 84% to 69% and 75%, respectively. Whereas for Ni/SiO_2, CH_4 and CO_2 conversions are reduced from 75% to 57% and 89% to 81%, respectively. The above results suggest that Ni/FA-CaO catalyst shows not only high activity but also a longer stability. The stability of Ni/FA-CaO is possible due to the fundamental component of aluminosillicate in fly ash. Aluminosillicate can probably prevent the formation of nickel aluminate and nickel silicate in calcination as happened in Ni/Al_2O_3 and Ni/SiO_2 systems. Nickel aluminate and nickel silicate were found to be inactive in reforming of methane reaction with carbon dioxide and are hard to be reduced to nickel metal. Some reports demonstrated that Ni supported on zeolites gave much higher catalytic activity and a long-term stability at 800 °C for this reaction [11,12]. Addition of CaO produces a protective layer of calcium aluminate and calcium silicate which also prevents the nickel sintering as well as carbon deposition [9], resulting in longer stability.

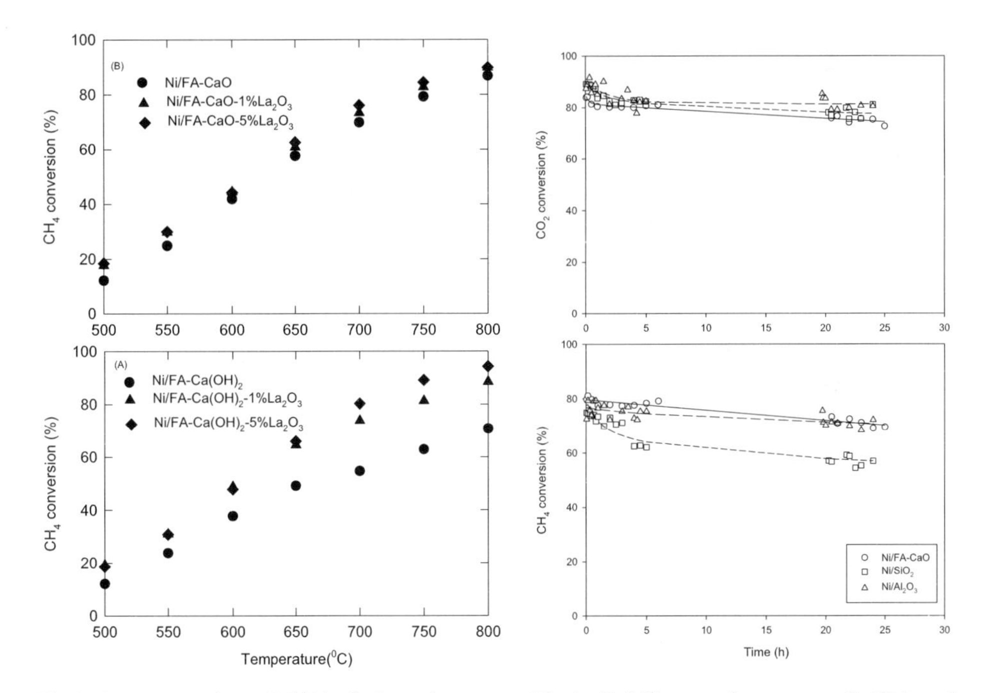

Fig.3 CH_4 conversion of Ni/FA-CaO catalysts.

Fig.4 Stability performance of Ni-based catalysts.

It has been believed that Ni sintering and carbon deposition are the two major causes of catalyst deactivation. Fig.5 presented the XRD patterns of fresh and used Ni/FA-CaO catalysts. It is seen that there is no Ni diffraction peak on the fresh catalyst. After 1 h reaction, the peak for Ni is still not significant. After 24 h reaction, Ni diffraction can be observed but the intensity is still small. These results suggest that the sintering of Ni on Ni/FA-CaO is not serious, which will partly contribute to the catalyst stability. From the figure, it is not able to see the graphite diffraction, which also indicates the much less carbon deposition on the catalyst. Our previous investigations show that Ni/Al_2O_3 and Ni/SiO_2 produce higher carbon deposition and Ni sintering under the same conditions after 1 h reaction [13].

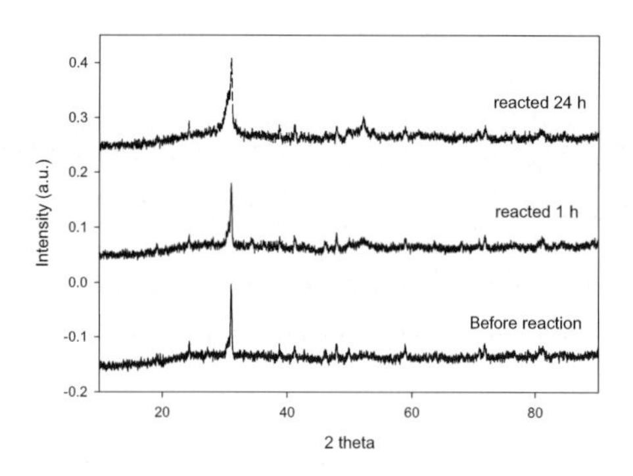

Fig.5 XRD patterns of Ni/FA-CaO before and after reaction.

Fig.5 XRD patterns of Ni/FA-CaO before and after reaction.

4. Conclusion

Fly ash, a solid waste, is a potential supporting material for Ni catalysts for carbon dioxide reforming of methane to synthesis gas. The treatment of the ash with a base, prior to the Ni loading, can greatly improve the catalytic activity and stability of the resultant catalysts. Such treatment influences the existence of nickel. Addition of CaO prevents the carbon deposition and nickel crystallite sintering, resulting in less deactivation. This kind of catalyst may be developed as an economical and potential catalyst for industrial application.

References

1. F. W. Chang, T. J. Hsiao, J. D. Shih, Ind. Eng. Chem Res. 37 (1998), 3838.
2. X. P. Xuan, C. T. Yue, S. Y. Li, Q. Yao, Fuel, 82 (2003), 575.
3. S. B. Wang, G. Q. Lu, G. J. Millar, Energy Fuels, 10 (1996), 896.
4. G. Q. Lu, S. B. Wang, Chemtech, 1999, 29 (1999), 37.
5. R. N. Bhat, W. M. H. Sachtler, Appl. Cataly. A. 150 (1997), 279..
6. U. L. Portugal, C. M. P. Marques, E. C. C. Araujo, E. V. Morales, M. V. Giotto, J. M. C. Bueno, Appl. Catal. A. 193 (2000), 173..
7. D. Halliche, O. Cherifi, A. Auroux, J. Therm. Anal. Calori. 68 (2002), 997-1002.
8. S. B. Wang, G. Q. Lu, Carbon, 36 (1998), 283-292.
9. S. B. Wang, G. Q. Lu, J. Chem. Technol. Biotechnol. 75 (2000), 589.
10. O. Yamazaki, T. Nozaki, K. Omata, K. Fujimoto, Chem. Lett. (1992), 1953.
11. Kim, G. J.; Cho, D. S.; Kim, K. H.;Kim, J. H., Catal. Lett. 28 (1994), 41.
12. J. S. Chang, S. E. Park, K. W. Lee, M. J. Choi, Stud. Surf. Sci. Catal. 84 (1994), 1587.
13. S. B. Wang, G. Q. Lu, Appl. Cataly. B. 16 (1998), 269..

Natural Gas Conversion VIII
F.B. Noronha, M. Schmal, E.F. Sousa-Aguiar (Editors)

Kinetic Depressing the Deposition of Carbon Species on Ni-Based Catalysts: La_2O_3 Adjusts the Reaction Rates of CO_2 Higher than CH_4 by Tuning the Ea and Pre-exponential Factor

Yuehua Cui, Huidong Zhang, Hengyong Xu*, Qingjie Ge, Yuzhong Wang, Shoufu Hou, Wenzhao Li

Dalian Institute of Chemical Physics, Chinese Academy of Sciences, Dalian 116023, P .R. China. E-mail: xuhy@dicp.ac.cn

The same amount of CH_4 and CO_2 was pulsed through the catalysts Ni/α-Al_2O_3 (A) and Ni/La_2O_3/α-Al_2O_3 (AL) in 773-873 K to study the deposition rates of carbon species CH_x ($0 \leq x \leq 3$) under the kinetic conditions. The pulse tests provided the possibility of non-deposition of carbon species on the catalyst AL. After elimination of the contribution of CO_2 rates in RWGS, the deposition of carbon species can be avoided on the catalyst AL for the reforming of equal molar of CH_4 and CO_2 under the kinetic conditions in 773-873 K.

1. Introduction

The deposition of carbon species on the Ni-based catalysts during the CH_4/CO_2 reforming decreases the catalytic activity and further completely deactivates the catalysts [1], great retarding the industrial application of CH_4/CO_2 reforming. Early thermodynamic studies [2] in CH_4/CO_2 reforming showed that high molar ratio of CO_2/CH_4 or high reaction temperature was necessary to avoid the carbon deposition on the Ni-based catalysts. The carbon deposition on Ni metal was mainly originated from CH_4 dissociation during the reforming [3]. Some approaches were designed to avoid carbon deposition in the CH_4/CO_2 reforming, including using special small grain support [4] or by addition of metal [5], basic metal oxidant [6] or La_2O_3 [7]. However, high temperature (above 1000 K) [5] or high molar ratio of CO_2/CH_4 (higher than 1) [8] was still needed for the depression of carbon species deposition.

However, the industrial application desires the reforming with an equal molar ratio of CO_2/CH_4 for the following F-T synthesis and the reforming at a relatively lower temperature for the economic standpoint. Therefore, this group directly studied the deposition rates of carbon species on Ni/α-Al_2O_3 (A) and Ni/La_2O_3/α-Al_2O_3 (AL) via the pulse of the same amount of CH_4 and CO_2 in 773-873 K under the kinetic conditions.

The formation and elimination of carbon species on Ni-based catalysts in the CH_4/CO_2 reforming could be summarized as Eqs. (1)-(4) [9,10]:

$CH_4 \rightarrow CH_x + (4\text{-}x)/2\ H_2 \quad (0 \leq x \leq 3)$ (1)

$CH_x + CO_2 \rightarrow 2\ CO + x/2\ H_2$ (2)

$2CO \rightarrow C + CO_2$ (3)

$CO + H_2 \rightarrow C + H_2O$ (4)

Under the kinetic conditions, the pulse reactions were kept in the initial stage, the Eqs. (3) and (4) could be basically avoided. The formation and elimination of CH_x were depended on the Eqs. (1) and (2). Therefore, this work investigated in detail the CH_4 dissociation rates and CO_2 reaction rates on the catalysts and the possibility of the depression of carbon species deposition on the catalyst under the optimized reaction conditions.

2. Experimental

The catalysts Ni (8 wt. %)/α-Al_2O_3 and Ni (8 wt. %)/La_2O_3 (2 wt. %)/α-Al_2O_3 were prepared [11] (denoted as A and AL). The transfer of mass and heat were firstly eliminated before the intrinsic kinetic studies [12,13].

The pulse tests were performed according to Ref. [13]. The pulse conversions of CH_4 or CO_2 ($Con_{pu}\%$) were kept far below the corresponding equilibrium conversions ($Con_{eq}\%$) of CH_4/CO_2 reforming.

The pulse reaction rates (R_P) of CH_4 or CO_2 were expressed as Eq. (5):

$R_P = Con_{pu}\% \times V / 22400 / t / m$ (5)

where V: the volume of the quantitative tube (0.3058 ml); $t = V_C/S_{V1}$ (s); V_C: the catalyst volume; S_{V1}: the flow rate of the carrier gas (ml/s); m is the mass of the catalyst (g); $R_{P,CH4}$ or $R_{P,CO2}$: the pulse rates of CH_4 or CO_2 through the catalyst (mol/s/g); and $R_{P,CO2/CH4}$ ($R_{P,CH4/CO2}$) indicates the pulse rates of CH_4 (CO_2) after pulsing CO_2 (CH_4) through the catalysts (mol/s/g).

The deposition rates of CH_x species ($R_{P,C}$) were expressed as Eq. (6):

$R_{P,C} = R_{P,CH4,\ max} - R_{P,CO2,\ max}$ (6)

Where $R_{P,CH4,max}$: a higher value between $R_{P,CH4}$ and $R_{P,CO2/CH4}$; So is for $R_{P,CO2,max}$. $R_{P,CO2,max}$ includes the CO_2 rates in the reaction with CH_x species and in the RWGS. $R_{P,C} > 0$ directly states the deposition of carbon species on the catalysts; while $R_{P,C} < 0$ possibly avoids the deposition of carbon species.

The activation energy (Ea) and the pre-exponential factor (A_0) were calculated by plotting ln R versus 1/T according to Eq. (7):

$\ln R_P = \ln A_0 - Ea/R_0 \times (1/T) + \ln(V/22400/V_b)$ (7)

where R_0: gas constant value 8.314 (J/K); T: pulse temperature (K); V_b: the preserve volume from the bottom of the quantitative tube to the entrance of the catalyst bed (3.14 ml).

The rates of CO_2 pulse without RWGS were expressed as Eq. (8):

$R_{P,CH4/CO2} - R_{P,RWGS} = R_{P,CH4/CO2} \cdot (1 - R_{S,RWGS}/R_{S,CO2})$ (8)

The steady kinetic reforming reactions were performed according to Ref. [13].

The rates of CO_2 reaction and CO formation in the reforming were expressed as Eq. (9) and (10):

$$R_{S,CO2} = (CO_{2,\,in} / N_2 - CO_{2,\,out} / N_2) \cdot (f_{CO2} / f_{N2}) \cdot Mol_{N2} / (V_C/S_{v2}) \quad (9)$$

$$R_{S,CO} = CO_{out} \cdot f_{CO} / (N_2 \cdot f_{N2}) \cdot Mol_{N2} / (V_C/S_{v2}) \quad (10)$$

Where subscript S: steady reforming reaction; subscript in and out: the peak area of the feedstock and product, respectively; f: the rectification factor; Mol_{N2}: the mole flow rate of N_2 (mol/s); S_{v2}: the total flow rate of the gases (ml/s).

The rates of the RWGS were calculated according to Eq. (11):

$$R_{S,RWGS} = 2 * R_{S,CO2} - R_{S,CO} \quad (11)$$

3. Results

3.1. Catalyst characterization

The XRD patterns of the catalysts A and AL shows the Ni particles on the catalysts. Based on the half-widths of the Ni peaks, the sizes of Ni particles on the catalysts A and AL were respectively estimated as 39.7 and 32.3 nm. The addition of La_2O_3 can promote the dispersion of Ni metal on the catalyst AL and resulted in smaller Ni particles on the catalyst AL than those on the catalyst A.

3.2. Elimination of transfer of mass and heat

The external diffusion could be eliminated when the contact time was controlled less than 0.556 s. When the particles of the catalysts were less than 0.175 mm (100-160 mesh), the inner diffusion can be eliminated for the catalysts A and AL. When the contact time was controlled less than 0.556 s, the temperature fluctuation of the pulse reaction was found within 0.5 K and the transfer of heat could be eliminated.

3.3. Kinetic pulse

The contact time was tuned to keep the pulse conversions of CH_4 or CO_2 (Con_{pu}%) far below the Con_{eq}% on every catalyst at different temperatures. The pulse reactions were controlled by kinetics. The pulse rates of CH_4 and CO_2 through the catalyst A and AL are shown in Fig. 1. The catalyst A showed the activity for CH_4 dissociation but not for the CO_2 conversion. $R_{P,CO2/CH4}$ was completely equal to $R_{P,CH4}$ at any temperature, indicating that the little amount of activated CO_2 on the catalyst A does not affect the CH_4 dissociation. However, $R_{P,CH4/CO2}$ were much higher than $R_{P,CO2}$, meaning the CH_x intermediates can greatly improve the conversion rate of CO_2. CO_2 was reacted with CH_x species according to 1:1 (molar ratio) to produce CO and H_2 via the reforming route under the kinetic conditions. The CH_x deposition rates ($R_{P,C}$) could be expressed as Eq. (6). The rates of CH_4 dissociation were always higher than those of CO_2 on the catalyst A, leading to the deposition of carbon species (CH_x) on the catalyst A.

The addition of La_2O_3 to the catalyst AL can improve CO_2 reaction activity, which can be due to the reaction of La_2O_3 with CO_2 on the catalyst AL [14]. $R_{P,CO2/CH4}$ was higher than $R_{P,CH4}$, meaning that the activated CO_2 by La_2O_3 improves the further dissociation of CH_4 on the catalyst AL. The CH_x intermediates formed on the catalyst during the first pulse of CH_4 also improved the dissociation rates of CO_2. $R_{P,CH4/CO2}$ were higher than $R_{P,CH4}$ or $R_{P,CO2/CH4}$ on the catalyst AL, which indicates that the deposition of carbon species may be possibly avoided on the catalyst AL under the kinetic conditions.

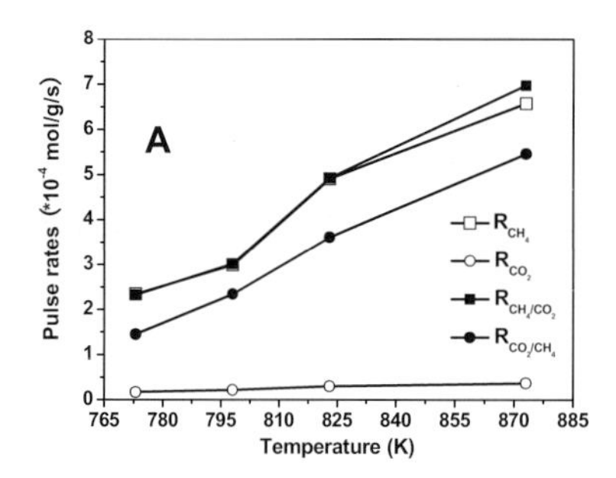

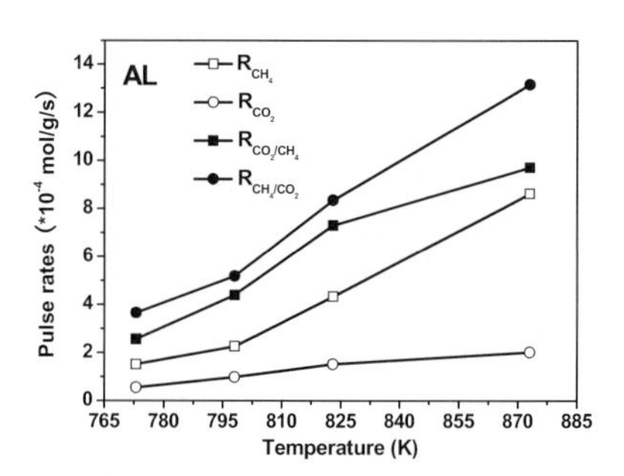

Fig. 1. The pulse rates of CH_4 and CO_2 through the catalysts A and AL.

3.4. The steady kinetics of CH_4/CO_2 reforming on the catalyst AL

In order to exactly investigate the possibility of the non-deposition of carbon species in the reforming reaction, the CH_4/CO_2 reforming was studied on the catalyst AL by the steady kinetic methods. Fig. 2 shows the reaction rates with the consideration of RWGS in the steady CH_4/CO_2 reforming (the left) and in the pulse tests (the right) under the kinetic conditions. CO_2 reacted with CH_4 in the reforming reaction and with H_2 in the RWGS. Therefore, $R_{S,CO2}$ were higher than $R_{S,CH4}$ on the catalyst AL in the reforming reaction (the left). After elimination of CO_2 reaction rates in the RWGS, the CO_2 conversion rates were almost equal to those of CH_4 in the reforming reaction on the catalyst AL. The carbon species deposition on the catalyst AL can be avoided in the equal molar CH_4/CO_2 reforming under the kinetic conditions in 773-873 K.

For the pulse tests of CH_4 and CO_2, the $R_{P,CH4/CO2}$ included the reaction rates of CO_2 with CH_x species and the reaction rates of CO_2 with H_2 originating from the first pulse dissociation of CH_4. $R_{S,RWGS}/R_{S,CO2}$ were calculated from the results shown in Fig. 2 (left). After elimination of CO_2 pulse rate in the RWGS according to Eq. (8), the remainder CO_2 pulse rate was slightly higher or equal to that of CH_4 pulse through the catalyst AL. These results further prove that the deposition of carbon species on the catalyst AL can be avoided for the reforming of equal molar of CH_4 and CO_2 in 773-873 K under the kinetic conditions.

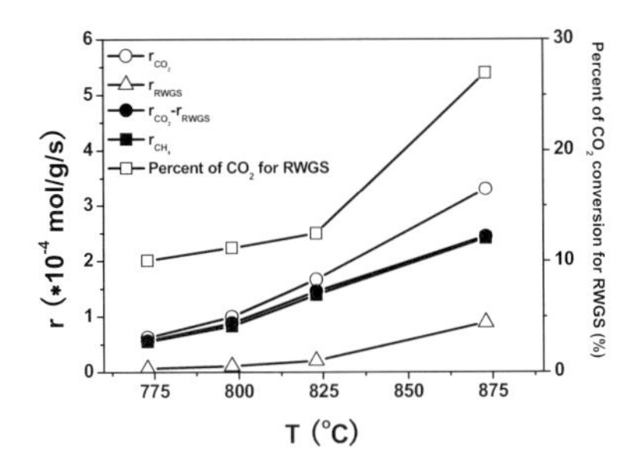

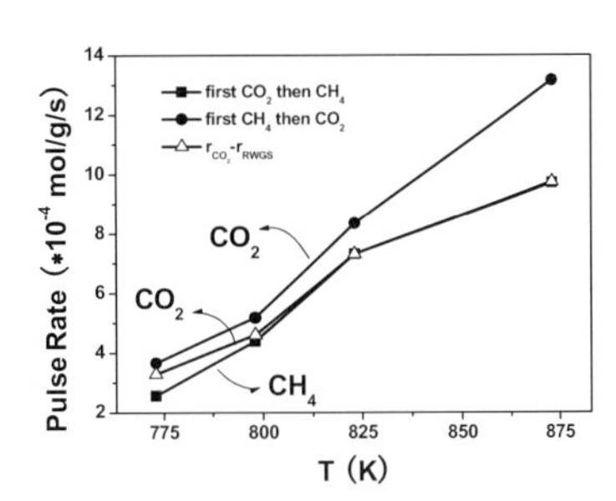

Fig. 2 The reaction rates of CH_4, CO_2, and RWGS in the steady reforming of CH_4 and CO_2 (the left) and in the pulse tests (the right) on the catalyst AL.

4. Discussion

What were the reasons for the depression of the deposition of the carbon species on the catalysts AL in the reforming of equal molar CH_4 and CO_2 below 1000 K under the kinetic condition? The effects of the La_2O_3 on the $R_{P,CH4,\ max}$ and $R_{P,CO2,\ max}$ were studied from the kinetic standpoints.

In fact, the deposition rate of the CH_x species on the catalysts was depended on the R_{CH4} and R_{CO2} in both forward and reverse directions due to the reversible reforming reaction [15]. When the reforming reaction was controlled in the kinetic region, the reverse direction of the reforming could be avoided. Therefore, $R_{P,C}$ can be expressed as Eq. (6).

Every rate was depended on their corresponding pre-exponential factor (A_0), activation energy (Ea), and the partial pressure (P) ($R = K_0 \cdot A_0 \cdot \exp(-Ea/(RT)) \cdot P$, K_0 is a constant value for CH_4 or CO_2 reaction). Therefore, higher partial pressure of CO_2 than CH_4 can lead to higher reaction rates of CO_2 than CH_4 in the reformation reaction, depressing the carbon deposition on the catalysts [8]. According to the industrial demands, the same molar of CH_4 and CO_2 were studied ($P_{CH4}=P_{CO2}$). The reasons for the non-deposition of carbon species on the catalyst AL can only be due to the tuned A_0 and Ea on the catalyst AL.

The Ea and A_0 for the pulses of CH_4 and CO_2 were respectively calculated based on $R_{P,CH4}$ and $R_{P,\ CH4/CO2}$ and are listed in Table 1. The addition of La_2O_3 on the catalyst A decreased the Ea for CO_2 reaction with CH_x by 21.4 kJ/mol and increased the Ea for CH_4 dissociation into CH_x by 18.4 kJ/mol, leading to higher Ea of CH_4 than CO_2. So are for A_0 values. The tuned Ea and A_0 led to higher rates of CO_2 than CH_4 on the catalyst AL, avoiding the deposition of carbon species on the catalyst AL. This work experimentally proved a proposal

[9] that the carbon deposition was avoided on the catalyst AL in 773-873 K under the kinetic conditions by tuning the Ea and A_0 for the CH_4 dissociation and CO_2 reaction with CH_x with the aid of La_2O_3.

Table 1 The Ea and A_0 for the pulse of CH_4 and CO_2 through the catalysts A and AL

Catalyst	Ea (kJ/mol)		A_0 (10^7)	
	CH_4	CO_2	CH_4	CO_2
A	93.4	98.9	9.7	15.9
A-L	111.8	77.5	117.7	1.4

5. Conclusion

The same amount of CH_4 and CO_2 was pulsed through the catalysts A and AL in 773-873 K to study the pulse rates of CH_4 dissociation into CH_x species ($0 \leq x \leq 3$) and the reaction of CO_2 with CH_x species under the kinetic conditions. The carbon species (CH_x) were deposited on the catalyst A in the pulses tests. $R_{P,CH4/CO2}$ were tuned to be higher than $R_{P,CH4}$ or $R_{P,CO2/CH4}$ on the catalyst AL with the possibility of non-deposition of carbon species. After elimination of CO_2 rate contribution in the RWGS, CO_2 remainder rates were almost equal to those of CH_4 dissociation. The deposition of carbon species were avoided in the reforming reaction of equal molar CH_4 and CO_2 on the catalyst AL under the kinetic conditions in 773-873 K. The reasons were attributed to the tuned Ea and A_0 for the CH_4 dissociation and CO_2 reaction with CH_x after the addition of La_2O_3.

References

[1] J.R. Rostrup-Nielsen, J.H.B. Hasen, J. Catal. 144 (1993) 38.
[2] A.M. Gadalla, B. Bower, Chem. Eng. Sci. 43 (1988) 3049.
[3] A.Jr. Sacco, F.W.A.H. Geurts, G.A. Jablonski, R.A. Gately, J. Catal. 119 (1989) 322.
[4] J.M. Wei, B.Q. Xu, J.L. Li, Q.M. Zhu, Appl. Catal. A 196 (2000) L167.
[5] Z. Hou, O. Yokota, T. Tanaka, T. Yashima, Appl. Surf. Sci. 233 (2004) 58.
[6] J.S. Chang, S.E. Park, J.W. Yoo, J.N. Park J. Catal. 195 (2000) 1.
[7] J.Z. Luo, Z.L. Yu, C.F. Ng, C.T. Au, J. Catal. 194 (2000) 198.
[8] X. Chen, K. Honda, Z.G. Zhang, Catal. Today 93-95 (2004) 87.
[9] M.C.J. Bradford, M.A. Vannice, Catal. Rev. Sci. Eng. 41 (1999) 1.
[10] M.C.J. Bradford, M.A. Vannice, Appl. Catal. A 142 (1996) 97.
[11] Y.H. Cui, H.Y. Xu, W.Z. Li, J. Mol. Catal. A: Chem. 243 (2006) 226.
[12] Y. Cui, H. Xu, Q. Ge, Y. Wang, W. Li, J. Mol. Catal. A 249 (2006) 53.
[13] Y.H. Cui, H.D. Zhang, H.Y. Xu, Appl. Catal. A: Gen. 2006, Accepted.
[14] Z.L. Zhang, X.E.Verykios, Catal. Today 21 (1994) 589.
[15] J.H. Edwards, A.M. Maitra, Fuel Process. Technol. 42 (1995) 269.

Natural Gas Conversion VIII
F.B. Noronha, M. Schmal, E.F. Sousa-Aguiar (Editors)

Investigation of unsteady-state redox mechanisms over ceria based catalysts for partial oxidation of methane.

C. Mirodatos[a], Y. Schuurman[a], A.C. van Veen[a], V.A Sadykov[b], L.G. Pinaeva[b], E. M. Sadovskaya[b]

[a]*Institut de Recherches sur la Catalyse, CNRS, 2 avenue Albert Einstein, F-69626, Villeurbanne Cedex, France.*
[b]*Boreskov Institute of Catalysis SB RAS, Pr. Lavrientieva, 5, 630090, Novosibirsk, Russia*

1. Introduction

Within the perspective of new domains of application like domestic hydrogen production from natural gas for fuel cell, the kinetic behavior of the most performing catalytic systems has to be properly taken into account under transient conditions and at short contact time since it corresponds to frequent operating procedures like process start-up, shut-down and fast changes in energy demand. To illustrate this new requirement, a transient kinetic approach of the partial oxidation of methane to syngas over Pt/ceria (ceria-zirconia)-based materials for which the authors have already carried out and reported various physico-chemical, kinetic and mechanistic investigations [1-5] is presented.
Ceria-based mixed oxides ($Ce_xM_{1-x}O_y$) are versatile solid oxygen exchangers. At high temperatures (400–800°C) the redox cycle $Ce^{3+} \Leftrightarrow Ce^{4+} + e^-$ facilitates oxygen storage and release from the bulk fluorite lattice. When combined with noble or non noble metal particles, this makes them ideal candidates for catalytic oxidation applications such as partial oxidation of methane into syngas (POM). However, the *surface* redox chemistry of ceria is highly sensitive to crystal structure defects, which can be tuned by substituting some of the Ce cations with ions of different size and/or charge [6], as illustrated in the present paper by using Zr as doping cation.

In the case of Pt catalysts supported on CeO_2 based solid mixed oxides, the rate of oxygen supply from the support may determine the rate of methane oxidation and especially the selectivity towards syngas [5]. In addition, the mixed oxide itself is capable to oxidize both CH_4 and CO even in the absence of gas phase oxygen, at least for a transient period [1,2,6]. This paper presents an overview of the main role of the metal and of the support as a function of the applied transient operating conditions, and the detailed kinetic calculation will be published in a forthcoming paper [9].

2. Experimental

$Ce_{0.5}Zr_{0.5}O_2$ mixed oxide was prepared by modified Pechini method and Pt (1.4 wt.%) was supported via incipient wetness [1,2]. 0.05 g of each sample was loaded into a tubular quartz reactor (ID=3 mm) and heated to 650°C under flowing He. Prior to each experiment the samples were oxidized in 2.2%O_2 + He flow (5 mL/s) for 30 min, leading to "pre-oxidized samples". The pre-oxidized samples could also be reduced under pure H_2 flow (2 L/h) at 650°C for 30 min, leading to "pre-reduced samples". Abrupt switches from He to the reaction mixture and vice versa (4.6% CH_4 (+ 2.2%O_2) in He at a flow rate of 5 ml/s) were performed and the outlet gas phase composition monitored by on-line mass spectrometer and chromatographic analysis [7].

3. Results and discussion

3.1. Switches from He to CH_4 (+ O_2) (POM) over CeO_2-ZrO_2.

When switching from He to 4.6%CH_4 + He, no detectable CH_4 conversion was observed at 650°C, either for the pre-oxidized or the pre-reduced samples.
When switching from He to 4.6%CH_4 + 2.5%O_2 + He (Fig.1A), the CH_4 conversion into total oxidation products -CO_2 and H_2O- establishes with transient dynamics depending on the sample pretreatment. For both catalyst pre-treatments, the steady state corresponds to a CH_4 conversion of about 24% with a full consumption of O_2 (Table 1). This matches closely the stoichiometry of the combustion: $CH_4 + 2O_2 \rightarrow CO_2 + 2H_2O$, where 2.5% of O_2 can fully oxidize 1.25% of CH_4, which means a conversion of 1.25/4.6 = 27%.

Table 1. Steady-state catalytic performances for CeO_2-ZrO_2 and Pt/CeO_2-ZrO_2.

Gas inlet composition	CH_4 conversion, CO and H_2 selectivity %	
	CeO_2-ZrO_2	Pt/CeO_2-ZrO_2
4.6% CH_4 +1.25% O_2 in He	nd	X_{CH4} = 41.6, S_{CO} = 82.6, S_{H2} = 100
4.6% CH_4 +2.5% O_2 in He	X_{CH4} = 24, S_{CO} = 0 S_{H2} = 0	X_{CH4} = 45.7, S_{CO} = 50.0, S_{H2} = 77

Note ; In all cases oxygen conversion is 100%

After catalyst pre-reduction (Fig.1A, full symbols), oxygen consumption is complete over the whole transient period (for reoxidizing the reduced ceria, then for oxidizing methane), but a short period of few seconds is required before methane starts to be oxidized. After catalyst pre-oxidation (Fig.1A, open symbols), oxygen is only partially consumed immediately after the switch to the reaction feed, before being fully consumed after about 40 s. Methane conversion increases more slowly before the steady state is achieved. CO_2 is immediately formed, but its concentration increases slowly (~60 s) to reach its steady- state value.

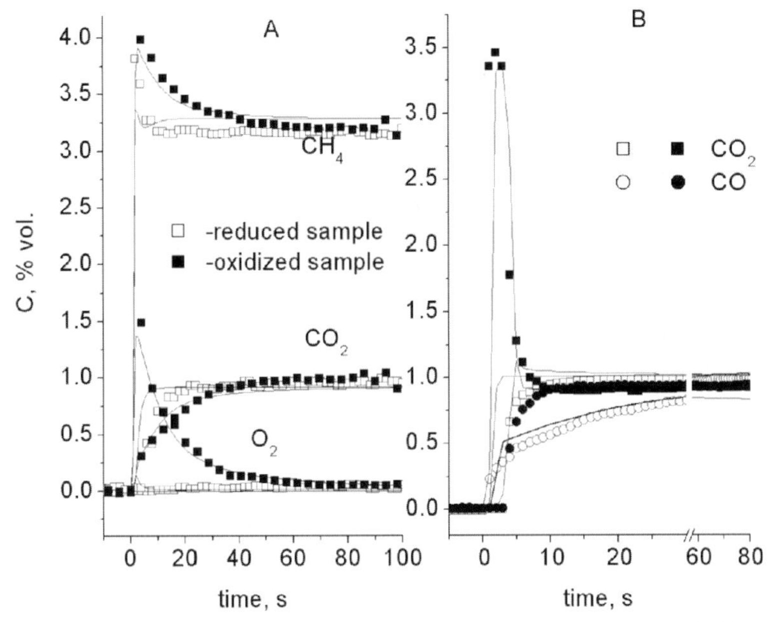

Figure 1 : Experimental (symbols) and calculated (lines) responses after switching from He to the reaction mixture $CH_4 + O_2$ in He over pre-reduced (open symbols) and pre-oxidized (full symbols) CeO_2-ZrO_2 (A) and 1.4% Pt/CeO_2-ZrO_2 (B) samples at 650°C.

As described elsewhere [2,5], the pre-oxidized surface would be saturated by the surface oxygen forms, while an even larger amount of oxygen may be dissolved within domain boundaries. Therefore, after the switch, the steady-state coverage of the surface by all oxygen forms is achieved, the respective transients of CH_4 combustion being affected by the oxygen transfer from domain boundaries to the surface. Along with domain boundaries, Frenkel-type defects present in the nanocrystalline Ce-Zr-O sample could also be responsible for a fast oxygen transfer from the bulk of particles to their surface [4]. Within the accuracy of the analysis, no carbon balance deficit was noted during the switch, either with pre-oxidized or pre-reduced sample. To explain a lower initial methane conversion observed on the pre-oxidized catalyst, intermediate slow steps for CH_4 activation involving participation of reduced sites (Z= Ce^{3+} or mixed oxide surface oxygen vacancies) that are scarce on the pre-oxidized surface should precede the fast oxidation of adsorbed methane by surface oxygen. For the pre-reduced surface, Z sites are available, but sufficient surface coverage by reactive oxygen species (ZO) should be attained before oxidation proceeds noticeably.

3.2. *Switches from He to CH_4 (+ O_2) (POM) over 1.4%Pt/CeO_2-ZrO_2*

For Pt-supported sample, the steady-state CH_4 conversion increases from 24% to 41÷46 %, depending on the oxygen inlet concentration, the oxygen being fully consumed (Table 1). Considering that 24% of the inlet methane was converted to CO_2 over the Pt-free support (surface area 65 m^2/g), it may be inferred that for the Pt-loaded system (33 m^2/g), within a simple additive scheme, the CH_4 conversion via its total oxidation on the support is about 2 times smaller, i.e. =12%. This corresponds to a consumption of about 0.53% of CH_4 and 1.06% of O_2. If one assumes that in parallel to the total oxidation on the support, only partial oxidation, i.e. $CH_4 + \frac{1}{2} O_2 \rightarrow CO + 2 H_2$, occurs on Pt sites, 1.14% of remaining O_2 would oxidize 2.28 % of CH_4, which would lead to a total methane conversion of (0.53+2.28)/4.6*100% = 61%. This value being higher than the experimental one (~ 42%), it comes that both complete and partial oxidation of methane occur on Pt sites.
In addition to this effect on the steady-state methane conversion, different dynamic transients were observed on the Pt -loaded sample (Fig.1B).
When switching either from He to 4.2%CH_4 or from He to 4.2%CH_4+ 2.2%O_2 + He over the pre-oxidized catalyst, a spike of CO_2 formation together with a spike of CH_4 conversion (not reported for sake of clarity of the Figure) are always observed either in the presence or in the absence of oxygen (Fig. 1B for the case : He to 4.2%CH_4+ 2.2%O_2 + He). This indicates that both pre-oxidation of Pt surface and accumulation of surface oxygen forms (ZO) and/or mobile oxygen dissolved within domain boundaries of support favor the complete oxidation of methane before the partial oxidation into syngas establishes. This delay observed for syngas formation clearly reveals that a certain reduction of the pre-oxidized Pt surface and of the ceria-zirconia support is required for syngas production, as reported in [1-5].
The amount of oxygen released as CO_2, CO and H_2O over the pre-oxidized catalyst during the transient reaction with pure CH_4 (obtained from transient curves integration) corresponds to the amount of highly mobile and reactive oxygen atoms (ZO). It was evaluated to be about $7*10^{20}$ atoms/g. This value corresponds approximately to a monolayer of the CeO_2-ZrO_2 material ($6.5*10^{20}$ atoms/g), calculated elsewhere from the surface area and also evaluated from chemical titration by CO oxidation [1-5].
Over the pre-reduced catalyst, in an oxygen-free feed, methane is converted exclusively to CO and H_2. The activation of CH_4 is generally assumed to be

dissociative and partly reversible [8], leading to surface CH_x fragments, stable enough to react in turn with oxygen atoms provided for the present case by the support. Here, the nature of intermediate adspecies leading to CO is not elucidated, except for the final carbonyl species, which were found to accumulate mostly as linear complexes on the Pt surface, as clearly demonstrated by in situ DRIFT spectra [8]. Note that in the absence of oxygen able to replenish rapidly the reduced sites, the carbonyl species adsorbed on Pt particles are also likely to react with hydroxyl groups of the support at interface sites, leading to formate adspecies formation [8]. The latter are slowly released into the gas phase as CO, as observed by the long tailing of CO formation when the pre-oxidized catalyst is contacted with CH_4 (Fig.1A).
On the pre-reduced catalyst in the presence of oxygen (Fig.1A), CO also appears initially, but is rapidly oxidized into CO_2 as oxidized sites (PtO and ZO) are formed. Again the delay in CO_2 appearance is assigned to surface hold up as carbonates till the surface/gas phase adsorption equilibrium is achieved.
On the basis of the above observations and a number of other ones which are reported in a forthcoming paper [9], a series of surface steps can be proposed for the partial oxidation of methane over the Pt/CeO_2-ZrO_2 system (Table 2).

Table 2. Elementary steps involved in the transient experiments over the Pt/CeO_2-ZrO_2 system

№	**CeO_2-ZrO_2**
1s	$[Z_1] + CO \leftrightarrow [Z_1CO]$
2s	$[Z_2] + H_2O \leftrightarrow [Z_2O] + H_2$
3s	$[Z_1CO] + [Z_2O] \rightarrow [Z_1] + [Z_2] + CO_2$
4s	$[Z_1CO] + [Z_1O] \rightarrow 2[Z_1] + CO_2$
$5s^{bis}$	$Obulk \rightarrow [Z_1O]$
$8s^{bis}$	$Z^* + 3\ CO \leftrightarrow [Z^*(CO)_3]$
9s	$O_2 + 2[Z_1] \leftrightarrow 2[Z_1O]$
$11s^{bis1}$	$[Z_1CH_2O(+2Z_1H)] + 3[Z_1O] \rightarrow CO_2 + 2H_2O + 6[Z_1]$
12s	$[Z_1CH_2O(+2Z_1H)] + 3/2O_2 \rightarrow CO_2 + 2H_2O + 3[Z_1]$
$10s^{bis}$	$CH_4 + 2[Z_1] + [Z_1O] \leftrightarrow [Z_1CH_2O\ (+ 2Z_1H)]$
	Platinum phase
2Pt	$PtO + CO \rightarrow CO_2 + Pt$
$3Pt^{bis}$	$Pt + [Z_2O] \rightarrow PtO + [Z_2]$
3Pt	$Pt + [Z_1O] \rightarrow PtO + [Z_1]$
6Pt	$2Pt + O_2 \rightarrow 2PtO$
7Pt	$PtO + CH_4 \rightarrow CO + 2H_2 + Pt$
8Pt	$4PtO + CH_4 \rightarrow CO_2 + 2H_2O + 4Pt$

Note : Two types of surface oxygen i) on-top highly reactive Z_1O and ii) more strongly bonded and less reactive Z_2O, are proposed for describing the CeO_2-ZrO_2 surface, for accounting for the observed transient curves under POM conditions. In addition, formates, carbonates and hydroxyls adspecies stored over the mixed oxide might affect the observed transients as well.

These simplified steps were used in the complete modeling of the transient curves, leading to a quite reasonable agreement with the experiment as exemplified in Fig. 1. All rate constants are given in [9] from which it could be derived that the step of oxygen transfer from the most active sites Z_1O to Pt (step 3Pt) was by order of magnitude faster than the other ones.

4. Conclusion

The dynamics of transient experiments carried out on nanocrystalline CeO_2-ZrO_2 and Pt/CeO_2-ZrO_2 materials in POM appear to be associated with i) at least two types of oxygen forms differing by their reactivity and ii) a fast surface/bulk oxygen diffusion, which might be determined by a high density of domain boundaries and Frenkel-type defects. The profile of these transients is clearly related to the state of the catalyst (preoxidized or pre-reduced). The fast oxygen transfer from the support to Pt sites favors preferential routes for POM reactions over Pt clusters, although the mixed oxide alone displays a reasonably high activity. Domain boundaries might be involved also in storage of carbonates and hydroxyls which affects observed transients as well.
All these mechanistic conclusions provide new trends for designing catalysts specifically adapted to applications involving essentially transient regimes.

5. References:

1. V.A. Sadykov, T.G. Kuznetsova, G.M. Alikina, Yu.V. Frolova, A.I. Lukashevich, et al Catal. Today 93-95 (2004) 45
2. V.A. Sadykov, T.G. Kuznetsova, S.A. Veniaminov et al, React. Kinet. Catal. Lett. 76 (2002) 83
3. V.A. Sadykov, T.G. Kuznetsova, Yu.V. Frolova-Borchert, G.M. Alikina, A.I. Lukashevich, V.A. Rogov, V.S. Muzykantov, L.G. Pinaeva, E.M. Sadovskaya, Yu.A. Ivanova, E.A. Paukshtis, N.V. Mezentseva, L.Ch. Batuev, V.N. Parmon, S. Neophytides, E. Kemnitz, K. Scheurell, C. Mirodatos, A.C. van Veen, Catal. Today 117 (2006) 475.
4. N.N. Bulgakov, V.A. Sadykov, V.V. Lunin and E. Kemnitz, React. Kinet. Catal. Lett. 76 (2002) 111.
5. T.G. Kuznetsova, V.A. Sadykov, S.A. Veniaminov, G.M. Alikina, E.M. Moroz, V.A. Rogov, O.N. Martyanov, V.F. Yudanov, I.S. Abornev, S. Neophytides. Catal. Today 91 – 92 (2004) 161
6. H. Borchert, Yu. Borchert, V. V. Kaichev, I. P. Prosvirin, G. M. Alikina, A. I. Lukashevich, V. I. Zaikovskii, E. M. Moroz, E. A. Paukshtis, V. I. Bukhtiyarov, and V. A. Sadykov, J. Phys. Chem. B. 109 (2005) 20077.
7. E.M. Sadovskaya, A.P. Suknev, L.G. Pinaeva, V.B. Goncharov, B.S. Bal'zhinimaev, C. Chupin, J. Pérez-Ramírez, C. Mirodatos, *J. Catal*, 225 (2004) 179.
8. E. Odier, Y. Schuurman, H. Zanthoff, C. Millet, C. Mirodatos. Proceedings "3rd International Symposium on Reaction Kinetics and the Development and Operation of Catalytic Processes" Ostende, Studies in Surface Science and Catalysis, Elsevier, Amsterdam, 2001, 133, 327-332.
9. L. G. Pinaeva, E. M. Sadovskaya, Yu. A. Ivanova, T. G. Kuznetsova, V. A. Sadykov, G. Grasso, Y. Schuurman, A.C. van Veen, C. Mirodatos, to be submitted.

Natural Gas Conversion VIII
F.B. Noronha, M. Schmal, E.F. Sousa-Aguiar (Editors)

Investigation by high throughput experimentation of ceria based catalysts for H_2 purification and CO_2 reforming of CH_4.

J. Despres, C. Daniel, D. Farrusseng, A.C. van Veen, C. Mirodatos

Institut de Recherches sur la Catalyse, CNRS, Villeurbanne, 69626 Cédex, France

1. Introduction

High-Throughput Experimentation (HTE) aims at i) increasing the probability for new materials discovery, ii) increasing speed and efficiency for process development and optimisation (faster access to optimised operating conditions through the development of micro-pilot technologies), iii) improving safety and environmental aspects (reduced reactor volumes and integrated mass and heat transfer improving the overall energy balance) [1,2].
Testing and comparing different catalytic compositions belonging to a given library, the HTE strategy is generally restricted to evaluate at simplified standard conditions as rapidly as possible catalytic performances such as reaction yield towards desired products. However, this fast testing procedure may eliminate good candidates which could have been revealed under more appropriate operating conditions, or select false hits which will turn out to be inactive after longer time on stream.
Parallel reactors when operated for fast primary screening can hardly be used for solving the above mentioned issues. However, in addition to primary information upon catalysts ranking, various ways exist to capture the essence of catalysts instability, improving considerably the selection process. Indeed catalysts instability in general refers to activation/deactivation processes occurring during steady-state operations. It also relates to catalyst behaviour under transient operating conditions which may correspond to permanently unsteady processes employing moving or fluidized bed, post-combustion, reverse flow reactors, domestic applications, etc.
In addition to using parallel reactors, an elegant and efficient way to have access to catalyst instability is to perform fast kinetic measurements under transient conditions, which may indeed add a leap in the acceleration of the investigation process [3].

This paper aims at illustrating how catalyst instability can be captured by HTE in the domain of natural gas conversion and hydrogen production. The first case study deals with the discovery of new ceria based material for hydrogen purification. Here, already published data are reinvestigated by focusing on the transient measurements of OSC and the use of parallel reactor for selecting performing new materials. New perspectives on this domain of HTE investigation of adsorption phenomena like the fast measurements of CO_2 adsorption are also presented. The second case study deals with the instability of Pt/doped ceria catalysts for CO_2 reforming of methane and unpublished results of an on-going HTE investigation are presented.

2. HTE for investigating OSC of new ceria based material for hydrogen purification. Perspectives for HT investigation of adsorption phenomena

Context. Ceria-based mixed oxides ($Ce_xM_{1-x}O_y$) are versatile solid oxygen exchangers. At high temperatures (400–800°C) the redox cycle $Ce^{3+} \Leftrightarrow Ce^{4+} + e^-$ facilitates oxygen storage and release from the bulk fluorite lattice. This makes them ideal candidates for catalytic oxidation applications such as the above mentioned hydrogen purification methods [5]. However, the *surface* redox chemistry of ceria is sensitive even at low temperatures to crystal structure defects, which can be tuned by substituting some of the Ce cations with ions of different size and/or charge [6,7].
In this first case study, the effects of doping the ceria support of platinum catalysts for the close reactions of SelOx, WGS and reverse water-gas shift (RWGS) are investigated using parallel synthesis and testing reactors, aiming at discovering new materials especially under dynamic conditions (start-up and shut-down) [8].

Methodology and results. A set of 12 catalysts containing 2 wt% Pt on $Ce_{0.9}M_{0.1}O_x$ was prepared (where M = Pb, Bi, Zr, V, W, Mo, Y, La, In, Sn, and two combinations of Zr/Bi, respectively) by impregnating platinum on various doped ceria supports. Details on this catalysts library and their characterization are given in [8]. Despite their nearly identical X-ray crystal structures, these catalysts were shown to display diverse redox/acidic/basic properties and different selectivities in the three test reactions. The best SelOx catalysts were also found to be the most active for WGS/RWGS equilibration. In addition to these conventional catalytic properties under steady-state regime, a major descriptor of these materials within the perspective of a domestic application is the oxygen storage capacity (OSC). The latter is involved during all the transient procedures of the targeted process, i.e. start-up and shut down of the hydrogen generator, according to the users needs. The OSC which controls the behaviour of the ceria based material under instable process conditions was evaluated by adapting a 16 channel parallel reactor to capture the breakthrough curves by admitting a mixture H_2/Ar =10/90 after catalyst oxidation. OSC was derived from the H_2 consumption of available oxygen of the ceria to form H_2O (Fig.1).

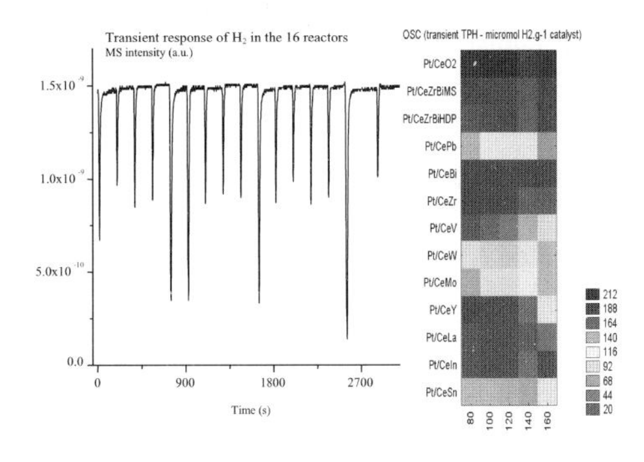

Figure 1: Evaluation of Oxygen Storage Capacity of catalysts (OSC) in transient mode for one generation of ceria based catalysts [8]

A 4.75 wt% Pt on high surface ceria catalyst was used as reference, displaying a very high OSC whatever the temperature. In contrast, 2% wt%Pt doped samples prepared for this study showed in general a lower OSC. Nevertheless, among 13 ceria additives, 4 (Pb, W, Mo and Sn) lead to increase the OSC by factors of 2 to 3 independently of the temperature. In contrast, V-doped ceria showed a continuous increase of the OSC with the temperature to attain an OSC of 120 micromols per gram of catalyst. Correlation between OSC and CO oxidation results were observed and discussed elsewhere [8].
As a general trend, it was found that the most performing samples exhibited the poorest OSC, indicating that excess of oxygen delivered by the system is quite detrimental for the selective oxidation of CO in the presence of hydrogen.
From a technical point of view, this method of transient OSC measurement at iso-temperature was proved to be efficient for assessing key features of 16 oxygen storing and oxygen conducting materials within a few hours.
Thus, by combining the information get from the HT measurements of OSC and the known effects of doping ceria with various heterovalent atoms [6], explanation of doped systems can be provided.

Perspectives for HT investigation of adsorption phenomena. Recently we extended this HTE facility to capture other types of adsorption phenomena. As such, isotherms of pure component and gas mixtures can be measured by means of parallel breakthrough curve experimentation. The capabilities of the testing units were demonstrated for the HT CO_2 adsorption on microporous materials [13]. As an example, Figure 2 presents the isotherms of CO_2 adsorption over a series of microporous materials acquired in a 16 channels parallel reactor. By using Langmuir models, the adsorption parameters were calculated. Thus, we found activation energy of 20 kJ/mol for the reference 13X zeolite in good agreement with literature data.

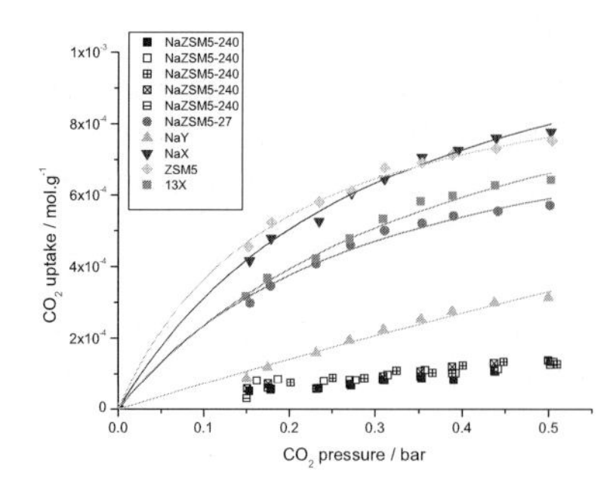

Figure 2: CO_2 adsorption isotherms at 120°C from a mixture $N_2/CO_2/He$ on zeolites. For the sample NaZSM5-240, five replicates have been studied [13].

3. Instability of Pt/doped ceria catalysts for CO_2 reforming of methane

Context. Crude natural gas may contain, in addition to methane, light alkanes and moisture, up to 30 vol. % of carbon dioxide, and up to 50% for the case of bio-gas. In order to convert these mixtures into syngas for a Gas-To-Liquid (GTL) process without prior separation, high performance catalysts for CH_4 and higher alkanes dry reforming are required, with a high activity and selectivity towards CO and H_2. Suitable catalysts are generally achieved with most group VIII based materials, but materials require also a high stability on stream ensured by a low coke formation, as coking constitutes the main drawback of the dry reforming process [9].

Accelerating this search of performing systems, HTE was recently implemented according different strategies. Maier et al. [10] used high-throughput synthesis and primary screening techniques for discovering and optimizing new formulas of catalysts able to be self reduced by the reacting feed of methane dry reforming. Several larges libraries have been synthesized with sol-gel methods using a synthesis robot and library design software. Those catalyst libraries were screened for catalytic activity and stability at 600°C in a simple high-throughput reactor system with micro-GC analysis. From several generations of catalysts, the best performing stable systems were found to be Ni on ceria doped with Al and Zr. The interest of the work lies in the fast access to catalyst instability in HTE, where either initial activation (linked to the in situ reduction of the active phases) or deactivation processes (related to carbon deposits and/or metal sintering) are revealed. However, only further conventional studies on selected formulas permitted a better understanding of these processes [10].

Methodology and results. In our laboratory, an on going medium throughput strategy was chosen to test rapidly the aging testing sequences for long period on stream with the fast characterization of carbon deposits by sequential H_2 and O_2 TPR sequences, able to rapidly elucidate the content and variety of coke deposits. Data may then be inserted into kinetic models designed to predict ageing processes, in view of foreseeing the state of the catalyst after virtual long term ageing experiments. A realistic ranking may thus be proposed in order to proceed towards a next generation of samples. In this paper, only results on HTE investigation of dry reforming aging are presented, the sequential H_2 and O_2 TPR measurements being kept for a full forthcoming paper.

Figures 3 and 4 illustrate how a series of Pt/ceria based catalysts tested in parallel reactors for the dry reforming of methane behaves as fresh and used systems, and the ranking which may be derived as a function of the dopant added to ceria, respectively.

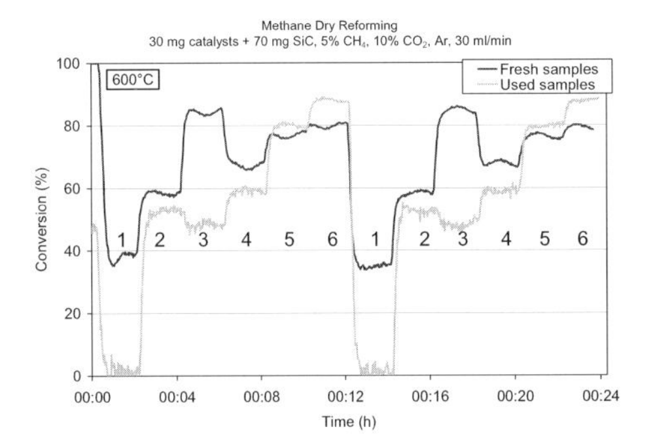

Figure 3: Time dependence of the methane dry reforming activity for a series of 6 Pt/ceria doped catalysts tested as fresh or used systems over series of 20 experimental cycles.

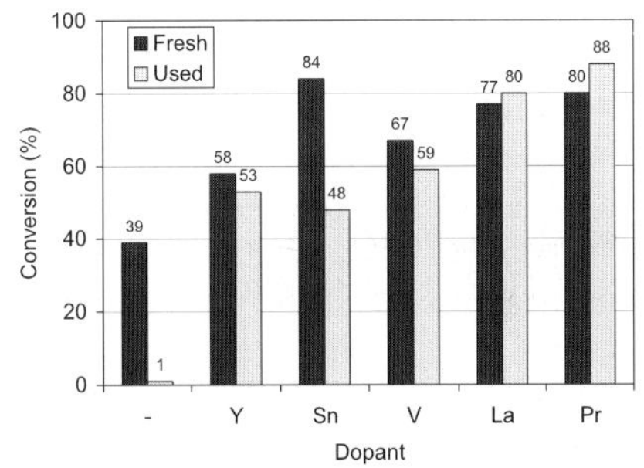

Figure 4 : Catalyst ranking according the nature of ceria dopant and the resistance to aging.

From these data it comes that the best resistance to coking was obtained for lanthanum and praseodymium doped materials, showing even a slight improvement in activity after aging cycles, while significantly lower performances and stability were observed for Y, Sn or V doped materials.
Referring to an advanced in situ characterization of lanthana doped Ni catalysts during methane dry reforming [11], it can be assessed that the specific role of the lanthanide additives is to protect the Ni particles with a layer of carbonates from emitting encapsulating carbon. This layer was shown to be built progressively with time on stream, which accounts for the improvement of performances upon aging. These carbonates in equilibrium with gas phase CO_2 act also as a continuous source of oxygen at Ni/ceria interface for oxidizing selectively into CO the CH_x fragments arising from methane decomposition [9,11]. For the case of V doped material, an increase in support acidity resulting from a spreading of V_2O_5 over the ceria surface (as shown for the close system VMgO [12]) might explain lesser performances, since the bi-functionality ensured by surface basicity is no more operating for this case. Finally, when an

easily alloying element like Sn is added to ceria, its likely migration onto the Ni particles might lead to a poisoning of the Ni phase since Sn is inactive for methane dissociation.

4. Conclusion

Through studies related to hydrogen purification and methane dry reforming, it is shown that high or medium throughput techniques can be adapted to capture unsteady processes in heterogeneous catalysis like ageing and dynamics of oxygen storage and CO_2 adsorption. However, a compromise has to be searched between fast screening of process parameters and deep understanding of the observed phenomena. The choice of the test reactions is therefore crucial. One option, as shown here, is to test sets of diverse catalyst formulations combined with real-time performance analysis and modeling, adapted to process instability.

Acknowledgements. This work was partly supported by the European research project "TOPCOMBI" (contract NMP2-CT2005-515792).

5. References

1. J. R.Engstrom, H. Weinberg, AiChE 46 (2000) 2.
2. D. Farrusseng, C. Mirodatos, in "High Throughput Screening in Chemical Catalysis. Technologies, Strategies and Applications", Hagemeyer, A, et al. (eds.), Wiley-CHV, Weinheim, 2004, p 239-269.
3. A.C. van Veen, D. Farrusseng, M. Rebeilleau, T. Decamp, A. Holzwarth, Y. Schuurman, and C. Mirodatos, J. Catal. 216 (2003)135.
4. R. M. Ormerod. Chem. Soc. Rev. 32 (2003)17.
5. S. D. Park, J. M. Vohs and R. J. Gorte. Nature 404 (2000) 265.
6. G. Rothenberg, E. A. de Graaf and A. Bliek. Angew. Chem. Int. Ed. 42 (2003) 3066.
7. J. A. Perez-Omil, S. Bernal, J.J. Calvino,J.C. Hernandez, C. Mira, M. P. Rodrıguez-Luque, R. Erni, and N. D. Browning, Chem. Mater. 17 (2005) 4282.
8. D. Tibiletti, E. A. (Bart) de Graaf, G. Rothenberg, D Farrusseng and C. Mirodatos, J. Catal. 225/2 (2004) 489.
9. L. Pinaeva, Y. Schuurman, C. Mirodatos, Environmental Challenges and Greenhouse Gas Control for Fossil Fuel Utilization in the 21st Century, M. Maroto-Valer, C. Song, Y. Soong, Eds, Kluwer Academic/Plenum Publishers, 2002, pp 313-327.
10. Kim, D. K., Maier, W. F., J. Catal, 2006, 238, 142-152.
11. A. Slagtern, Y. Schuurman, C. Leclercq, X. Verykios, C. Mirodatos, J. Catal. 172 (1997) 118-126.
12. A. Pantazidis, C. Mirodatos, ACS Symposium Series,Vol. 638, p. 207. Am. Chem. Soc.,Washington, DC, 1996.
13. G. Morra, A. Desmartin-Chomel, C. Daniel, U. Ravon, D. Farrusseng, M. Krusche, C. Mirodatos, to appear in Chem. Eng. J., 2007.

F.B. Noronha, M. Schmal, E.F. Sousa-Aguiar (Editors)

A Carbon Reaction Pathway for Dimethyl Ether, Methanol and Methane from Syngas on an Alumina Supported Palladium Catalyst

Masood Otarod,[a] Sami H. Ali[b]

[a]*University of Scranton, Scranton, PA 18510, USA*
[b]*Kuwait University, Kuwait*

1. Abstract

A novel modeling procedure was employed to identify a carbon pathway mechanism for the conversion of syngas to dimethyl ether over an alumina supported palladium catalyst. The concentration of adsorbed carbon monoxide, CO(a), and hydroxycarbon intermediates, CHO(a), CH_2O(a) and CH_3O(a), were estimated by comparing the model to the ^{13}CO tracing of the transient response of naturally occurring dimethyl ether, CH_3OCH_3. The path of methanation was identified by estimating the concentration of the adsorbed carbon, C(a), and the combined concentration of adsorbed hydrocarbonaceous species CH_x(a), x=1,2,3 from the isotope transient of methane.

The results indicate that methane, methanol and dimethyl ether adsorb on the surface of the catalyst at high exchange rates and that the formation of the methoxy intermediate appears to be the rate controlling step in the synthesis of dimethyl ether. A comparison of the estimated concentrations of adsorbed carbon monoxide as estimated from the dimethyl ether and methanation data suggests that the amount of adsorbed carbon monoxide may not be equally accessible in its entirety to all reaction pathways.

2. Introduction

The conversion of syngas to dimethyl ether over Pd/Al proceeds according to the following overall reactions

$$CO + 2H_2 \leftrightarrow CH_3OH \quad (1)$$

$$CO + H_2O \leftrightarrow CO_2 + H_2 \quad (2)$$

$$2CH_3OH \leftrightarrow CH_3OCH_3 + H_2O \quad (3)$$

$$CO + 3H_2 \leftrightarrow CH_4 + H_2O \quad (4)$$

While the characterization of the palladium based catalysts for the conversion of syngas to methanol, methane and dimethyl ether has been the subject of extensive research [1-4], the mechanism of conversion of syngas to methane, methanol and dimethyl ether over alumina supported palladium has not yet been fully studied. Ali and Goodwin [5] conducted a ^{13}CO isotope transient tracing of this reaction system on Pd/Al and reported a set of isotope transient tracing rate data for the naturally occurring CH_4, CH_3OH and $(CH_3)_2O$. Their data provided an opportunity to probe this reaction system and this article is dedicated to the examination of a mechanism for the conversion of syngas on Pd/Al.

3. Modeling and Data Analysis

If a step function of an isotopically labelled reactant is superimposed on a reaction system running at steady state, transients will develop in the rates of the effluents from which the concentration of the intermediates and the step velocities can be inferred. The mathematics of the analysis of the isotope transient tracing data is simplified if the experiments are conducted in a gradientless CSTR or a differential diffusion-free PFR.

^{13}CO tracing of the overall reactions (1-4) produces transient rates for methanol, dimethyl ether and methane. For dimethyl ether they are $(CH_3)_2O$, $^{13}CH_3OCH_3$ and $(^{13}CH_3)_2O$. The transients of methane are CH_4 and $^{13}CH_4$. Similarly, for methanol they are CH_3OH and $^{13}CH_3OH$. Modelling of this reaction system under steady state isotope tracing involves the identification of a mechanism, derivation of the material balance equations and the estimation of the parameters by comparing the model to the data. These procedures are detailed as follows.

3.1. Mechanism

The following mechanism was selected.

$$H_2 + 2\,(a) \longrightarrow 2H(a) \quad (5)$$

$$CO + (a) \longrightarrow CO(a) \quad (6)$$

$$CO(a) + H(a) \longrightarrow CHO(a) + (a) \quad (7)$$

$$CHO(a) + H(a) \longrightarrow CH_2O(a) + (a) \quad (8)$$

$$CH_2O(a) + H(a) \longrightarrow CH_3O(a) + (a) \quad (9)$$

$$CH_3O(a) + H(a) \leftrightarrow CH_3OH + 2(a) \quad (10)$$

$$CH_3O(a) + CH_3O(a) \longrightarrow (CH_3)_2O(a) + O(a) \quad (11)$$

$$(CH_3)_2O(a) \leftrightarrow (CH_3)_2O + (a) \quad (12)$$

$$CO(a) + (a) \longrightarrow C(a) + O(a) \quad (13)$$

$$C(a) + xH(a) \longrightarrow CH_x(a) + x(a) \quad (14)$$

$$CH_x(a) + (4\text{-}x)H(a) \leftrightarrow CH_4 + (5\text{-}x)(a) \quad (15)$$

The character (a) indicates a vacant adsorption site or an adsorbed surface species.

Carbon tracing does not provide direct information concerning the adsorption of hydrogen. However, almost all investigators agree that hydrogen dissociatively adsorbs on Pd and reacts with other species. Steps (6-10) are adopted from the work of Poutsma et al [6] where they suggested that the formation of methanol on palladium involves the addition of adsorbed hydrogen to adsorbed carbon monoxide. Steps (11) and (12) draw on the ideas proposed by Schiffino and Merrill [7]. Steps (13-15) are a variation on the mechanism for methanation reported by Vannice and coworkers [8]. The letter x is used generically to balance Eqs. (14) and (15) because under ^{13}CO tracing hydrogenation sequence of steps cannot be identified and as such it has no direct impact on the modelling of carbon tracing data. Consistent with these works, equilibrium was assumed for Eqs. (10), (12) and (15).

3.2. Data

Transient fractional concentrations for CH_3OCH_3 and CH_4 were adopted from the work of Ali and Goodwin [5]. The experiments were conducted on a Pd/Al precipitated from $Pd(NO_3)_2$ precursor. The mode of the operation was differential with an overall CO conversion of 0.57% and no evidence of mass transfer was observed. The reactor had a charge of $W = 0.05$ grams of 5% Pd catalyst supported on γ-Al_2O_3.

3.3. Mass conservation equations

The detail of the derivation of the material balance equations for isotope tracing in a differential plug-flow reactor is given by Happel [9]. They are partial differential equations with a space variable w and time t. A comprehensive exposition of these equations to modelling isotopic transients in a differential PFR are given by Otarod et al [10].

To model the transient tracing data, it is assumed that the catalyst is uniformly distributed throughout the reaction chamber. The space variable w is chosen to be the amount of the catalyst swept by the reacting gases from the beginning of the bed. Thus, if W is the total amount of the catalyst, then $0 \le w \le W$.

For a typical gas phase species A with a steady state volumetric flow rate of $F^A(w)$, and an adsorbed intermediate A(a) with a steady state surface concentration of $C^{A(a)}$, they assume the following form.

$$\frac{\partial}{\partial w}[F^A(w)z^A(t,w)] + \frac{\tau}{W}\frac{\partial}{\partial t}[F^A(w)z^A(t,w)] = R_A \tag{16}$$

$$C^{A(a)}\frac{\partial}{\partial t}[z^A(t,w)] = R_{A(a)} \tag{17}$$

Where, for the mechanism described by Eqs. (6-15), A = CO, $(CH_3)_2O$, CH_4, CH_3OH, and A(a) = CO(a), CHO(a), CH_2O(a), CH_3O(a), $(CH_3)_2O$(a), C(a), CH_x(a). $z^A(t,w)$ and $z^{A(a)}(t,w)$ represent, respectively, the fractional concentration of the gas phase species A and adsorbed species A(a) with respect to the sum of their concentration and that of their isotopically labelled cohorts.

R_A and $R_{A(a)}$ are the rate functions for A and A(a) respectively which are derived on the basis of the power law in lieu of the elementary mechanistic steps described by Eqs. (5-15). τ is the space time of the reactor.

Under steady state tracing and the hypothesis of the differential PFR mode $F^A(w)$, the flow rate of A, depends on the bed length w, but $C^{A(a)}$ the concentration of adsorbed intermediate A(a), is independent of w.

For a tubular reactor these equations are to be solved subject to the following boundary and initial conditions.

$$z^{A}(t<0,w)=1, \quad z^{A(a)}(t<0,w)=1$$
$$z^{A}(t\geq 0,w=0)=0, \quad z^{A(a)}(t\geq 0,w=0)=0$$

4. **Results and Discussion**

Fig. 1 is the result of the least squares curve fitting procedure of dimethyl ether transient and methane data to the mass balance equations corresponding to the mechanism. The agreement between the model and the data is remarkable.

Adsorbed intermediates can only be resolved to the extent that their presence is detectable by the experimental data. Dimethyl ether reaction pathway is identified by the concentrations of CO(a), CHO(a), CH_2O(a), CH_3O(a) and $(CH_3)_2O$(a), and the step velocities. ^{13}CO tracing cannot distinguish among CHO(a), CH_2O(a) and CH_3O(a) complexes. Yet, CH_3O(a) is distinguishable from the other two species as, according to the model, it is in an exchange reaction with methanol. Thus, under ^{13}CO tracing, the steps involving CHO(a) and CH_2O(a) were combined and a combined concentration was estimated.

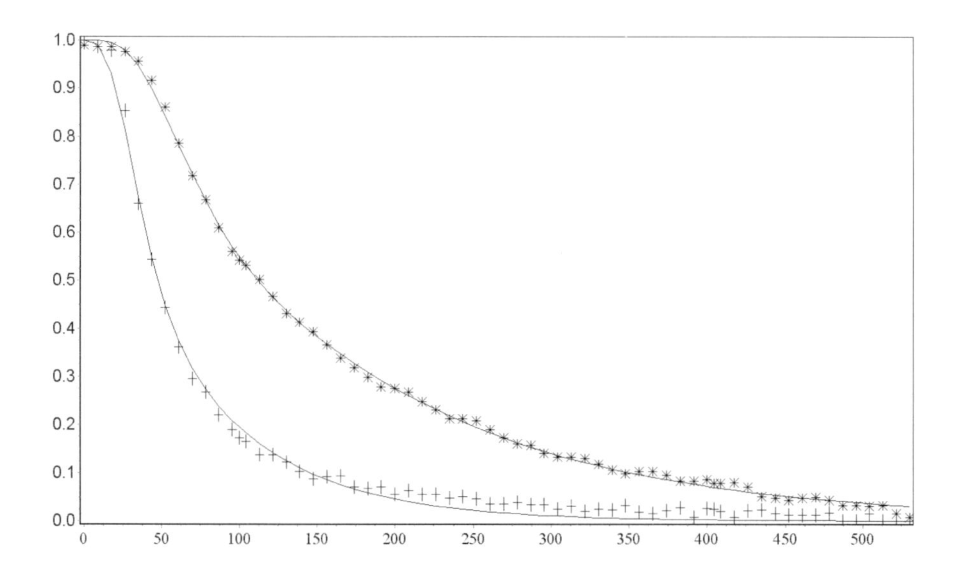

Figure 1. Fractional marking level in the rates of dimethyl ether and methane versus time (second). * = dimethyl ether data, + = methane data, − = model..

Equilibrium was assumed for Eq. (10) since data correlation indicated that its steady state forward and backward rates were large.

Dimethyl ether was found to adsorb on the catalyst at a concentration of 6.71 μ mol/g catal. This result confirms the findings of Schiffino and Merrill [7] where they point out that dimethyl ether adsorbs weakly on alumina. The concentration of $CH_3O(a)$ was estimated to be 47.6 μ mol/g catal. and that of $CHO(a)+CH_2O(a)$ was 5.05 μ mol/g catal.

It was also found that methanol adsorbs on the catalyst and equilibrates with methoxy at a very high rate. Schiffino and Merrill [7] have presented a detailed mechanism concerning the dehydration of methanol on γ-alumina catalyst involving the interaction of methoxy/methoxy and methanol/methoxy to form dimethyl ether. While kinetically significant, these interactions are inconsequential to our modeling approach as these steps are indistinguishable under ^{13}CO. The estimated concentration of adsorbed carbon monoxide from the path of dimethyl ether was 6.33 μ mol/g catal.

The concentrations of methanation intermediates were estimated by modeling the transient of methane. It was found that C(a) was adsorbed at 3.39 μ mol/g catal. and $CH_x(a)$ was present at 2.4 μ mol/g catal. Only a small amount of CO(a) (0.01 μ mol/g catal.) was used for methanation. This result seems to confirm the assertion by Vannice and coworkers [8] that the ``large surface concentrations of ir-active CO which exist on these Pd catalysts do not represent the form of adsorbed CO participating in the RDS ..."

Thus, the concentration of hydroxycarbon adsorbed species plus that of the adsorbed carbon monoxide amount to a total surface coverage of 71.53 μ mol/g catal. While the concentration of the adsorbed noncarbonaceous species cannot be estimated from the present tracing data, the total holdup of carbon containing species seems to be within a comparable range of the data reported from the CO chemisorption experiments [5].

5. Conslusion

The results obtained concerning the reversibility of the elementary steps and the concentration of adsorbed intermediates suggest that the mechanism proposed here satisfactorily describes the surface reactions on Pd(NO)/Al as concerns the carbon pathway.

Since $CH_3O(a)$ is the most abundant intermediate in a path of the reaction running at steady state, its formation seems to be the slowest step in the mechanism, and Eq. (9) is likely to be the rate determining step in the dimethyl ether reaction pathway.

Further studies involving isotopes can be helpful in elucidating the mechanism of the synthesis of this very important compound. As an adjunct to carbon pathway analysis, ^{18}O tracing can be particularly useful for probing the path of dimethyl ether synthesis as carbon and oxygen seem to follow the same path from CO to dimethyl ether.

Reference:

1. W.-J. Shen, M. Okumura and M. Haruta, Appl. Catal., A: General, 213 (2001) 225.
2. A.L. Bonivardi, D.L. Chiavassa, C.A. Querini and M.A. Baltanás, Stud. Surf. Sci. Catal. 130 (2000) 3747.
3. F.A.P. Cavalcanti , A. Yu. Stakheev and W.M.H. Sachtler WMH, J. Catal. 134 (1992) 226.
4. M. A. Vannice, S.-Y. Wang and W.H. Moon, J. Catal. 71 (1981) 152.
5. S.H. Ali and J.G. Goodwin, J. Catal., 176 (1998) 3.
6. M.L. Poutsma, L.F. Elek, P.A. Ibarbia, A.P. Risch and J.A. Rabo, J. Catal. 52 (1978) 157.
7. R.S. Schiffino and R.P. Merrill, J. Phys. Chem. 97 (1993) 6425.
8. S.-Y. Wang, W.H. Moon and M. A. Vannice, J. Catal. 71 (1981) 167.
9. J. Happel, Isotopic Assessment of Heterogeneous Catalysis, Academic Press, 1986.
10. M. Otarod, Y. Lecourtier, E. walter and J. Happel, Chem. Eng. Comm., 116 (1992) 127.

Natural Gas Conversion VIII
F.B. Noronha, M. Schmal, E.F. Sousa-Aguiar (Editors)

Low Emission Conversion of Natural Gas to Hydrogen

D. Trimm[a], I. Campbell[b], Y.Lei[c], N. Assanee[c] and N.W.Cant[c]

[a] *CSIRO Petroleum, Melbourne, VIC3168, Australia*
[b] *CSIRO Energy Technology, Sydney, NSW2234, Australia*
[c] *The University of New South Wales, Sydney, NSW2052, Australia*

Abstract

The decomposition of methane to hydrogen and carbonaceous residues under conditions suitable for solar heating (ca. 750°C) has been studied. Although nickel and iron-containing catalysts provide decomposition at appropriate temperatures, deposition of carbon leads to deactivation. Regeneration is found to be slow. However, scrap material from a steel mill, containing iron as a major catalytic component, was found to give high yields of hydrogen. Carbon was deposited on the scrap material and the solid product is suitable for recirculation to a steel mill.

1. Introduction

Although the production of synthesis gas and of hydrogen by the reforming of natural gas are well established industrial processes [1], there are disadvantages associated with the reactions. Both steam reforming and dry reforming are endothermic reactions which require significant heating, usually by combusting part of the natural gas [1]. The combustion produces carbon dioxide, an unwanted greenhouse gas.
Hydrogen may also be produced by the decomposition of methane, a process that can be adjusted to avoid the production of greenhouse gases [2].

$$CH_4 = C + 2H_2 \qquad (1)$$

Non-catalysed and catalysed pyrolysis yields hydrogen and carbonaceous residues, some of which can be economically attractive. The generation of greenhouse gases as a result of heating natural gas to pyrolysis temperatures can be avoided by solar heating, but this process also introduces some difficulties.

Consideration of the design of solar powered furnaces reveals that the maximal irradiated area can be achieved for a temperature of about 750°C, which necessitates the use of catalysts for the pyrolysis. Metallic and zeolitic catalysts have been recommended [2], but the separation of the catalyst from deposited carbon and catalyst regeneration introduces new problems. As a result, the present studies have been focused on identifying disposable catalysts that can facilitate pyrolysis at ca 750°C.

Various forms of iron oxides have been considered for catalysts for methane decomposition [3] and the steam-iron process [4]

$$CH_4 + Fe_2O_3 = CO + 2H_2 + 2FeO \quad (2)$$

$$CO + Fe_2O_3 = CO_2 + 2FeO \quad (3)$$

$$2FeO + H_2O = Fe_2O_3 + H_2 \quad (4)$$

Although complicated by the intermediate production of carbon oxides and the possibility of carbide formation, the generation of pure hydrogen was found to be possible at temperatures of the order of 600-800°C. Thus iron was seen to be a possible catalyst for the relatively low temperature solar decomposition.

2. Experimental

Comparisons of the efficiency of various catalysts were assessed using a flow reactor system. Natural gas (>98%methane) was monitored using Brooks 5850 mass flow controllers and the mixed gases were passed to a tubular reactor (300mm long, 10mm ID) held in an electrically heated furnace. Exit gases were separated using a 1.5m Carbosphere column temperature programmed from 80 to 230°C and a 2.4m Chromosorb 102 column, temperature programmed from 40 to 120°C and analysed using TCD detectors. Flow rates were controlled between 15 and 90 ml/min and passed over 0.5 – 2 g of catalyst. Where necessary, the catalysts were reduced at 650°C or 850°C in pure hydrogen (130 ml/min) before use.

Experiments to study the reduction and oxidation of iron based catalysts were carried out using a flow reactor, with water injected at a controlled rate using an Isco pump. Product analysis was completed using a mass spectrometer (Balzer Thermostar).

Various types of iron scrap were obtained from BlueScope Steel. Analysis results are shown in Table 1. Due to the nature of the material, average (within 2%) values are reported.

Table 1: Analysis of scrap materials (wt %)

Sample	Mg	Al	Si	P	Ca	V	Mn	Fe
BS1	2.15	0.3	0.8	0.06	5	0.06	0.65	50
BS6	0.8	1.9	1.9	0.3	3.4	0.02	2.1	50
BS7	3.2	0.9	3.5	0.45	13.5	0.73	2.3	50

3. Results and Discussion

Initial experiments showed the effect of inert surfaces on the pyrolysis of methane to be negligible. Experiments with various carbons, as suggested in the literature [5], was found to favour pyrolysis at 850°C but required expensive high surface carbons. XRF analysis of the carbons showed only traces of potentially catalytic inorganic material and the amounts of hydrogen were found to follow a logarithmic relationship with surface area as reported by Muradov [5,6]. Results reported in Table 2 show that up to ca 35% hydrogen could be produced over activated carbons, but only at 850°C.

Table 2: Comparison of carbon catalysts at 850°C

Catalyst	Surface Area m^2/g	Space Velocity (h^{-1})	Hydrogen, %
Metallurgical coke	0 - 10	4,800	0.29
Graphite/coke	0 - 10	10,650	0.05
Graphite bar	0 10	1,000	0.4
Graphite rod	0 - 10	490	1.1
Carbon black N660	36	1,000	8.5
Carbon black N774	29	1,000	9.5
PICACARB	1200	980	36.4
PICACTIF	1364	990	37.8
Coarse GAC 3560	2230	990	28.4

The study of various metal and alloy catalysts was then initiated. In general, attention was focused on scrap material such as Incoloy (a Ni-Fe-Cr alloy) and Monel (a Ni-Cu alloy) turnings, with comparison experiments being carried out with conventional catalysts (Table 3). Up to 60% of hydrogen in the exit stream could be produced at 650°C but the catalyst rapidly deactivated as a result of carbon formation.. Somewhat surprisingly, metal oxide catalysts were found to be less active than metal – at least in the case of magnetite - suggesting that methane decomposition rather than the reduction of iron oxides with methane (reaction 2) was operative. At higher temperatures, reduction was found to dominate.

Testing of alternative materials that could promote the pyrolysis included scrap material that could be buried (together with the deposited carbon) or recycled to a steel blast furnace. Although carbon dioxide would be produced in the furnace, the necessity of adding carbon to the steel furnace would be avoided. To this end various scrap materials, all containing iron, were tested. The samples were ground to -0.71mm+0.50 mm and reduced before use. As seen from Table 4, these produced some of the best results, with up to ca 60 % of hydrogen being produced at 850°C. Most of the iron was converted to iron carbide and a large excess of free carbon was produced. Despite pre-reduction, carbon oxides were also present in the exit gases, but only in small amounts (Figs.1a and 1b). XPS analysis of the surface of the scrap materials showed a lack of homogeneity, with iron based materials mixed with Ca based flux particles. The relative efficiency of hydrogen production may reflect coverage of the iron surface by a film of flux.

Table 3: Hydrogen production from metal and alloy catalysts

Catalysts	Space velocity (h^{-1})	% H_2 in exit gas			% CH_4 converted for highest hydrogen
		650°C	750°C	850°C	
Monel turnings	400	0.16	2.0	14.3	7.2
Monel turnings	2,080		0.29	2.2	2.3
Monel turnings	10,540	0.08	0.12		
Incoloy turnings	400	0.2	0.25	3.0	1.4
Incoloy turnings	2,120	0.02	0.05	0.62	0.1
Ni Cement	10,520	1.7	2.6	5.5	2.5
Steam Reforming catalyst	10,400	62.7	14.2		45.3
HT WGS catalyst[1]	10,560	43.4[2]	43[3]		26.7
Magnetite catalyst	10,570		0.02[2]	37.5[2]	44.8
Magnetite catalyst (after 850° above)	10,480	2.0	17.1		10.7
Reduced Magnetite catalyst	10,500	36.3			22.4

[1] High temperature water gas shift catalyst containing iron oxides and chromia.
[2] Accompanied by carbon oxides since the catalyst was not pre-reduced.
[3] Result achieved after ca 3 hours operation at 650°C

Although this result was very promising, the presence of carbon oxides in the exit gases was unwelcome and an in-depth study of a model system was carried out. The catalyst has been described by Takenada et al. [3] and consists of iron oxide promoted by 5% nickel with chromia (8%) added to stabilise the surface area at higher temperatures. The use of more expensive materials led to consideration of a carbon formation–carbon removal cyclic operation (reactions 1-4 above), carbon removal being attempted using water as a gasifying agent. Typical results obtained with fresh catalyst are shown in Fig. 2a and Fig. 2b.

Table 4: Hydrogen production from steel scrap *

Steel scrap	Space velocity (h^{-1})	% H_2 in exit gas*			% CH_4 converted	Surface area m^2/g
		650°C	750°C	850°C		
BS1 catalyst reduced 650°C	400	14.8	62	94	89.0	29.6
BS1 catalyst reduced 850°C	950	14.5	58.3	88	79	29.6
BS6 catalyst reduced 650°C	380	6.2	16.7	42.8	26.0	6.2
BS6 catalyst reduced 650°C	960	2.4	7.5	25.4	14.2	6.2
BS7 catalyst reduced 650°C	380	11.9	34.1	64.9	50.4	3.4
BS7 catalyst reduced 650°C	950	7.4	22.2	35.4	18.5	3.4

* Small amounts of carbon oxides were produced in all cases even though the catalysts were pre-reduced.

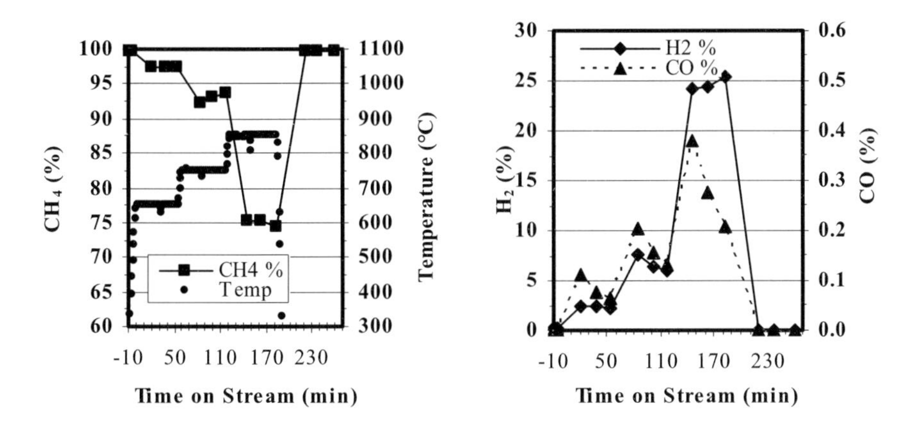

Fig. 1: (a) Typical CH_4 decomposition over BS6 (Space Velocity: 960 – 1120 ml/hr/gcat); (b) H_2 and CO formation during CH_4 decomposition over BS6 (Space Velocity: 960 – 1120 ml/hr/gcat)

It is clear that methane conversion is initiated at 470°C with the production of significant amounts of carbon monoxide through oxide reduction. As the catalyst was reduced, the amount of carbon monoxide dropped and hydrogen, originating from methane decomposition, increased to ca 55% of the exit gases at ca 620°C. Once again, it proved impossible to totally remove carbon monoxide, and this seemed to be due to the slow diffusion of oxygen from the inside of the catalyst particle to the external surface where reaction with methane occurred. Removal of deposited carbon by steam at 650°C was found to be a very slow process (Fig. 2b), with small amounts of hydrogen and carbon monoxide being generated for long periods.

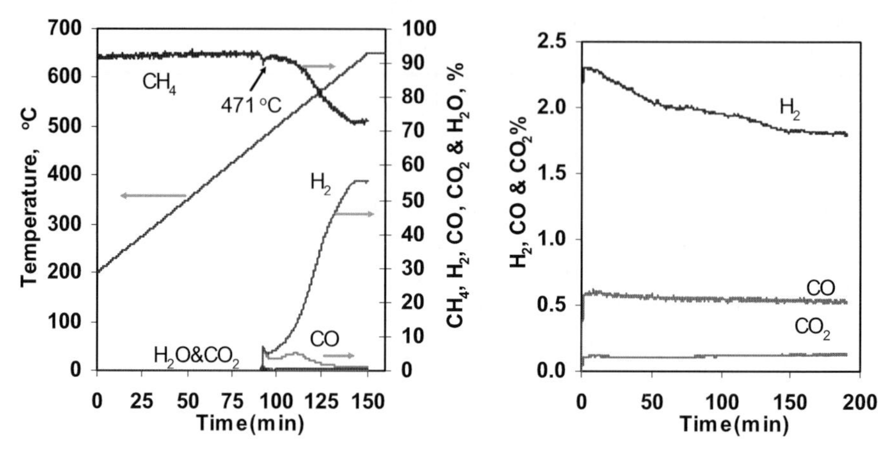

Fig. 2: (a) Temperature Programmed CH_4 decomposition over promoted Fe_2O_3
(b) Isothermal reoxidation of reduced iron oxide in water vapor (2.3% H_2O) at 650°C

Complete combustion of carbon on the catalyst after steam gasification revealed that 75% of the methane decomposed and deposited on the catalyst remained after gasification with steam. As a result, it was not surprising that an attempt to carry out a cyclic reduction-oxidation gave very poor yields on the second cycle.

Thus it would seem that, under the conditions pertinent to optimised solar heating, a cyclic reduction/decomposition-oxidation process is unsatisfactory, due to encapsulation of the catalyst by coke. Catalytic decomposition using scrap materials from a steel mill is preferred, but – under some conditions – small amounts of carbon monoxide are produced. If necessary, these small traces may be removed by methanation or selective oxidation [7].

References

[1] M.V.Twigg (ed), Catalyst Handbook, Wolfe Scientific Texts, London, 1989.
[2] T.V. Choudhary, E. Aksoylu,and D.W. Goodman, Catal. Rev. Sci. Eng., 45 (2003) 151.
[3] S. Takenada, N. Hanaizumi, V. T.D. Son and K. Otsuka, J. Catal., 228 (2004) 405.
[4] V. Hacker, R. Fankhauser, G. Faleschini, H. Fuchs, K. Friederich, M. Muhr and K. Kordesch, J. Power Sources, 86 (2000) 531.
[5] N. Muradov, F. Smith, C. Huang and A. T-Raissi, Catal. Today, 102-103 (2005) 225.
[6] N. Muradov, Internat.J.Hydrogen Energy, 26 (2001)1168
[7] D.L. Trimm, Appl. Catal. A: General, 296 (2005) 1.

Natural Gas Conversion VIII
F.B. Noronha, M. Schmal, E.F. Sousa-Aguiar (Editors)

SiC as stable high thermal conductive catalyst for enhanced SR process

F. Basile[a*], P. Del Gallo[b], G. Fornasari[a], D. Gary[b], V. Rosetti[a], A. Vaccari[a]

[a] *Dipartimento di Chimica Industriale e dei Materiali, Università degli Studi di Bologna, Viale Risorgimento, 4, 40136 Bologna (Italy),*
[b] *Air Liquide, Centre De Recherche Claude-Delorme, 1, chemin de la Porte des Loges – B.P. 126 - Les Loges-en-Josas, 78354 Jouy-en-Josas Cedex (France)*
**corresponding author: basile@ms.fci.unibo.it*

1. Abstract

Silicon carbide has been chosen as a support for steam methane reforming (SMR). In fact, its good conductive properties may improve the temperature profile, while decreasing the high ΔT caused by the high endothermicity of the reaction and, at the same time, increasing the heat transfer from external furnace, reactor wall and catalyst particles. 10 wt % Ni was deposited on the SiC support by incipient wetness impregnation. The sample was calcined at different temperatures in order to study both the chemical-physical properties and the interaction between the support and active phases. The sample calcined at 700°C was tested in a SMR laboratory plant under different operative conditions, in order to evaluate the activity and stability with time-on-stream. On a laboratory scale, the catalyst shows good results, although, at very high temperatures (960°C) the support shows a slight SiO_2 formation.

2. Introduction

Currently steam reforming of hydrocarbon, especially steam methane reforming (SMR) is the most used process for syngas and hydrogen production [1].
The process for syngas and hydrogen production consists of sulphur removal, feed gas preheating, pre-reforming, reforming and, eventually, H_2/CO shift and CO_2 removal. Natural gas reacts with steam on a Ni-based catalyst in a primary

reformer at 800-1000°C and 15-30 bar to produce syngas at residence time of several seconds, thus obtaining a H_2/CO ratio around 3. Excess steam is introduced to avoid carbon deposition, and the feed H_2O/CH_4 (S/C molar ratio) is typically 2-5, depending on the syngas end use. Ni is chosen because it is cheap, if compared to noble metals catalysts, and still highly active in steam reforming; ceramic-supported nickel catalysts are commonly used in steam reforming in the form of pellets, ring or tablets [1, 2]. Due to severe reaction conditions, such as high temperature, pressure and hydrogen content, industrial plants register problems linked to materials: tube resistance to high temperatures in SR conditions, catalyst deactivation and diffusion limitation through the catalyst. In particular, the temperature in the tube may range from 600°C to 950°C because high heat fluxes are necessary, due to the endothermicity of the reaction. In these conditions, the temperature difference between the external- and inner-tube wall can be higher than 60°C for a thickness of 15 mm: due to this strong stress, the tube may crack during operation. The catalyst deactivation increases this problem since higher temperatures are required to achieve the same results. Reaction rate is strongly affected by the heat transfer of the ceramic pellets, and the residence time is necessarily high (2-4 s). To avoid all these problems, a catalyst support having high thermal conductivity could be used. This kind of support can increase the heat transfer radially and along the bed, thus making it possible to operate at lower residence times, while smoothing the thermal profile both inside the reactor tube and at the interface between the tube wall and catalyst. A silicon carbide supported catalyst was tested in a research activity focused on preparing differently shaped supported catalysts (pellets, monoliths, foams) characterized by high thermal conductivity, such as ceramic carbide, metal gauze and alloy. Silicon carbide exhibits a high thermal conductivity and mechanical strength, a low specific weight, and chemical stability: these features make it a promising candidate as a catalyst support in place of the traditional insulator supports (Al_2O_3, $MgAl_2O_4$) either in several highly endothermic or exothermic reactions or in aggressive reaction media [3,4]. Leroi et al. [3,4] and Sun et al. [5] tested Ni/SiC catalysts in the partial oxidation of methane, and found that they present excellent catalytic activity and stability, even though carbon formation was registered [3, 4, 5].
The aim of this paper is to illustrate the features of a catalyst containing 10 wt % of Ni deposited on SiC; this catalyst was prepared by incipient wetness impregnation, and its activity in SRM was carried out in reaction conditions similar to those used in industrial plants. Different calcination temperatures of the impregnated precursor (500, 700 and 900°C) were investigated.

3. Experimental

The 10 wt % Ni/SiC catalysts were prepared by incipient wetness impregnation using both $Ni(NO_3)_2$ as precursor and pellets of SiC (0.84 - 1.41 mm) as support by grinding cylinder on 8-14 mesh sieves. The SiC cylinders (type b-SiC) were

produced by SICAT France. Three samples were prepared and calcined at 500 and 700°C for 2 h, and at 900°C for 12 h. Powder X-ray diffraction analyses (PXRD) were carried out using a Philips PW1050/81 diffractometer equipped with a graphite monochromator in the diffracted beam ($\lambda = 0.15418$ nm). A 2θ range from 10° to 80° was investigated at a scanning speed of 70°/h. Temperature programmed reduction and oxidation (TPR/O) were carried out with H_2/Ar 5/95 v/v for TPR and O_2/He 5/95 v/v for TPO (total flow rate 20 ml/min) at 10 °/min in the 100-950°C temperature range using a ThermoQuest CE instrument, TPDRO 1100. The BET surface areas were determined by N_2 absorption using a Carlo Erba instrument, Sorpty model 1750.
In order to study the activity of the catalysts, a laboratory plant was assembled, with a fixed bed reactor operating up to 20 bar. In this plant the feed is provided by both a cylinder of pure CH_4 and a HPLC pump for the water. The water is vaporized using a heater at 400°C and then mixed with methane before entering the reactor. The reactor is made of a special Incoloy 800HT material, with a wall thickness of 9 mm and inner diameter of 12 mm. The reactor is heated with an electric oven, and 10 ml of catalyst (10 cm in bed length) are placed in the isothermal zone, while filling the bottom and the top section of the tube with an inert material (beads of corundum or quartz pellets). Inside the catalytic bed, along the axial direction of the flow, a metal wire is placed in the centre of the bed, where a thermocouple moves to measure the axial thermal profile during reaction. The reaction products are sampled on-line with an automatic valve, after water condensation, and tested by two gas chromatographs equipped with HWD and Carbosieve SII columns. In these tests He is used as the carrier gas for the analysis of CH_4, CO, and CO_2 and N_2 as the carrier gas for the H_2 analysis.

4. Results and discussion

The samples of Ni/SiC catalyst calcined at three different temperatures (500, 700 and 900°C) showed only SiC and NiO phases (Fig. 1). It can be observed that the sample calcined at 900°C showed an amorphous band between 20-30°2θ, attributable to SiO_2 formation during the thermal treatment at such high temperature. Furthermore, the XRD and the BET results showed that the calcination temperature does not significantly affect the NiO crystal size (20-30 nm) nor the surface area values of the support (18-22 m^2/g), which did not show any significant change even after impregnation and thermal treatment.
After reduction at 800°C for 2 h, the sample calcined at 900°C showed the presence of metallic Ni together with a Ni_2Si phase; while the samples calcined at lower temperatures (500 and 700°C) showed the Ni_2Si phase only (Fig. 1). The presence of this phase may be due to the direct reaction between metallic Ni (formed at 500°C) and SiC, a reaction that is thermally activated over 600°C, while Ni_2Si is stable up to 950°C [6-8].

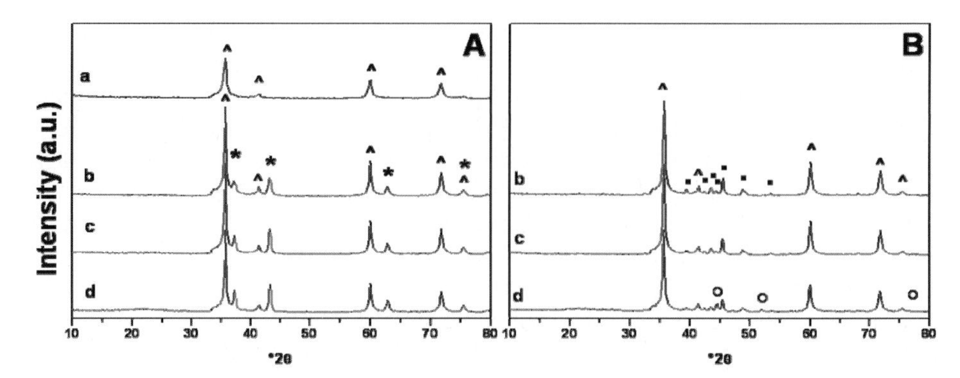

Figure 1. PXRD patterns of (A) calcined samples and (B) samples calcined at 900°C and reduced at 800°C. (a) SiC support; (b) calcination at 500°C, (c) 700°C and (d) 900°C. [(^) SiC; (*) NiO; (°) Ni; (■) Ni_2Si].

Considering that Ni_2Si is mainly observed in the reduced samples which were previously calcined at 500 and 700°C, it is possible to state that in the sample calcined at 900°C, in which the formation of SiO_2 was observed, the Ni_2Si formation was inhibited by the SiO_2 layer, because the latter prevented the contact between SiC and metallic Ni, partly passivating the surface of the support.

TPR analyses (Fig. 2) carried out on the support did not show any reduction peaks. For Ni samples a main reduction peak at 400 - 450°C with a shoulder at 500 - 550°C was clearly visible. These peaks correspond to the NiO reduction, which differently bounded with the support.

The oxidation analysis carried out on the support showed an oxidation shoulder starting at 750°C and related to the superficial oxidation of SiC. The oxidation increased at 850 - 900°C due to bulk SiC oxidation. As for Ni samples, the oxidation of the Ni occurred at 500 - 600°C. Moreover, TPO analyses showed a further oxidation peak centered at 750°C, before the support oxidation. This peak may be attributed to the oxidation of Ni_2Si phase obtained by TPR at 950°C, which partially decomposes Ni_2Si forming Ni^0.

Catalytic activity in SMR reaction was carried out for the 10 wt % Ni/SiC sample while maintaining the outlet temperature of the catalytic bed constant at 870°C or 960°C by varying the oven set temperature. The methane conversion and the syngas selectivity are in agreement with the thermodynamic equilibrium prediction calculated by changing both the S/C ratio and the outlet temperature. In terms of catalyst stability, tests at 870°C show very constant values in the syngas composition during the time-on-stream (Fig. 3). At 960°C the composition of the syngas is still stable but the oxygen balance (in/out) starts to go above 100%, probably due to the formation of SiO_2. This is also confirmed by the TPO analysis (Fig. 2) where at 950°C the oxygen uptake becomes significant, therefore showing a limit in the process temperature of the catalyst.

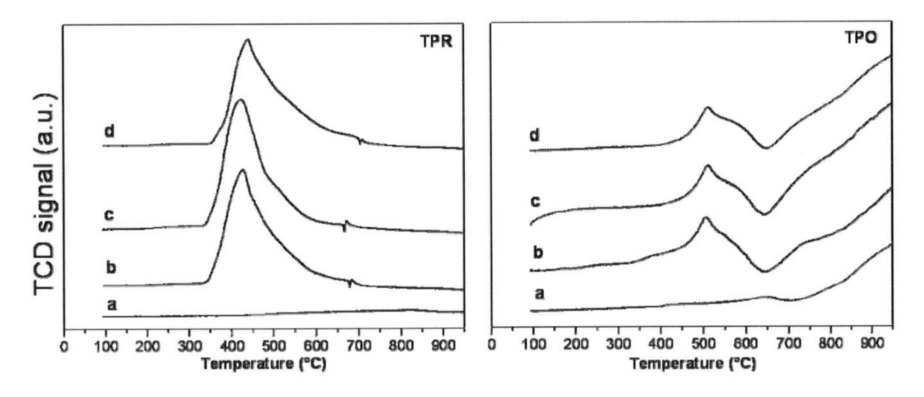

Figure 2 Temperature programmed reduction (TPR) and oxidation (TPO) analyses carried out on (a) SiC support; 10 % Ni /SiC calcined at (b) 500°C, (c) 700°C and (d) 900°C.

Table 2. Catalytic results of the 10 wt % Ni / SiC calcined at 700°C (CT = 1.1 s).

T_{out}	870°C	870°C	960°C	960°C
S/C	1.7	2.5	1.7	2.5
Conv. CH_4	79.0	79.4	81.7	95.9
Yield H_2	77.5	78.6	79.9	94.8
Sel. CO	76.3	72.4	85.2	68.8
Sel. CO_2	23.7	27.6	14.8	31.2

On the other hand, an interesting feature is the ΔT between oven temperature and outlet temperature (≈ 10°C), which is an indication of a smoothed axial profile, whereas for commercial catalysts tested in the same conditions a larger difference was observed. There was also a different and peculiar thermal behavior along the bed, where a smoothed axial profile was observed (Fig. 3). This behavior makes it possible to work with a lower oven temperature, thus overcoming some of the previously stated problems.
After catalytic tests, the used sample presents a slight decrease in surface area with respect to the fresh catalyst (fresh sample: 20 m^2/g, used sample: 17 m^2/g); while the XRD analysis reveals the presence of metallic Ni as the predominant phase instead of Ni_2Si which is present in the reduced sample, probably due to the higher temperature reached by the sample in the reactor (960°C) (Fig. 4). In fact, only the reduced sample, previously calcined at 900°C, showed metallic Ni besides the Ni_2Si phase. In the used sample, the presence of the SiO_2 amorphous phase is much more evident with a broad band between 15-30° 2θ. It is probably formed by the reaction with water occurring at very high temperature and during the tests at 960°C.

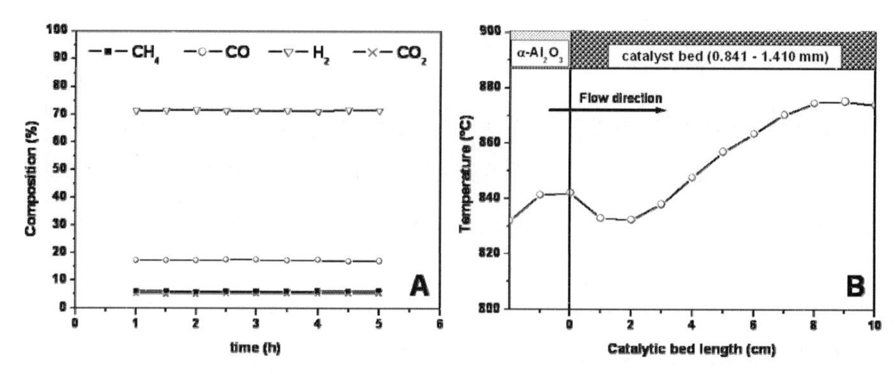

Figure 3. Dry syngas composition (A) and axial thermal profile (B) measured during steam reforming reaction on 10 % Ni/SiC calcined at 700°C (P 10 bar; CT 1.1 s; S/C 1.7 mol/mol).

5. Conclusions

Ni/SiC samples may be good catalysts for the steam reforming of methane in lab scale, due to the high conductivity of the support. The limit has been registered in high temperature tests (> 960°C) due to SiC oxidation to SiO_2, with a corresponding decrease in the surface area. This effect increases while operating on pilot scale conditions, in which the formation of SiO_2 dramatically increases. A study on the improvement of the support stability either by passivation at high temperature or by coating with an inert submicronic oxide is in progress.

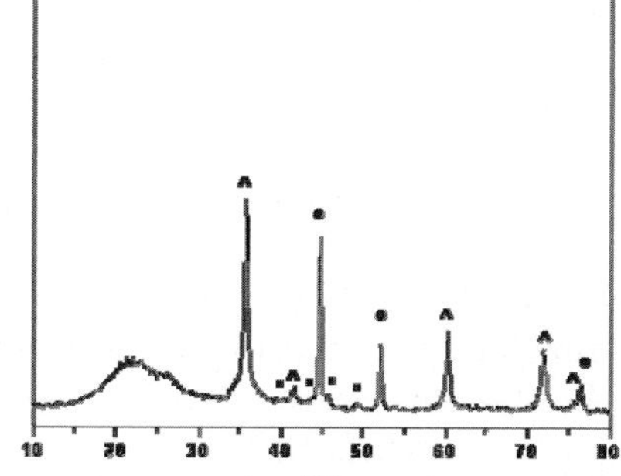

Figure 4. PXRD patterns of the used sample. [(^) SiC; (•) Ni; (■) Ni_2Si].

References

1. S. Rakass, H. Oudghiri-Hassani, P. Rowntree, N. Abatzoglou, *J. Power Sources,* **158** (2006) 485-496.
2. M. A. Peña, J. P. Gómez, J.L.G. Fierro, *Appl. Catal. A: Gen.,* **144** (1996) 7-57.
3. P. Leroi, B. Madani, C. Pham-Huu, M. J. Ledoux, S. Savin-Poncet, J. L. Bousquet, *Catal. Today*, **91-92** (2004) 53-58.
4. P. Leroi, Etude du reformage du methane en gaz de synthèse sur cataliseurs à base de carbur de silicium, PhD French Thesis, National N° 03STR13080 (2003).
5. W-Z. Sun, G-Q. Jin, X-Y. Guo, *Catal. Comm.,* **6** (2005) 135-139.
6. R. Roccaforte, F.la Via, V. Ranieri, P. Musumeci, L. Calcagno, G. G. Condorelli, *Appl. Phys A.,* **77** (2003) 727-833.
7. K.W. Richter, K. Chandrasekaran, H. Ipser, *Intermetallics,* **12** (2004) 545-554.
8. Y. Du, J. C. Schuster, *Metall. Mater. Trans. A,* **30A** (1999) 2409-2418.

Natural Gas Conversion VIII
F.B. Noronha, M. Schmal, E.F. Sousa-Aguiar (Editors)

Catalytic partial oxidation of CH_4 and C_3H_8: experimental and modeling study of the dynamic and steady state behavior of a pilot-scale reformer

Ivan Tavazzi, Alessandra Beretta, Gianpiero Groppi, Alessandro Donazzi, Matteo Maestri, Enrico Tronconi, Pio Forzatti

Dipartimento di Chimica, Materiali e Ingegneria Chimica "G. Natta"
Politecnico di Milano, piazza Leonardo da Vinci 32, Milano 20133, Italy

1. Abstract

An investigation which combines kinetic tests, adiabatic reactor tests and mathematical modeling has been performed to gain insight on Catalytic Partial Oxidation (CPO) of hydrocarbons. A pilot-scale honeycomb reactor, equipped with sliding thermocouples, was assembled for testing the CPO of methane over a 2% (wt/wt) Rh/α-Al_2O_3 catalyst at atmospheric pressure. Both the cold start-up and the steady state performances of the reactor are presented and discussed. A one-dimensional (1D) mathematical model of the reactor was applied to the quantitative analysis of the observed performances. The model analysis was extended to the simulation of CPO of propane.

2. Introduction

The present study deals with the production of synthesis gas via the CPO process suitable for innovative applications including on board and distributed generation of H_2. On previous works the authors have addressed several aspects of the CPO process. Catalyst preparation and deposition methodologies were optimized. A kinetic investigation was developed by performing catalytic tests in an annular reactor under quasi-isothermal and out of thermodynamic control conditions. A molecular indirect-consecutive kinetic scheme was obtained [1]. A 1D heterogeneous, dynamic model of an adiabatic reactor implementing the molecular reaction scheme was developed and applied for simulating the

behavior of CPO reformers with different geometry of the support [2]. Model predictions were validated with catalytic tests performed in a lab scale packed bed of spheres reactor [3]. The observed thermal efficiency of the reactor strongly depended on the flow rate and high Gas Hourly Space Velocity (GHSV) operating conditions were identified to achieve a near-adiabatic behavior of the reactor. The observed start-up dynamics were consistent with the occurrence of an indirect reaction sequence. The model analysis of the axial temperature profiles revealed that in the packed bed reactor the thermocouple was equilibrated with the gas phase temperature.
Starting from the previous experience, in this work a pilot-scale honeycomb reformer was developed. Experiments of CH_4 CPO were run over a 2% (wt/wt) $Rh/\alpha\text{-}Al_2O_3$ coated honeycomb monolith and were aimed to characterize the dynamic response of the system, the thermal efficiency and steady state performances of the reactor. The measured trends were quantitatively analyzed by applying a 1D, heterogeneous, single channel model of the adiabatic reactor which implemented a molecular kinetic scheme independently derived. The model analysis was extended to simulate the performances of the reactor in the case of CPO of C_3H_8, regarded as a model species for higher hydrocarbons.

3. Reactor and catalyst design

Methane and air ($O_2/CH_4 = 0.54$) were fed from high pressure cylinders by high capacity mass flow controllers. For safety reasons the two streams were preheated separately and mixed in a Sulzer® static device just upstream from the reactor. The total flow rate was 1.26 $Nm^3 \cdot h^{-1}$ (corresponding to a GHSV equal to $3.8 \cdot 10^6$ $Ncm^3 \cdot g_{cat.}^{-1} \cdot h^{-1}$, referred to the washcoat weight).
The reactor was a thermally insulated stainless steel cylinder (Fig. 1) designed to host and separate three monolith modules (a catalyst coated monolith, whose features are reported in Table 1, in between two inert monoliths acting as thermal shields). As shown in Figure 1 the inlet and outlet gas temperatures were measured at fixed axial position, while the temperature within the catalyst was measured by a thermocouple sliding in an open channel of the monolith.

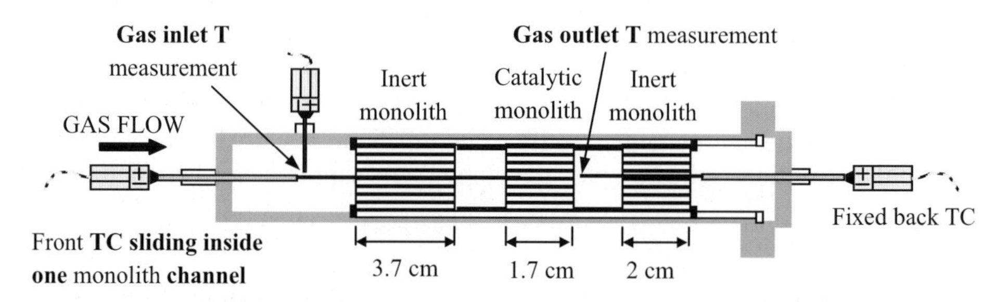

Figure 1. Sketch of the honeycomb reformer; (TC = Thermocouple).

Table 1. Geometrical and physical properties of the honeycomb reactor.

Reactor diameter [m]:	0.018	Specific heat [$kJ \cdot kg^{-1} \cdot K^{-1}$]:	0.925
Channel diameter [m]:	0.0011	Thermal conductivity [$W \cdot m^{-1} \cdot K^{-1}$]:	2.5
Channels per square inch [$inch^{-2}$]:	400	Inert bed length [m]:	0.037
Monolith density [$kg \cdot m^{-3}$]:	2300	Catalytic bed length [m]:	0.017
Monolith void fraction [-]:	0.75	Washcoat density [$kg \cdot m^{-3}$]:	1.38

The $Rh/\alpha\text{-}Al_2O_3$ washcoating of the cordierite honeycomb was realized by dipping the monolith into a slurry of the catalytic powders, prepared from a $Rh(NO_3)_3$ precursor, followed by blowing of the excess catalyst. A catalyst loading of 330 mg was estimated by weigth difference before and after coating. Operation of the rig consisted of first analyzing the reactant mixture, by-passing the reactor initially kept under N_2 stagnant atmosphere at room temperature. The preheated feed stream was then injected to the reformer at time zero of the dynamics. The dynamics of product distribution and temperature were acquired with an ABB® AO2000 continuous analyzer. C, H, O mass balances based on GC analyses at steady state were closed within 2%.

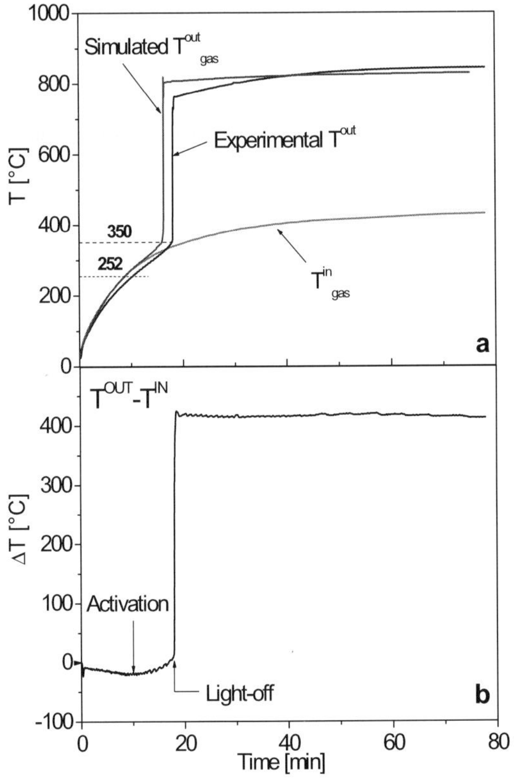

Figure 2. Temperature start-up dynamics.

4. Results

4.1. Start-up

Figure 2a reports the experimental dynamics of the inlet gas temperature and of the temperature measured inside the channel at the back end of the Rh-coated monolith upon injection to the reactor of the reactant mixture preheated at 524 °C. Heat losses along the reactor assembly were not a major issue at the operating high flow rate. Indeed, only a temperature decrease lower than 20 °C was initially measured between the front and the back thermocouple after injection of the feed mixture in the absence of reaction. By subtracting the dynamics of the inlet gas temperature (Fig. 2b) one can appreciate that the outlet gas temperature started increasing at about 250 °C when the catalyst started being active toward the reaction of CH_4 total oxidation. Once reached the temperature of 350 °C at the outlet of the monolith, with inlet temperature of 338 °C, ignition rapidly occurred. In about 20 seconds the outlet temperature reached 760 °C after which continued to increase following the slow dynamics of the inlet gas temperature. Upon ignition CH_4 and O_2 conversion (not shown) had a steep increase; the outlet concentration of H_2O and CO_2 passed through a maximum while the net production of H_2 and CO increased monotonically, consistently with the occurrence of an indirect reaction scheme. An interpretation of the mechanism governing the light-off was given in [3] and will be discussed in better detail in a following work.

4.2. Steady state

The measured axial temperature profiles were characterized by the presence of a maximum (Fig. 3), consistently with the occurrence of an exo-endothermic reaction sequence. At the inlet gas temperature equal to 431 °C the observed conversion of CH_4 and selectivity to syngas (see table in Fig. 3) were 2% and 1% lower, respectively, than equilibrium values calculated under adiabatic conditions. Accordingly, because of the governing role of endothermic reactions on the steady state performances of the reactor [3], the measured temperature of the outlet gas was higher than the adiabatic equilibrium temperature: 834 °C vs. 766 °C, respectively. The thermal efficiency of the reformer was estimated as ratio (α) of the observed temperature rise across the reactor to the temperature increase expected under adiabatic conditions according to the measured product composition [3]. A value of α equal to 1.08 was estimated which confirms the close approach to the adiabatic behavior achieved by combining good thermal insulation and high flow rate. The slight excess of enthalpy is consistent with experimental uncertainties on gas composition and radial temperature and velocity profiles.

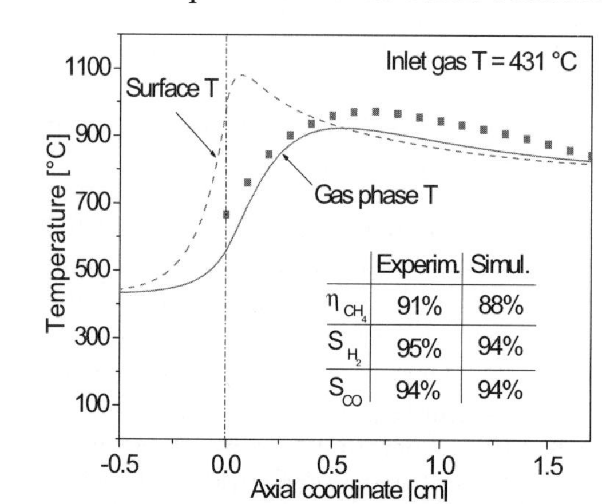

	Experim.	Simul.
η_{CH_4}	91%	88%
S_{H_2}	95%	94%
S_{CO}	94%	94%

Figure 3. Measured (symbols) and predicted (lines) axial temperature profiles at steady state; percent conversion (η) and selectivity

5. Model analysis

The behavior of the reformer was simulated by integrating a molecular kinetic scheme into a 1D, heterogeneous, single channel

model of the adiabatic reactor, previously developed [2]. The contribution of homogeneous reactions was neglected. The proposed indirect reaction scheme [1] consists of six molecular stoichiometries including a primary step of total oxidation of CH_4 followed by steam and CO_2 reforming responsible for the formation of syngas, the water gas shift reaction, and H_2 and CO post-combustion. The corresponding rate equations (not shown here) have been reported in Table 2 of [1] along with the kinetic parameters estimated over a 0.5% (wt/wt) Rh/α-Al_2O_3 catalyst. The kinetics of CH_4 CPO over catalysts with higher Rh loading have been investigated by the authors and will be discussed in a forthcoming work. Table 2 reports the kinetic parameters adopted in the present work. Kinetic tests of C_3H_8 CPO were also performed and could be preliminarily described by invoking a three-fold activity of propane combustion and reforming with respect to the kinetics of CH_4.
According to findings reported in the literature [4] the thermocouple sliding inside the monolith channel was expected to be equilibrated with the temperature of the gas phase, due to the small ratio of the thermocouple (0.5 mm) to the channel diameter (1.1 mm). Figure 3 reports the axial gas and surface temperature profiles predicted at steady state. A satisfactory match was evidenced between the predicted temperature of the gas phase and the experimental measurements which provided a validation of the reactor model. The residual discrepancies were likely associated to minor deviations of the pilot reactor from the ideal 1D, adiabatic behavior. The predicted surface temperature profiles were characterized by the presence of a hot spot near the catalytic monolith entrance, with a peak temperature of 1080 °C.
Concerning the start-up dynamics, the calculated trend of the outlet gas temperature is reported in Figure 2a. At low temperature the predicted back end temperature coincided with the temperature of the incoming gas, due to the perfect adiabaticity assumed in the model. When the inlet temperature reached 325 °C the ignition of the reactor was predicted at the outlet gas temperature of 350 °C, in very good agreement with observations. Upon light-off the predicted gas temperature was about 60 °C higher than the experimental one and after ignition had a weaker dependence on the inlet gas temperature.

Table 2. Arrhenius-type kinetic (k_{kin}) and adsorption ($K_{ads.}$) constants at 873 K for CH_4-CPO.

Surface reaction	**TO**	**SR**	**DR**	**RWGS**	**Ox, H_2**	**Ox, CO**	Surface adsorption	**O_2**	**H_2O**	**CO**
$\mathbf{k_{kin}}^{873K}$	0.0405	0.0283	0.0283	2.545	113	9.69	$\mathbf{K_{ads}}^{873K}$	30.8	0.0345	55.4
$\mathbf{E_{act}/R}$	10100	10000	10000	6640	6500	9150	$\mathbf{-\Delta H_{ads}/R}$	7710	19100	5770

$[k_{kin}] = [mol \cdot g_{cat.}^{-1} \cdot s^{-1} \cdot atm^{-1}]$; $[K_{ads}] = [atm^{-1}]$; $[E_{act}/R] = [-\Delta H_{ads.}/R] = [K]$. TO: Total Oxidation; SR: Steam Reforming; DR: Dry Reforming; RWGS: Reverse Water Gas Shift; Ox,CO (H_2): CO (H_2) oxidation.

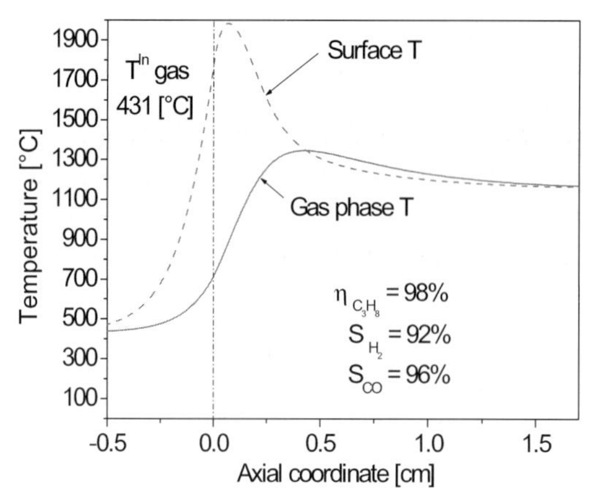

Figure 4. Steady state temperature profiles predicted in simulations of C_3H_8 CPO.

Figure 4 summarizes the predicted steady state performances of the reformer in C_3H_8 CPO run at the same O_2/C ratio and operating conditions adopted in the CH_4 run. The outlet products composition and temperature equaled the adiabatic equilibrium values.

The qualitative features of the temperature profiles corresponded to those discussed about CH_4. Still, in the case of C_3H_8 CPO peak surface temperatures as high as 2000 °C were predicted as a result of the combined effect of the higher overall exothermicity of C_3H_8 with respect to CH_4 CPO and of the lower diffusion coefficient of C_3H_8 which limits the reaction rate of endothermic reactions in the hottest catalyst zone. Albeit under such temperature conditions neglecting the contribution of homogeneous reactions undermine the accuracy of the simulations these results pose a warning with respect to catalyst stability and enlighten the necessity of the identification of suitable operating conditions for realizing the process of CPO of propane.

6. Conclusion

CH_4-CPO experiments were run in a pilot-scale honeycomb reactor over a $Rh/\alpha\text{-}Al_2O_3$ catalyst. Near adiabatic conditions were realized by means of an efficient radial insulation of the partial oxidizer and operating at high flow rate. The observed cold start-up dynamics revealed that Rh activated at 250 °C and ignition occurred at 350 °C with instant production of syngas. Both the observed start-up dynamics and the steady state temperature profile were consistent with the occurrence of an indirect reaction scheme. 1D model analysis well captured the features of the reformer behavior. Simulations of the CPO of C_3H_8 evidenced that a purely heterogeneous indirect reaction process leads to extremely high surface temperatures.

References

[1] I. Tavazzi, A. Beretta, G. Groppi, P. Forzatti, J. Catal. 241 (2006) 1-13.

[2] M. Maestri, A. Beretta, G. Groppi, E. Tronconi, P. Forzatti, Catal. Today 105 (2005) 709-717

[3] I. Tavazzi, M. Maestri, A. Beretta, G. Groppi, P. Forzatti, AIChE J 52 (2006), 3234-3245

[4] A. J. Rankin, R.E. Hayes, S.T. Kolaczkowsky, Trans I. Chem. E. A, 73 (1995), 110-121

Natural Gas Conversion VIII
F.B. Noronha, M. Schmal, E.F. Sousa-Aguiar (Editors)

Additive effect of O_2 on propane catalytic dehydrogenation to propylene over Pt-based catalysts in the presence of H_2

Changlin Yu, Hengyong Xu[1]*, Qingjie Ge, Wenzhao Li

Applied Catalysis Laboratory, Dalian Institute of Chemical Physics, Graduate School of the Chinese Academy of Sciences, 457 Zhongshan Road, Dalian 116023, China

1. Introduction

The catalytic dehydrogenation of light alkanes for the production of short-chain olefins is the important route for the exploitation of natural gas as raw material for highly priced and clean chemicals. Propane dehydrogenation is of increasing importance because of the growing demand for propylene.The catalytic dehydrogenation greatly suffers equilibrium limit and huge energy consumption. Oxidative dehydrogenation, while exothermic and not equilibrium limited, suffers from low selectivity at high conversions with currently known catalysts and is therefore not practiced commercially [1]. With the idea to supply heat by in situ combustion of some of the hydrogen and drive the equilibrium towards the product side but not at the expense of propylene selectivity, we performed the experiments to explore the possibility of feeding a small quantity of oxygen to the conventional catalytic dehydrogenation system.

2. Experimental

2.1. Catalysts preparation

A series of new M promoted Pt-Sn/M-γ-Al_2O_3 catalysts (M: Ce, Zn, La, Cr, Fe, Zr and Mn) were prepared by continuous impregnation method as described in literatures [2-3]. For M promoted Pt-Sn/M-γ-Al_2O_3 catalysts, Sn and M were first deposited by co-impregnating $SnCl_2$ and $M(NO_3)_x$ ethanol solution, then dried and calcined, finally the Pt component was deposited, then

*Corresponding author. Tel.: +86(411) 84581234; fax: +86(411) 84581234.
E-mail address: xuhy@dicp.ac.cn (H.Xu)

dried and calcined again. The loadings of Pt and Sn were 0.3 and 0.9 wt.%. The molar ratio of Pt/Sn/M was 1:5:5.

2.2. Catalytic test

The reactions were studied in a fixed bed quartz micro-reactor. The reactions were carried out at atmospheric pressure. The catalyst loading was 0.3 g for all experiments. All catalysts were previously reduced under flowing pure H_2 (12.6 ml/min) at 576 °C for 2.5 h. The reactor was placed in an electrical furnace, and the temperature controlled using external and internal thermocouples. The reactant mixture with varying compositions of propane, oxygen, hydrogen and helium, was premixed and fed to the reactor using electronic mass flow controllers (Bronkhurst). The product gas was analyzed for hydrocarbons and for gases (O_2, CO, CO_2) by Shimadzu GC14-C equipped with a TCD (Carbosieve SII packed column) and a FID (Porapak-Q packed column). From the composition data the fractional conversion of propane was calculated. The selectivities were calculated based on moles propane converted to CO/CO_2 (C-selectivity).Water and hydrogen in the exit gas was not analyzed by the GC.

3. Results and discution

The effect of small quantity of oxygen on the catalytic performance of this series of catalysts was firstly examined. Reaction results are show in Figure1. It

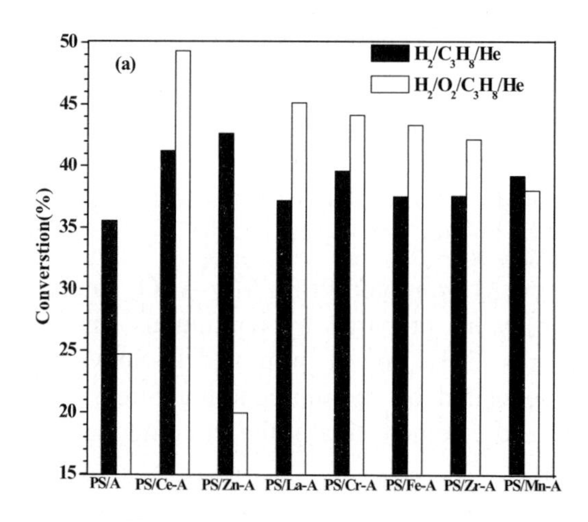

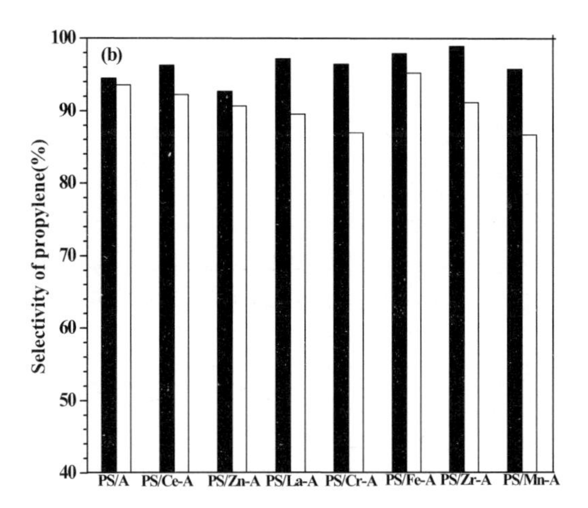

Figute 1. Effect of gas medium on the catalytic performance of Pt-Sn/M-Al catalysts in propane dehydrogenation. Reaction conditions: T=576°C; $H_2/C_3H_8/He$ =1:1:5(molar ratio), total GHSV=3800 h^{-1}; $H_2/O_2/C_3H_8/He$ =1:0.3:1:5 (molar ratio), total GHSV=3960 h^{-1}; Reaction time: 60 min. P: Pt, S: Sn, A: γ-Al_2O_3.

can be seen that the presence of O_2 greatly changes the catalytic behavior. An obvious conversion decrease over Pt-Sn/Al and Zn promoted Pt-Sn/γ-Al_2O_3 catalysts is observed. In $H_2/C_3H_8/He$ system, the conversions of propane over Pt-Sn/Al and Pt-Sn/Zn-Al catalysts are 35% and 43% respectively. But in $H_2/O_2/C_3H_8/He$ system conversions decrease to 24% and 18% respectively. While for Cr, Fe, and Zr, especially for Ce and La, promoted Pt-Sn/γ-Al_2O_3 catalysts, a notable increase in conversion is obtained. Due to the presence of oxygen, over Cr, Fe, Zr, Mn, Ce and La promoted Pt-Sn/γ-Al_2O_3 catalysts, conversions of propane increase from 39, 37, 38, 39, 41 and 37% to 44, 43, 42, 40, 50 and 46% respectively. As for the selectivity to propylene, the introduction of O_2 leads to a slight decrease which depends on the catalyst. Over Pt-Sn/Ce-γ-Al_2O_3 and Pt-Sn/La-γ-Al_2O_3 catalysts, the selectivities towards propylene decrease from 96 and 97% in $H_2/C_3H_8/He$ system to 92 and 90% in $H_2/O_2/C_3H_8/He$ system. The results of experiments suggest that it is possible to feed small quantity of oxygen to catalytic dehydrogenation system to increase the yield of propylene but not obvious at the expense of propylene selectivity over some promoted Pt-Sn/γ-Al_2O_3 catalysts.

The effect of temperature on the performance of Pt-Sn/La-γ-Al_2O_3 catalyst in different gas medium was tested. The results are shown in Figure2. Figure2

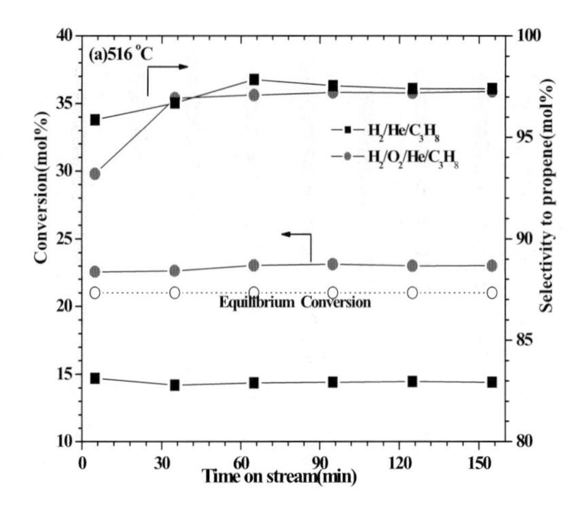

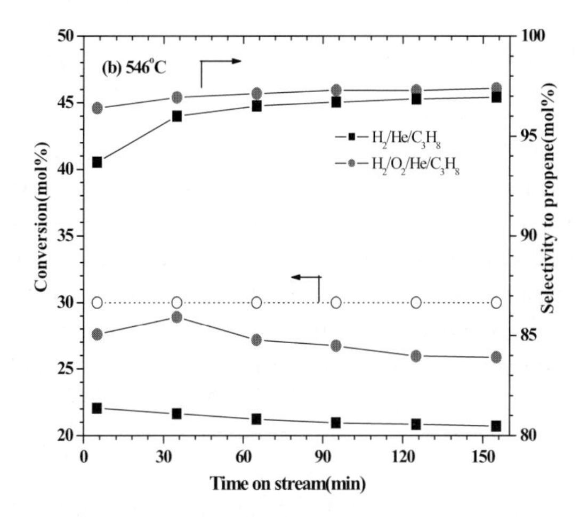

Figure 2. Propane dehydrogenation to propylene over Pt-Sn/La-Al catalyst in different gas medium at different temperatures. Dashed line represents equilibrium conversion of propane in catalytic dehydrogenation. Reaction conditions: $H_2/C_3H_8/He$ =1:1:3(molar ratio), total GHSV=5000 h^{-1}; $H_2/O_2/C_3H_8/He$ =1:0.3:1:3 (molar ratio), total GHSV=5400 h^{-1}.

shows that, at 516°C, the conversion of propane is about 15% in H_2/He atmosphere and it increases to 23% in $H_2/O_2/He$ atmosphere. Only a slight decrease of the initial selectivity of propylene (from 97 to 96%) can be found in the presence of oxygen. No O_2 is detected by GC in products and the selectivity towards CO_x is about 0.7%. It shows that more than 99% of O_2 is selectively converted to H_2O. Obviously, O_2 existence dose not lead to an obviously extra loss in propylene selectively. The mean yield of propylene in the case of O_2 presence is about 22% which is much higher than that (14%) in H_2/He atmosphere and even exceeds the equilibrium value of propane catalytic dehydrogenation (about 21%). At 546°C, a similar picture has been given. The conversion of propane increases from 21 to 27% due to the presence of oxygen. However, the CO_x selectivity shows a slight increase (about 1.2%) and the mean yield of propylene reaches to 25%. At 546 °C, the conversion of propane (27%) in O_2 atmosphere is slight below the equilibrium value (about 30%).

The main finding reported here is that under certain conditions, feeding a small quantity of oxygen to conventional catalytic dehydrogenation system will greatly favor promoting the yield of propylene. The results are dependent on the catalyst. In oxygen atmosphere Pt-Sn/Al and Pt-Sn/Zn-Al catalysts are much less stable than in H_2/He atmosphere. This suggests that oxygen or steam formed gives an adverse effect on the stability of those catalysts, which accelerates the decay of Pt active sites. Previous studies [4-5] have shown that oxygen or steam adsorption over platinum surface can distinctly reduce the interactions between metal and support, thus promoting the sintering of platinum particles. Another adverse factor of steam is that steam could destroy active platinum tin clusters supported by γ-Al_2O_3 [6].This adverse factor of oxygen or steam could make the catalysts less stable. But for some other M promoted Pt-Sn/Al catalysts, this effect is almostly negligent.

What is the role of oxygen in this $H_2/O_2/C_3H_8$/He system? To our knowledge, the reaction conditions are much different from those existent literatures report [7-9]. From the effects of temperature on the performance of catalyst, we can see that at low temperatures the role of oxygen becomes more distinct, which may suggest that oxygen could mainly benefit the dehydrogenation reaction at low temperatures. Oxygen may act the following functions. Hydrogen is selectively combusted by oxygen through platinum catalysis or through the lattice oxygen in promoter metal oxides and the reaction equilibrium is driven toward the products side. However, even 100% of O_2 is consumed by hydrogen the amount of consumption hydrogen may not enough to increase the conversion of propane to such an extent, especially at lower temperatures. So, another function of oxygen is perhaps to take part in the reaction steps of propane dehydrogenation. The limiting step of propane dehydrogenation on platinum catalysts is considered to be the β H elimination of the adsorbed propane [10].The active H produced over the surface of platinum may react with O^{2-} in M promoter, thus yielding hydroxyl groups. The hydroxyl groups could react with β H of adsorbed propane, thus significantly increase the rate of dehydrogenation. A similar phenomenon has been reported by Kogan [11] that in steam atmosphere a faster dehydrogenation rate than in hydrogen atmosphere was observed over Pt-In catalyst supported on corundum. They ascribed the effect of steam to the fact that the hydroxyl groups adsorbed on the Pt bimetallic clusters reacted with β H of propane, thus significantly increase the rate of dehydrogenation.The formation of hydroxyl groups would be related to the temperature. The elevated temperatures may destroy the hydroxyl groups and do not benefit the dehydrogenation in oxygen atmosphere. It has been reported that treating CeO_2 with hydrogen leads to the formation of hydroxyl groups as is apparent from FT-IR measurements [12]. The intensity of those υ (OH) bands increase with increasing reduction temperature up to over 400 °C. At even higher temperature these bands disappear.The activity of catalysts in oxygen atmosphere could be related the lability of lattice oxygen in M promoters. The the high lability of lattice oxygen could result in the high activity, as indicated in ceria promoted Pt-Sn/Al catalyst. The other role of M promoters may be related the stabilization of active PtSn phases. The special stabilization of promoters to active PtSn phases can decrease the adverse effect of oxygen or steam and further maintain the high performance in $H_2/O_2/C_3H_8$ system.

4. Conclusions

Feeding a small quantity of oxygen to propane catlytic dehydrogenation system catalyzed by metal oxide promoted Pt-Sn/Al catalysts can greatly promote the yield of propylene. Hydrogen is selectively consumed by oxygen through platinum catalysis or through the lattice oxygen in promoter and the equilibrium is driven toward the products side. Another role of oxyge is perhaps to take part in the reaction of dehydrogenation.

References

1. M. Baerns, O. Buyevskava, Catal., Today, 45 (1998) 13.
2. C. L. Yu, Q.J. Ge, H.Y. Xu, W. Z. Li, Journal of Fuel Chemistry and Technology (Chinese), 34(1) (2006) 209.
3. C.L. Yu, Q. J. Ge, H.Y. Xu, W.Z. Li, Appl. Catal. A. Gen., xxx (2006) xxx
4. E. Ruckenstein, X. D. Hu, J. Catal., 100(1986)1-16
5. H. Glassl, R. Kramer, K. Hayek. J. Catal., 68(1981)388-396
6. W. S. Dong, H. J. Wang, X. K. Wang, and etal, J. Mol. Catal., (China) 13(3) (1999)181.
7. M.Huff, L.D.Schmidt, J. Catal., 149(1994)127-141.
8. J. G.Tsikoyiannis, D. L.Stern, R. K. Grasselli, J. Catal.184(1999)77.
9. L. Låte, J.-I. Rundereim, E.A. Blekkan, Appl.Catal. A: Gen.,262 (2004) 53.
10. P. Biloen, F. M. Dautzenberg, W. M. H. Sachtler, J. Catal, 50(1977)77-86.
11. S. B. Kogan, M. Herskowitz, Catal. Commun., 2(2001) 185.
12. S. Bernal, J. J. Calvino, G. A. Cifredo, and et al,J. Chem. Soc. Faraday Trans., 89 (1993) 3499.

Natural Gas Conversion VIII
F.B. Noronha, M. Schmal, E.F. Sousa-Aguiar (Editors)

Au/TiO_2 as a catalyst for the selective hydrogen combustion (SHC) applied to the catalytic dehydrogenation of propane

Hilde Dyrbeck[a], Edd A. Blekkan[a]

[a]*Department of Chemical Engineering, Norwegian University of Science and Technology, NO-7491 Trondheim, Norway*

Abstract

Selective catalytic oxidation of hydrogen in the presence of hydrocarbons was studied in a fix bed quartz reactor, over 3wt%Au/TiO_2 and 5wt%Au/TiO_2 catalysts. The reaction can be coupled with catalytic dehydrogenation, providing *in situ* heat to the dehydrogenation reaction and removing produced hydrogen. Avoiding the nonselective combustion of the hydrocarbons in the mixture is essential to the process. Both 3wt%Au/TiO_2 and 5wt%Au/TiO_2 are able to combust hydrogen, but in a gas mixture with propane and oxygen the selectivity is dependent upon the feed ratio of hydrogen and oxygen. At 550 °C, with propane present, no carbon oxides are formed when the H_2:O_2 ratio is 4, but at lower H_2:O_2 ratios some CO and CO_2 is formed.

Keywords

Selective hydrogen combustion; Au/TiO_2; propane dehydrogenation

1. Introduction

The catalytic dehydrogenation of light alkanes is an important reaction, much due to the growing demand for light alkenes as propene and ethene as feedstock in chemical industries.

The catalytic dehydrogenation, e.g. of propane (1), is a strongly endothermic reaction favoured by low pressures. This, combined with a tendency towards coke formation on the catalyst are important factors influencing the reactor design, which has to allow for heat transfer to the reaction and frequent regeneration of the catalyst. Conventional catalytic dehydrogenation of light alkanes is practiced commercially worldwide, and several commercial processes are available. These differ mainly in reactor arrangement (fixed, moving or fluidized beds) and catalyst (either based on supported Pt or chromia-alumina catalyst) [1]. Both Pt and Cr catalysts deactivate rapidly, and require frequent oxidative regenerations [2].

$$C_3H_8 \rightarrow C_3H_6 + H_2 \qquad (\Delta H^0_{298} = 124 \text{ kJ/mole}) \qquad (1)$$

A high concentration of hydrogen will limit the yield of product alkene, and removing the hydrogen is thus favourable for the equilibrium yield. This can be done by continuously removing H_2 using membranes, or through a selective reaction with co-reactant such as O_2, thus removing the hydrogen as water [3,4]. In a new process concept the produced hydrogen can be combusted catalytically to provide *in situ* heating, and if the combustion is coupled with the dehydrogenation reaction an autothermal process can be achieved, thus removing some restraints in the reactor design. At 500 °C the heat of combustion of hydrogen is approximately twice the heat required for propane dehydrogenation. Hence, to get a heat-balanced process it is necessary to combust approximately half the hydrogen produced in the dehydrogenation.

$$H_2 + \tfrac{1}{2}O_2 = H_2 \qquad (\Delta H^0_{298} = -241.83 \text{ kJ/mole}) \qquad (2)$$

This has been previously tested in a staged cofeed process mode [5], or in a system where the SHC and DH catalysts are physically mixed (the so-called DH+SHC redox mode [6]). Pt based catalysts are known to be active for both dehydrogenation and combustion, and has therefore been the focus of our initial work. Here we report on the use of Au/TiO_2 as a SHC catalyst.

2. Experimental

The gold-based catalysts reported here are 3 and 5 wt% Au, supported on TiO_2. They were prepared using a sol-gel technique giving small Au particles, as described by Duff et al. [7]. The Au sol was prepared from reduction of chloroaurate(III) ions by partially hydrolysed tetrakis(hydroxymethyl)-phosphonium chloride (THPC) solution. A calculated amount of 0.2 M NaOH and THPC (1.2 ml of 80 wt.% THPC in water diluted to 100 ml) was added to the desired amount of continuously stirred distilled water. After 2 minutes a 43mM $HAuCl_4$ solution was added. While pouring the gold(III) solution into the alkaline THPC mixture a rapid colour change from yellow to dark orange occurred, indicating the formation of the gold sol. The TiO_2 (commercially available, 30% rutile and 70% anatase, Sg = 30 m^2/g) was dispersed in 50 ml water and the Au sol was added dropwise. The pH was adjusted to 2 by 0.2 M HNO_3 and the solution was stirred for 1 hour. The mixture was filtrated and no colour could be observed for the filtrate. The sample was washed with ethanol until no Cl- was detected in the filtrate using $AgNO_3$.

The catalysts were characterized using EXAFS and N_2 adsorption (BET). The catalysts have a mean Au particle size below 2 nm and correspondingly low coordination numbers. The particle size is determined from the first Au-Au coordination number assuming spherical particles. Both samples had surface areas close to 35 m^2/g. Further details of the preparation and characterization are given elsewhere [8].

All catalysts were tested in a fixed-bed quartz reactor (U-shaped) with an inner diameter of 3.5 mm. The reactor was placed inside an electrical furnace, and both external and internal thermocouples were used to monitor and control the temperature. Gases were metered using electronic mass-flow controllers, and the product gas composition of the dry gas was measured using an on-line gas chromatograph. The total flow of feed gas was 100 ml/min; the amounts of H_2 and C_3H_8 were fixed (2 ml/min and 10 ml/min respectively) while the flow of O_2 was varied to achieve different H_2/O_2 ratios, the balance being He. In the experiments without propane, helium was added to maintain the same total flow and partial pressures of H_2 and O_2. The experiments were done at temperatures up to 550 °C and the pressure was 1 atm., with a loading of 0.15 g catalyst. The reported selectivities are based on moles carbon converted to $CO/CO_2/C_3H_6$ (C-selectivity). Further reaction details can be found in [3].

3. Results and discussion

The Au catalysts were active in hydrogen combustion in the absence of hydrocarbons. The results are summarized in Table 1.

The activity was slightly higher for the 5wt%Au/TiO_2 sample, and for both samples the temperature needed to achieve 50% conversion decreased when the concentrations of hydrogen and oxygen were increased, indicating an overall positive reaction order. There was no discernible deactivation when the hydrogen combustion was performed at low to moderate temperatures, but for catalysts exposed to hydrocarbons at higher temperatures some of the activity was lost. This could be due to sintering of the gold particles, or coke formation on the surface.

Table 1. Activity for hydrogen combustion over Au/TiO_2 catalysts reported as the temperature necessary for 50% conversion (T_{50}).Conditions: 0.15 g catalyst, 1 atm., total flow 100 ml/min (using He as the diluent).

Gas flows (ml/min)			T_{50} (°C)	
O_2	H_2	Ratio H_2/O_2	$3Au/TiO_2$	$5Au/TiO_2$
1	4	4	90	63
2	4	2	100	73
0,5	2	4	100	70
1	2	2	130	100
2	2	1	160	120

The SHC-selectivity was tested with feed gas mixtures containing 10% propane. With excess hydrogen, the SHC selectvity was high. For example, when using a H_2/O_2-ratio of 4 in the feed, no CO or CO_2 were formed over the entire temperature range 20-550 °C over the $3Au/TiO_2$ catalyst. The temperature necessary for full conversion of hydrogen is considerably higher when propane is present, for all three different H_2/O_2 ratios of feed gas composition that was tested. From this it can be suggested that in the presence of propane not only hydrogen and oxygen are adsorbed on the catalyst surface, but also propane or other carbon-containing components.

Figure 1 shows results from the SHC experiments with the Au/TiO_2 catalysts at 550 °C. The conversion of oxygen is always complete at these conditions, and Fig. 1a shows the propane conversion. The conversion is low, and increasing with increasing amount of oxygen in the flow. The selectivity to the SHC-reaction depended on the gas composition.

The 5wt%Au/TiO_2 is slightly more active than the 3wt%Au/TiO_2 at higher H_2/O_2 ratios. The lowest flow gives the lowest conversion and the corresponding highest selectivity to propene (Fig. 1b), indicating that the propane converted at these conditions is mainly dehydrogenated according to Eq. (1). The activity for propane conversion is as expected low for these catalysts, and the selectivity to propene for both catalysts drops dramatically with increasing oxygen flows.

Figure 1 c and d show the carbon selectivity to CO_2 and CO for different H_2/O_2 ratios. The CO_2 selectivity is similar for the two catalysts, while there is a difference in the trend for the CO formation. The 5wt%Au/TiO$_2$ is 100 % SHC selective at substoichiometric O_2 amounts, while the 3wt%Au/TiO$_2$ produces some CO even at these conditions. At higher O_2 flows both catalysts produce

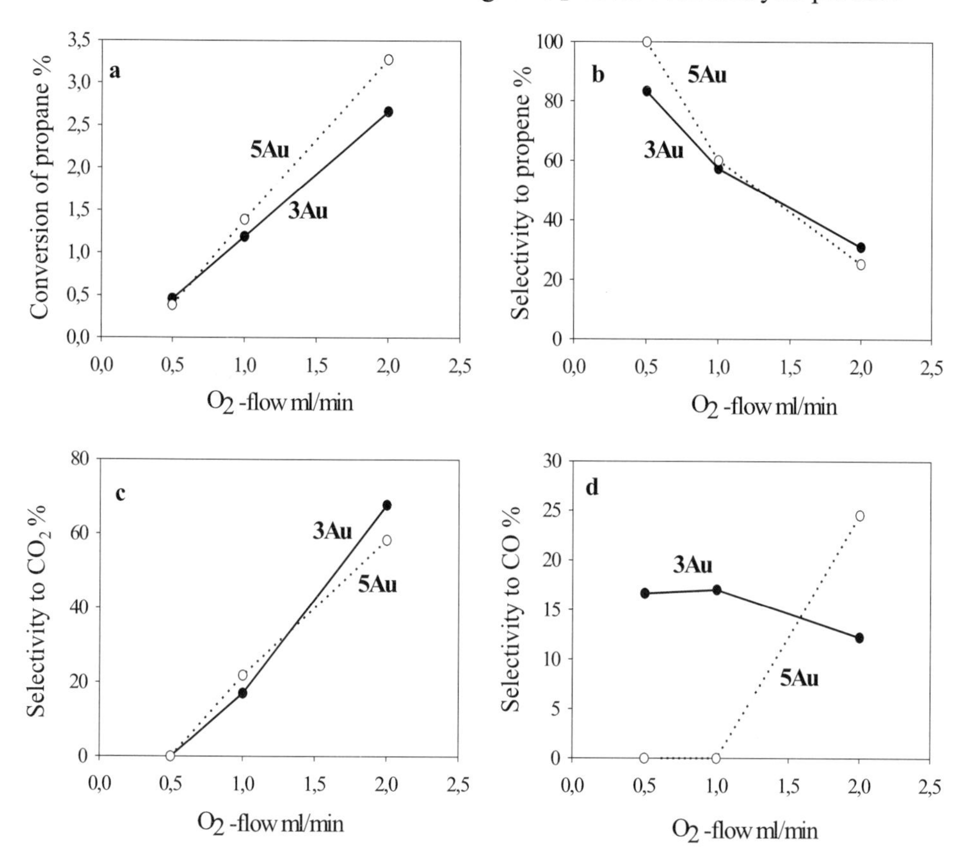

Figure 1. SHC over 5wt%Au/TiO$_2$ and 3wt%Au/TiO$_2$ at 550 °C; conversion of propane (a), selectivity to propene (b), CO_2 (c) and CO (d).

carbon oxides, which is similar to the behavior of Pt/SiO_2, PtSn/SiO_2 and Sn/SiO_2 as described previously [3].

4. Conclusions

The Au/TiO_2 system is a potential catalyst for the SHC reaction. As long as the $H_2 : O_2$ ratio exceeds 2 (stoichiometry), the oxygen reacts preferably with the H_2 to H_2O. The CO and CO_2 selectivity increases when the ratio is reduced. We can only speculate on the mechanism allowing the SHC. Both hydrogen and hydrocarbon combustion are assumed to be surface catalyzed reactions. At the conditions where the hydrogen is selectively combusted it is reasonable to assume that adsorbed hydrogen dominates the surface. However, hydrocarbons must also be adsorbed, since we observe some propene formation.

Acknowledgements
We thank the Research Council of Norway for financial support through the KOSK programme. We also thank NTNU for strategic funding for the promotion of equal opportunities.

References

[1] R. K. Graselli, Catal. Today, 49 (1999) 141.
[2] T. Waku, J. A. Biscardi and E. Iglesia, J. Catal., 222 (2004) 481.
[3] L. Låte, J.-I. Rundereim and E. A. Blekkan, Appl. Catal. A, 262 (2004) 53.
[4] J. G. Tsikoyiannis, D. L. Stern and R. K. Graselli, J. Catal., 184 (1999) 77.
[5] T. Imai: U.S Patent 4,435,607 (1984).
[6] R. K. Graselli, D. L. Stern and J. G. Tsikoyiannis, Appl. Catal. A, 189 (1999) 9.
[7] D. G. Duff, A. Baiker and P. P. Edwards, J. Chem. Soc. Chem. Commun. (1993) 96.
[8] N. Hammer, I. Kvande, D. Chen, W. van Beek and M. Rønning, Accepted, Topics in Catalysis.

Natural Gas Conversion VIII
F.B. Noronha, M. Schmal, E.F. Sousa-Aguiar (Editors)

Oxidative dehydrogenation of ethane at short contact times

Silje F. Håkonsen, Anders Holmen

Department of Chemical Engineering, Norwegian University of Science and Technology (NTNU), Sem Sælands vei 4, NO-7491 Trondheim, Norway

ABSTRACT

Oxidative dehydrogenation of ethane at short contact times has been examined in a continuous flow reactor over PtSn and $LaMnO_3$ monoliths. The effect of the PtSn catalyst is observed to decrease substantially when increasing the furnace temperature from 700 – 800 °C. PtSn was shown to give higher selectivity to ethene than $LaMnO_3$ when sufficient amounts of hydrogen was added to the feed.

1. INTRODUCTION

Ethene is an important feedstock for many chemical processes. Traditionally ethene is produced by steam cracking. However, steam cracking suffers from several limitations. One promising alternative to this process could be oxidative dehydrogenation of ethane which offers advantages such as short contact time, adiabatic operation and reduced reactor volumes [1]. It has been shown that high yields of ethene can be obtained by oxidative dehydrogenation of ethane at short contact times using Pt-Sn monoliths [2-5] and $LaMnO_3$ monoliths [6, 7].
The present study is focused on the production of ethene via oxidative dehydrogenation of ethane at short contact times over PtSn and $LaMnO_3$ monoliths. The product distribution has been investigated both with and without PtSn catalyst at different furnace temperatures. In addition, the loading of Pt on the monoliths was varied from 1.0 wt% to 0.1 wt% at a furnace temperature of 700 °C. The effect that PtSn and $LaMnO_3$ has on the product distribution was also compared.

2. EXPERIMENTAL

Cylindrical pieces of extruded Cordierite ($2MgO \cdot 2Al_2O_3 \cdot 5SiO_2$; Corning) were used as support. The monolith pieces (l=10 mm, d=15 mm, 62.2 cells/cm^2) were washcoated by a dispersion of Disperal P2 (286 m^2/g) (Condea). The washcoated monoliths were then dried (120 °C, 4 h) and calcined in a flow of air (550 °C, 4.5 h). To obtain Pt-Sn catalysts the washcoated monoliths were first impregnated with an aqueous solution of $Pt(NH_3)_4(NO_3)_2$. The impregnated monoliths were then dried and calcined at 550 °C for 4.5 hours. After that they were impregnated by an aqueous solution of $SnCl_2 \cdot 2H_2O$ and finally dried and calcined in two steps: First for 30 minutes at 100 °C and then for 90 minutes at 700 °C. The Pt-Sn monoliths were reduced *in situ* prior to the experiment in a flow of hydrogen at 600 °C for 0.5 h.
The γ-alumina washcoated monoliths were also used as support for the $LaMnO_3$-catalysts. First the monoliths were stabilized with 5 wt% La_2O_3 by impregnating them with a solution of $La(NO_3)_3 \cdot 6H_2O$. Then they were impregnated with an equimolar aqueous solution of $La(NO_3)_3 \cdot 6H_2O$ and $Mn(CH_3COO)_2 \cdot 4H_2O$. Subsequently the monoliths were dried and calcined. The latter impregnation was repeated until 30 wt% perovskite with regard to the washcoat was obtained.
The oxidative dehydrogenation of ethane was studied at close to atmospheric pressure and in the temperature range of 600–900 °C in a continuous flow apparatus consisting of a quartz reactor with an inner diameter of 15 mm. The catalyst was placed between two inert monoliths acting as radiation shields. Temperature profiles were measured by a movable thermocouple placed in a small quartz tube inside the monoliths. The furnace temperature is measured between the quartz reactor and the inside of the furnace. Two condensers at the reactor outlet removed any water from the product gas. The dry samples were analyzed by two on-line gas chromatographs.

3. RESULTS AND DISCUSSION

3.1. Effect of the catalyst at different temperatures

Fig. 1A) and Fig. 1B) show the conversion of ethane and the selectivity to ethene as a function of the C_2H_6/O_2-ratio at a furnace temperature of 700 °C and 800 °C, respectively. The figures illustrate the difference between experiments run with and without PtSn catalyst.
At a furnace temperature of 700 °C a large effect of the PtSn catalyst was observed on the conversion of ethane as well as on the selectivity to ethene. For the experiment performed over the PtSn monolith, much higher temperatures were measured through the reactor bed than during the experiment without any

catalyst. This is most probably caused by a larger extent of combustion over the catalyst. The oxygen conversion is almost complete over the monolith containing PtSn, whereas oxygen is detected in the product gas in relatively large amounts over the washcoated monolith containing no PtSn.

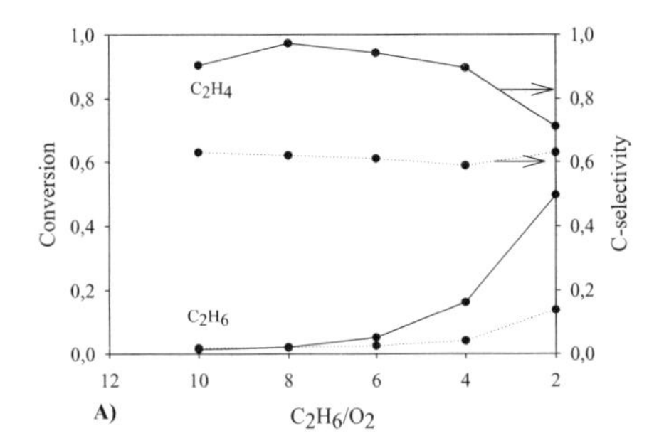

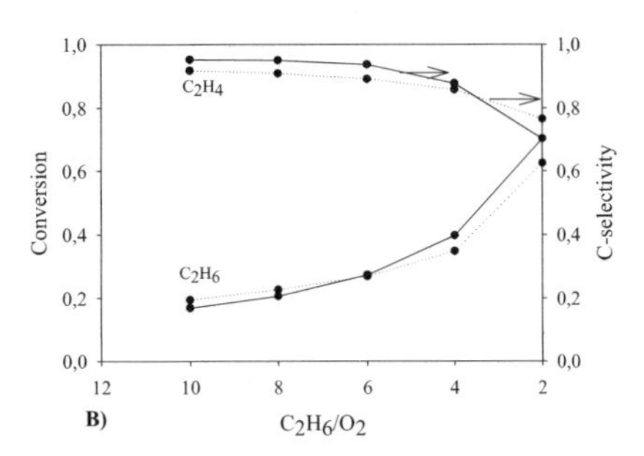

Fig. 1. A comparison between the use of PtSn catalyst on alumina (solid lines) and alumina alone (dotted lines). A) Experiments performed at a furnace temperature 700 °C. B) Experiments performed at a furnace temperature of 800 °C. Feed [Nml/min]: C_2H_6 (308), air (147-733), H_2 (154) and Ar (1391-805). Total flow rate: 2000 Nml/min.

The experiments shown in Fig. 1A) and Fig. 1B) illustrate that at a furnace temperature of 700 °C the PtSn catalyst has a significant effect on the oxidative dehydrogenation of ethane, while already at a furnace temperature of 800 °C the gas phase reactions seem to be dominating.

3.2. Effect of platinum loading

The effect of platinum loading on the conversion of ethane and the selectivity to ethene is shown in Fig. 2. The same experimental conditions as for the experiment shown in Fig. 1A) were used. A Sn/Pt-ratio of 2 (wt/wt) was kept constant for all four Pt loadings.

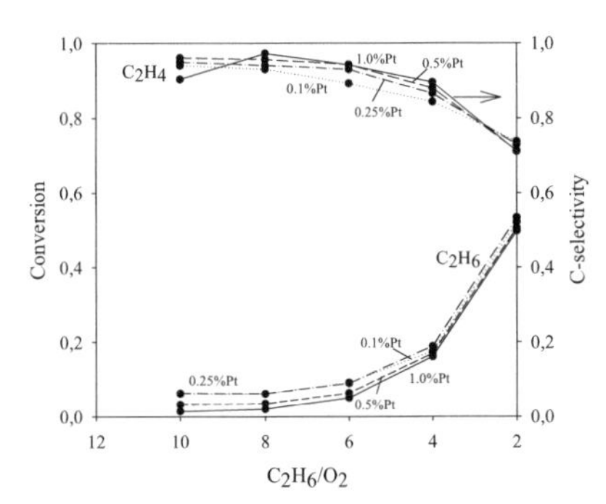

Fig. 2. The effect of Pt loading. 1.0% Pt (solid lines), 0.5% Pt (dashed lines), 0.25%Pt (dashed-dotted lines) and 0.1% Pt (dotted lines). Sn/Pt-ratio = 2 (wt/wt). Feed [Nml/min]: C_2H_6 (308), air (147-733), H_2 (154) and Ar (1391-805). Total flow rate: 2000 Nml/min.

The results in Fig. 2 indicate only small effects of changing the content of Pt. When decreasing the weight percent of platinum on the monoliths, the conversion of ethane increases, especially at higher C_2H_6/O_2-ratios. This increase in conversion results in a slight reduction of the ethene selectivity. The lower selectivity is mainly a result of increased production of CO. The conversion of oxygen was >93% for all four platinum loadings.

3.3. Comparing PtSn with $LaMnO_3$

Fig. 3A) shows the ethane conversion and the selectivity to ethene as a function of furnace temperature over PtSn and $LaMnO_3$. The selectivity to the carbon oxides as a function of ethane conversion is shown in Fig. 3B). Fig. 3A) shows that the ethane conversion is almost similar for the two catalysts while PtSn gave higher ethene selectivity in the whole temperature interval and in particularly at temperatures below 700–750 °C. Fig 3B) indicates that the lower ethene selectivity is mainly due to the large production of CO_2 over the $LaMnO_3$ catalyst compared to the PtSn catalyst. The oxygen conversion and selectivity to the other main carbon based by-products are almost the same for the two catalysts. It therefore seems that PtSn is much better in inhibiting total combustion than is the perovskite.

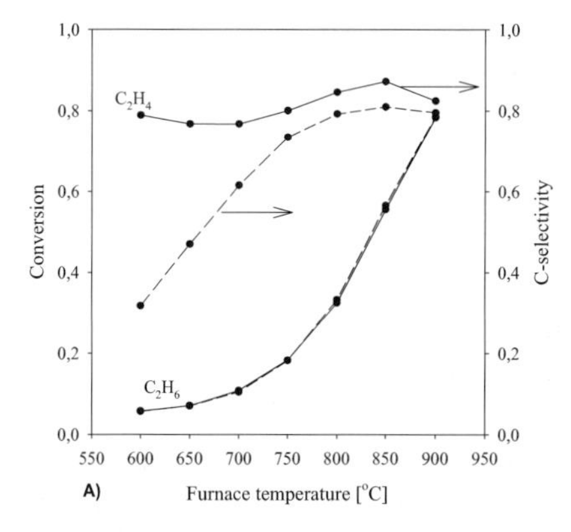

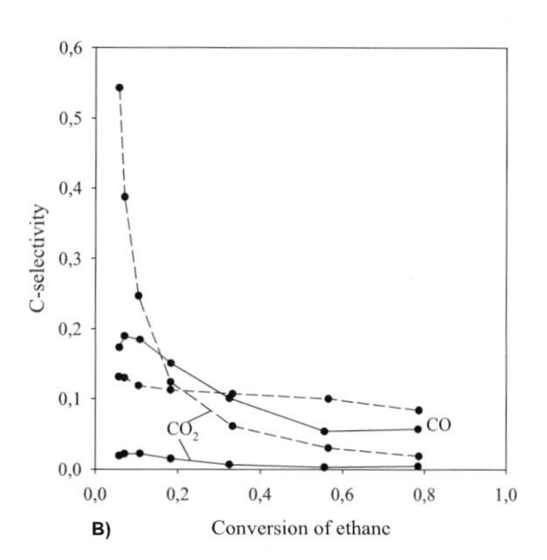

Fig. 3. A comparison between the use of PtSn catalyst (solid lines) and $LaMnO_3$ (dotted lines). Feed [Nml/min]: C_2H_6 (308), air (367), H_2 (154) and N_2 (1171). Total flow rate: 2000 Nml/min.

Fig. 4 shows the effect of adding hydrogen to the reaction mixture both over PtSn (solid lines) and $LaMnO_3$ (dotted lines). When no hydrogen was added $LaMnO_3$ gave higher ethene selectivity due to large amounts of CO_x formed over the PtSn catalyst. However, by increasing the amount of hydrogen in the feed, the ethene selectivity is greatly enhanced over PtSn by reducing the formation of CO_x. This effect is not observed to the same extent over $LaMnO_3$ where only the production of CO_2 is reduced upon hydrogen addition. No such reduction was observed for the CO selectivity.

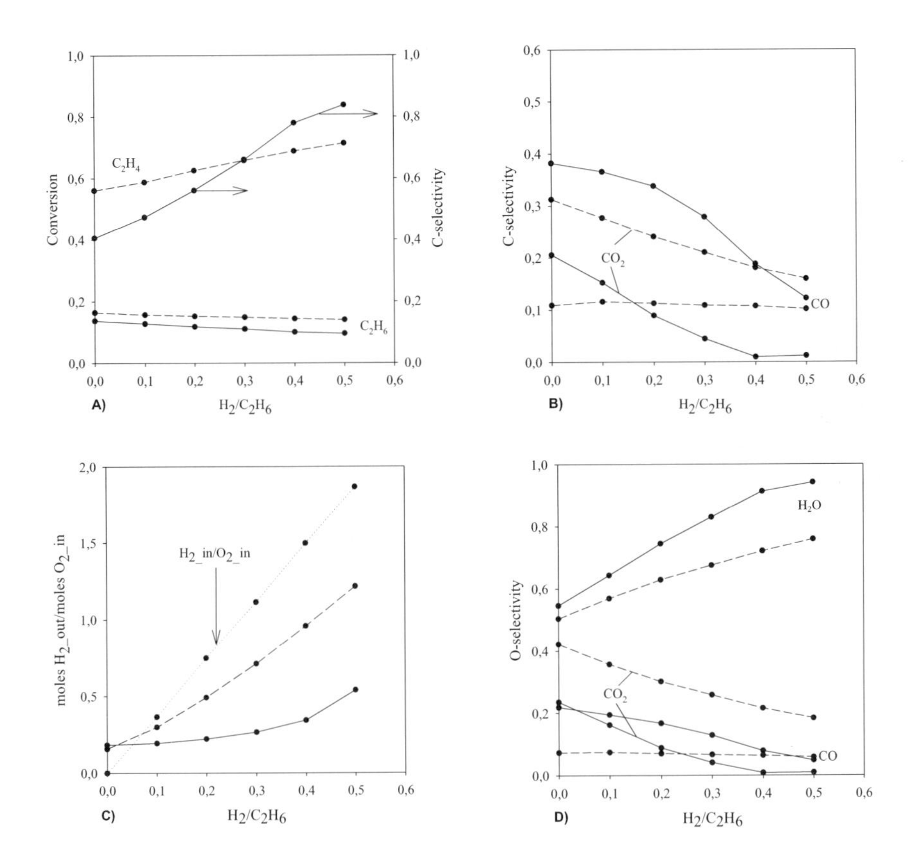

Fig 4. A comparison between the use of PtSn catalyst (solid lines) and $LaMnO_3$ (dotted lines). Effect of hydrogen addition on A) conversion of ethane and selectivity to ethene, B) selectivity to the CO_x, C) amounts of hydrogen in the product compared to hydrogen in the feed and D) O_2-selectivity. Feed [Nml/min]: C_2H_6 (308), air (367), H_2 (0-154) and N_2 (1171-1325). Total flow rate: 2000 Nml/min. Furnace temperature: 700 °C

The amount of hydrogen in the product stream compared to the amount in the feed and the oxygen selectivity are shown in Figs. 4C) and D). It is clearly seen that the consumption of hydrogen is larger over the PtSn catalyst than over $LaMnO_3$. However, when co-feeding hydrogen there is no net production of hydrogen over any of the catalysts at a furnace temperature of 700 °C. Fig. 4D)

shows that as more hydrogen is fed to the reaction, more water and less CO_x is formed as product. Nevertheless, this effect is more evident over PtSn than for $LaMnO_3$. The oxygen conversion is almost the same over the two catalysts (>93%).
It seems that PtSn has a greater capacity of inhibiting oxidation reactions of hydrocarbons when hydrogen is present in the feed than is the $LaMnO_3$ catalyst. As a consequence, PtSn produces water rather than carbon oxides to a greater extent than $LaMnO_3$, thereby giving a higher selectivity to ethene.

4. CONCLUSIONS

The catalytic effect of PtSn impregnated on washcoated monoliths was evident at a furnace temperature of 700 °C, but already at a furnace temperature of 800 °C the gas phase reactions are dominating. Reducing the Pt loading on the monoliths appeared to increase the ethane conversion and decrease the selectivity to ethene only slightly. However, even very small loadings of the PtSn catalyst did not seem to lower the ethene yield much. By comparing PtSn to $LaMnO_3$, it is evident that PtSn gives much higher ethene selectivity when sufficient amounts of hydrogen were added to the feed. In the presence of hydrogen, oxygen preferably reacts with hydrogen to produce water over the PtSn catalyst. This effect was not as evident over $LaMnO_3$ where larger amounts of CO_2 were observed as a result of total combustion.

ACKNOWLEDGEMENT

The financial support from the Norwegian Research Council through the KOSK program and the Norwegian University of Science and Technology is greatly acknowledged.

REFERENCES

[1] S.F. Håkonsen and A. Holmen, in: Handbook of heterogeneous catalysis, eds. G. Ertl, H. Knözinger, F. Schüth and J. Weitkamp (Weinheim, 2007) In press.
[2] C. Yokoyama, S.S. Bharadwaj and L.D. Schmidt, Catal. Lett. 38 (1996) 181.
[3] A. Bodke, D. Henning and L.D. Schmidt, Catal. Today 61 (2000) 65.
[4] D.A. Henning and L.D. Schmidt, Chem. Eng. Sci. 57 (2002) 2615.
[5] B. Silberova, J. Holm and A. Holmen, in: Studies in Surface Science and Catalysis, 147, eds. X. Bao and Y. Xu (Elsevier, Amsterdam, 2004) pp. 685.
[6] F. Donsì, R. Pirone and G. Russo, J. Catal. 209 (2002) 51.
[7] F. Donsì, R. Pirone and G. Russo, Catal. Today 91-92 (2004) 285.

Natural Gas Conversion VIII
F.B. Noronha, M. Schmal, E.F. Sousa-Aguiar (Editors)

Study of synthesis gas production over structured catalysts based on LaNi(Pt)O_x- and Pt(LaPt)-CeO_2-ZrO_2 supported on corundum

S. N. Pavlova, N. N. Sazonova, V. A. Sadykov, G. M. Alikina, A. I. Lukashevich, E. Gubanova, R.V. Bunina

Boreskov Institute of Catalysis, Siberian Branch of the Russian Academy of Sciences, pr. Lavrentieva, 5, 630090, Novosibirsk, Russia

Study of partial oxidation of methane (POM), steam (SR), authothermal (AR) and dry (DR) reforming of methane over catalysts containing $LaNiO_3$ (pure or promoted by Pt) or $LaPtO_x/CeO_2$-ZrO_2 supported on the fragments of corundum monolith reveals that the most effective catalysts contain $LaNiO_3$ and $LaPtO_x$ with La excess. The reduction pretreatment as well as addition of Pt to $LaNiO_x$ facilitates the formation of syngas at a lower temperature, furthermore, Pt favors a high catalyst stability preventing catalyst coking.

1. Introduction

The catalytic partial oxidation of methane at short contact times is now considered as an attractive technology for the small-scale and distributed production of syngas in the stationary and mobile fuel processing [1]. Thus, integration of POM into stage combustors of methane for gas turbines favoures combustion stabilization and decline of the nitrogen oxide content in exhausts [2]. Syngas may also be added to the conventional vehicle internal combustion engine in order to decrease pollutant emissions especially during the start-up period [3]. Realization of POM at short contact times (< 0.1 sec) allows to reduce the size of the fuel processor and requires application of monolithic catalysts having a low pressure drop [4,5]. A sequence of reactions was shown to proceed along the monolithic catalytic bed during POM: exothermic partial or complete oxidation of methane at the inlet part and endothermic methane steam and dry reforming in the rear zone [5]. This leads to the appearance of hot spots with the temperature >1000°C and substantial temperature gradient along the catalyst bed. Hence, in the inlet zone, evaporation of expensive noble metals contained in known monolith catalysts occurs, while in the outlet zone deactivation due to carbon build up is observed [4]. To solve these problems, the use of different catalysts along the catalytic bed which ensure a high activity and stability in corresponding reactions could be a promising approach.

The catalysts based on Pt-promoted $LaNiO_x/Ce\text{-}ZrO_x$ were shown recently to be very efficient in POM [5-7]. In this work, the comparative study of POM, steam (SR), authothermal (AR) and dry (DR) reforming of methane over catalysts containing $LaNiO_3$ (pure or promoted by Pt) or $LaPtO_x/CeO_2\text{-}ZrO_2$ supported on the fragments of corundum monolith was performed. The catalyst performance has been studied at extremely short contact times ~ 1- 6 ms.

2. Experimental

As supports, separate triangular channels of $\alpha\text{-}Al_2O_3$ monolithic substrate with length of 10-20 mm were used [5,6]. To prepare catalysts, first, $Zr0_{0.8}Ce_{0.2}0_2$ and, subsequently, $LaNiO_x$ ($LaNi_{0.9}Pt_{0.1}$) or Pt ($LaPtO_x$) were supported by the incipient wetness impregnation as in [6]. Composition of the catalyst active components is presented in Table 1.

Temperature–programmed POM reaction for 10 mm catalyst channels subjected to different pretreatments were studied as described in [6].

The testing of the catalysts in POM, SR and DR was carried out at atmospheric pressure in a plug-flow quartz reactor and temperatures in the range of 550-900 °C. The catalysts were pretreated *in situ* at 900°C for 1 h in O_2 (H_2) flow. The reaction mixture (POM: 3.5 –12 vol.% CH_4, 1.75-6% O_2, N_2 – balance, CH_4/O_2~2; SR: 3.5 –12 vol.% CH_4, $H_2O:CH_4$ =1- 3, He – balance; DR: 7 vol.% CH_4, $CO_2:CH_4$ =1, N_2 – balance) was fed with the flow rate in the range of 5.5-36 l/h, i.e. at 1-6 ms. The outlet temperature of the catalyst channel was used to characterize the activity. The reagents and reaction products were analyzed by GC and on-line IR absorbance analyzer PEM-2M equipped with an electrochemical H_2 sensor.

3. Results and discussion

3.1. Temperature-programmed POM reaction over Ni-containing catalysts.

Typical curves of reagent consumption and product evolution during POM temperature-programmed reaction over LN and LNP catalysts are shown in Fig. 1. The reaction starting temperature depends on the catalyst composition and pretreatment. Thus, for LNP, after all pretreatment types, the formation of syngas is observed at temperatures ~ 130-200°C lower as compared to the LN sample without Pt. For both catalysts, pre-reduction with hydrogen or methane (the data are not shown for brevity) decreases the starting temperature of syngas evolution as well. It is now well known that the deep oxidation of methane mainly proceeds over Ni or Pt oxides whereas metallic state is responsible for the synthesis gas formation [7,8]. A lower temperature of syngas formation over reduced catalysts is due to more facile pyrolysis of CH_4 over metallic Ni and Pt. For oxidized catalysts, a higher temperature of syngas evolution is explained by the fact that decomposition of a precursor – Ni-containing perovskite to form metal Ni (Pt) particles has to occur in situ under the reaction mixture action [6]. Thereby, the reduction pretreatment of fresh catalysts is necessary to ensure the rapid start-up of syngas generators at low temperatures.

In all cases, first, products of complete methane oxidation - CO_2 and water are observed. For oxidized and H_2-reduced catalysts, H_2O evolution followed by CO_2 appearance is observed in the presence of oxygen in the gas phase that implies oxidation of hydrogen formed as a result of CH_4 pyrolysis. Then, at increasing temperature, CO always appears in the gas phase even in the presence of oxygen, whereas H_2 is observed only after nearly complete consumption of oxygen except for the reduced LNP catalyst (Fig.1). For the latter case, appearance of H_2 when oxygen consumption is not complete means that syngas formation can occur via the direct route including pyrolysis of methane in which CO and H_2 are formed as primary reaction products. This path is supposed to proceed in parallel with the indirect route involving complete oxidation of CH_4 to CO_2 and water followed by steam and dry reforming of residual methane yielding CO and H_2.

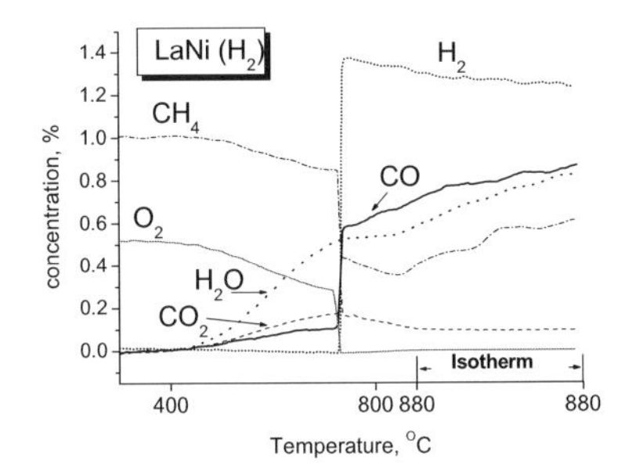

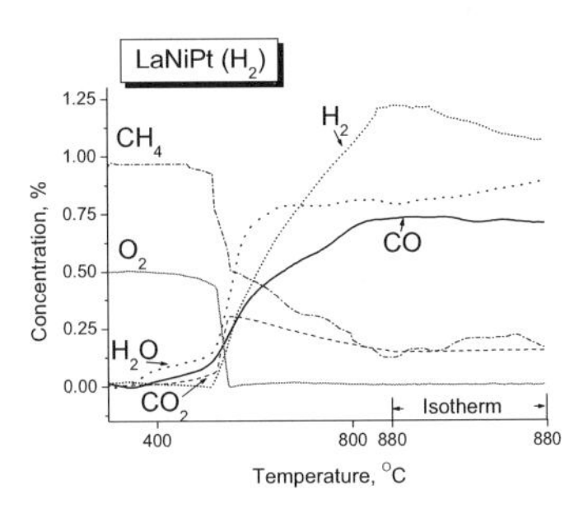

Fig. 1. Temperature-programmed POM reaction over 10 mm channel of H_2-reduced catalysts $LaNiO_x/CeO_2$-ZrO_2/α-Al_2O_3 and $LaNiPtO_x/CeO_2$-ZrO_2/α-Al_2O_3.

Table 1. Efficient first-order rate constants for POM, SR, ATR and DR at 750°C and 7%CH_4 for the catalysts with different active components

Catalyst	Active component composition[5]	Rate constants, s^{-1}				
		POM	SR (3)[1]	SR (1)[1]	ATR[3]	DR
LN	$LaNiO_x$	110	89[2]	0		
LNP	LaNiPt (0.4%Pt)	63	39	16		62[4]
0.4P	0.4%Pt	40	24	19	38	5
0.4PL7	0.4%Pt+La(1:7)	94	35	44	79	44
1.8P	1.8%Pt	41	8.6	4	17	1
1.8PL1	1.8%Pt+La(1:1)	47	14	5	19	9

[1] $H_2O:CH_4$; $H_2O:CH_4$ =2; [3]- 7%CH_4, 21% H_2O, 1.8%O_2; [4]20 mm channel; 7%CH_4, $CH_4:CO_2$=1:1; [5] –all catalysts contained ~7% $Zr0_{0.8}Ce_{0.2}0_2$

3.2. Catalyst activity in dry, steam and autothermal reforming.

Data on SR, DR and ATR for studied catalysts are presented in Table 1 and Fig.2-3. For SR, a methane conversion and rate constant increase at increasing temperature up to 800°C and ratio from 1 to 3 (Table 1). The exception is the catalyst 0.4PLa7 (Fig.2) for which the methane conversion and rate constant decrease in some extent with increasing $H_2O:CH_4$ ratio. Moreover, at the lowest water concentration, this catalyst ensures the highest SR activity among all catalysts (Table 1). At the peak water concentration, the highest rate constant is observed for LN. However, for this catalyst, the increase of methane concentration up to 12% results in the strong decrease of the SR rate constant as compared with that of Pt-promoted catalyst LNP (Fig.4). Such an activity drop for the catalyst containing pure $LaNiO_x$ is caused by profound Ni coking at a high CH_4 concentration whereas addition of Pt hinders it [9]. For SR over all

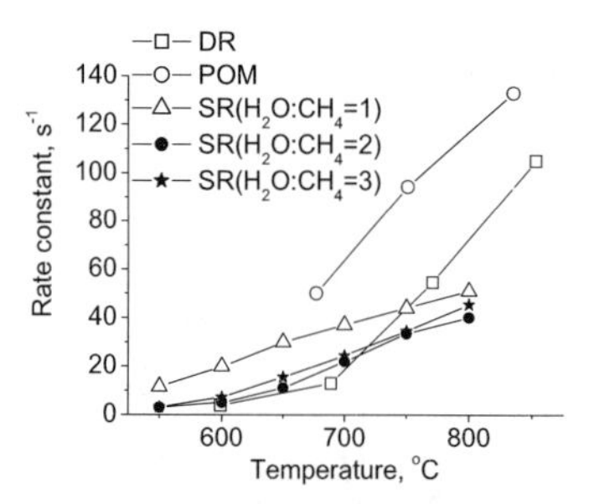

Fig.2. Temperature dependence of the rate constants of POM, DR and SR at different H_2O concentration for 0.4PL7 catalyst.

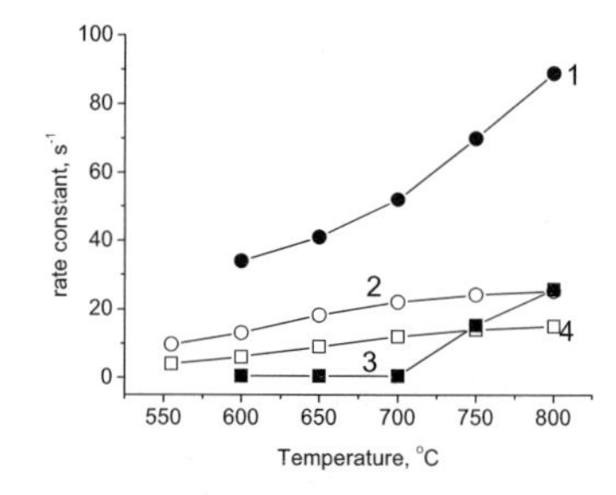

Fig.3. The rate constant of SR versus temperature for LN(1,3) and LNP(2,4) at different CH_4 concentrations: 1, 2 – 7% CH_4; 3, 4 - 12% CH_4. $H_2O:CH_4$ = 2.

catalysts, the addition of oxygen results in an appreciable increase of the rate constant (Table 1), however, syngas selectivity declines especially at a low temperature.

As concerning the catalyst performance in DR, the high constant rates are observed for LNP and 0.4PL7. A lower activity of 1.8P and 1.8PL as compared with La-unpromoted 0.3P could be due to a lower dispersion of Pt in the former catalysts. DR activity of pure Pt-containing samples significantly increases by addition of La, especially in excess. Furthermore, for DR, the effect of La is more pronounced as compared to SR (Table 1). Previously, the enhanced DR activity of Ni/La_2O_3 [10] and $Pt/La\text{-}ZrO_2$ [11] was proposed to be related to decoration of the Ni(Pt) crystallites with lanthanum oxycarbonates species which favor removal of carbon formed under the DR reaction. Therefore, a high activity of LNP and 0.4PL7 catalysts in DR could be due to this function of lanthana.

3.3. Steady-state catalyst activity in POM.

For all catalysts, the conversion of reactants and products selectivity increase with the temperature (Fig.4-5). The oxygen conversion is incomplete only for experiments with the diluted feed containing 3.5% CH_4 at contact time ~ 1 ms and temperature below 650°C.

Values of methane conversion strongly depend on the catalyst composition. Addition of Pt to $LaNiO_x$ decreases methane conversion and syngas selectivity. For Pt-only active components as well as for sample 1.8PtL1, methane conversion does not exceed 35%, and these catalysts show a low syngas selectivity (Fig.4-5) due to a high combustion ability caused by Pt. This results also in decline of the overall methane conversion. For the catalyst 0.4PLa7, a steep rise in methane conversion and syngas selectivity is observed with increasing temperature above 700°C whereas the catalyst LN ensures a high methane conversion and syngas selectivity in the whole temperature range. High activity and selectivity of these catalysts in POM correlate with their effective performance in SR and DR favoured by scavenging of carbon at the periphery of Ni(Pt) metal particles by lanthanum oxycarbonate species. The values of dry and

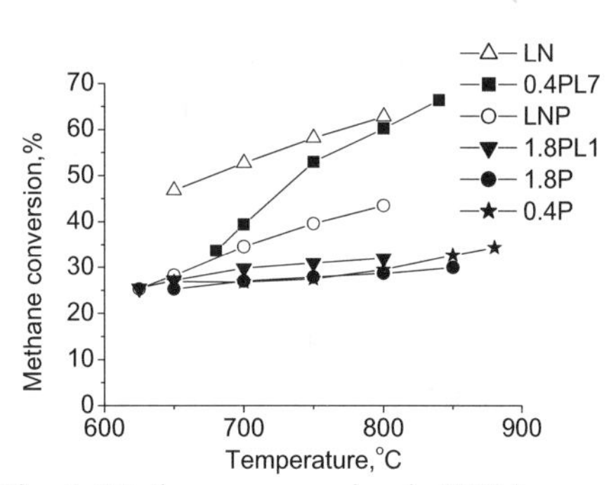

Fig. 4. Methane conversion in POM over catalysts of different composition. 7%CH_4 +3.5%O_2 in N_2. Contact time 3.2 ms

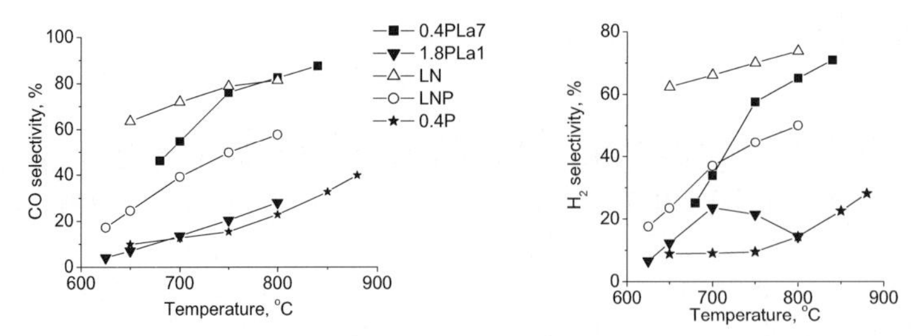

Fig. 5. CO and H_2 selectivity in POM over catalysts of different composition. $7\%CH_4 + 3.5\%O_2$ in N_2. Contact time 3.2 ms.

steam reforming rate constants are comparable by an order of magnitude with those of partial methane oxidation (Table 1). This implies that at least a part of methane could be converted into syngas via the indirect route involving combustion of methane followed by subsequent reforming reactions of remaining methane molecules.

4. Conclusion

The data of POM TPR show that the reduction pretreatment as well as addition of Pt to $LaNiO_x$ facilitates the formation of syngas at a lower temperature, furthermore, Pt favors a high catalyst stability preventing catalyst coking. Comparative study of POM, steam (SR) and dry (DR) reforming of methane over the catalyst channels with supported $LaNiO_3$ (pure or promoted by Pt), $LaPtO_x$ and CeO_2-ZrO_2 supported on the corundum monolith reveals that the most effective catalysts contain $LaNiO_3$ and $LaPtO_x$ with La excess. Therefore, at the inlet part of the catalyst bed, $LaNiO_3$-containing catalysts stable to evaporation in the presence of oxygen at high temperatures should be placed, while, in the rear part, $LaPtO_x$ stable to coking could be used.

References

1. C. Song, Catal.Today 77 (2002) 17.
2. M. Lyubovsky, L. L. Smith, M. Castaldi, et al. Cat. Today 83 (2003) 71.
3. C.G. Bauer, and T.W. Forest, Int. J. Hydr. En. **26** (2001) 50.
4. D. A. Hickman and L. D. Schmidt, J. Catal., 138 (1992) 267.
5. V.A. Sadykov, S,N. Pavlova et al. Kinetics Catal. 46 (2005) 227.
6. S. Pavlova, N. Sazonova, V. Sadykov, et al. Catal. Today 105 (2005) 367.
7. A. P. E. York, T. Xiao, and M. L. H. Green, Topics in Catalysis 22 (2003) 345.
8. O.P.Aghalayam, Y.K. Park, Vlachos, J. Catal. 213 (2003) 23.
9. I. Alstrup, B.S. Clausen, C. Olsen, R.H.H. Smits, J. R. Rostrup-Nielsen, Stud. Surf. Sci. Catal. 119 (1998) 5.
10. V. A. Tsipouriari, Z.Zhang, X.E. Verykios, J.Catal. 179 (1998) 283.
11. S. M. Stagg-Williams, F. B. Noronha, G. Fendley, D. E. Resasco, J. Catal. 194 (2000) 240.

Natural Gas Conversion VIII
F.B. Noronha, M. Schmal, E.F. Sousa-Aguiar (Editors)

DIRECT SYNTHESIS OF PROPANE/BUTANE FROM SYNTHESIS GAS

Kaoru Fujimoto,[a] Hiroshi Kaneko,[b] Qianwen Zhang,[a] Qingjie Ge,[a] Xiaohong Li[a]*

[a]*Faculty of Environmental Engineering, The University of Kitakyushu, 1-1, Hibikino, Wakamatsu-ku, Kitakyushu, Fukuoka 808-0135, Japan*
[2]*The Japan Gas Synthesis Co. LTD.,1-12-10 , Minato-Ku,Tokyo, 105-0003, Japan*

Abstract

Efficient conversion of syngas into LPG over a hybrid catalyst is a novel method for the synthesis of clean fuel from natural gas, coal or biomass. Hybrid catalysts based on methanol synthesis catalyst and zeolite were investigated in a fixed bed reactor. Experimental results demonstrated that the modified Cr-Zn methanol catalyst and β-zolite increased the activity, selectivity and stability of hybrid catalyst. The newly developed catalyst showed one through CO conversion of 81% and selectivity of 78% and 15% for LPG and naphtha, respectively at 370 °C and 5.1 MPa. Bench plant is now under operation.

1. Introduction

Syngas (H_2/CO) production from natural gas or coal is the established technology. The LPG synthesis from syngas would be most promising but remain unexploited technology. Fig. 1 shows relationship between above technologies.

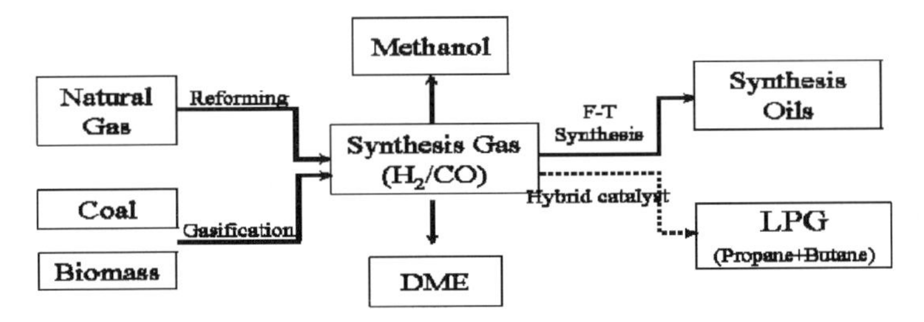

Fig. 1 Three Routes for LPG

The syngas, which is commercially produced by reforming of natural gas or gasification of coal or biomass, can be converted into hydrocarbons (LPG)

through methanol or DME as an intermediate in above processes.
As we have reported [1-4], LPG was directly synthesized from syngas over a hybrid catalyst which composed of methanol synthesis catalyst and zeolite. The hybrid catalyst that composed of Cu-Zn methanol synthesis catalyst and Y-type zeolite showed a good initial activity and selectivity for synthesis of LPG. However, its stability remained to be improved because the Cu-Zn catalyst deactivated under an atmosphere of CO_2 and water at high temperature. The hybrid catalyst consisted of Pd-based and β-zeolite showed a higher stability and activity than Cu-based catalyst. High catalyst cost is caused by high content of Pd in methanol catalyst would be obstructive to commercial application of this new process.
In this work, the performance of hybrid catalyst based on the modified Cr-Zn and β-zeolite by palladium was investigated, and a highly stable and active hybrid catalyst for directly synthesis of LPG from syngas was presented.

2. Results and discussion

2.1. The performance of hybrid catalysts for synthesis gas to LPG

Differently from F-T synthesis reaction mechanism[5-7], LPG was directly synthesized from syngas through methanol or DME as an intermediate over a hybrid catalyst. The hybrid catalyst exhibits two functions: methanol synthesis and selective conversion of methanol into hydrocarbons. Because of the synergistic effect of methanol synthesis catalyst and zeolite, the coexistence of both steps, methanol synthesis and methanol conversion in the single reactor, remove the limitation of thermodynamic equilibrium of methanol formation from syngas and CO conversion reaches high level (almost 100%) at low reaction pressure.
Usually the reaction temperature of conversion of methanol or DME into hydrocarbon was higher than 300°C, and at that high temperature Cu-Zn methanol synthesis catalyst suffer from the deactivation with time on stream. Supported palladium catalyst for methanol synthesis on the other hand show higher stability but lower activity than Cu-Zn methanol catalyst, and addition of calcium into supported palladium promoted its activity.
Chromium and zinc oxide as methanol catalyst requires high temperature and pressure because of low activity for methanol synthesis. But it is more stable than the catalyst based Cu-Zn oxides at high temperature and high partial pressure of CO_2 and H_2O[8,9].
The hybrid catalyst employed is a mixture of Cr-Zn methanol synthesis catalyst and β-zeolites with a ratio of silica to alumina between 37 and 400.
Zeolite containing of Pd, was prepared by ion exchange method.
A pressurized flow type reaction apparatus with a fixed bed tubular reactor with an inner diameter of 6mm was used for this study. Details of the reaction

procedures and the product analysis have been described elsewhere.
Table 1 shows the promotion of LPG synthesis by the modification with small amount of palladium into Cr-Zn oxide. As Table 1 shows, Pd-free catalyst showed a low activity, because Cr-Zn had low activity for methanol synthesis at 2.1MPa. Added palladium in Cr-Zn resulted in the high activity for LPG synthesis, while keeping the LPG selectivity ((C_3+C_4) % in hydrocarbons) almost constant. That demonstrated that the small amount of Pd on Cr-Zn and β-zeolite benefited the activity and selectivity of hybrid catalyst for synthesis of hydrocarbons.

Table 1: The performance of hybrid catalysts as a function of palladium content in Pd/Cr-Zn

Catalyst	A	B	C	C	D
Reaction pressure (MPa)	2.1	2.1	2.1	5.1	5.1
CO Conversion (%)	21.4	23.9	33.9	66.8	72.0
Product Yield (C%)					
Hydrocarbons	12.5	12.6	20.6	40.3	46.8
DME	0.6	0.0	0.0	0.03	0.11
CO_2	8.9	11.2	13.3	26.5	25.1
Hydrocarbon Distribution (C%)					
C_1	5.0	3.9	3.8	6.4	2.3
C_2	11.5	10.0	8.9	8.0	3.0
C_3	45.1	45.2	48.2	47.4	37.4
C_4	31.2	33.9	31.9	31.3	40.6
C_5	5.4	5.5	5.4	5.2	9.7
C_{6+}	1.8	1.5	1.6	1.7	6.0
LPG(C_3+C_4)	76.3	79.1	80.2	78.7	78.0

Note: W/F=9.0g·h/mol, 2.1MPa, 375℃, feed: Ar/H_2/CO/CO_2=3/65/24/8,
Catalyst A no Pd, Catalyst B,C,D contains small amount of Pd

2.2. The optimization of reaction conditions over hybrid catalysts

Compared with Cu-Zn, Cr-Zn showed a lower activity for methanol synthesis from syngas. The increase in the reaction pressure would benefit the formation of methanol. As Fig.2 shows, a high reaction pressure promoted the CO conversion, yield of hydrocarbons, and selectivity for LPG. Also, the yield of methane decreased at high reaction pressure. This is favorable because methane is the most unfavorable product in this process. Therefore, the medium reaction pressure (3.0-6.0MPa) was suitable for the LPG synthesis from syngas, because the high catalytic activity is essential for the commercial application.

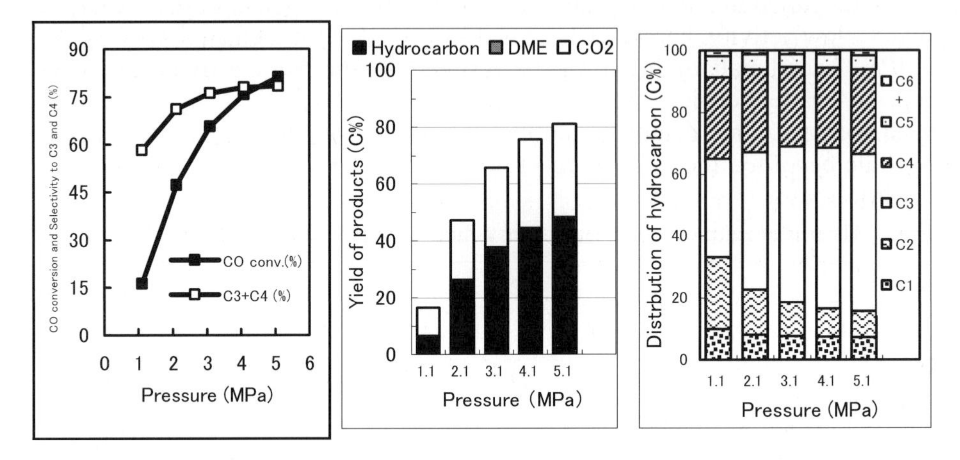

Fig. 2 Effect of reaction pressure
Reaction conditions: 400°C, 8.9 mol/(h.g catalyst), Syngas: Ar: CO: H2 = 3: 32: 65, Catalyst: 1.0g

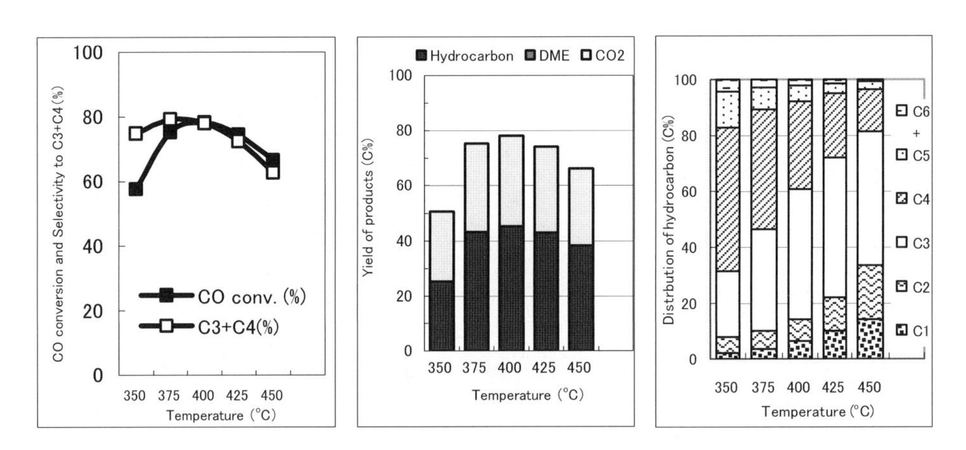

Fig. 3 Effects of reaction temperature
Reaction conditions: 5.1MPa, 8.9 mol / (h.g catalyst), Syngas: Ar:CO:H_2=3:32: 65, Catalyst: 1.0g

Fig.3 shows the CO conversion and the hydrocarbon distribution as a function of reaction temperatures. When the reaction temperature was raised, the selectivity of C_1 and C_2 hydrocarbons increased while C_4, C_5 and C_6+ hydrocarbons decreased. The CO conversion reached the maximum at 400℃ while the selectivity for LPG kept at high level. However, the high reaction temperature was not favored by equilibrium yield of methanol from syngas, and the rate of methanol formation become the rate control for hydrocarbon synthesis from synthesis gas over hybrid

catalyst. Low concentration of methanol and DME, as intermediates from synthesis gas to hydrocarbons, would restrict the rate of hydrocarbon formation. Thus high reaction temperature decreased the CO conversion. Temperature also showed a worked effect on the distribution of hydrocarbon products. At low temperature, the average molecular weight of the product hydrocarbons was larger than that at high temperature, where direct hydrogenation of CO into methane or the hydro-cracking of higher hydrocarbon catalyzed by Pd may result in the high yield of methane. Also, high temperature enhanced cracking rate of C_4-C_6+ hydrocarbons, resulting in the decrease in the heavy hydrocarbons selectivity, which was deleterious for the activity of zeolite. So the appropriate temperature of reaction is 370-400°C.
The flow rate of syngas played a main role in the CO conversion and distribution of product hydrocarbons. As Fig.4 shows, the CO conversion increased with an increase of W/F as usual while keeping the total selectivity for LPG almost constant. The selectivity for light hydrocarbons (C_1, C_2 and C_3) increased while the selectivity for the heavy hydrocarbons (C_4, C_5 and C_6+) decreased when the contact time(W/F) increased, probably because of secondary reaction of C_5+ hydrocarbons. That the low W/F also resulted in a low CO conversion and low the selectivity for LPG fraction was attributed to the formation rate of methanol from synthesis gas was the rate-determining step in the total process of synthesis of hydrocarbons.

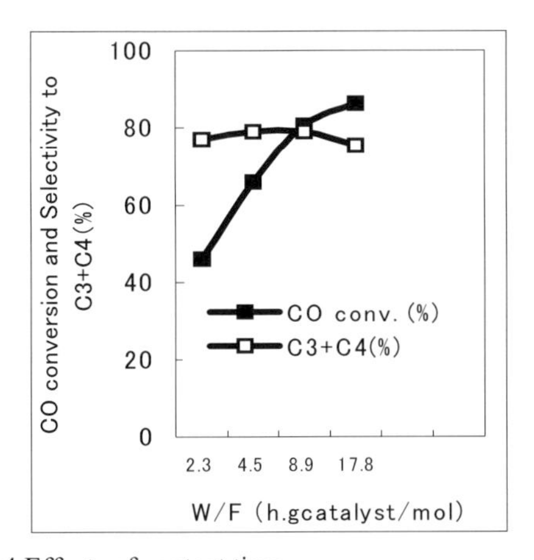

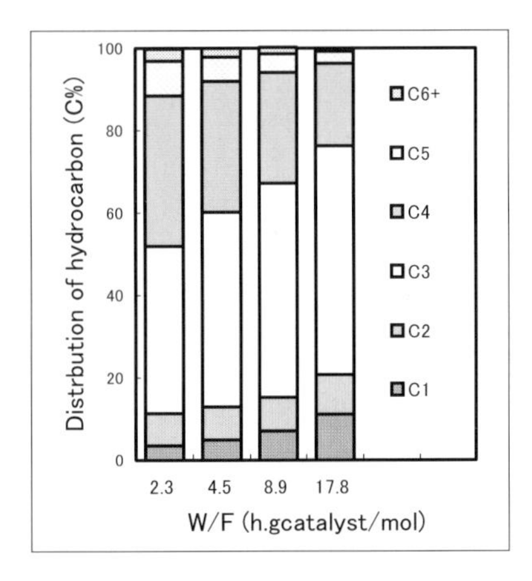

Fig.4 Effects of contact time.
Reaction conditions: 400°C, 5.1MPa, Syngas: Ar: CO: H_2 = 3: 32: 65, Catalyst: 1g

2.3. The catalytic stability of hybrid catalysts

The stability experimental results (Fig. 5) exhibit that the Zn-Cr hybrid catalyst show the good catalytic stability, the CO conversion and LPG selectivity could keep >65% and 75%, respectively, during 150 hours operation. The high selectivity to LPG over low content Pd-modified hybrid catalyst was attributed to three dimensions structure and large pore size of β -zeolite. Olefins (propylene and butenes) in product hydrocarbons formed in the cage of zeolite would easily leave the active site and diffused to the surface of palladium for hydrogenation reaction. Hydrogenation of olefins would stop the growth of carbon chain and decrease the yield of heavy hydrocarbons and coke. There was no olefin found in the product, because the intermediate olefins produced from methanol or DME were hydrogenated in situ catalyzed by Pd.

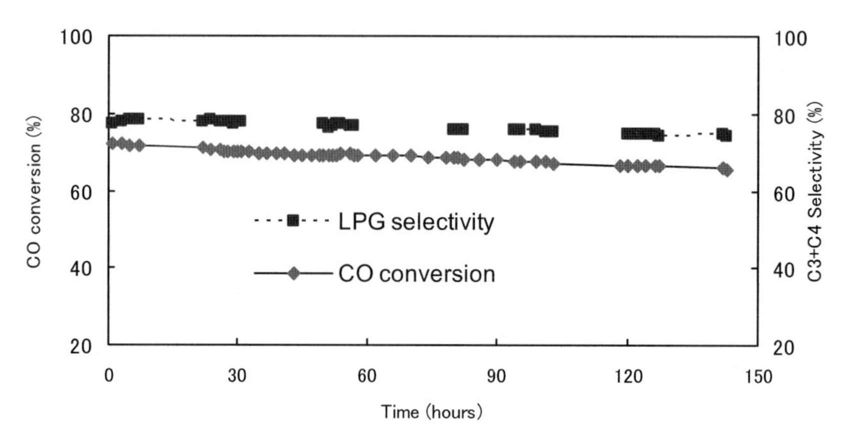

Figure 5. The stability experiments of Zn-Cr hybrid catalyst at 375°C, 5.0MPa, 9 g·h/mol.

3. Summary

The modification of Cr-Zn catalyst by Pd increased the activity and selectivity for LPG synthesis and modification of β -zeolite increased stability of the hybrid catalyst. The medium reaction pressure (3.0-5.0MPa) was suitable for LPG synthesis from syngas using the hybrid catalyst giving 78% CO conversion and 79% LPG and 15% light naphtha selectivity. Based on these results, a bench scale plant whose capacity is 1Kg/day is now under operation.

Reference

1. Q. Zhang, X. Li, K. Asami, K. Fujimoto, Catal. Lett., 2005, 102: 52
2. Q. Zhang, X. Li, K. Asami, S. Asaoka, K. Fujimoto, Fuel Processing Technology, 2004, 85: 1139
3. K. Asami, Q. Zhang, X. Li, S. Asaoka, K. Fujimoto, Stud. Surf. Sci. Catal, 2004, 147: 427
4. Q. Zhang, X. Li, K. Asami, K. Fujimoto, Catal. Today, 2005, 104: 30
5. L. Fan, K. Fujimoto, Appl. Catal. A., 1993, 106: 1.
6. M. Stocker, Micropor. Mesopor. Mat., 1999, 29: 3.
7. L. Jean-Paul, Catal. Today, 2001, 64: 3.
8. K. Fujimoto, Y. Yu, Stud. Surf. Sci. Catal., 1993, 77: 393
9. I. Melian-Cabrera, M. Lopez Grannados, L. G. Fierro, J. Catal., 2002, 210: 285.

Natural Gas Conversion VIII
F.B. Noronha, M. Schmal, E.F. Sousa-Aguiar (Editors)

LPG Synthesis from DME with semi-direct method

Wenliang Zhu[1], Xiaohong Li [1]*, Hiroshi Kaneko[2], Kaoru Fujimoto[1]

1.Department of Chemical Processes and Environments, Faculty of Environmental Engineering,University of Kitakyushu, Kitakyushu, Fukuoka, 808-0135, Japan
E-mail: lixiaohong@env.kitakyu-u.ac.jp Tel.: +81 93 695 3378; Fax: +81 93 695 3376
2. The Japan Gas Synthesis Co. LTD

Abstract

The study of liquefied petroleum gas (LPG) synthesis from syngas via dimethyl ether (DME) was studied in a fixed-bed reactor. ZSM-5 and Beta zeolite were tested for selectively converting DME to LPG fraction. Results demonstrate that both zeolites have good activity for the conversion of DME, but posssess different hydrocarbon distribution. Moreover, the ratio of SiO_2/Al_2O_3 has great effect on the performance of Pd-Beta zeolite in the conversioon of DME to LPG and Pd-Beta zeolite with high ratio of SiO_2/Al_2O_3 possesses very good LPG fraction. At the optimized reaction conditions, good catalytic performances of Pd-Beta are obtained with 100% DME conversion and around 85% LPG selectivity, especially only around 0.3% COx in products.

Key words: LPG synthesis, DME conversion, ZSM-5, Beta zeolite, syngas.

1. Introduction

Since the decrease of petroleum reserve and the increase of its price accordingly, energy security has been receiving high attention recently. Besides the exploration of more advanced energy technologies (e.g., fuel cells), conversion of natural gas, biomass, or coal into high-quality fuels is an effective way to reduce or eliminate the dependency on petroleum. At present, the conversion of natural gas or coal via syngas (H_2+CO) to liquid fuels, i.e. gas-to-liquid (GTL), is successful [1, 2], but to LPG is a new project. The technology of LPG synthesis from syngas is great promising but remains unexploited.

There are three ways for LPG synthesis from syngas, The first one is the synthesis of LPG from syngas directly; the second one is semi-direct synthesis of LPG from syngas via methanol/DME which are produced from syngas; the third one is indirect method through olefins from methanol or DME. This work is LPG synthesis from DME with semi-direct method.

DME can be used as a clean high-efficiency fuel and is considered as a promising alternative of LPG. However, although its physical properties are similar to those of liquefied petroleum gas, the infrastructures for its application should be modified when it is used for household cooking and heating. Therefore, it is impossible to be popularized on a large scale in a short term. Since methanol or DME have been commercially produced from syngas, it become feasible for LPG synthesis in large scale from syngas via DME.

In our previous works, LPG can be synthesized from DME over a hybrid catalyst consisting of hydrogenation catalyst and zeolite [3]. However, in those cases, LPG selectivity was low and much COx was produced simultaneously, which would reduce the yield of LPG. In this work, a novel catalyst based on Beta zeolite with palladium was developed and good performances of high activity and selectivity for LPG synthesis from DME was presented.

2. Experimental

ZSM-5 and Beta zeolites with different ratios of SiO_2/Al_2O_3 were used in the study and catalysts were prepared by ion-exchanged method in palladium nitrate solution. All predecessors of catalysts were dried usually at 120℃ and calcined at 550 ℃ for 3 hours. A tubular reactor with 6 mm inner diameter was used in the reactions and 0.3 g catalyst with 20/40 mesh was packed in the middle of reactor. The feedstock was the mixture of DME and hydrogen. Before the introduction of the feedstock into reactor, catalysts were activated in pure hydrogen at 385 ℃ for 3 h. The products in gaseous state were analyzed by two on-line GC with FID and TCD detector respectively. Argon in the feedstock was used as an internal standard for calculation.

3. Results and discussion

With respect to the investigations of methanol-to-hydrocarbons reaction, the medium-pore zeolite ZSM-5 has been an excellent catalyst in methanol-to-gasoline (MTG) reaction [5, 6]. Small-pore zeolite SAPO-34 with middle acidity is considered as a very promising catalyst for the light olefins production [7, 8]. As in the study of methanol to hydrocarbons, zeolite-type materials are considered as a favorable alternative catalyst for LPG synthesis from DME. Since LPG is thc mixture of propane and butane, medium-pore zeolite is considered as a very promising candidate for LPG synthesis from DME. As pointed out by Jin in the investigation of LPG Synthesis from methanol/DME over various kinds of H-form zeolite, different zeolites possessed different hydrocarbon distributions and medium-pore ZSM-5 was suggested as a promising catalyst for the production of desired (C3+C4)-enriched hydrocarbons [4]. Therefore, ZSM-5 was first used as a catalyst to investigate LPG synthesis from DME.

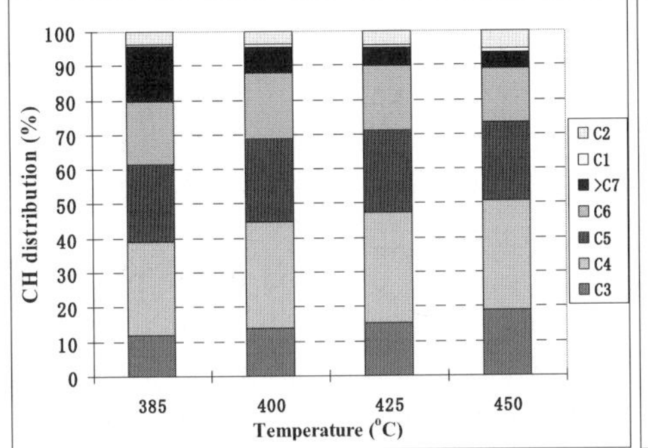

Fig.1. the effect of temperature on CH distribution with H-ZSM-5 as catalyst

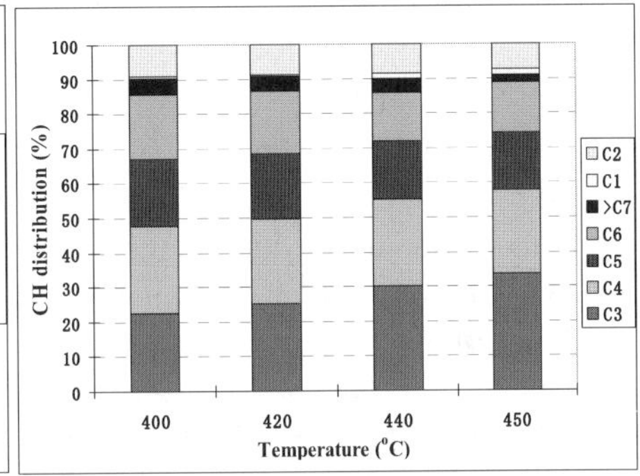

Fig.2. the effect of temperature on CH distribution with 0.2%Pd-ZSM-5 as catalyst

Fig. 1 is the temperature dependence on the hydrocarbon distribution in the DME conversion reaction over H-ZSM-5 zeolite at the pressure of 2.0MPa, H_2/DME of 10 (the molar ratio of H_2/DME) and W/F of 1.0 (residence time, g cat. h/mol). Since there was no hydrogenation catalyst in the reaction, $C_3^=$ and $C_4^=$ olefin are taken as LPG. It can be seen that LPG selectivity (C3+C4) increases from 40% to 50% with the increase of reaction temperatures from 385 to 450 ℃ respectively, but not so high. This is due to that too much C5, C6 and C7 were produced in the DME conversion. However, the selectivity of C1+C2 is low, around 2-3%.

It has been clear that the main steps of methanol to hydrocarbons pass through the intermediates of lower carbonic olefins [13]. If olefin is hydrogenated over hydrogenation catalysts in time, the compositions in the products will be just paraffin. The schematic reaction of DME to LPG is as follows:

CH_3OCH_3 == olefin + paraffin +H_2O

Therefore, besides pore structures of zeolites which hydrocarbon distribution depends strongly on, the hydrogenation function of catalyst is another important factor for hydrocarbon distribution in LPG synthesis. If the hydrogenation function is too strong, lower carbonic olefins would be hydrogenated into paraffin before not binding together or only binding together into C2 carbonic olefins and thus main composition in the products would be methane or ethane.

On the contrary, when hydrogenation function is weak, the molecular binding between intermediate hydrocarbon species cannot stop in time, heavy hydrocarbons, even aromatic serials would be produced. In such case, hydrocarbons in products would be paraffin, olefins and aromatic, which accorded with the results in the methanol to gasoline. Since the objective of this study is the synthesis of LPG, the distribution of hydrocarbons should be very narrow and especially focus on C3 and C4 paraffin. Therefore, it is very important for the catalyst to possess a proper hydrogenation function.

Fig. 2 shows the temperature dependence of carbon distribution over 0.2%Pd-ZSM-5. Compared with the results over H-ZSM-5, hydrocarbon distribution changes limited. The selectivity of LPG is still not so high, only 48% to 58% with the increase of temperatures from 385 ℃ to 450 ℃ respectively. Moreover, too much C5 and C6 are still detected in the products. However, the selectivity of heavy hydrocarbons (Carbon number >7) drop and, on the contrary, ethane selectivity increases obviously. Although the production projects towards either light olefins or motor fuels could be carried out by regulating catalysts and engineering conditions in terms of the different targets [11], the selectivity of LPG in this case is low and there is a need to find new catalyst.

At present, studies about light hydrocarbons production has moved away from large-pore molecular sieves considering the effect of shape selectivity in zeolites [9, 10]. LPG is the mixture of propane and butane and it is better to be synthesized over small-pore or medium-pore zeolite. Indeed, there are a few studies were reported over large-pore zeolite, such as USY, SAPO-5 or Beta-type zeolites for light hydrocarbons production [1, 3, 4, 12]. Now, we try to synthesize LPG over large pore Beta zeolite. Catalyst Beta-A and Beta-B with different ratio of SiO_2/Al_2O_3 were prepared by ion-exchange method in the palladium nitrate solution with same concentration.

Fig. 3 and Fig. 4 show the results of DME conversion over Pd-Beta catalysts A and B as a function of the temperature at the pressure of 2.0MPa, H_2/DME of 10 and W/F of 1.0. They demonstrate very different catalytic performance. Since there is some palladium on the catalyst, no olefin was detected in the products.

Figure 3 is the results of catalyst Pd-Beta-A with low SiO_2/Al_2O_3 ratio. The results show that, at the low temperature of 385 ℃, the main products was methane, around 97.5% and LPG (C3 + C4) is only 2.2%. However, with the increase of temperature from 385 to 430 ℃, the selectivity of methane drops sharply, from 97.5% to 27.3%; simultaneously, LPG selectivity increases from 2.2% to 60%.

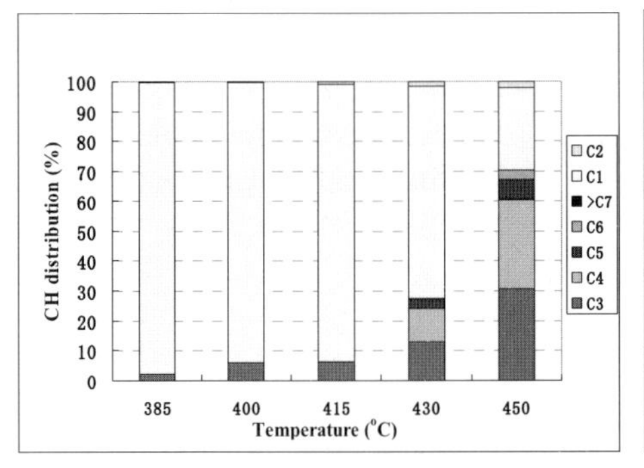

Fig.3. the effect of temperature on the CH distribution over Pd-Beta-A as catalyst

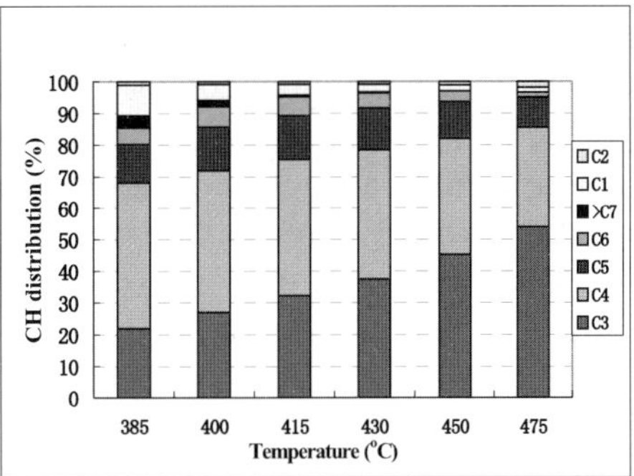

Fig.4. the effect of temperature on the CH distribution over Pd-Beta-B as catalyst

In order to further study the catalytic performance of Pd-Beta zeolite, the temperature dependence of hydrocarbon distribution over Pd-Beta-B with high ratio of SiO_2/Al_2O_3 was investigated and results is shown in Fig.4. Compared with the results over Pd-Beta-A, the catalytic performance of Pd-Beta-B changed greatly. Even at low temperature of 385 ℃, methane selectivity was very low, only 4.1%, and continuously dropped as the increase of reaction temperatures. At 450 ℃, methane selectivity was only 2.4%. In addition, the selectivity of LPG increased in the whole range of temperature. At 475 ℃, it was 85.3% with 100% DME conversion. When the temperatures are above 450 ℃, the selectivity of methane increases a bit again. This phenomenon attribute to the strong cracking performance of Beta zeolite for heavy hydrocarbons at high temperatures.

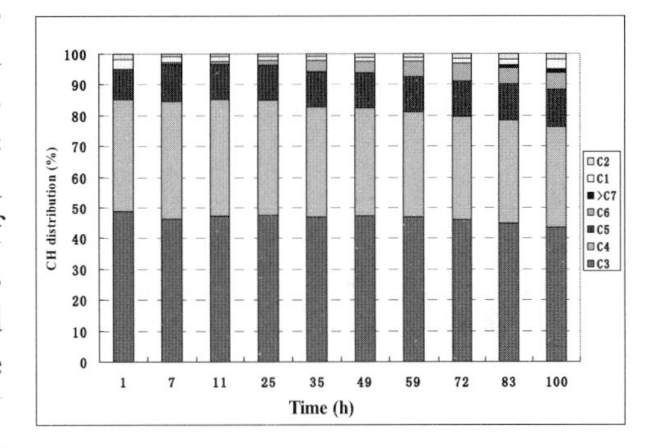

Fig. 5. 100 h results of LPG synthesis from DME over Pd-Beta-B zeolite

From above results, there is a need for further studies to explore the reasons why LPG selectivity increases sharply with the increase of reaction temperature over Pd-Beta zeolites and why Pd-Beta zeolites with different ratio of SiO_2/Al_2O demonstrates such different catalytic performances. Maybe all of these link to the strong acid side and large pore of Beta zeolite.

Figure 5 is the results of long-term test of Pd-Beta-B zeolite for the conversion of DME to LPG. During 100 h run, the reaction is relative stable at the optimized reaction conditions with temperature of 450 ℃, pressure of 1.0Mp, W/F of 2.0 and H_2/DME of 19. 100% DME conversions and 86–78% LPG selectivity are achieved, and especially, only around 0.3% CO_x in products.

4. Conclusion

The catalytic performance of ZSM-5 and Beta zeolite were evaluated as LPG synthesis catalysts in the DME conversion reaction. ZSM-5 catalysts demonstrate low LPG selectivity even at high reaction temperatures. Beta zeolites with different ratios of SiO_2/Al_2O_3 show very different catalytic performances at different reaction temperature. Furthermore, catalyst Pd-Beta-B zeolite exhibits a good activity and a favorable stability during 100 h run. In addition, there is very small COx detected in products.

Acknowledgements

The Japan Gas Synthesis financially supported this study and the authors gratefully acknowledge their financial support.

Reference

[1] S. Michael, Micropor. Mesopor. Mater. 29 (1999) 3.
[2] K. Cyril, Catal. Today 71 (2002) 437.
[3] Kenji Asami, Catal. Today 106 (2005) 24
[4] Yingjie Jin, Sachio Asaoka et.al., Fuel process technol, 85 (2004) 1151.
[5] M. Kang, J. Mol. Catal. A, Chem. 150 (1999) 205.
[6] T.C. Xiao, L.D. An, H.L. Wang, Appl. Catal. A, Gen. 130 (1995) 187.
[7] M. Kang, M.H. Um, J.Y. Park, J. Mol. Catal. A, Chem. 150 (1999) 195.
[8] M. Kang, C.T. Lee, J. Mol. Catal. A, Chem. 150 (1999) 213.
[9] M. Stocker, Microporous Mesoporous Mater. 29 (1999) 3.
[10] F. Salehirad, M.W. Anderson, J. Catal. 29 (1996) 117.
[11] G.F. Froment, W.J.H. Dehertog, A.J. Marchi, Catalysis 9 (1992) 1.
[12] Ø. Mikkelsen, S. Kolboe, Microporous Mesoporous Mater., 29 (1999) 173
[13] Michael Sto¨cker, Microporous Mesoporous Mater., 29 (1999) 3–48

Natural Gas Conversion VIII
F.B. Noronha, M. Schmal, E.F. Sousa-Aguiar (Editors)

Performance of monolithic catalysts with complex active component in partial oxidation of methane into syngas: experimental studies and modeling

V. Sadykov[a], S. Pavlova[a], Z. Vostrikov[a], N. Sazonova[a], E. Gubanova[a], R. Bunina[a], G. Alikina[a], A. Lukashevich[a], L. Pinaeva[a], L. Gogin[a], S. Pokrovskaya[a], V. Skomorokhov[a], A. Shigarov[a], C. Mirodatos[b], A. van Veen[b], A. Khristolyubov[c], V. Ulyanitskii[d]

[a]*Boreskov Institute of Catalysis SB RAS, pr. Lavrentieva, 5, Novosibirsk, 630090, Russia*
[b]*Institut de Recherches sur la Catalyse - CNRS, 2 av. Albert Einstein, 69626 Villeurbanne Cedex – France*
[c]*VNIIEF, Sarov, 607190, Russia.*
[d]*Lavrentiev Institute of Hydrodynamics, 630090, Novosibirsk, Russia.*

1. Introduction

Partial oxidation of methane (POM) on monolithic catalysts at short contact times is a promising process for design of compact syngas generators. As was demonstrated for Rh- or Pt- supported catalysts [1-3], optimization of their performance requires process modeling based upon a detailed elementary step reaction mechanism verified for pure metals. For more complex active components such as Pt-promoted $LaNiO_3$ /Ce-Zr-O etc [4], elucidation of such detailed elementary kinetics would require too extensive research. This work presents a verification of more simple approach to modeling of both steady-state and start-up performance based upon using the rate constants for the reactions of methane selective oxidation and reforming reactions estimated for small separate units of monolithic catalysts (channels etc) in nearly isothermal conditions.

2. Experimental

Two types of honeycomb monolithic substrates based on corundum (a hexagonal prism with a side of 40 mm and triangular channels with wall thickness of 0.2– 0.3 mm) or fechraloy foil (cylindrer 50 mm diameter, 200 - 400 cpsi, 20 μm foil thickness) were used. The metal surface is protected by a thin (~10 μm) nonporous layer of corundum supported by the dust blasting [4]. The active component comprised of mixed $LaNiO_3$/Ce-Zr-La-O oxides (up to 15 wt.%) and Pt (up to 0.5 wt.%) was supported via washcoating and/or impregnation procedures followed by drying and calcination [4].
The monolith catalysts (50 mm length) along with the front and back-end thermal shields (2 cm each) were placed into a tubular stainless-steel reactor. The axial temperature at selected points along the monolith was scanned by thermocouples located in the plugged central channel. The linear velocity of the feed was varied in the range of 0.5-6 m/s. The feed was comprised of natural gas (NG) - 22-29 vol.%, air – balance. To ignite the process, the air-NG mixture preheated up to 400°C was fed to the reactor warmed to the same temperature in air. The gas composition was monitored by GC, MS and IR absorbance gas analyzers.
Separate structured elements (triangular channels of corundum monolith, a roll of Fe-Cr gauze (wire diameter 0.2-0.3 mm, square mesh ~ 0.5x0.5 mm, length 10 mm) or a wire spiral (wire diameter 0.5 mm, external spiral diameter 3 mm, length 16 mm) with supported protective corundum sublayer and the same type of mixed active component Pt/$LaNiO_3$/Ce-Zr-La-O were tested in a quartz reactor of 4 mm inner diameter at contact times 1-15 ms using feeds containing 1-15% of methane and required amount of oxidants (O_2, CO_2 and H_2O) both in the stationary and temperature-programmed modes. Overall loading of active components on the gauze was ~ 5 wt. %, and on spiral ~2 wt. %, i.e. several times smaller than that for corundum channels (~18 wt. % [5]).

3. Modeling

Reactor was assumed to be adiabatic and of a plug-flow type. One-dimensional model containing detailed mass and energy balances for the gas phase and the monolithic catalyst [6] was used. The kinetic model of partial methane oxidation includes stages of complete methane oxidation, steam reforming and water gas shift reaction. The rate constants and activation energies for these reactions were estimated from results of isothermal experiments with triangular corundum-based channels using rate equations of the type suggested in [7-9]. In POM modeling, the reaction rates were calculated using the surface concentrations of reagents and the catalyst temperature. A set of non-linear algebraic equations for the gas and surface concentrations of components was solved by a stabilization method. Both steady-state and transient concentration and temperature profiles were obtained and compared with experimental data for the real-size monoliths.

4. Results and discussion

4.1. Performance of separate elements

At 600-700 °C for all types of elements, the oxygen in the CH_4+O_2 feed was not completely consumed. In these conditions a higher CH_4 conversion and syngas yield were observed for the corundum-supported element (Fig. 1) with a big loading of the active component distributed within porous walls. This feature can be explained by syngas generation via steam reforming (SR) in pores where O_2 is depleted. At high (~900 °C) temperatures where all O_2 is consumed in the inlet part of an element, the highest CH_4 conversion and syngas yield were obtained for elements on metallic substrates with a low loading of active components comprised of thin (~10 microns) layers This suggests an important role of the heat transfer from the inlet part (where O_2 is consumed) into the rear part where endothermic reactions of CH_4 reforming occur. Similarly (Fig. 2), a high-temperature performance of the same active component in the CH_4 dry reforming (DR) is higher when it is supported onto a gauze, which helps to avoid heat-transfer limitations. When O_2 is added to the feed, performance strongly declines for gauze-supported active component due to suppression of CH_4 activation. This phenomenon is less strongly expressed for porous corundum-based structural element due to more reduced state of the active component located within pores of substrate. For structural elements based upon corundum channels, rather close rate constants for POM, SR and

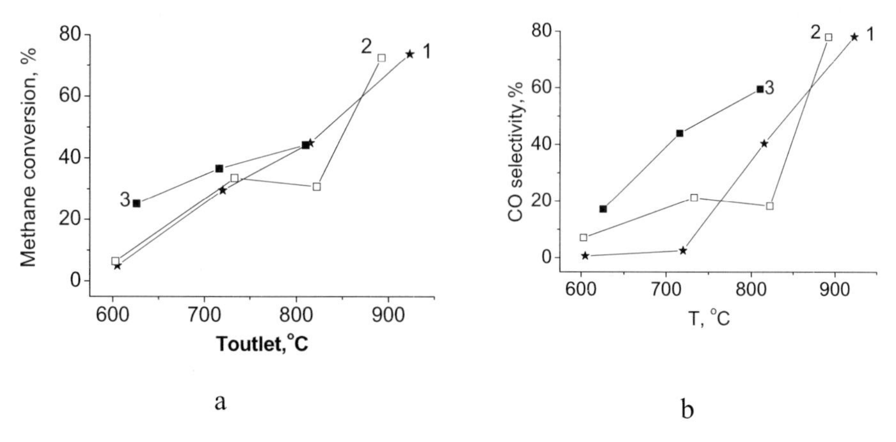

Fig. 1. CH_4 conversion (a) and CO selectivity (b) vs T in POM on spiral (1), gauze (2) and corundum channel-based (3) catalytic elements. Contact time 8 ms, feed 7%CH_4 +3.5%O_2

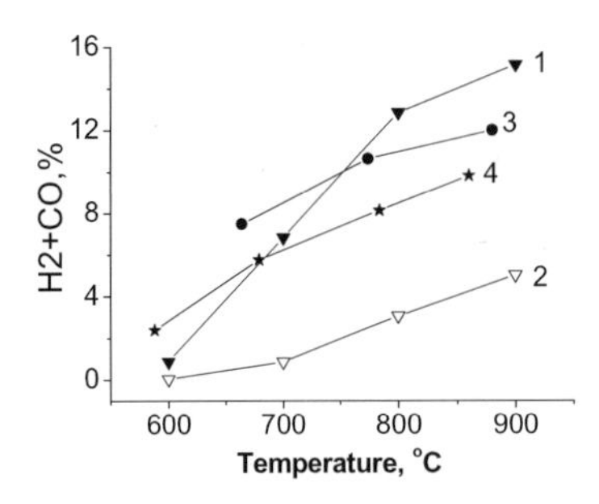

Fig. 2. Syngas content in converted feed vs T for structural elements based upon gauze (1,2) or corundum channel (3, 4) in CH_4 DR (1, 3) or oxy-DR (2, 4). Inlet feed 7%CH_4+7%CO_2 (1,3) or 7%CH_4+7%CO_2 + 1.75% O_2 (2,4), contact time 8 ms.

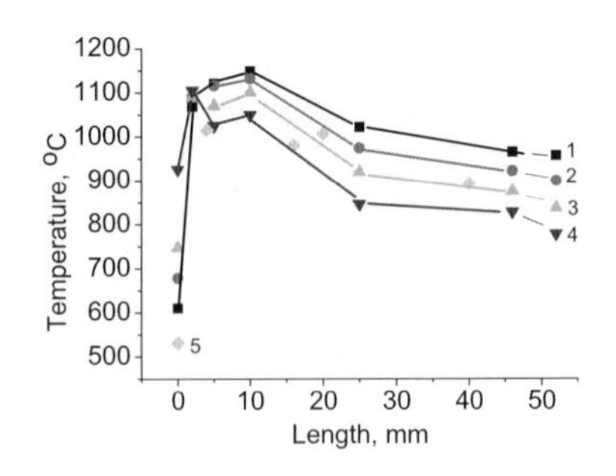

Fig. 3. Temperature profiles for monolithic catalysts on corundum (1-4) or fechraloy (5) substrates. Contact time (s): 1-0.02, 2-0.04, 3-0.06, 4-0.1, 5-0.03. 27 % NG in air, no back thermal shield, T at 52 mm measured in the gas phase after monolith.

DR of CH_4 were obtained varying in the range of 50-200 s^{-1} at temperatures in the range of 650-800 °C [5]. Since at temperatures ~ 600 -800°C the mass transfer coefficient for the triangular channel was estimated to be ~ 600-800 c^{-1}, this suggests that for POM the impact of the mass transfer could be appreciable.

4.2. Steady-state performance of monolithic catalysts and its modeling

In the autothermal mode, performance of catalysts on corundum and fechraloy foil substrates is practically identical despite a lower content of the active component for the latter case. At the inlet gas temperature of 400°C and the linear velocity in the range of 0.5-2 m/s, methane conversion and CO + H_2 yield are nearly constant being equal to ~85-90% and 50-53 vol.%, respectively. These performance characteristics are comparable to those for much more expensive Rh-containing monolithic catalysts [2,3]. No deactivation of performance with time –on-stream was observed.The temperature profiles along the catalyst length (Fig. 3) are rather similar for both types of monolithic substrates. In the narrow inlet part of the monolith highly exothermal reaction of methane combustion occurs leading to a steep rise of temperature. After reaching the maximum at a position depending upon the space velocity (Fig. 3), the temperature of the catalyst declines due to the heat consumption by endothermic steam and dry reforming reactions. The outlet temperature increases with the space velocity reflecting overall increase of the heat generated within the reactor in the nearly adiabatic mode. Less steep temperature profile (from T_{max} to T_{end}) for the catalyst on the fechraloy substrate implies some impact of the heat transfer by conduction along the metal foil. As revealed by modeling (Fig. 4), for the main part of the monolith the catalyst temperature exceeds that of the gas phase thus demonstrating importance of the heat and mass transfer processes. The peak temperature is correctly reproduced only with a due regard for steam reforming occurring even in the presence of gas-phase oxygen, apparently within pores of corundum substrate (vide supra). This reaction is also favored by a big difference between the concentration of reagents (especially, oxygen) in the gas phase and at the surface of the channel due to the effect of the diffusion-controlled mass transfer (Fig. 5).

4.3. Ignition characteristics and their modeling

Under conditions of high-temperature POM process, complex perovskite-like $LaNi(Pt)O_3$ oxide is reduced yielding small highly reactive Ni^o particles promoted by Pt [4,5]. This decreases typical ignition temperature of POM for mildly reoxidized catalyst to ~ 400 °C from 500-600 °C for as-prepared sample. Ignition usually starts at some intermediate point within the monolith (Fig. 6). The ignition point moves downstream with decreasing the contact time (increasing space velocity). After ignition, the hot spot moves towards the monolith inlet. Modeling describes reasonably good experimental temperature profiles within the monolith. The peak of deep oxidation products concentration $-H_2O$ and CO_2 (Fig. 7) is achieved within 30 s from start-up, when the temperature in the middle of monolith increases up to ~ 750 °C. After attainment of the steady-state temperature profile within all monolith, concentration of deep oxidation products decreases and that of CO increases to nearly constant levels, while a slow increase of H_2 content continues. This suggests that CO formation is mainly controlled by the primary process of CH_4 transformation within the inlet part, while for H_2 formation a slower stage of CH_4 SR is important as well. These dynamic features agree with those earlier observed for Rh/ alumina monolith [2].

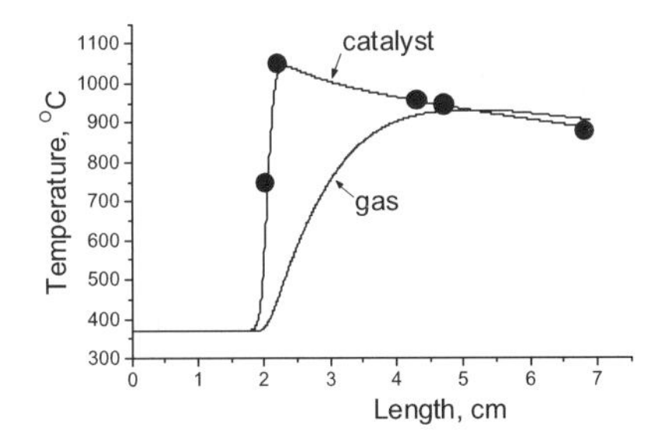

Fig. 4. Experimental (points) and calculated (line) temperature profiles within the monolithic catalyst and in the gas phase. 28.5% of NG in air, contact time 0.05 s.

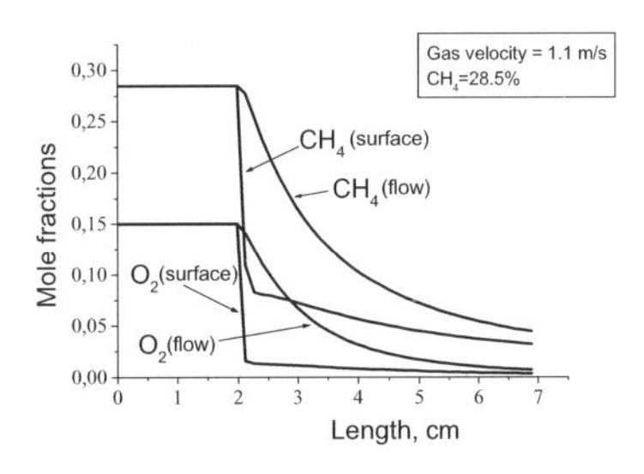

Fig. 5. Calculated profiles of gas-phase and surface concentrations of reagents. Contact time 0.05 s, inlet T 370 °C, 28.5% of NG in air.

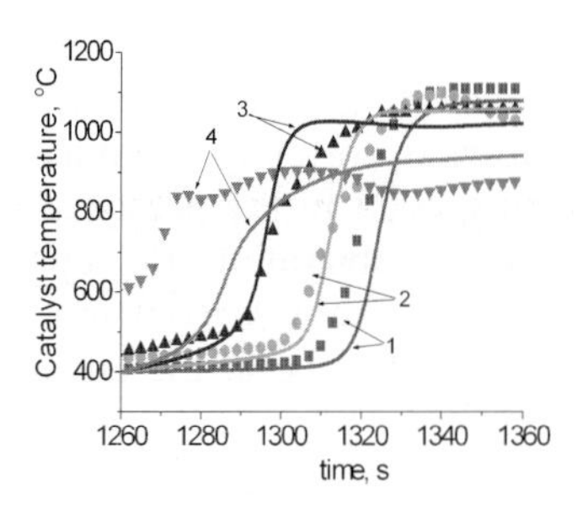

Fig. 6. Comparison of experimental (points) and calculated (lines) dynamics of temperature variation at selected points (1-2, 2-5, 3-10, 4-25 mm) in monolithic catalyst on corundum substrate during start-up. Contact time 0.11 s, 25.5% NG +15.7 %O_2 in N_2

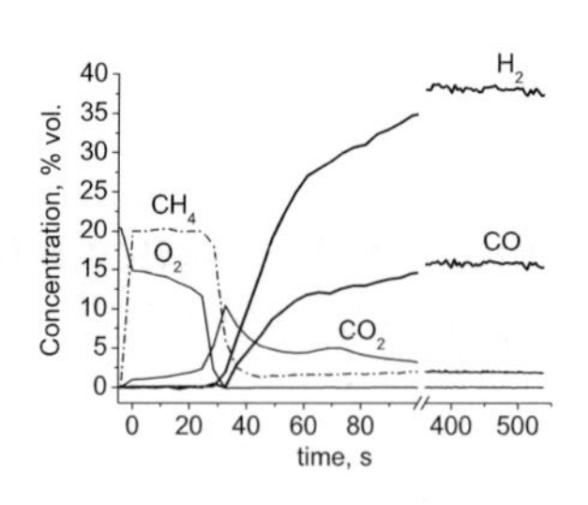

Fig. 7. Typical dynamics of the concentration of reagents and products variation during start-up for monolithic catalyst on corundum substrate. Start-up by switching the stream of air through catalyst preheated to 400 °C for feed 27% NG in air.

At the same contact time, the temperature profile stabilizes more rapidly for the catalysts based on metallic substrate as compared to corundum ones due to a higher thermal conductivity of the metal foil.

5. Conclusions

Experimental studies and modeling demonstrated importance of the heat and mass transfer for the process of the natural gas partial oxidation into syngas at short contact times on monolithic catalysts with a complex active component Pt/$LaNiO_3$/Ce-Zr-La-O. A simplified approach for modeling based upon using rate constants for the reactions of methane transformation into syngas estimated for separate structural elements of monolithic catalysts in nearly isothermal conditions was successfully verified.

This work was carried out in frames of European Associated Laboratory on Catalysis. Support by ISTC 2529 Project is gratefully acknowledged.

References

1. G. Veser, J. Frauhammer, Chem. Eng. Sci. 55 (2000) 2271.
2. R. Schwiedernoch, S. Tischer, C. Correa, O. Deutschmann, Chem. Eng. Sci. 58 (2003) 633.
3. R. Horn, K.A. Williams, N.J. Degenstein, L.D. Schmidt, J. Catal. 242 (2006) 92.
4. V.A. Sadykov, S.N. Pavlova et al. Kinetics Catal. 46 (2005) 227.
5. S. Pavlova, N. Sazonova, V. Sadykov, S. Pokrovskaya, V. Kuzmin, G. Alikina, A. Lukashevich, E. Gubanova, Catal. Today 105 (2005) 367.
6. G. Groppi, E. Tronconi, P. Forzatti, Catal. Rev.-Sci. Eng., 41 (1999) 227.
7. D.L. Trimm, C.-W. Lam, Chem. Eng. Sci. 35 (1980) 1405.
8. J.Wei, E. Iglesia, J. Catal., 224 (2004) 370.
9. T. Numaguchi, K. Kikuchi, Chem. Eng. Sci. 43 (1988) 2295.

Natural Gas Conversion VIII
F.B. Noronha, M. Schmal, E.F. Sousa-Aguiar (Editors)

CO_2 reforming of methane to syngas over Ni/SBA-15/FeCrAl catalyst

Kai Wang,[a] Xiujin Li,[b] Shengfu Ji,[*a] Si Sun,[a] Dawei Ding,[a] Chengyue Li[a]

[a] *State Key Laboratory of Chemical Resource Engineering, Beijing University of Chemical Technology, 15 Beisanhuan Dong Road, P.O.Box 35, Beijing, 100029, China*
[b] *Department of Environmental Eenineering, Beijing University of Chemical Technology, 15 Beisanhuan Dong Road, 100029, China*

Abstract: A series of Ni/SBA-15/FeCrAl catalysts with different Ni content were prepared using the FeCrAl alloy foils as support. The catalytic activity and the stability for carbon dioxide reforming of methane were tested at atmospheric pressure. The structure of the catalysts was characterized by XRD and TPR measurements. The results indicated that the Ni/SBA-15/FeCrAl catalysts showed the higher conversion of CH_4 and CO_2 under the reaction conditions. The 8.0% Ni/SBA-15/FeCrAl catalyst had an excellent catalytic activity and stability at 800℃ for 1400 h. There were the interaction among the NiO, the SBA-15 and the FeCrAl in the Ni/SBA-15/FeCrAl catalysts.

Key words: FeCrAl alloy, Ni catalyst, methane reforming, synthesis gas

1. Introduction

Recently, the monolithic catalysts have gained considerable attention with lower pressure drop, smaller size of reactor, lower gradient of temperature and suitable order flow channels in many forms [1-3] compared with conventional fixed-bed reactors with pellet or powder catalysts. The monolithic catalysts have great promising application in the reactions with high space velocity and heat effect such as carbon dioxide reforming of methane to synthesis gas reaction. The nickel catalyst has comparative activity and selectivity of the noble metal catalysts. However, they are rapidly agglomerated and deactivated on the transitional supports such as Al_2O_3 and SiO_2. This disadvantage might be avoided when the nickel catalyst has been supported on mesoporous silica and the FeCrAl alloy foils. In our previews work, the higher stable Ni/SBA-15 catalyst was prepared and the catalyst had a higher conversion of CH_4 and CO_2

[*]Corresponding authors. Tel.: +86 10 64412054; Fax: +86 10 64419619.
E-mail addresses: jisf@mail.buct.edu.cn (S. Ji)

under the reaction conditions [4]. In this paper, a series of the Ni/SBA-15/FeCrAl metallic monolithic catalysts were prepared. The structure was characterized using XRD and TPR techniques. The catalytic activity for carbon dioxide reforming of methane was investigated.

2. Experimental

The Ni/SBA-15 was prepared by impregnating an appropriate amount of the nickel nitrate solution into the SBA-15 at room temperature overnight, followed by drying at 110℃ for 24 h and calcination at 550℃ for 5 h. The mass fraction of Ni in the Ni/SBA-15 was in the range of 3.0wt% ~ 16wt%. The metallic monolithic catalysts were prepared using the FeCrAl alloy foils (OC404, Sandvik Steel, Sweden) as support. The FeCrAl alloy foils were rolled into cylinders in different diameter and 50 mm in length and calcined at 950℃ for 15 h in air. Then these supports were immersed in the γ-Al_2O_3 slurry, with a withdrawal velocity of 3cm/min, dried at room temperature, and then calcined at 500℃ for 4 h. Mixing slurry of the Ni/SBA-15 and γ-Al_2O_3 was prepared by wet milling. The calcined monolithic supports were dipped into the slurry, withdrawn at a constant speed of 3cm/min, dried at room temperature in air, and then calcined at 500℃ for 4 h, the Ni/SBA-15/FeCrAl metal support monolithic catalysts were obtained. The weight of the Ni/SBA-15 in the Ni/SBA-15/FeCrAl catalyst was ca. 10wt%. XRD of the catalysts was recorded using a Rigaku D/Max 2500VB2+/PC diffractometer with Cu Kα radiation at 40 kV. TPR experiments were performed with a Thermo Electron Corporation TPD/R/O 1100 series catalytic surfaces analyzer. The catalytic activity was tested in a quartz flow reactor (i.d., 6 mm) at atmospheric pressure. Reactant gas of CH_4/CO_2 (50/50 v/v) was introduced into the reactor with a gas hourly space velocity (GHSV) of 20,000 ml/g(cat+coat)·h. Before reaction, the catalysts were reduced in hydrogen flow at 750℃ for 3.5 h. The outlet products were measured with online GC with a TC detector.

3. Results and discussion

3.1 XRD of the samples

The XRD patterns of FeCrAl and Al_2O_3/FeCrA support are presented in Fig. 1. The characteristic peaks of FeCr are observed at 44.33° and 64.6° (JCPDS 34-0396, Fig. 1a). After heat treatment at 950℃ for 15 h, the additional peaks (25.5°, 35.0°, 37.6°, 43.2°, 52.4°, 57.3°, 66.3°, 68.0°) are observed in Fig. 1b, suggesting that α-Al_2O_3 (JCPDS 88-0826) is formed on the FeCrAl surface due to the oxidation of aluminum. The formation of the α-Al_2O_3 layer can improve the combination ability between the Al_2O_3 washcoat layer and the FeCrAl support [5-6]. The characteristic peaks of γ-Al_2O_3 (36.9°, 45.4° and 67.5°) are observed when FeCrAl support has been coated γ-Al_2O_3 slurry (Fig. 1c). The

XRD patterns of Ni/SBA-15/FeCrAl catalysts are presented in Fig. 2. The characteristic peaks of NiO phase are observed at 37.2°, 43.2°, 62.8° and 75.3° with upon 4.0% Ni content, whereas no characteristic peaks of NiO phase can be observed with below 4.0% Ni content in the catalysts. This indicates that NiO is highly dispersed on the surface of the FeCrAl support. Furthermore, the intensity of the NiO peaks increases gradually with increasing Ni content.

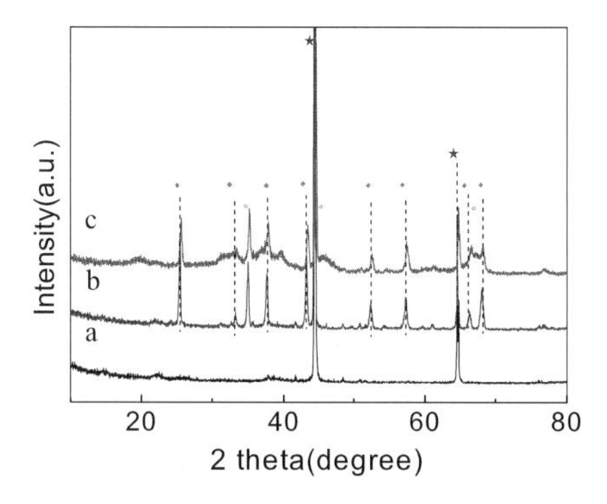

Fig. 1 XRD patterns of FeCrAl samples
(a) FeCrAl; (b) FeCrAl pre-Oxidized at 950℃;
(c) Al_2O_3/FeCrAl;
℃ FeCr; ℃ γ-Al_2O_3; ● α-Al_2O_3

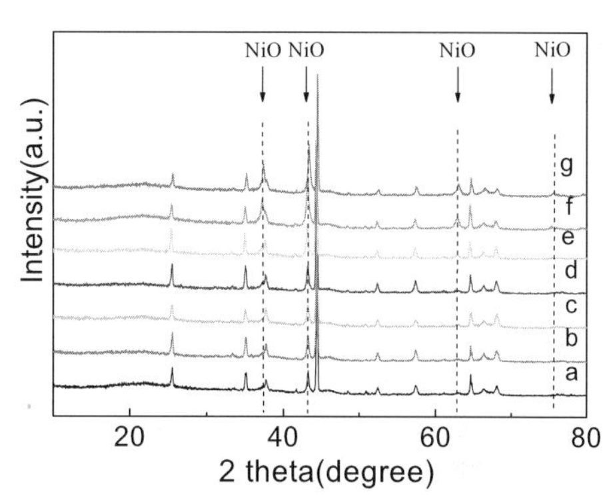

Fig. 2 XRD patterns of Ni/SBA-15/FeCrAl catalysts
(a) 3.0%Ni/SBA-15/FeCrAl; (b) 3.5%Ni/SBA-15/FeCrAl; (c) 4.0%Ni/SBA-15/FeCrAl; (d) 5.0%Ni/SBA-15/FeCrAl; (e) 8.0%Ni/SBA-15/FeCrAl; (f) 12%Ni/SBA-15/FeCrAl; (g) 16%Ni/SBA-15 /FeCrAl

The small angle XRD patterns of SBA-15 and Ni/SBA-15/FeCrAl catalysts are presented in Fig. 3.

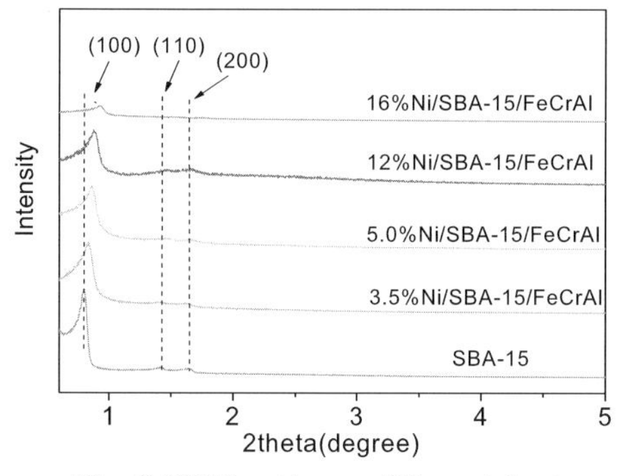

Fig. 3 XRD patterns of the catalyst

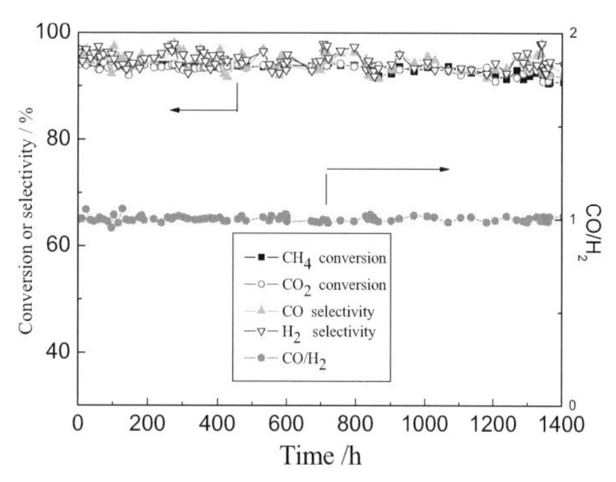

Fig. 4 Catalytic stability of 8.0%Ni/SBA-15/FeCrAl catalyst
GHSV = 20,000ml/g(cat.+coat)h; $V_{(CH4)}/V_{(CO2)} = 1$; T = 800℃

All Ni/SBA-15/FeCrAl catalysts show well-resolved patterns with a prominent diffraction peak (100), two additional weak diffraction peaks indexed to (110) and (200) reflections, which is similar to those of SBA-15. This indicates that the hexagonal mesoporous structure of SBA-15 is still present in the Ni/SBA-15/FeCrAl catalysts. However, the (100) diffraction peak of the Ni/SBA-15/FeCrAl samples shift to higher 2θ values compared with that of SBA-15, which might attribute to the constriction of their frameworks with Ni content during the calcination procedure. In addition, a gradual attenuation in the intensity of the (100) peak is observed with increasing Ni content, implying that partial blocking of the hexagonal pore walls of the SBA-15 materials occurs upon Ni introduction, especially at high Ni content.

3.2 Catalytic activity of the samples

The catalytic activity for carbon dioxide reforming of methane to synthesis gas between 700°C to 850°C is shown in Table. 1. For the Ni/SBA-15/FeCrAl catalysts, the conversion of CH_4 and CO_2 as well as the selectivity of CO and H_2 increases with increasing Ni content at the same reaction temperature. When Ni contents are from 4.0wt% to 16wt%, the metallic monolith Ni/SBA-15/FeCrAl catalysts exhibit a good catalytic activity at atmospheric pressure.

Table 1 Catalyst performance of the Ni/SBA-15/FeCrAl catalysts

Catalysts	T	Conv.(%)		Sel.(%)		Catalysts	T	Conv.(%)		Sel.(%)	
	(℃)	CH_4	CO_2	CO	H_2		(℃)	CH_4	CO_2	CO	H_2
3.0%Ni	700	48.6	46.7	74.5	82.0	8.0%Ni	700	76.8	80.5	87.4	94.2
/SBA-15	750	58.1	58.9	85.1	89.3	/SBA-15	750	85.9	88.0	89.2	96.4
/FeCrAl	800	69.4	66.2	87.3	94.4	/FeCrAl	800	94.8	91.8	93.0	97.6
	850	75.6	75.3	90.2	98.9		850	97.2	94.4	94.4	98.1
3.5%Ni	700	56.8	57.3	74.6	85.7	12%Ni	700	76.6	78.0	91.4	93.7
/SBA-15	750	63.4	63.7	83.8	93.1	/SBA-15	750	87.0	88.0	94.1	97.5
/FeCrAl	800	72.2	73.9	86.7	98.4	/FeCrAl	800	94.2	92.7	94.5	97.7
	850	82.5	81.8	84.6	98.8		850	97.5	94.8	93.8	98.6
4.0%Ni	700	74.8	74.6	81.0	88.5	16%Ni	700	83.9	81.6	89.9	95.7
/SBA-15	750	82.9	84.8	84.5	95.6	/SBA-15	750	88.6	90.2	91.5	97.6
/FeCrAl	800	87.4	87.7	86.1	95.6	/FeCrAl	800	95.3	94.4	91.5	98.0
	850	94.3	92.1	87.0	97.0		850	98.6	96.5	93.0	98.3
5.0%Ni	700	76.3	74.5	82.1	90.7						
/SBA-15	750	82.0	84.6	85.6	94.6						
/FeCrAl	800	88.6	89.4	87.5	96.5						
	850	92.6	92.0	88.6	98.6						

As far as Ni content is concerned, 16%Ni/SBA-15/FeCrAl shows higher conversion and selectivity compared with other Ni content. However, there is slight increase in catalytic activity from 8.0% to 16%. It is a better choice of 8% Ni/SBA-15/FeCrAl catalyst for the stability test. The relationship between the catalytic activity and stability of the 8% Ni/SBA-15/FeCrAl catalyst under the 800℃ is shown in Fig. 4. At the beginning of the reaction, the CH_4 and CO_2 conversion are 94.5% and 94.1%, respectively. After reaction 1400 h, they decrease down to 91.7% and 92.5%, respectively. In addition, the CO and H_2 selectivity decrease down to 92.5% and 94.1% from 96.8% and 96.9%. The catalytic activity and the stability of the Ni/SBA-15/FeCrAl catalysts are better than that of the Ni/SBA-15 catalysts [4].

3.3 TPR of the samples

H_2-TPR curves of the Ni/SBA-15/FeCrAl catalysts are presented in Fig. 5.

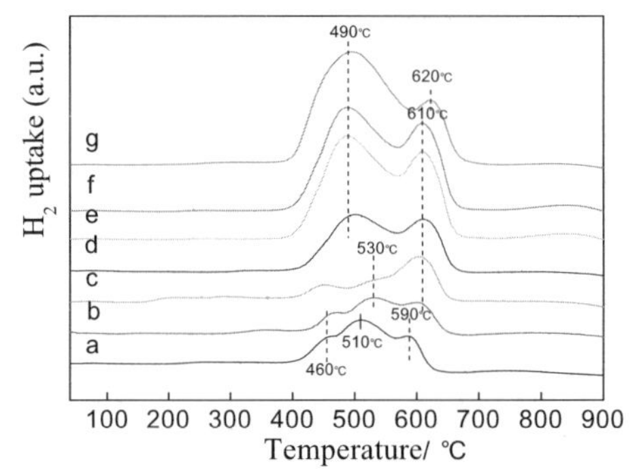

Fig 5 TPR patterns of Ni/SBA-15/FeCrAl.
(a) 3.0%Ni/SBA-15/FeCrAl; (b) 3.5%Ni/SBA-15/FeCrAl; (c) 4.0%Ni/SBA-15/FeCrAl; (d) 5.0%Ni/SBA-15/FeCrAl; (e) 8.0%Ni/SBA-15/FeCrAl; (f) 12%Ni/SBA-15/FeCrAl; (g) 16%Ni/SBA-15 /FeCrAl

There are no obvious reduction peaks of the FeCrAl alloy foil and SBA-15 in the temperature range of 25℃ ~ 900℃. In general, for the Ni-based catalysts, the reduction peaks at ~ 470℃ and over 600℃ are attributed to the reduction of large NiO particles and the reduction of NiO in intimate contact with the oxide support, respectively [7-8]. In the Ni/SBA-15/FeCrAl catalysts, when Ni content is 3.0%, the three reduction peaks appear at 460°C, 510°C, and 590°C, respectively. And the reduction peaks at 510°C and 590°C shifts to higher temperature with increasing of Ni content (Fig. 5-a, -b, -c). However, when Ni content is above 5.0%, there are two reduction peaks at 490°C and 610°C, respectively (Fig. 5-d, -e, -f). And the reduction peak at 610°C shifts to 620 °C when Ni content is 16% (Fig. 5-g). These indicate that, in the Ni/SBA-15/FeCrAl catalysts, the reduction property of the NiO particles depends upon both the Ni content and SBA-15 as well as the FeCrAl. Simultaneity, the intensity of reduction peaks increased with increasing Ni content. These indicate that there are the interaction among the NiO, the SBA-15, and the FeCrAl.

4. Conclusions

When Ni contents are from 4.0wt% to 16wt%, the metallic monolith Ni/SBA-15/FeCrAl catalysts exhibit a good catalytic activity for carbon dioxide reforming of methane at atmospheric pressure. And the 8wt%Ni/SBA-15/FeCrAl catalyst have an excellent catalytic activity and stability at 800℃ for 1400 h. XRD results indicates that in Ni/SBA-15/FeCrAl catalysts the hexagonal regularity mesoporous structure of the SBA-15 was sustained. The nickel species are highly dispersed on the surface of the FeCrAl support. TPR results indicate there are the interaction among the NiO, the SBA-15 and the FeCrAl.

Acknowledgements

Financial funds from the Chinese Natural Science Foundation (Project No: 20473009), the Beijing Natural Science Foundation (Project No: 8062023), the National Basic Research Program of China (Project No. 2005CB221405) and the National ''863'' Project of China (No. 2006AA10Z425) are gratefully acknowledged.

Reference

1. E. Włoch , A. Łukaszczyk , B. Sulikowski, et al. Catal. Today, 114 (2006) 231.
2. S. Roy, A. K. Heibel, W. Liu, et al. Chem. Eng. Sci., 59 (2004) 957.
3. G. Groppi, W. Ibashi, E. Tronconi, et al. Chem. Eng. J., 82 (2001)57.
4. M. Zhang, S. Ji, L. Hu, et al. Chin. J. Catal, 27 (2006) 777
5. X. Wu, D. Weng, Su Zhao, et al. Surf. Coat. Technol., 190 (2005) 434.
6. S. Zhao, J. Zhang, D. Weng, et al. Surf. Coat. Technol., 167 (2003) 97.
7. R. Molina, G. Poncelet, J. Catal., 173 (1998) 257.
8. A. M. Diskin, R. H. Cunningham, R. M. Ormerod, Catal. Today, 46 (1998) 147.

Natural Gas Conversion VIII
F.B. Noronha, M. Schmal, E.F. Sousa-Aguiar (Editors)

Scale-up Challenges in Synthesis Gas Production

Søren Gyde Thomsen[a], Olav Holm-Christensen[a], Thomas S. Christensen[a]

[a]*Haldor Topsøe A/S, Nymøllevej 55, 2800 Lyngby, Denmark*

1. Introduction

It has for a long time been realised among plant owners and investors that large petrochemical complexes have far better economics than small complexes. Therefore, to harvest the benefit of this economics of scale effect, the present trend is to design and build still larger plants – such as for instance very large GTL facilities comprising multiple lines each with a high single line capacity or huge methanol and ammonia plants.

It is obvious that in order to exploit the full potential of these plants, all individual pieces of process equipment must function as expected – in terms of process performance, but equally important also in terms of reliability over their expected life time. The malfunction or an unexpected break-down of one single piece of equipment may significantly hamper the plant economics, thus it is of utmost importance that the scaling-up issues are addressed properly each time a critical piece of equipment is designed to exceed the capacity of previous industrial references.

The present paper will specifically discuss the issues and challenges relevant to scale-up plants for conversion of natural gas to a synthesis gas production for Fischer-Tropsch based GTL, methanol, ammonia, etc. A single line of a typical synthesis gas production for the above products comprises the following: a number of desulphurisation vessels, the reformer section and the process gas waste heat cooling section. In some cases, an additional process gas heated reformer may be included as well. In order to assess and mitigate any risks associated with a scale-up, each of these process steps and equipment items need to be evaluated in detail.

A number of other issues need attention for the large plants of the future such as transportation restrictions, limitations in the capacity of rotating equipment, etc.

Although similarly important, these are considered beyond the scope of this paper.

2. Use of CFD in Scaling Process

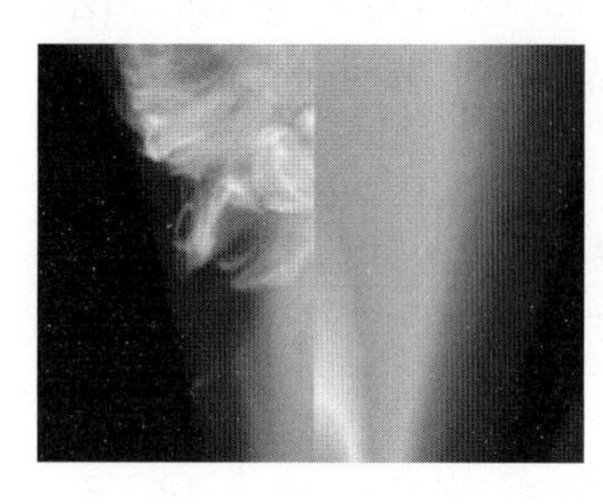

Figure 1

Computational Fluid Dynamics (CFD) is an invaluable tool for design purposes in as much as it involves no scaling assumptions – it works equally well for geometries of small and big physical scale. CFD is, however, not based on first principles alone; it involves empirical models for description of turbulence and combustion. It is therefore essential to validate the models used against experiments and to do so, a number of test campaigns have been carried out on test units scaled from industrial size.

One of these campaigns aimed at testing a second order closure model for swirling flows. Several Haldor Topsøe reactor designs involve gas burners and some of these generate highly complex flow fields characterised by intensive swirl and central vortex break down. A flow field of this kind is hard to capture by simple turbulence models and the use of second order closure models for this kind of simulations is required. It is particularly difficult for the turbulence models to predict the strength of the central back flow formed by the vortex break down correctly. An experiment was therefore designed to measure the axial velocity in a plane downstream the opening of the gas burner. Laser Doppler Anemometry (LDA) was used to measure local flow velocities in the non-combusting flow field, and four experiments were made to ensure repeatability of the results. In the graph,

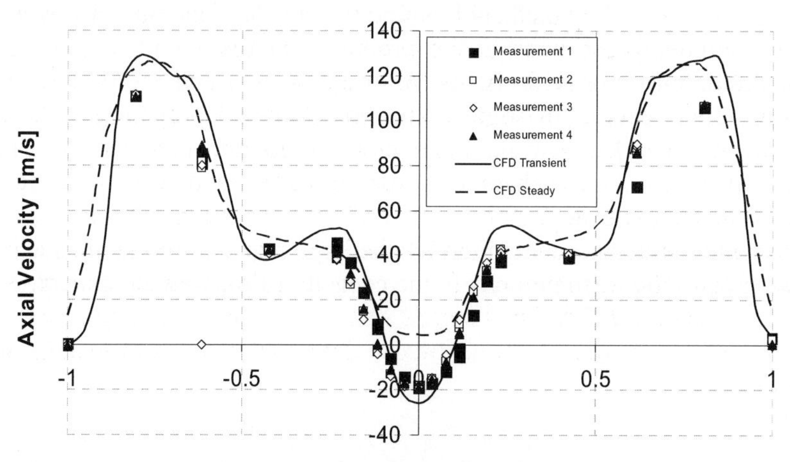

Figure 2

(Figure 2) the markers indicate the measurements taken as a function of the relative radial distance from the centre axis of the burner, and the curves describe the output of the CFD simulations with the advanced turbulence model mentioned. The experiments show that the velocity profile is symmetric around the centre line (radius = 0), and very close to the centre we find axial velocities with negative sign indicating backflow in this region. This corresponds to our expectations and what the CFD simulations must be able to accurately capture.
The dashed curve represents the results of a steady state simulation and the solid one that represents the average of output over different points in time generated from a simulation that allows the flow field to vary with time – a transient simulation in CFD terminology. The back-flow involves a dynamic behaviour that can only be correctly predicted if time dependence is included in the model. This can be visualised by comparing the 2 curves shown in the graph above – the time averaged transient results are almost identical to the measurements while the steady results fail to predict any back-flow.
The experimental investigation demonstrates that application of the correct turbulence model used in transient mode leads to highly credible results. As mentioned, CFD is not based on scale assumptions of any kind and the tool can therefore with this turbulence model be used for burners of all (relevant) sizes.

3. Adiabatic Reactors in Large Scale Plants

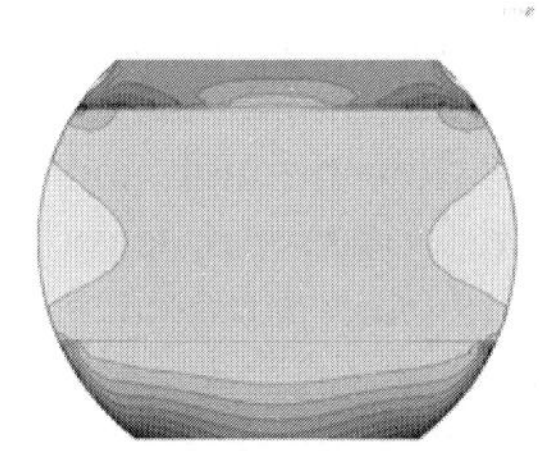
Figure 3

Catalytic processes such as desulphurisation and pre-reforming are performed in adiabatic vessels. The pressure drop essentially limits the catalyst bed height and for very large single line capacity plants, the bed height to diameter ratio becomes very small. In order to make optimal use of the volume in such reactors, it is tempting to make the reactor spherical rather than cylindrical with spherical heads. The concern would, however, be the flow distribution in the catalyst bed. Again, CFD comes in as an indispensable tool. Figure 3 shows the result of a simulation of the flow through a spherical reactor with a catalyst bed height of approx. 50% the internal diameter. The figure depicts the flow distribution in a vertical plane through the reactor. The results indicate a non-uniform distribution of the flow, and studies of residence time distributions confirm that the catalyst utilisation is significantly impaired with the spherical design compared to the traditional cylindrical design. The implications of the above considerations are that there is a limit to the economical size, i.e. diameter of the adiabatic catalytic reactor. At this limit, it will be a better option to choose 2 parallel reactors instead of one mega reactor.

4. Some Aspects in the Reforming Section

In large scale plants, the Autothermal Reformer (ATR) is the heart of the conversion of natural gas to synthesis gas. CFD is intensively used to model the ATR burner and reactions in the combustion chamber and catalyst bed to ensure that mixing, combustion, chemical reaction as well as temperature distribution in the combustion chamber are correctly controlled.

In the ATR, heat is provided by sub-stoichiometric combustion of the feed gas. This fuel-rich combustion involves the risk of incomplete combustion to soot particles. The ATR process is designed for soot-free operation, and it is achieved through careful design of burner, combustion chamber and catalyst bed as well as by the selection of the operating conditions (pressure, temperature and steam-to-carbon ratio). The chemical composition of the synthesis gas at the outlet of the ATR reformer is determined through use of heat and mass balance calculations, equilibrium thermodynamics and chemical reaction kinetics. Two tools used in conjunction are necessary for the scale-up process, i.e. pilot testing [1] and reactor modelling, and these together with extensive industrial feed-back ensures reliable scaling to even larger plants.

5. Process Gas Waste Heat Section

In the process gas waste heat section of a synthesis gas plant, the very hot gas leaving the reformer is cooled in either a Waste Heat Boiler (WHB) alone or in a combination of a WHB and a steam superheater (SSH). Again, more careful design is required for both pieces of equipment because the flow through and thereby the duty and size is increased. Moreover, the demand for better process efficiencies dictates more aggressive gas compositions, which further strengthen the requirements for good design supported by precise analyses.

The inlet system to the WHB in which the hot and aggressive synthesis gas is admitted into the boiler tubes is considered the most critical part of the equipment. Thorough thermal analyses using both CFD and the Finite Element (FE) method are required to assess the temperature field in the complex system comprising tube sheet refractory lining, ceramic and metallic ferrules, tube sheet and tubes. Such analyses are to take into account radiation as well as all fluid dynamic effects generated by contraction and expansion of the synthesis gas as it passes into the boiler tube. The analyses must verify that the temperatures of the metallic parts exceed neither the mechanical design temperature nor the temperature at which metal dusting corrosion becomes a concern. At the same time, it must be verified that the heat release in the boiler at the inlet section is well below the critical heat flux at which the steam/water circulation becomes insufficient to cool the boiler tube.

The tube sheet, the tube sheet refractory and the tube sheet to shell junction of the WHB pose a specific challenge because the plant is scaled-up. The steam side of the boiler in a synthesis gas plant works at high pressures in the range 45 to 120 bars and with increasing shell diameters, the only viable boiler types are those with flexible tube sheet design where the tube sheets are supported by the boiler tubes. This design can only be made with a rigorous FE based stress analysis and in each case the tube sheet to shell junction must be shaped to ensure sufficient flexibility of the tube sheet and at the same time, make the parts strong enough to sustain the pressure. Figure 4 shows an example of a tube sheet to shell transition subject to pressure and thermal loads.

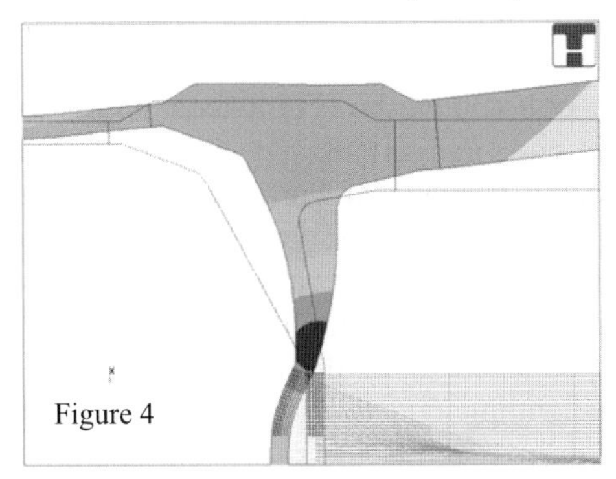
Figure 4

Steam superheating is often an attractive option to utilise some of the sensible heat in the process gas. It does, however, impose severe requirements on the material used for the SSH. On one side, metal dusting resistant materials are to be used for high metal temperatures and on the other side, materials that are insensitive to stress corrosion cracking are to be used in the colder end of the SSH. As these two requirements are to some extent contradictory, it is found that the best option is to split the SSH into two parts or even two vessels. This has the added advantage, that the size of the SSH can be kept reasonable even in a very large size methanol plant.

6. Process Gas Exchange Reformer Option

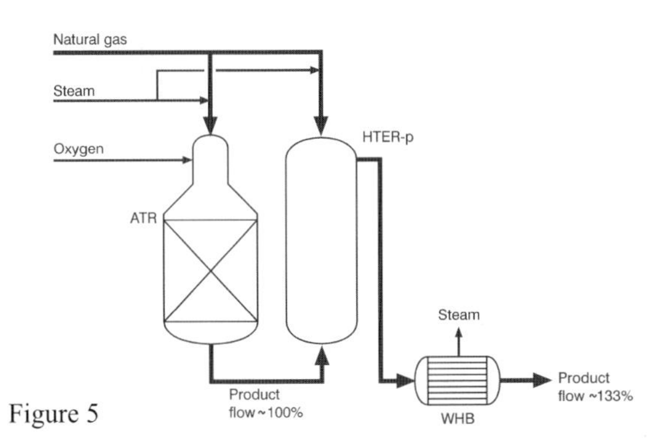

Figure 5

Increasing the capacity of a plant would normally imply larger pieces of equipment. However, sometimes it is possible to improve the capacity radically by introducing additional equipment and keep the remaining equipment at a size that is within the proven range. In steam reforming processes as performed in the ATR, the very hot process gas is traditionally cooled in a steam producing process gas boiler. However, this is not an optimal exploitation of the energy,

because more steam than needed for the process is produced. If the energy in the hot process gas is instead utilised to generate more reforming reaction in a gas-gas heat exchange reformer, it will be possible to increase the capacity of a plant without increasing the energy needed by the process and without increasing the size of the original reformer. Such a scheme is shown in Figure 5, where the Haldor Topsøe Exchange Reformer (HTER-p) has been integrated.
In 2003 one of the reforming lines at Sasol Synfuels in South Africa was converted from simple autothermal reforming to include an HTER-p. The combined throughput from the HTER working in parallel with the ATR was increased by 33% (equivalent to 1,250 MTPD of ammonia). A more thorough description of the successful implementation of this HTER can be found in [2].
The HTER technology is also applicable for other technologies than production of synthesis gas for GTL. In [3], it is described how the HTER technology can be adapted for an ammonia plant front-end, and in this way reach a plant capacity of 3300 MTPD without increasing the size of the primary reformer beyond the size of a conventional and well proven reformer. In this way, the scale-up issue for the plant has been transferred from the primary reformer to the HTER. The HTER is, however, easily scalable, since it is made up of a number of tube assemblies that can be adjusted to fit the required size. The mechanical design of the HTER is challenging due to the presence of high temperatures and the fact that the gas is corrosive, but the successful operation of the above-mentioned plant at Sasol for more than 3 years has proven the technology.

7. Conclusion

A number of challenges in connection with scale-up to large size units for production of synthesis gas have been identified. In order to address these properly, a rigorous design process relying heavily on tools like CFD and FE analysis is necessary. At the same time, it is necessary to verify that the methods used have the ability to precisely predict the conditions prevailing in the process equipment.

References

1. Christensen, T. S, Østberg, M. Bak Hansen, J-H, "Process Demonstration of Autothermal Reforming at Low Steam-to-Carbon Ratios for Production of Synthesis Gas", Ammonia Safety Symposium, (46), 2001
2. Loock, S., Ernst, W.S, Thomsen, S.G, Jensen, M.F.: "Improving carbon efficiency in an auto-thermal methane reforming plant with Gas Heated Heat Exchange Reforming technology", the 7th World Congress of Chemical Engineers, Scotland, 2005.
3. Thomsen, S. G, Han, Pat, Loock, Suzelle, Ernst, Werner:"The first Industrial Experience with a Hador Topsøe Exchange Reformer", Ammonia Safety Symposium, (51), 2006.

Natural Gas Conversion VIII
F.B. Noronha, M. Schmal, E.F. Sousa-Aguiar (Editors)

Studies of Carbon Deposition over Rh-CeO_2/Al_2O_3 during CH_4/CO_2 Reforming

Rui Wang,[a,b] Xuebin Liu,[a] Qingjie Ge,[a] Wenzhao Li,[a] Hengyong Xu[a*]

a Dalian Institute of Chemical Physics,Chinese Academy of Sciences, Dalian 116023, China
[b] *Graduate University of Chinese Academy of Sciences, Beijing 100049, China*

1. Abstract

Three types of carbon species (α, β, γ-carbon) exist on highly dispersed Rh, Rh crystals, and Lewis acid sites of Al_2O_3 support over Rh/Al_2O_3 during CH_4/CO_2 reforming, respectively. When CeO_2 was added, the more α-carbon has been observed, which was suggested to be located on the newly formed Rh-CeO_2 interface, and as an active intermediate is related to the reforming activity. The β-carbon was remarkably decreased and most likely responsible for the deactivation of catalyst. The γ-carbon formation was greatly suppressed, which is usually considered not to participate in the reforming reaction.

2. Introduction

Methane reforming by carbon dioxide has received much attention because of the suitable utilization of both CH_4 and CO_2, two major greenhouse gases. This reaction generates synthesis gas with a H_2/CO ratio of unity, which can be adjusted by combining with steam reforming to be suitable for the production of methanol and the Fischer-Tropsch. The addition of oxides or noble metal as promoters [1-3] has been found to improve the activity and coke deactivation of catalyst. There is, however, a limited amount of fundamental research concerning the coking process during CH_4/CO_2 reforming. Zhang et al. [4] have found that three kinds of carbonaceous species, C_α, C_β and C_γ species exist on

* Corresponding author: Tel: +86-411-8458-1234 E-mail: xuhy@dicp.ac.cn

the Ni/Al_2O_3 catalyst. The active C_α species is suggested to be responsible for CO formation, while the less active β-C and γ-C ones contribute to catalyst deactivation. M. A. Goula et al. [5] have suggested that two kinds of carbon species were mainly found over $Ni/CaO-Al_2O_3$ catalysts, where the amount and reactivity of them are influenced by the CaO/Al_2O_3 ratio. Recently, we have reported that the $Rh-CeO_2$ interaction could cause the coexistence of Ce^{4+}/Ce^{3+} and $Rh^0/Rh^{\delta+}$ redox couples over $Rh-CeO_2/Al_2O_3$, which is found to not only improve the reforming activity, but also enhance the coke resistance of Rh/Al_2O_3 catalyst [6]. In the present work we further studied the amount, the site location and reactivity of carbon species on $Rh-CeO_2/Al_2O_3$ catalyst, with which both reforming activity and coke resistance of the catalyst was further correlated.

3. Experimental

3.1. Catalyst Preparation

An Rh/Al_2O_3 catalyst was prepared by impregnating $\gamma-Al_2O_3$ with an aqueous solution of $RhCl_3 \cdot nH_2O$. An $Rh-CeO_2/Al_2O_3$ was prepared by co-impregnation of $\gamma-Al_2O_3$ with a mixed solution of $RhCl_3 \cdot nH_2O$ and $Ce(NO_3)_3 \cdot 6H_2O$ (the mol ratio of Rh/Ce= 1.2), and Rh/CeO_2 catalyst was prepared by impregnation of CeO_2 with solution of $RhCl_3 \cdot nH_2O$. The resulting solids were dried at 100°C for 24 h and calcined at 500 °C for 2h in air. CeO_2 support was obtained after calcinations of $Ce(NO_3)_3 \cdot 6H_2O$ at 500 °C in air for 6h and $\gamma-Al_2O_3$ was commercial catalyst. The Rh content was 1 wt% in all of the prepared catalysts.

3.2. Catalytic Studies

Catalytic activities were investigated in a tubular quartz reactor with an inner diameter of 8 mm. 150mg of catalyst was reduced in H_2 (20ml/min) at 500 °C for 30 min and then heated to 700 °C in N_2 (50ml/min). The reactions were performed at 700°C with a CO_2:CH_4 mole ratio of 1:1 and a flow rate of 100 ml/min. The effluents were analyzed on line by a TCD gas chromatography with two columns of TDX-01 and MS-8A. The conversion of reactants was evaluated according to the carbon balance.

3.3. Catalyst Characterazation

Temperature-programmed oxidation (TPO) and CO_2-temperature-experiments were performed with an AutoChem 2910 apparatus. 150mg of catalysts were loaded into a U-type quartz reactor and a thermocouple was placed at the top of the catalyst layer. The catalysts first reduced in H_2 for 1h at 500 °C and then heated to 700 °C in He for 1h. The feed gas of CH_4/CO_2 mixture (1:1) or pure CH_4 pulses (300.6 μl) was fed to the reactor for required time. Then, the catalysts were kept in He for 0.5h and cooled to room temperature, and subsequently exposed to 5% O_2/He or 5% CO_2/He (15 ml/min) by ramping at a heating rate of 5 °C/min from ambient temperature to 900 °C. The effluents were measured on line with a mass spectrometer (OminStar 300).

CO_2 contributes a fragment to the signal at *m/e*= 28 (11.4%), and this was subtracted from the total intensity at *m/e*= 28 to get the amount of CO.

4. Result and disscussion

4.1. Catalytic Performance

Table 1. Reaction and Data over Rh-based Catalysts in CH_4/CO_2 Reforming at 700 °C [a]

Catalyst	Conversion $CH_{4\ equ.}$ [b]	Initial CH_4 Conversion [c] (%)	CO/H_2 Molar ratio	Stability [d]	Carbon [e] ($\mu mol.g^{-1}$)
Rh/Al_2O_3	72.79 %	52.2	1.21	88.9	31.2
$Rh\text{-}CeO_2/Al_2O_3$	-	59.93	1.27	94.6	19.97
Rh/CeO_2	-	39.9	1.34	94.8	4.25
Al_2O_3	-	<1%	-	-	370.23
CeO_2	-	<1%	-	-	0

[a] CH_4/CO_2 = 1, GHSV = 40000 $ml{\cdot}h^{-1}{\cdot}g^{-1}$, T= 700 °C; m= 150 mg;
[b] Thermadynamic equilibrium data at 700 °C ; [c] CH_4 conversion at 10 min on stream;
[c] the ratio between the CH_4 conversion at 25 h and that at 10 min on stream;
[d] determined by successive pulsing O_2 at 800 °C

Table 1 displays the reaction data of the non-promoted and ceria-promoted Rh catalysts, with the bare supports as references. For bare Al_2O_3, there was no significant CH_4 conversion but 370.23μmol/g carbon was formed. Obviously, the Lewis acid sites of Al_2O_3 facilitate the cleavage of C-H bonds of methane, leading to an enhanced carbon accumulation at the high temperature [7]. For Rh/Al_2O_3, methane can be activated on Rh and converted to CH_x species, as a very active intermediate (Equ. I), which readily react with CO_2 (Equ. II). Only a small part of them could be further dehydrogenated into coke precursors. The molar ratio of CO/H_2 yield is larger than 1, demonstrating the simultaneous occurrence of the reverse water-gas shift reaction ($CO_2 + H_2 \rightarrow CO + H_2O$) during the reaction.

$$CH_4 \rightarrow CH_x + (2\text{-}x/2)\, H_2 \qquad (I)$$

$$CO_2 + CH_x \rightarrow 2CO + x/2H_2 \quad (II)$$

In the case of bare CeO_2, there were nearly no CH_4 conversion and carbon deposition. This indicates CeO_2 itself could not convert CH_4 to coke precursors, however, methane can be partially oxidized by oxygen species of ceria to form CO and H_2. When Rh was loaded on CeO_2, the initial CH_4 conversion (39.9%) was lower than that of Rh/Al_2O_3, due to the lower surface area of CeO_2 compared to Al_2O_3. When CeO_2 as promoter was added to Rh/Al_2O_3, the initial CH_4 conversion $Rh\text{-}CeO_2/Al_2O_3$ was found to be higher than that over Rh/Al_2O_3, and about 95% of the initial CH_4 conversion was maintained after 25h of the reaction time. In addition, the molar ratio of CO/H_2 yield decreased in the order $Rh/CeO_2 > Rh\text{-}CeO_2/Al_2O_3 > Rh/Al_2O_3$. The higher molar ratio of CO/H_2 yield favors the occurrence of the reverse water-gas shift reaction. The production of

H_2O facilitates the carbon removal, which suppresses the carbon deposition and improves the stability of catalysts. These results are in a good agreement with the measurement of both stability and carbon deposition during the CH_4/CO_2 reforming reaction.

4.2. TPO studies after CH_4/CO_2 reforming

The TPO measurements of the carbon species formed during the CH_4/CO_2 reforming over the various catalysts are shown in Fig. 2, in which only the CO_2 signals of mass spectra (MS) are involved.

The TPO profile of bare Al_2O_3 (Fig. 2a) consisted of two deconvoluted peaks at 570 °C and 625 °C, respectively, corresponding to the carbon species formed on Lewis acid sites but two different strength. Moreover, no H_2O signal was detected by MS during TPO experiment. Therefore, the carbon species are probably the graphitic carbon (coke)[8], which seems not to participate in the reforming reaction [7].

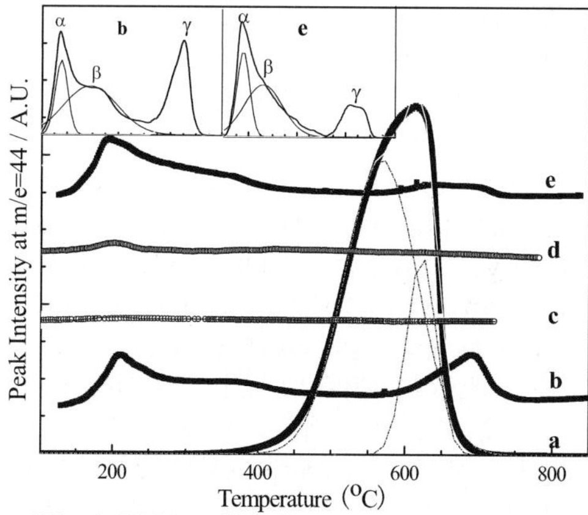

Fig. 2 TPO profiles for (a) Al_2O_3 (b) Rh/Al_2O_3 (c) CeO_2 (d) Rh/CeO_2 (e) $Rh\text{-}CeO_2/Al_2O_3$ after 25h of CH_4/CO_2 reforming reaction at 700 °C

For Rh/Al_2O_3 (Fig. 2b), three types of carbon species were formed, designated as α, β and γ-carbon according to the oxidation temperature. It is interesting to note that when the α and β-carbon peak appeared, a distinct water signal was detected in MS spectra simultaneously. This reveals that both α-carbon (CH_x) and β-carbon (CH_y) are hydrogen-containing carbon species but with different hydrogen content ($x > y$). It is generally accepted that methane activation mainly occurs on metal surface. Therefore, α-carbon would be formed on highly dispersed Rh, which provides more active centers for CH_4 decomposition and generate more active intermediates (CH_x). While β-carbon would be formed on bulk Rh crystals not directly contacted with support, which possess relatively a strong dehydrogenation ability, resulting in the formation of CH_y species with less H during methane decomposition. β-carbon is less reactive for CH_4/CO_2 reforming than α-carbon and would be possibly further dehydrogenated into inactive coke precursors with time of reaction. These hydrogen-containing carbon species, as the active intermediates species for the CH_4/CO_2 reforming reaction, they could react with CO_2 to form CO and H_2. The γ-carbon appeared at 690 °C, the least reactive carbon species, is usually assigned to graphitic carbon deposited on support, which is remarkably smaller than that over bare Al_2O_3.

No carbon deposition was observed on bare CeO_2 under CH_4/CO_2 reforming conditions (Fig. 2c). In the case of Rh/CeO_2 (Fig. 2d), the peak for α-carbon is at 190 °C , at least 20 °C lower than that of Rh/Al_2O_3.

When CeO_2 was added to Rh/Al_2O_3 (Fig. 2e), the α-carbon was increased while the β, γ -carbon was remarkably decreased. The distribution of the three forms of carbon species depends on the formation of the novel Rh-CeO_2 interface. The unique interaction between Rh and CeO_2 facilitates the activation of both CH_4 and CO_2 molecules [6], which enhances the conversion of CH_4 to CH_x species and remarkably suppresses the complete dehydrogenation of CH_4 to γ -carbon.

4.3. TPO studies after CO_2-TPSR

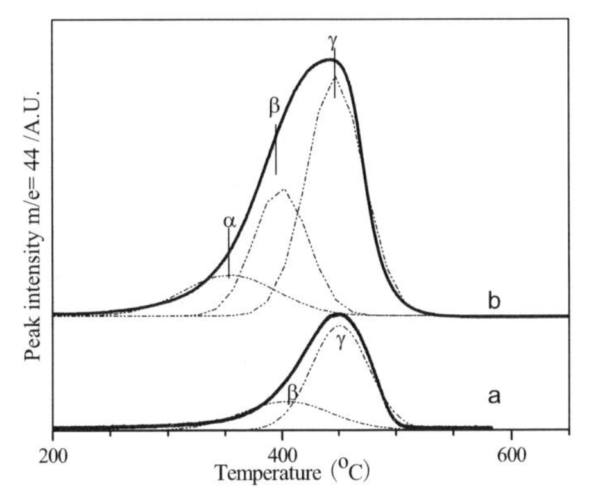

Fig. 3 TPO profiles for Rh/Al_2O_3 after 72 pulses of CH_4 (a) followed by CO_2-TPSR to 700 °C (b) without CO_2-TPSR

To further clarify the role of different carbon species in reforming reaction, the CO_2-TPSR measurement up to 700 °C for 30 min under CH_4 pulses (dominates carbon formation) over Rh/Al_2O_3 was performed. In order to check the carbon remained on the catalyst after CO_2-TPSR, TPO experiment was carried out subsequently (Fig. 3a). It is found that two deconvoluted peaks CO_2 peak appeared at 400 and 450 °C respectively, corresponding to β, γ-carbon described in Fig 2b. This suggests that the carbon species could not completely react with CO_2. A direct TPO profile of Rh/Al_2O_3 after CH_4 pulses but no CO_2-TPSR procedure, as reference, is shown in Fig. 3b. It is observed that besides the peak at 400 and 450 °C (β, γ-carbon), an additional peak appeared at 350 °C (α-carbon). The absence of α-carbon peak in Fig 3a is due to its complete removal by CO_2 during the previous CO_2-TPSR peocedure, This just suggests that the α-carbon as the active intermidiates has a high reactivity with CO_2 during the reforming reaction. The difference of β and γ-carbon between Fig. 3a and Fig. 3b demonstrates that the less reactive carbon (β-carbon) could partially react with CO_2, but another part of it would be further dehydrogenated to γ-carbon, which seems not to participate in reforming reaction. Therefore, we conclude that the β-carbon is most likely responsible for deactivation of the catalyst.

Finally, a mechanistic model of the carbon deposition over Rh-CeO_2/Al_2O_3

catalyst at 700 ^{o}C is shown in Fig. 4. The newly formed Rh-CeO_2 interface remarkably slows down the coking process, which can be explained by our previously suggested cycle mechanism of $Rh^0/Rh^{\delta+}$ and Ce^{4+}/Ce^{3+} redox couples over Rh-CeO_2/Al_2O_3 [6]. In pathway 1, CH_4 decomposes on highly dispersed Rh into α-carbon (CH_x), as the active intermediates, which could completely react with CO_2 to CO and H_2. $Rh^0/Rh^{\delta+}$ redox couples promote CH_4 decomposition into CH_x intermediates and H_2 on Rh-CeO_2 interface. On the other hand, Ce^{4+}/Ce^{3+} redox couples facilitate CO_2 activation and dissociation into CO and surface oxygen, and surface oxygen accelerates the oxidation of CH_x intermediates to CO and hydrogen. Pathway 2 leads to the formation of β-carbon on Rh crystals, which would be further dehydrogenated to coke precursors and be associated with the deactivation of the catalyst. CeO_2 in the vicinity of Rh crystals favors the reaction of carbon species with CO_2, which decreases the amount of β-carbon. In pathway 3, CH_4 is dehydrogenated completely on Lewis acid sites of Al_2O_3 to ɣ-carbon (coke), which could not react with CO_2. The presence of CeO_2 enhances the conversion of CH_4 to the active imtermediates, and consequently decreases the complete dehydrogenation of CH_4 to the coke on Al_2O_3. In CH_4/CO_2 reforming reaction, when many of Rh crystals are covered by carbon species, the Rh-CeO_2 interface is still available for the reaction. Thus, the Rh-CeO_2/Al_2O_3 catalyst exhibits a better stability than Rh/Al_2O_3.

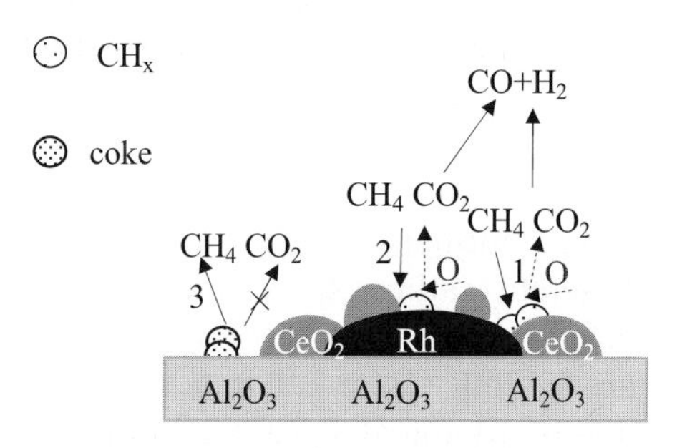

Fig. 4 A mechanistic model of the carbon deposition over Rh-CeO_2/Al_2O_3 catalysts

Acknowledgements

Financial support by the the Natural Basic Research Program of China (2005cb221401), is gratefully acknowledged.

References

1. S.-H. Seok, S.H. Choi, E. D. Park, S. H. Han, J.S. Lee, J. Catal. 209 (2002) 6.
2. Y.-Z. Chena, B.-J. Liawb, W.-H. Lai, Appl. Catal. A 230 (2002) 73.
3. H.S. Roh, H.S. Potdar, K.-W. Jun, Catal. Today 93–95 (2004) 39.
4. Z.L. Zhang, X.E. Verykios, Catal. Today 21 (1994) 589.
5. N. C. Triantafyllopoulos, S. G. Neophytides, J.Catal. 239 (2006) 187.
6. R. Wang, H.Y. Xu, X.B. Liu, Q.J. G, W.Z. Li, Appl. Catal. A (2006) 305.
7. K. Nagaoka, K. Seshan, K. Aika, J.A. Lercher, J. Catal. 197 (2001) 34..
8. T. Koerts, M.J.A.G. Deelen, R.A. van Santen, J. Catal. 138 (1992) 101.

Natural Gas Conversion VIII
F.B. Noronha, M. Schmal, E.F. Sousa-Aguiar (Editors)

Development of CO_2 Reforming Technology

Fuyuki Yagi[a], Shinya Hodoshima[a], Shuhei Wakamatsu[a], Ryuichi Kanai[a], Kenichi Kawazuishi[a], Yoshifumi Suehiro[b] and Mitsunori Shimura[a]

[a]*Chiyoda Corporation, 3-13 Moriya-cho, Kanagawa-ku, Yokohama-City, Kanagawa, 221-0022, Japan*
[b]*Japan Oil and Metals National Corporation, 1-2-2 Hamada, Mihama-ku, Chiba-City, 261-0025, Japan*

Abstract

A CO_2 reforming catalyst and process have been developed. The commercial scale CO_2 reforming catalyst has demonstrated its activity and stability in pilot plant tests under operating conditions for the manufacture of synthesis gas with a H_2/CO ratio of 2.0 for approximately 7,000 hours. In these tests, synthesis gas was generated by CO_2 and H_2O reforming with high energy efficiency: low CO_2/carbon and low H_2O/carbon ratios in feed gas composition. Our catalyst exhibited high carbon removal characteristics with steam from the catalyst surface, which prevents carbon deposition on our CO_2 reforming catalyst.

Introduction

CO_2 reforming process has recently attracted much attention as an effective technology for the utilization of CO_2 and natural gas. This technology is important for potential use in environmental preservation and energy diversification, because CO_2 (greenhouse gas) is available as a portion of raw gas, and clean fuels (FT oil, DME, etc.) are produced from product synthesis gas.

In the GTL processes, the capital cost of the synthesis gas production section accounts for approximately 60% of the total capital cost[1]. Therefore the development of more compact processes for syngas production with high energy efficiency is required. Such operations are available for CO_2 and H_2O

reforming under low feed CO_2/carbon and H_2O/carbon ratios. However, carbon deposition will be a difficult problem under these conditions[2-6]. So the development of a catalyst resistant to coking will be one of the most important issues for CO_2 and H_2O reforming technology. Effective carbon removal with steam on the catalyst surface will suppress carbon deposition[4]. We have taken part in the Japan Oil, Gas and Metals National Corporation (JOGMEC) GTL National project since 1999, in our continuing effort towards the establishment of this new CO_2 and H_2O reforming catalyst and synthesis gas production process on a pilot plant scale.

Experimental

Catalyst: Commercial size ring catalysts with an outer diameter of 15.9mm, an inner diameter of 6.4mm, and a length of 15.9mm were prepared by the loading of noble metals (Rh and/or Ru) on MgO carriers[7,8]. MgO carriers were highly crystallized, and surface areas of the catalysts were less than 5m^2/g[7,8]. The average side crashing strength of the catalysts was 500N/piece, and estimated pressure drop was 0.011MPa/m (gas linear velocity : 2.2m/s).

Bench test : An accelerated catalyst life test (CO_2 reforming) was employed in the bench scale test. The commercial size ring catalyst was crushed to 2-4mm size, and 50 cm^3 of the catalyst was loaded into the reactor. The reaction test using a conventional Ni loaded catalyst was also executed. The CO_2/CH_4 ratio of the feed gas was 1.0, and the reaction conditions were 1123K, 2.0MPaG, and GHSV=6000 h^{-1}.

Pilot test : For pilot plant operations, a commercial size catalyst batch of 0.12m^3 was charged into a single-tube reformer of 12.0m length and 11.0cm inner diameter and was reduced at around 1073K under atmospheric pressure before the experiments. Synthesis gas with H_2/CO=2.0 was generated directly by CO_2 and H_2O reforming of natural gas. The H_2O/CO_2/carbon ratio was 1.15-1.64/0.40-0.60/1, and the reaction conditions were 1138-1163K, 1.5-1.9MPaG and GHSV=3000 h^{-1}. Natural gas composition, $C_1/C_2/C_3/C_4/C_5/C_6/N_2$, was 85.8/8.3/3.1/1.3/0.1/0.1/1.3 vol%, respectively.

Carbon removal test: CH_4 diluted with N_2 was fed through the catalyst to make carbon deposits on the catalyst surface. The deposition condition was 1123K, 0.2MPaG of CH_4 partial pressure, 2.0MPaG of total pressure and GHSV=3000 h^{-1}. Deposited carbon was removed with pulsed steam (0.5 μL, 15pulses) at 1123K under atmospheric pressure. The quantities of removed carbon were analyzed as CO and CO_2 by mass spectroscopy. Five samples utilized in this test were as follows; one MgO carrier, three types of Ru/MgO catalysts and one conventional γ-Al_2O_3 carrier, respectively. The surfece area of γ-Al_2O_3 was 156m^2/g. Regarding the three Ru/MgO catalysts, the relative weights of Ru loaded on the MgO carriers were 1.0, 3.0 and 18.3, and they were represented as Ru1.0/MgO, Ru3.0/MgO and Ru18.3/MgO.

Results and Discussion

Figure 1 compares reaction conditions to produce various kinds of synthesis gases (1123K, 2.0MPaG). Similar carbon limit curves are utilized to design steam reformers[2,3]. The horizontal axis indicates feed O/C ratio, and the vertical axis shows feed H/C ratio. Oblique axes represent the feed gas composition (CO_2/CH_4 ratio and H_2O/CH_4 ratio), and the four dotted lines indicate feed gas composition to obtain various synthesis gases with different H_2/CO ratios.

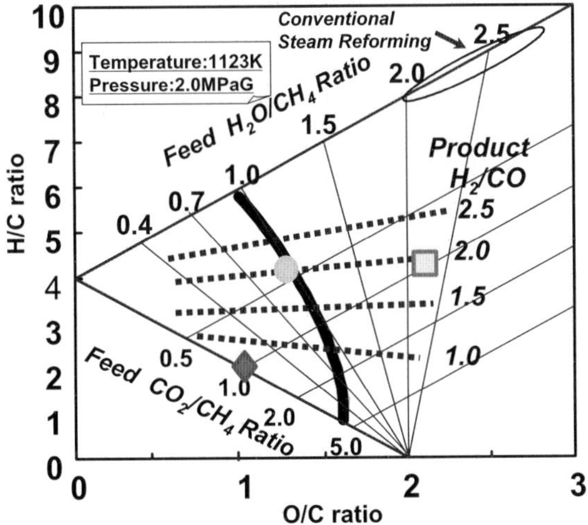

Fig. 1 Relation between Feed Gas Composition and H_2/CO Molar Ratios of Product Synthesis Gases.
Reaction conditions under which carbon deposition will occur are also indicated in this figure.

Nielsen et. al. defined carbon activity (Ac) as an indicator of the possibility of carbon formation[2,3]. For example, with regard to Boudouard reaction ($2CO \rightarrow C + CO_2$), Ac is expressed as the following equation.

$Ac = K \cdot (P_{CO})^2/(P_{CO_2})$

K; equilibrium constant

P_{CO}; partial pressure of CO

P_{CO_2}; partial pressure of CO_2

If this value exceeds 1.0, carbon is expected to appear in the gas phase. In the area that is enclosed by the bold curve and two oblique axes, product gases have high thermodynamic potential for carbon deposition because their carbon activity is higher than 1.0.

As shown in Fig. 1, conventional steam reforming operations are being executed under reaction conditions whose H_2O/CH_4 ratios are higher than 2.0 to avoid carbon deposition. To obtain synthesis gas with a ratio of 2.0 which is suitable for FT synthesis and methanol production by using a conventional steam reforming catalyst, the addition of CO_2 with an approximate ratio = 1.0 will be required. This reaction condition is shown in Fig. 1 as a square plot. Because H_2O reforming and CO_2 reforming are large endothermic reactions,

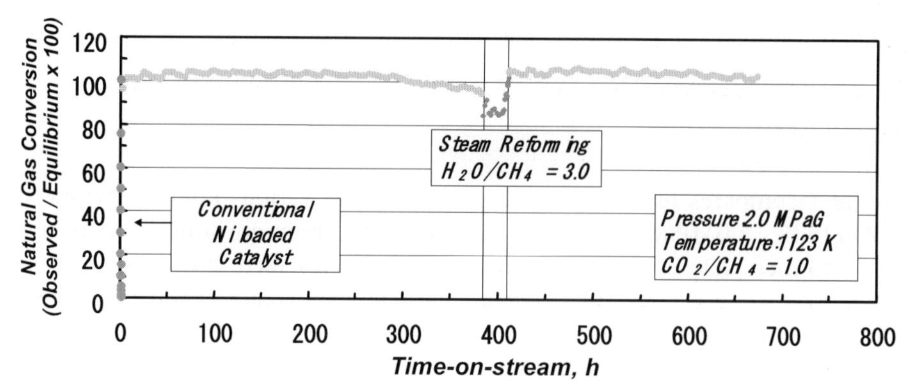

Fig. 2 Result of Catalyst Life Test for CO_2 reforming (H_2/CO=0.7) .

reduction of feed H_2O and CO_2 is required for an operation with higher energy efficiency. Our target condition for the production of synthesis gas with a ratio of 2.0 is shown as a circle plot. It will be impossible for conventional steam reforming catalysts to operate under this condition because of obstinate carbon deposition, though operation with low energy consumption is possible. In Fig. 1, a lozenge plot shows the reaction condition of CO_2 reforming. Carbon deposition will easily occur because carbon activity at the outlet of the catalyst bed is 2.6 on our CO_2 reforming experiment.

Accelerated catalyst life tests (CO_2 reforming) were employed in the bench scales. The typical results of the CO_2 reforming are shown in Fig. 2. In this figure, the conversions of natural gases are presented by comparison with the percentage of the equilibrium values in the same reaction conditions. A conventional Ni loaded catalyst was immediately deactivated under this condition. Ru3.0/MgO maintained stable activity for approximately 300 hours. These results supported the robustness of the catalyst against carbon deposition. After around 380 hours

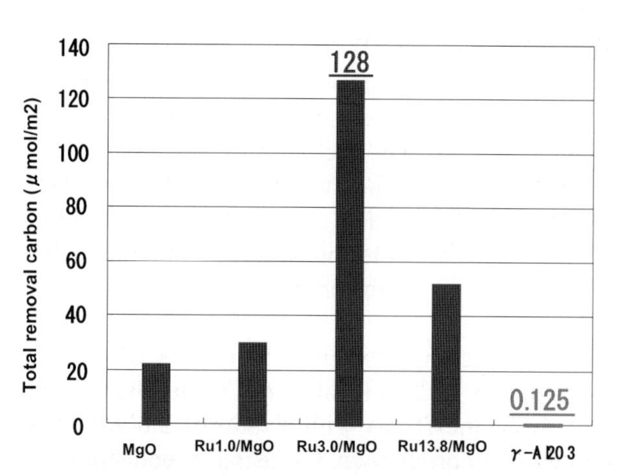

Fig. 3. The Quantities of Removed Carbon with Steam Pulses

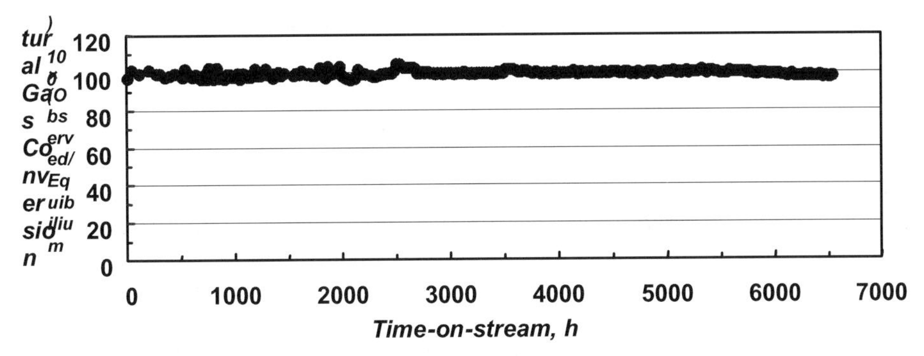

Fig. 4 Result of Pilot Plant Operation (H_2/CO=2.0).

on stream, the catalyst regeneration test was executed by steam reforming with H_2O/CH_4=3.0, and it was confirmed that the catalyst could be effectively regenerated. This result indicates that the morphology of carbon deposited on the catalyst surface seems to be a material like an atomic carbon that is easily converted into carbon monoxide by steam. The result of the regeneration test indicated that the steam played a significant role in removing the deposited carbon from the catalyst surface.

So the quantities of carbon removed from the catalyst surfaces with steam were evaluated for five samples. In Fig. 3, the amounts of removed carbon were shown as the values per total surface area of the samples installed into an analyzer. As shown in Fig. 3, our MgO showed an extremely high carbon removal characteristics in comparison with conventional γ-Al_2O_3. Nielsen et. al reported that the dissociation rate of steam was fast on the MgO surface[4]. The MgO itself has a high potential to activate steam on its surface[4], and thus activated steam is considered effective in removing deposited carbon in our experiment. Furthermore, the crystallized MgO can maintain a stable structure under reforming reaction conditions in which high pressure steam is present. Carbon removal activity was enhanced by the addition of metals, which indicates that active sites for carbon removal were newly generated on metal support interfaces. The decline of carbon removal activity on Ru18.3/MgO was considered to be due to the decrease of metal dispersion.

We also confirmed that our catalysts exhibited strong metal support interaction by TEM and CO adsorption. The metal particle size estimated by CO adsorption was extremely larger than the one estimated by TEM. The Boudouard reaction ($2CO \rightarrow C + CO_2$) will be prohibited because CO produced during the reforming reaction is easily removed from the catalyst surfaces.

The results of the pilot plant operation are shown in Fig. 4. In this test, an oparation was executed under the target conditions as explained in Fig. 1. Ru3.0/MgO was used in this test. Although raw natural gas contains higher hydrocarbons, as explained in experimental section, pre-reformer which is usually set to convert them to methane was not installed in our plant, and therefore raw natural was fed directory into the reformer. As shown in this figure, stable operation for approximately 7,000 hours was attained under the target reaction conditions. During the operation, little change in temperature and pressure drop of the catalyst bed was observed. After the operation, the catalyst was discharged from the reformer and carbon content of the catalyst was measured. The quantity of carbon deposited on the catalysts which had been loaded during the operation was less than 0.1wt%. These results demonstrated carbon-free operation.

We have plans to establish our technology with a GTL demonstration plant whose capacity will be 500BPD on the next phase of the GTL national project. Our GTL process is being developed by sharing development items: Chiyoda Corporation is in charge of the development with regard to the synthesis gas production section, Nippon Steel Corporation is taking charge of the development of the Fischer-Tropsch section, and Nippon Oil Corporation is carrying out the development of the hydro-cracking section. This will be the first GTL demonstration plant (from natural gas to FT oil) in Japan. After successful completion the new GTL process that we develop will be a promising candidate for the leading GTL process throughout Japan.

Conclusions

CO_2 reforming catalysts and a process that can generate synthesis gas by CO_2 and H_2O reforming with low CO_2/carbon and low H_2O/carbon ratios in feed gas composition have been developed. Catalyst performance was confirmed in pilot plant tests for approximately 7,000 hours. Our catalyst exhibited high removal activity regarding the removal of deposited carbon with steam.

References

1. P. J. Lakhapate, V. K. Prabhu, *Chemical Engineering World,* 2000, **35**, 77.
2. N. R. Udengaard, J.-H. B. Hansen and D. C. Hanson, *Oil & Gas Journal, March 9* (1992) 62.
3. J. R. Rostrup-Nielsen and J.-H. B. Hansen, *J. Catal.,* 1993, **144**, 38.
4. I. Alstrup, B.S. Clausen, C. Olsen, R. H. H. Smith and J. R. Rostrup-Nielsen, *Stud.Surf. Sci. Catal.*, 1998, **119**, 5.
5. K. Tomishige, Y-G. Chen, K. Fujimoto, *J. Catal.*, 1999, **181**, 91.
6. F. Yagi, R. Kanai, S. Wakamatsu, R. Kajiyama, Y. Suehiro and M. Shimura, *Catal. Tod.*, 2005, **104**, 2.
7. World Patent, WO98/46523
8. World Patent, WO98/46524

Natural Gas Conversion VIII
F.B. Noronha, M. Schmal, E.F. Sousa-Aguiar (Editors)

The optimization of preparation, reaction conditions and synthesis gas production by redox cycle using lattice oxygen

Xiaoping Dai [a, b], Changchun Yu [a*], Ranjia Li [a], Zhengping Hao [b*]

[a] *The Key Laboratory of Catalysis CNPC, University of Petroleum Beijing, Beijing 102249, China*
[b] *Research Center for Eco-Environmental Science, Chinese Academy of Sciences, Beijing 100085, PR China*

ABSTRACT

The aim of this work was to optimize the catalyst preparation and reaction conditions, and produce syngas by redox cycle over $LaFeO_3$ oxide using lattice oxygen instead of gaseous oxygen. Catalyst preparation methods, pretreatment processes and reaction conditions were investigated and characterizations were conducted by XRD and BET. The research results showed that $LaFeO_3$ oxide, prepared by SGW method and calcinated at 1173 K for 5 h, exhibited the best catalytic performance for the CH_4 partial oxidation without gaseous oxygen. It was proved possibly that the lattice oxygen instead of molecular reacted with methane to produce synthesis gas by sequential redox cycle.

1. INTRODUCTION

Production of synthesis gas from methane and conversion of synthesis gas to a range of fuels and chemicals could become increasingly of interest as the reserves of crude oil are depleted and/or the price of crude rises [1]. The catalytic partial oxidation of methane to synthesis gas was an exothermic reaction and gave a H_2/CO ratio of 2, which served as a feedstock of many chemical processes, e.g., Fischer-Tropsch and methanol syntheses.

A novel method for synthesis gas production by lattice oxygen instead of molecular oxygen has been proposed by our group [2]. The present paper reported the effects of catalyst preparation and reaction conditions on the

* Corresponding authors: Changchun Yu. Tel: 86 10 89707447, Email: daixpcup@126.com
Zhengping Hao Tel: 86-10-62923564, Email: zpinghao@rcees.ac.cn

catalytic performance for the methane partial oxidation to synthesis gas over $LaFeO_3$ catalyst in the absence of gaseous oxygen, and sequential redox cycles were investigated for the production of synthesis gas.

2. EXPERIMENTAL

The perovskite oxide $LaFeO_3$ catalysts were prepared by sol-gel method (ethanol (SGE) or water (SGW) as solvent), complexation method (CW), and pyrolysis method (PW) method. The details of the preparation method were previously described [3]. The catalyst precursors were calcined at 1173 K for 5 h. The characterizations of catalyst were carried out by XRD (ShiDU XRD-6000 diffracometer) and BET (NOVA1200).

The catalytic performance was tested by pulse reaction, continuous flow reactions and sequential redox cycles in a fixed-bed quartz micro-reactor filled with 200 mg sample. Detailed procedure was previously described [3].

3. RESULTS AND DISCUSSION

3.1. The optimization of preparation conditions

The tested catalysts exhibited more difference in activity among various samples by CW, SGW, PW and SGE methods, as shown in Fig. 1.

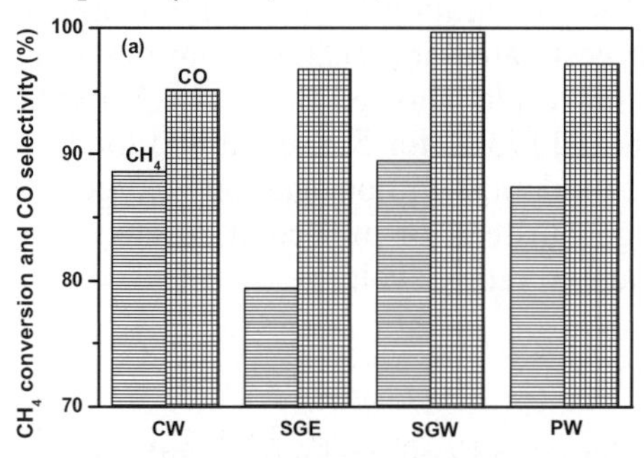

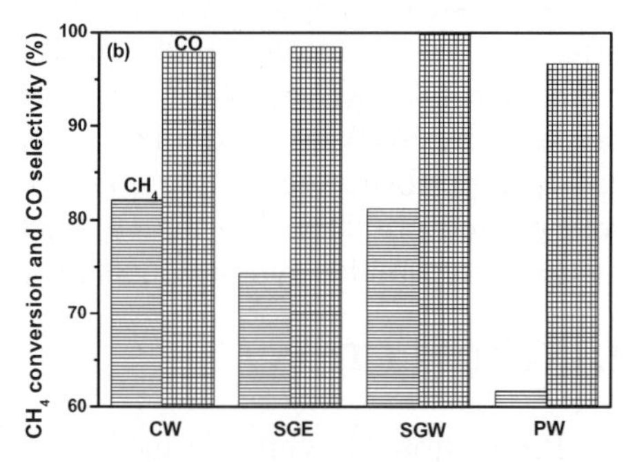

Fig. 1 The reaction results by different preparation methods for methane oxidation at 1173 K using lattice oxygen (a) Fresh catalyst (b) Recovery catalyst partially with gaseous oxygen

The SGE sample displayed the lowest activity for methane conversion, while the CO selectivity was the highest over SGW sample over fresh samples. But the difference of CO selectivity was not distinct over these samples in recovered oxygen partially, and the activity declined at different degree over all these samples. The declined trend was clearer over PW sample for methane conversion. Obviously, the amount of oxygen can not only be ascribed to the surface oxygen, because the amount of oxygen at the first surface layer on the basis of low BET surface area (ca. 6 m^2/g oxide calcined at 1173 K). Therefore, it was clear that the bulk lattice oxygen of $LaFeO_3$ oxide should participate in the partial oxidation of CH_4 to synthesis gas.

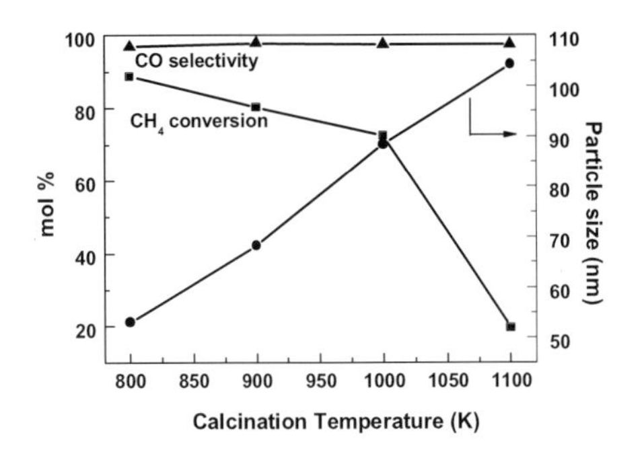

Fig. 2 The CH_4 conversion, CO selectivity and particle size vs. calcination temperature

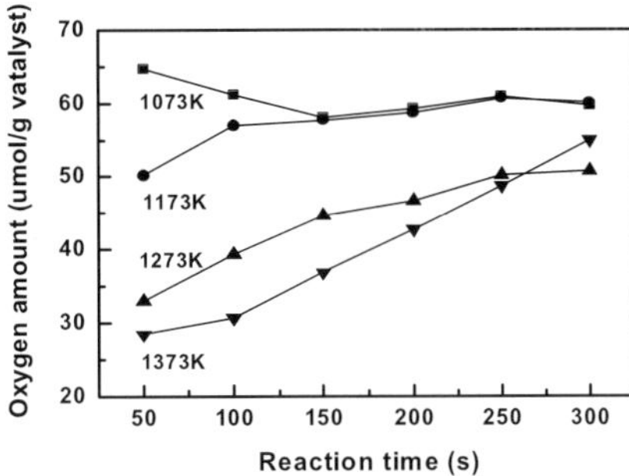

Fig. 3 The oxygen exchange capacity of 50 s cumulative data at different calcination

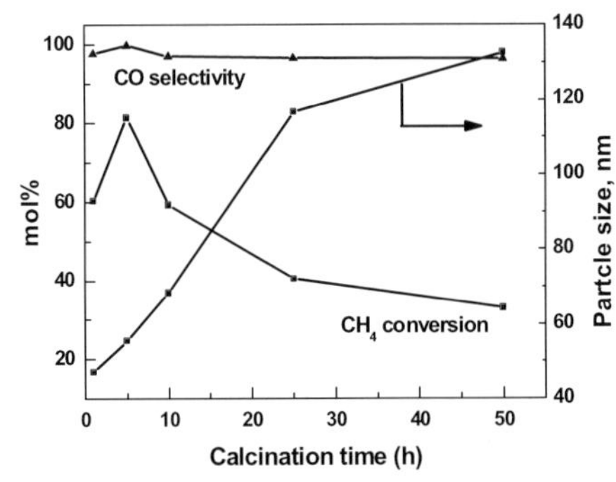

Fig. 4 The CH_4 conversion, CO selectivity vs. calcination time at 1173 K

Fig. 2 presented the results of the methane partial oxidation experiments performed by continuous flow reaction over $LaFeO_3$ catalyst. It was observed that the lower the calcination temperature, the higher the methane conversion. Especially, at the higher calcination temperature, the declined trend was more quickly. CO selectivity kept high level of 95%, and increased slightly when the calcination temperature increased. With the calcination temperature increased, the surface particles were highly agglomerated and the size was increased. The effect of calcination temperature for partial oxidation of methane was twofold. On one hand, it was favorable to produce more oxygen vacancies in which active oxygen was formed at higher temperature. On the other hand, the methane conversion declined because of the decrease of surface area at higher calcination. From Fig. 3, the oxygen exchange capacity was highest at lower calcination temperature. With the reaction time increased, the oxygen exchange capacity increased. Exposure of $LaFeO_3$ oxides to a reduced atmosphere could generate oxygen vacancies due to the loss of lattice oxygen and reduction of the oxidative stated Fe ion simultaneously, providing pathways of oxygen transport through the lattice [4], which decreased the resistance of oxygen migration and lead to enhancement of oxygen diffusivity.

The methane conversion increased firstly with calcination time, and reach it's maximum at 5 h, then declined with calcination time, as shown in Fig. 4. The CO selectivities kept the level of 95% among the tested calcination time, while particle sizes increased sharply with calcination time prolongated.
The preparation conditions were optimized by above investigations, and the catalyst, prepared by SGW method, calcination temperature 1173 K for 5 h, exhibited the best catalytic performance.

3.2. The effluence of reaction conditions

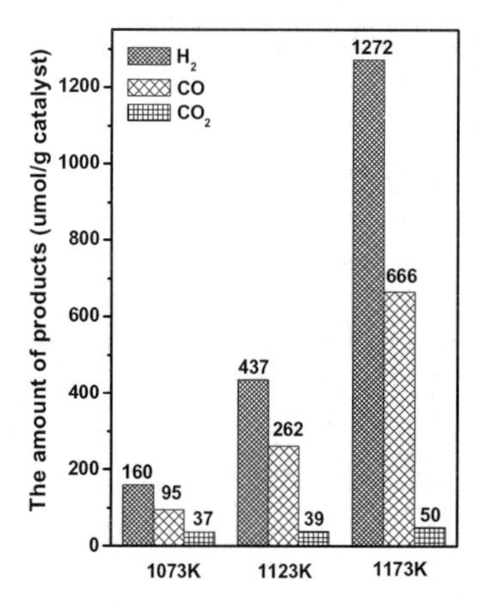

Fig. 5 The total amount of products in the 15 pulse reactions over $LaFeO_3$ oxide at different temperature

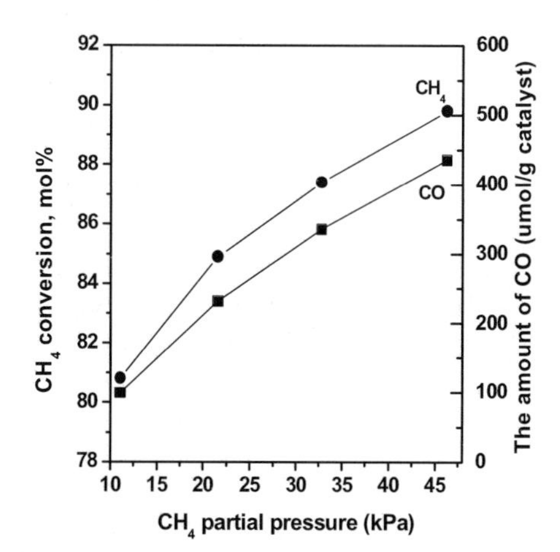

Fig. 6 Methane conversion and CO selectivity as a function of methane partial pressure at 1173 K

From Fig. 5, the total amount of CO and H_2 increased sharply when reaction temperature increased from 1073 K to 1173 K, due to the higher diffusion rate at high temperature. The total selectivity of CO increased with reaction temperature, and much more of CH_4 was selectively oxidized to CO and H_2 at higher temperature. With the oxygen species on the surface of $LaFeO_3$ oxide declined, the increased concentration of oxygen vacancies provided pathways of oxygen transport through the lattice, and lead to enhancement of oxygen diffusivity. Compared with literature [5, 6], the carbon resistance was improved, and the amount of oxygen species for synthesis gas formation increased.

With the increase of methane partial pressure, the methane conversion and the amount of CO increased, as shown in Fig. 6. This was due to the fact that, with the increase of methane partial pressure, it was favorable to increase the chance of contact with active site. On the other hand, with the increase of CH_4 conversion, the exothermic reaction could lead to the temperature of catalyst bed increased, which was often observed in methane partial oxidation [7, 8]. This was favorable to the migration of oxygen species and methane conversion.

3.3. The syngas production by redox cycle over $LaFeO_3$ catalyst

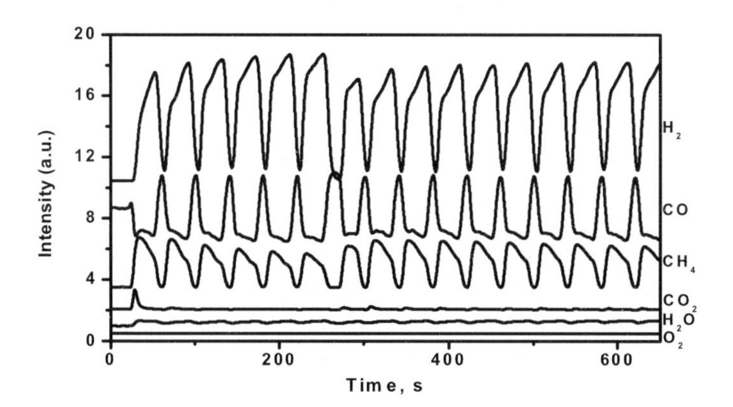

Fig. 7 Transient responses by redox cycles between O_2/Ar and CH_4/He over $LaFeO_3$ oxide at 1173K

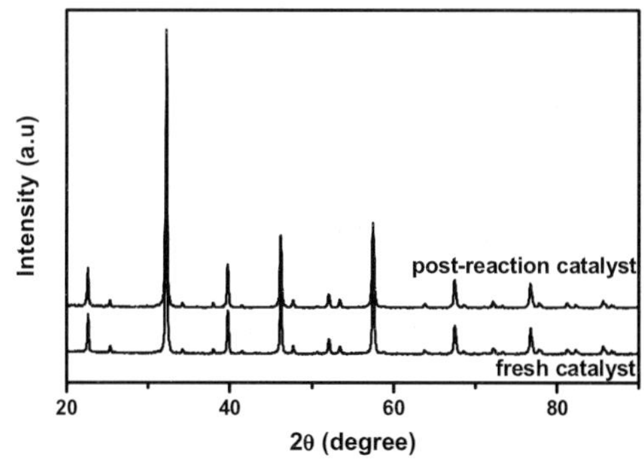

Fig. 8 XRD profiles over fresh and post-reaction $LaFeO_3$ catalyst

The results by sequential redox cycle in Fig. 7, showed the reactant and product responses between O_2/Ar oxidation for 10 s and CH_4/He reduction for 30 s over the $LaFeO_3$ catalyst at 1173 K. CO_2 was formed firstly, then CO and H_2 were detected for the first cycle. However, from the second redox cycle to sixth redox cycle, only a small quantity of CO_2 was produced, while CO and H_2

were the dominant products. With the increase of cyclic number from the seventh redox cycle, the amount of CO_2 increased, which may be due to the change of the redox properties of Fe.

From the 15 redox cycles at 1173 K, the $LaFeO_3$ oxide maintained relatively high catalytic activity (Fig. 7) and structural stability from XRD characterization (as shown in Fig. 8), but the products provides evidence of the appearance of deactivation phenomena. When the lattice oxygen consumed at reductive atmospheres (CH_4/He) was partially restored at oxidative atmospheres for 10 s, the oxidation level of the catalyst may not reach the initial state, but the most reactive oxygen species can be recovered, which can react with CH_4 to form synthesis gas. The results of catalyst reoxidation at redox cycle indicated that the oxide had a high reactivity towards the oxygen and the oxygen can be consumed completely. It was possible that the lattice oxygen instead of molecular reacted with methane to synthesis gas by sequential redox cycle.

4. CONCLUSION

Catalyst preparation and reaction conditions were optimized by pulse reaction and continuous flow reactions. The $LaFeO_3$ oxide, prepared by SGW method and calcinated at 1173 K for 5 h, exhibited the best catalytic performance for the CH_4 partial oxidation without gaseous oxygen. The rate of oxygen migration during the CH_4 reaction with $LaFeO_3$ was strongly affected by the reaction temperature and methane partial pressure, and increased with raising temperature and methane partial pressure. Methane can be converted selectively to synthesis gas, under the appropriate reaction conditions by sequential redox cycles. The oxygen species of selective oxidation CH_4 to synthesis gas can be recovered by re-oxidation using gaseous oxygen.

ACKNOWLEDGEMENTS

Financial funds from the Chinese Natural Science Foundation (project No. 20306016, 20322201) are gratefully acknowledged.

REFERENCES:

1. Dry Mark E, Catal. Today, 2002, 71 (3-4): 227.
2. Li R J, Yu C C, Dai X P, Shen S K, Chinese J. Catal., 2002, 23 (6): 549.
3. Dai X P, Wu Q, Li R J, Yu C C, Hao Z P, J. Phys. Chem. B, 2006, 110 (51), 25856.
4. Ramos T, Atkinson A, Solid State Ionics, 2004, 170 (3-4): 275.
5. Yaremchenko A A, Kharton V V, Veniaminov S A, Belyaev V D, Sobyanin V A, Marques F M B, Catal Commun., 2007, 8 (3) 335.
6. Li R J, PhD thesis, China University of Petroleum (Beijing) (2002).
7. Ji Y Y, Li W Z, Xu H Y, Chen Y X, Appl. Catal. A, 2001, 213 (1): 25.
8. Liu S L, Xiong, G X; Dong H, Yang W S, Appl. Catal. A, 2000, 202 (1): 141.

Natural Gas Conversion VIII
F.B. Noronha, M. Schmal, E.F. Sousa-Aguiar (Editors)

Effect of Rh addition on activity and stability over Ni/γ-Al2O3 catalysts during methane reforming with CO2

Marco Ocsachoque[a], Claudia E. Quincoces[a], M. Gloria González[a]

a Centro de Investigación y Desarrollo en Ciencias Aplicadas Dr. J.J.Ronco (CINDECA) (CONICET, UNLP), 47 Nro 257, 1900 La Plata, Argentina.

ABSTRACT

The effect of low Rh contents was studied in Ni/γ-Al_2O_3 catalysts. Activity, stability and carbon deposition on these catalysts were analyzed for methane reforming with CO_2. Catalysts were characterized by XRD, TGA, TPR and flow-reaction.
Results indicated that the catalyst promoted by Rh showed a higher activity and a similar carbon deposition than the Rh and Ni monometallic catalysts. Rh addition favors an interaction compound in the Rh-Ni/γ-Al_2O_3 that would improve the methane reforming activity of the supported Ni-based catalysts.

1. INTRODUCTION

In last decades, methane reforming with CO_2 to produce synthesis gas has acquired special attention due to appraisement of natural gas and CO_2, both of them with impact on the environment and energy resources [1,3]. This reaction is attractive because it can be employed in areas where the water is not available and it produces syngas with lower H_2/CO ratios convenient for Fisher-Tropsch synthesis [4] and can contribute to the use of natural gas fields containing considerable CO_2 amount.

Reaction conditions (high temperatures, CH_4/CO_2 ratio) used in methane reforming are the principal deactivation causes of supported Ni catalysts employed at industrial scale. Deactivation produced by carbon deposition, metallic sintering and poisoning by sulfur affects their catalytic properties. At the same time, supported catalysts of noble metals presented high catalytic activity for dry reforming and resistance to deactivation by carbon deposition. However, it is important to develop Ni catalysts resistant to deactivation considering the low availability and high cost of noble metals. One of the most important options for enhancing activity and stability in methane reforming reactions is the use of bimetallic catalyst. Several studies [5-11] report that addition of noble metals to Ni- based catalyst increases the thioresistance and reduces carbon deposition. Rostrup-Nielsen reported that Rh and Ru decrease the formation of filamentous carbon due to low carbon solubility in the metal. Ruckennstein and Wang [12] stated that supports as γ-Al_2O_3, SiO_2, La_2O_3 and MgO have important effect on the metallic Rh formation; this metallic Rh is proposed as active site for reforming with CO_2. These results suggest that the addition of noble metals in the formulation of Ni/γ-Al_2O_3 catalysts can improve stability and resistance to catalyst deactivation, and in this way they are more convenient at industrial scale.

The aim of this work was analized the effect of the addition of low Rh contents in Ni/γ-Al_2O_3 catalysts. Activity, stability and carbon deposition on these catalysts were studied for methane reforming with CO_2.

2. EXPERIMENTAL

Ni and Rh monometallic catalysts were prepared by impregnation at incipient wetness of γ-alumina with solutions of $Ni(NO_3)_2.4H_2O$ and $RhCl_3.4H_2O$, respectively. The corresponding bimetallic samples were prepared by impregnating first Rh and then Ni, in nominal contents of 0.5% Rh and 5% Ni onto the support. Then, Ni and Rh-Ni catalysts were dried and calcined at 500°C for 1 h. The Rh monometallic sample was calcined at 380°C. Before the catalytic test, catalysts were reduced in situ with a pure hydrogen flow at 650°C for 1h.

Fresh and used catalysts were characterized by X Ray Diffraction, Temperature Programmed Reduction and Thermal Gravimetric Analysis in order to analyze physical and physicochemical characteristics and their effect on catalytic properties, with special emphasis in deactivation problems.

Identification of the crystalline phase by XRD was determined in a Philips PW 1740 equipment, with CuKα radiation in the range 2θ = 20-80° and with scan speed of 0.02°min^{-1}.

TPR analyses were carried out in a conventional equipment (Quantasorb QS JR-2) with samples of 0.1g heated from RT to 1000°C at a rate of 10°C min^{-1} using a 10% (v/v) H_2/N_2 gas flow rate of 20 cm^3 min^{-1}.

The amount of carbonaceous deposits was determined by thermogravimetric analysis (TGA) in a Shimadzu TGA-50H under air stream (20 cm^3 min^{-1}). The temperature was increased from 25 to 800°C at a heating rate of 10°C min^{-1}. Experiments were performed over samples extracted from the reactor after 20 reaction hours.

Catalytic properties were determined in a flow system with fixed bed reactor connected in series to a Perkin Elmer chromatograph containing a 4 m Porapack Q column, a thermal conductivity detector and He gas (40 cm^3 min^{-1}) as carrier. The reactor was fed with 100 cm^3 min^{-1} of a CH_4/CO_2=1.2 mixture and He as balance. A mass of 0.05 g catalyst was loaded into the quartz reactor. Catalytic reaction was performed at 650°C and atmospheric pressure in conditions of chemical control. The sulfur resistance was measured by adding 0.3 ppm of H_2S to the reactor feed.

3. RESULTS AND DISCUSSION

Catalytic test

Table 1 shows the effect of Rh addition on the initial activity and stability after 15 reaction hours. The deactivation degree (R_D) is defined as the ratio between initial conversion to conversion at 15 h in reaction.

Results indicate higher activity for the bimetallic catalyst and a similar deactivation degree for tested samples. The investigation will be continued under more severe temperature conditions to analyze the behavior of Rh-Ni catalysts with respect to carbon deposition.

Table 1. Catalytic activity and stability.

Catalyst	$\%X_{CH4(1h)}$	$R_D^{(1)}$
Rh/γ- Al_2O_3	36.6	0.95
Ni/γ- Al_2O_3	37.0	0.90
Rh-Ni/γ- Al_2O_3	42.7	0.90

Thermogravimetric analysis, in air current, shows that carbon contents in Rh-Ni catalyst (2.5%) and Ni catalyst (2.1%) were similar (Fig.1a). dTGA as a function of temperature for these catalysts is represented in Fig. 1b, where it is observed that the Ni catalyst presents an only broad peak between 370 and 700°C while the bimetallic catalyst show two signals, at 400 and 680 °C.

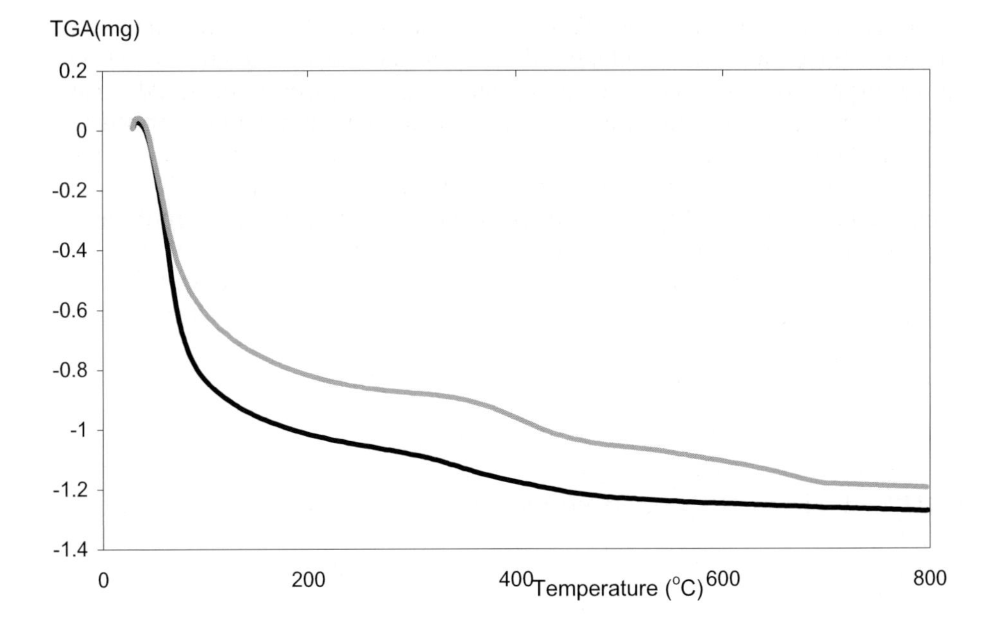

Fig.1a. Thermogravimetric diagrams of Ni monometallic (black curve) and Rh-Ni bimetallic (red curve) supported catalysts.

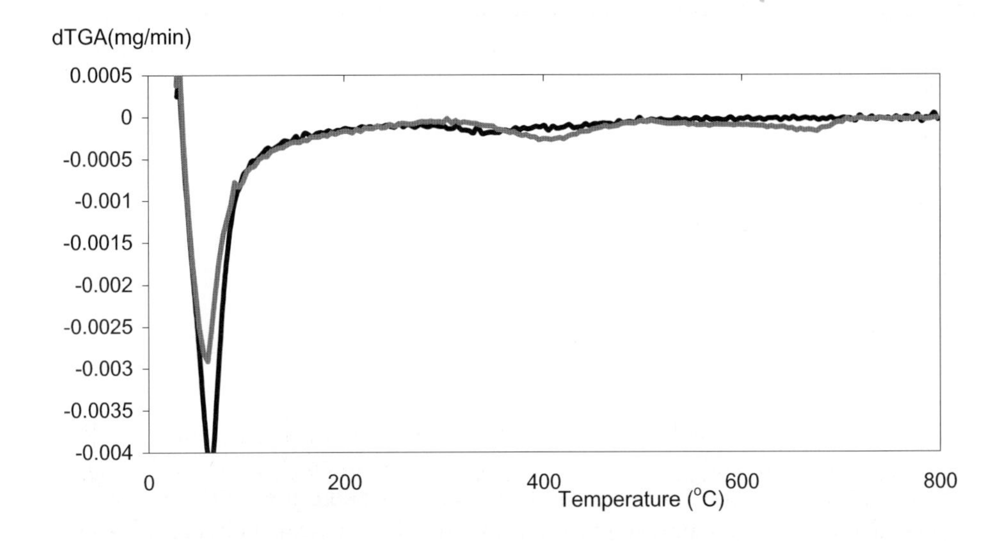

Fig.1b. dTGA for Ni (black curve) and Rh-Ni (red curve) catalysts.

The signal at 400°C may be related with a superficial carbide and the second species above 600°C is identified as whisker-like filamentous carbon. According to literature [13], species at 400°C may be reaction intermediate and their reactivity can be correlated with catalytic activity and with CO formation. The whisker- like filamentous carbon corresponds to absorbed carbon atoms derived from methane decomposition and CO dissociation.

Concerning the stability, the bimetallic catalyst showed high resistance to carbon formation, instead, this catalyst was sensitive to poisoning by sulfur presenting low thioresistance which leads to a decrease of its activity to 35% of its initial activity, after 23 hours.

Characterization of catalysts

Figure 2 shows TPR profiles of mono and bimetallic catalysts. For the monometallic Ni catalyst, two peaks at 420 and 640°C can be attributed to Ni interaction with the support. The peak around 640°C had a long tail toward higher temperature and a significantly stronger intensity than that at 420°C. These results indicate that at least two types of NiO species were present in the Ni catalyst. One type was NiO, which was weakly interacted with γ- Al_2O_3. The other NiO species was assignable to a strong interaction with the support surface by a solid-state reaction between NiO and γ-Al_2O_3. For Rh-Ni catalyst, a broad peak from 300 to 493°C can be observed and this is attributed to interaction of reduced Rh with the support [14]. It is well known that Rh and alumina interact by Rh diffusion into subsurface and bulk

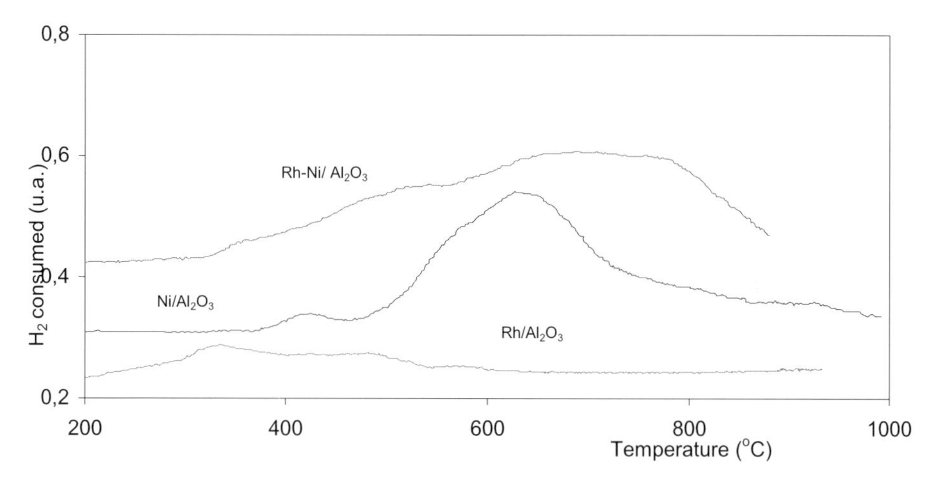

Fig.2. Temperature Programmed Reduction profiles of mono and bimetallic catalysts.

of γ-Al_2O_3 when the catalyst is calcined in an oxidizing atmosphere. This makes it possible that Rh oxide is less reducible [15,16].

Spectra of Rh-Ni bimetallic catalyst show a broad peak at 750°C. This signal is attributed to an Rh-Ni interaction compound that implies an intimate contact between Ni and Rh.

X-Ray diffractograms of reduced Ni-Rh/γ-Al_2O_3 do not show the Ni° signal, which is observed in the Ni/γ-Al_2O_3 sample. XRD results would indicate the presence of small Ni particles in the bimetallic catalyst and would suggest that the Rh addition favors Ni dispersion on the support surface.

CONCLUSIONS

Conclusions of this work can be summarized as follows:

-The addition of 0.5% Rh in Ni/γ-Al_2O_3 catalysts improves the catalytic activity by modifications of the active phase.

-TPR results for Rh-Ni/γ-Al_2O_3 catalysts show the existence of a species that is reduced at high temperature assigned to an interaction compound between the two metallic species. By XRD, the disappearance of the Ni signal in the bimetallic catalyst suggests that the Rh presence favors Ni dispersion.

-The addition of Rh favors the formation of more reactive carbonous species that are correlated with catalytic activity.

-The interaction between Rh and Ni and the Ni dispersion would be the responsible for the activity increase in the bimetallic catalyst.

References

1.M.C.J. Bradford, M.A. Vannice, Catal. Rev. Sci. Eng. 41 (1) (1999).
2. S.Wang, G.W. Lu, Appl. Catal. A 169 (1998) 271.
3.J.A. Lercher, J.H. Bitter, W. Hally, W. Niessen, K. Seshan, Stud. Surf. Sci. Ctal.101 (1996) 463.
4. A.M. Gadalla, B. Bower, Chem. Eng. Sci 43 (1988) 3049.
5. J.J. Strohm, J. Zheng, C. Song, J. Catal. 283 (2006) 309.
6. Z. Hou, P. Chen, H. Fang, X. Zheng, T. Yashima, Int. J. Hydrogen Energy (in press).
7.- J.R. Rostrup-Nielsen, T.S. Chistensen, I. Dybkjaer, Stud. Surf. Sci Catal 113 (1998) 81.
8.-J.H.Bitter, S. Seshan, J.A.Lercher, J. catal. 176 (1998) 93.
9.-S. Yokota, K. Okumura, M. Niwa, Catal. Lett. 84 (2002) 131.
10.- Z.L.Zhang, V.A.Tsipouriari, A.M. Efstathiou, X.E. Verykios, J. Catal. 158 (1996) 51.
11.-H.Y. Wang, E.Ruckenstein, Appl. Catal. 204 (2000) 143.
12.- E. Ruckenstein and H.Y. Wang, J. Catal. 187 (1999) 151.
13.- Y-G.Chen, J. Ren, catal. Lett. 29 (1994) 39.
14.- M. Ferrandon and T. Krause, Appl. Catal. A 311 (2006) 135.
15.- H.C. Yao, S. Japar, M. Shelef, J. Catal. 50 (1977) 407.
16.- R. Burch, P.K. Loader, N.A. Cruise, Appl. Catal 147 (1996) 375.

Natural Gas Conversion VIII
F.B. Noronha, M. Schmal, E.F. Sousa-Aguiar (Editors)

Slurry Phase DME Direct Synthesis Technology -100 tons/day Demonstration Plant Operation and Scale up Study-

Yotaro Ohno[a], Hiroshi Yagi[a], Norio Inoue[b], Keiichi Okuyama[b], Seiji Aoki[b]

[a] JFE Holdings Inc., 1-2 Marunouchi 1-chome, Chiyoda-ku, Tokyo 100-0005, Japan
[b] DME Development Co., 3-9 Shinbashi 3-chome, Minato-ku, Tokyo 105-0004, Japan

1. Introduction

Dimethyl ether (DME) is an innovative clean fuel which can be used for various sectors: household, transportation, power generation, etc. How to transport and utilize DME is extensively investigated in Japan, because DME could be a promising option to diversify energy resources in the future. DME has been used mainly as a propellant for spray cans. Recently, in China, they have started to use it as a LPG substitute. Approximately 200,000 tons/year are produced worldwide by a dehydration reaction of methanol. In order to use DME as a fuel, it must be produced at low cost in large quantities.

As methanol, itself, is produced from synthesis gas (hydrogen and carbon monoxide), it would be more efficient to produce DME directly from synthesis gas. We have carried out the technical development of DME direct synthesis process from synthesis gas for the past 15 years.

In the fundamental research, a new efficient catalyst and slurry phase process for DME synthesis were developed [1]. After pilot plant (5 tons/day) testing [2], five years development project using 100 tons/day demonstration plant has started in 2002 by DME Development Co., Ltd. which is a consortium of nine Japanese companies and one French company: JFE, Taiyo Nissan Corporation, Toyota Tsusho Corporation, Hitachi Ltd., Idemitsu Kosan Co., Ltd, Marubeni Corporation, INPEX Corporation, LNG Japan Corporation, Japex Co., Ltd and Total, with funds provided by the Ministry of Economy, Trade and Industry, Japan [3].

2. Characteristics of DME direct synthesis and synthesis gas production

2.1. DME synthesis reaction and equilibrium conversion

DME synthesis reaction (1) from synthesis gas (H_2, CO) is composed of three reactions: methanol synthesis reaction (2), methanol dehydration reaction (3), and water gas shift reaction (4). The overall reaction is exothermic and the reaction heat at methanol synthesis step is dominant.

$$3CO+3H_2 \rightarrow CH_3OCH_3+CO_2 \quad -246.0\text{kJ/DME-mol} \quad (1)$$
$$2CO+4H_2 \rightarrow 2CH_3OH \quad -181.6\text{kJ/DME-mol} \quad (2)$$
$$2CH_3OH \rightarrow CH_3OCH_3+H_2O \quad -23.4\text{kJ/DME-mol} \quad (3)$$
$$CO+H_2O \rightarrow CO_2+H_2 \quad -41.0\text{kJ/DME-mol} \quad (4)$$

The equilibrium conversion of synthesis gas (CO conversion plus H_2 conversion) for DME synthesis reaction (1) is much higher than that for methanol synthesis reaction (2) and has its maximum peak where H_2/CO ratio corresponds to the stoichiometric value, that is, with H_2/CO ratio of 1.0.

2.2 Slurry phase reactor for DME synthesis and catalyst system

As the reaction of DME synthesis is highly exothermic, it is more important to control the reaction temperature than that in the case of methanol synthesis, because the higher equilibrium conversion of DME synthesis gives much more reaction heat, and hot spot in the reactor could damage the catalyst.

In the slurry which is composed of an inert solvent containing fine catalyst particles, the reactant gas forms bubbles and diffuses into the solvent, and chemical reaction takes place on the catalyst. The reaction heat is quickly absorbed by the solvent, which has a large heat capacity, and thanks to the high effective heat conductivity, the temperature distribution in the slurry could be homogeneous.

In the slurry phase reactor, as catalyst particles are surrounded by the solvent, mass transfer mechanism of reactants and products is different from that in the fixed bed reactor. In this connection, a catalyst system adequate to the slurry phase reactor has been developed. In order to enable a large scale test, a catalyst mass production technology has been also developed.

There is an additional merit that the slurry phase reactor for DME synthesis is simpler than that of FT synthesis. It is because DME synthesis reaction products are all in gas phase at the reactor outlet and separation of liquid product and catalyst is not required.

2.3. Separation of products from unreacted synthesis gas

By-product of reaction (1) is CO_2, actually with methanol and a very small amount of water, separation of liquid product and unreacted synthesis gas is efficiently done at chilled temperature(–39°C). CO_2 can be removed by dissolving in product DME.

In recent SRI report [4], an evaluation of so called JFE process is introduced. However, unfortunately the process described as JFE process is quite different from our actual process, especially for CO_2 separation, maybe due to lack of information.

JFE's basis is to use a physical absorption based on DME - CO_2 vapor liquid relations. On the other hand, SRI uses a chemical absorption based on MDEA. In SRI's process, due to the chemical absorption, neither methanol nor DME is allowed in synthesis gas for the MDEA absorber, the separation of the product such as DME and methanol from synthesis gas is very strict and therefore, the synthesis gas should be chilled less than –51°C. Two distillation columns are provided for this separation.

The separation sequence is also different. In SRI's process, the sequence is the synthesis gas-product separation, the water removal and the DME-methanol separation, while JFE's the synthesis gas-product separation, the DME purification and the methanol-water separation and through this sequence, the methanol-water separation becomes smaller than that of SRI's. In total, SRI's process requires six columns in the separation and purification section and the very severe separation, while JFE's requires only three columns, which leads to less energy and investment cost.

2.4. Newly developed autothermal reformer with recycled CO_2 for synthesis gas production

The synthesis gas ($H_2/CO=1$) adequate for DME synthesis can be produced in an autothermal reformer (ATR) with O_2 and recycled CO_2 from DME synthesis itself by the following reaction;

$$2CH_4+O_2+CO_2 \rightarrow 3CO+3H_2+H_2O\downarrow \quad -30.4\text{kJ/DME-mol} \qquad (5)$$

The burner top structure and furnace inner profile were designed to enhance mixing of feed in order to realize both higher thermal efficiency and much lower soot formation and residual methane at the same time. Water cooled burner and well designed refractory works are applied to keep its robustness.

Reforming catalyst bed is located at the lower part of ATR to have role of completing reforming and having better gas distribution. Synthesis gas going out of ATR is rapidly quenched to prevent Boudouard reaction. Synthesis gas production (5) and DME synthesis reaction (1) give the following overall reaction:

$$2CH_4+O_2 \rightarrow CH_3OCH_3+H_2O \quad -276.4\text{kJ/DME-mol} \qquad (6)$$

which indicates that DME and water is generated by a kind of partial oxidation of methane.

2.5. Comparison of Direct DME synthesis and Indirect DME synthesis

Table 1 shows a comparison of "Direct synthesis" and "Indirect synthesis". Indirect synthesis is two step process of methanol synthesis by reaction (2) from synthesis gas and DME production by dehydration reaction (3).

The synthesis gas ($H_2/CO=2$) adequate for methanol synthesis can be produced similarly in ATR with O_2 and steam by the following reaction;

$$2CH_4+O_2+H_2O \rightarrow 2CO+4H_2+H_2O\downarrow \quad -71.4\text{kJ/DME-mol} \qquad (7)$$

The combination of (7),(2) and (3) gives the same overall reaction formula (6).

Table 1. Characteristics of Two DME process

Process	Direct (JFE)	Indirect (Two Step)	
		Methanol	Dehydration
Reaction pressure(MPa)	5	8-10	1-2
Reaction temperature(°C)	240-280	180-270	300-340
One through conversion(%)	50	38	70
Reaction by-product	CO_2	-	Water
(Water+MeOH)/DME (molar ratio)	0.1	-	1.9
Reactor	Slurry Phase	Fixed Bed	Fixed Bed
Cold gas efficiency(%)	71(83)	57(83)	
( ) – Theoretical cold gas efficiency(%)		66(84)	87(98)

The methanol synthesis is done actually at pressure 8-10MPa because the equilibrium conversion of the methanol synthesis is low at pressure 5MPa.

Although conversion of dehydration reaction is high, around 70%, the energy requirement for evaporation of liquid methanol and separation of three components, DME, by-product water and unreacted methanol is very large. Taking account of this energy requirement, the cold gas efficiency of dehydration process goes down to 87% and the overall cold gas efficiency of DME production from natural gas is estimated as about 57%.

The calculated theoretical value of cold gas efficiencies of the DME production from methane, the methanol production from methane, and the dehydration of methanol is shown in Table 1. In this table, estimated values for actual plant are also indicated; an evaluated value from operation results of an actual plant of the methanol synthesis, a predicted value by process simulation for the methanol dehydration and the direct DME synthesis.

It is understood that the direct synthesis could have big advantages of the lower synthesis pressure and the higher cold gas efficiency over the indirect synthesis.

3. Operation results of 100 tons/day demonstration plant

3.1. DME 100 tons/day demonstration plant

Figure 1 shows a simplified process flow diagram of 100 tons/day demonstration plant located at Kushiro in the north of Japan.

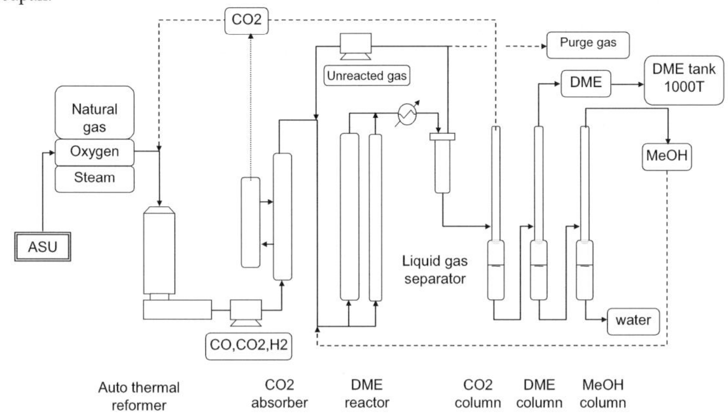

Figure 1. Process Flow Diagram of 100 tons/day DME Synthesis Plant

Natural gas is reformed in an autothermal reformer at the condition; the outlet temperature is 1000-1200°C and the inner pressure is 2.3MPa with oxygen, steam and carbon dioxide recycled from a carbon dioxide removal unit and a purification unit to give synthesis gas of $H_2/CO=1$. The synthesis gas is cooled, compressed and carbon dioxide is removed by a methanol absorption and supplied into DME synthesis reactor. There are two reactors in parallel; main reactor is 2.3m in inner diameter and 22m in height, small reactor is 0.65m in inner diameter and 28m in height. The small reactor is equipped to get various engineering data with higher gas velocity conditions. Reaction heat is removed by internal heat exchanger coils to generate steam. The standard reaction condition is temperature: 260°C and pressure: 5MPa. The effluent of the reactor is cooled and chilled to separate DME from unreacted gas, which is recycled to the reactor. DME is purified in two distillation columns and stored in pressurized tank (1,000tons). By-product methanol is recycled to the DME synthesis reactor after water removal to be converted into DME.

3.2. Operation results of 100 tons/day demonstration plant

The design, manufacturing and construction of 100 tons/day demonstration plant started in 2002 and completed in November 2003.

As shown in Table 2, six test runs were carried out during 2003-2006 to demonstrate the viability of the technology and to obtain various engineering data aimed to establish the scale-up technology to commercial plant.

Table 2. Results of Test Operation

RUN NO.	Period	Duration (day)	DME production(t)
RUN100	2003/12/12-2004/1/26	43	1,240
RUN200	2004/6/20-7/31	39	2,500
RUN300	2004/10/6-12/16	72	4,230
RUN400/500	2005/6/15-11/13	152	9,070
RUN600	2006/4/1-5/12	40	2,480
(Total)		346	19,520

Although the first run had some mechanical troubles, after that, the operations were very stable for both synthesis gas production and DME synthesis. The total operation time was almost one year (346 days). The cumulative DME production reached to about 20,000 tons. The product DME was supplied for various development projects of DME utilizations. The fourth run and the fifth one were conducted continuously to get the long term performance data of the catalyst.

The cold gas efficiency (Calorific value of produced DME /Calorific value of feed natural gas as raw materials) of overall process was evaluated as 69.4% from the operational data. This value is sufficiently high for demonstration plant and that of commercial scale plant can be estimated as 71%, which satisfies the assumed value in Table 1. The cold gas efficiency of ATR section and DME synthesis section was 84.9% and 81.7%, respectively.

3.3. ATR performance

At the start-up of ATR, it was supplied with preheated natural gas, steam and oxygen, and the gas temperature at the outlet of ATR was controlled mainly by steam addition. After DME synthesis started up, by-product CO_2 was recycled to ATR and, replacing steam, H_2/CO ratio of synthesis gas was adjusted to 1.

Under the condition of m =0.34 (m is oxygen ratio, m=1 is defined for complete combustion), S/C=0.25, outlet gas temperature is 1,050°C at 2.3MPa, 90% of reforming reaction took place before the catalyst bed, the amount of soot formation and residual methane was very low; 0.2mg/Nm3 (target: 10mg/Nm3) and 0.15%, respectively. No damage of burner and refractory were found after one year operation.

3.4. DME synthesis reactor performance

Before start-up, the catalyst slurry was charged into the reactor by pump. The reactor was heated up by steam and the catalyst was first subjected to pre-reduction by synthesis gas. The reactor temperature was controlled stably by steam drum pressure up to 280°C. The temperature distribution in the slurry was very homogeneous and there was no hot spot observed. The following performances were demonstrated; Total conversion: 96% (target: 95%), DME/(DME+MeOH) (carbon mole ratio): 91 % (target: 90%) and Production rate: 109tons/day (target: 100tons/day). The purity of produced DME was higher than 99.8% which satisfied the target purity (higher than 99%: Tentative specification proposed by Japan DME Forum).

Figure 2 shows an evolution of the catalyst activity for 152days in the fourth and fifth runs continuously operated. The catalyst activity change was high for the first 1,000 hours, but later looks like stabilized. The catalyst life is expected at least one year for the commercial scale plant.

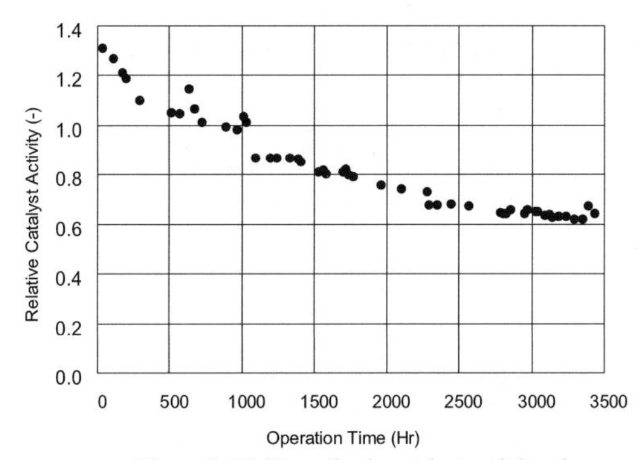

Figure 2. DME synthesis catalyst activity change

4. Scale up of slurry phase reactor

4.1. Process parameters of DME synthesis slurry phase reactor

Figure 3 shows process parameters governing the DME synthesis slurry phase reactor.

Various operation parameters affect the phenomena inside the reactor, and finally the reactor performances such as feed gas consumption, reactor capacity and control system.

The operation data of conversion were compared with those of smaller scale plant including autoclave. The conversion varies as a function of W/F, W: Charged catalyst weight in the reactor, and F: Gas flow rate through the reactor. As shown in Figure 4, the data of conversion for very wide range of reactor scale are almost the same for the same W/F. Plant scale has little effect on conversion. It means that overall reaction is mainly controlled by chemical reaction and the production rate is proportional to the catalyst weight in the reactor.

By this principle, the volume of 100 tons/day reactor had been decided almost 20 times larger than 5 tons/day pilot plant reactor. This simple principle was confirmed by the above operation results, and can be applied to the commercial scale reactor deign.

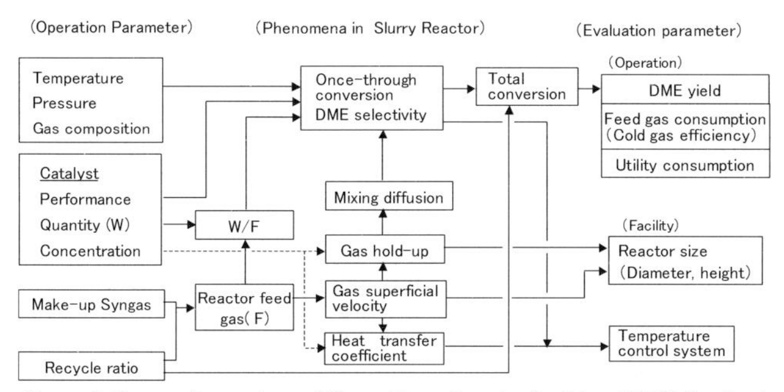

Figure 3. Process Parameters of Slurry Phase Reactor for Direct DME Synthesis

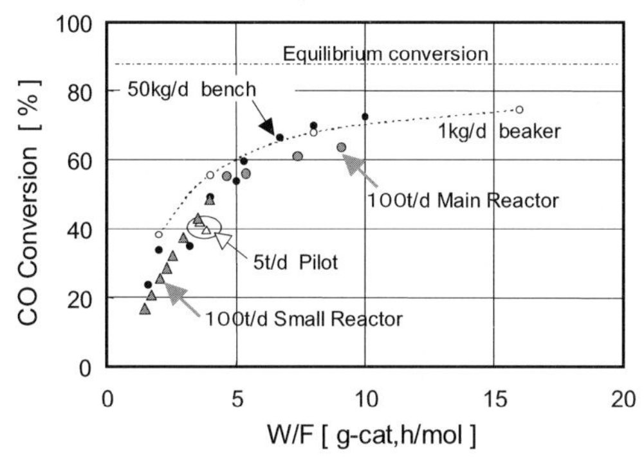

Figure 4. Effect of W/F on Once-through CO Conversion

4.2. Fluid dynamic phenomena in slurry phase reactor

The gas hold up (εg) was measured in increasing gas velocity (Ug) up to 40 cm/s. Within this range, the gas hold up increased smoothly without any sudden change as shown in Figure 5. As for the synthesis reaction with higher gas velocity, DME production yield by catalyst weight is almost constant up to 40 cm/s.

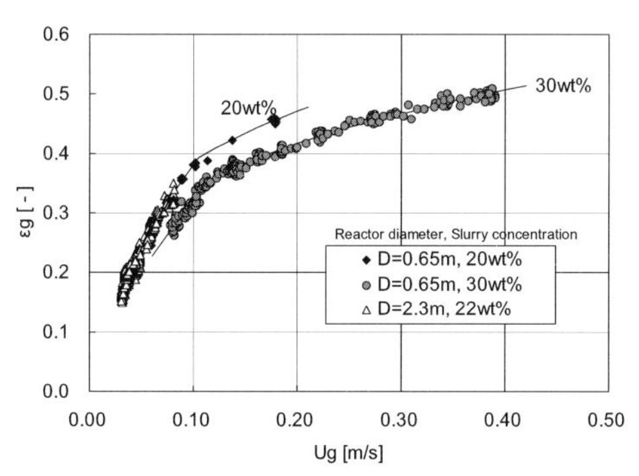

Figure 5. Effect of Superficial Gas Velocity on Gas Hold-up

The gas hold-up decreases by 0.05 with slurry concentration increasing 20% to 30%. There is no clear dependence of the gas hold up on the reactor diameter. According to Krishna [5], if the ratio of gas bubble size (db) to reactor diameter (D) is smaller than 0.125, reactor diameter has no effect on the gas holdup. In our case, the ratio (db/D) is less than 0.05; for D=0.65-2.3m, db=5-30mm (estimated by a fluid dynamic simulation with increasing Ug).

The mixing diffusion coefficient in liquid phase was estimated by one dimensional reactor model simulation and also by a fluid dynamics model simulation for liquid and gas bubbles movement.

In this one dimensional simulation, the reaction and heat transfer was simulated in changing a factor for the mixing diffusion coefficient calculated by Deckwer [6] equation. With factor of 0.5-1.0, the simulated temperature distribution and conversion at the outlet is similar to the measured ones.

The heat transfer coefficient from the slurry to heat exchanger tubes was also evaluated from heat balance data of the reactor. Deckwer equation for heat transfer coefficient agrees with the operational data with some allowance.

Based on these data, the commercial scale production of 3,000tons/day could be realized by a single reactor with 7m in diameter and 44m in height. Figure 6 shows an example of simulation for a commercial scale reactor, the temperature distribution is homogeneous and the synthesis conversion is sufficiently high.

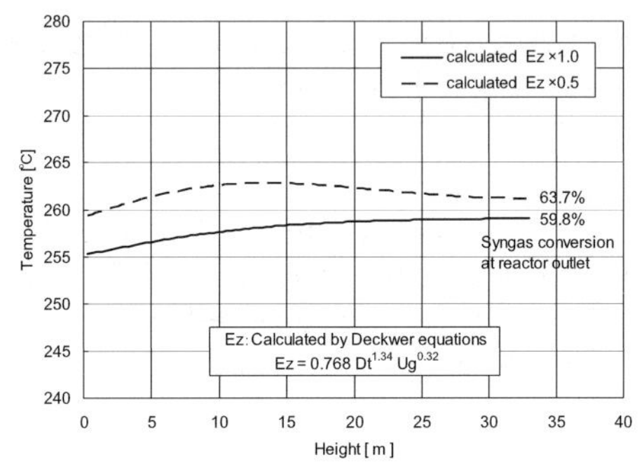

D=7m, Ug=0.2m/s, W/F=5.2kg h/kmol,
Inlet gas temp. 220°C, Cooling water temp. 213°C, Slurry conc. 30wt%

Figure 6. Example of 1-D Reactor Simulation for commercial scale plant

5. Conclusion

DME is an excellent clean fuel, easy to transport, and could be used extensively in various sectors as household, transportation and power generation. DME can be produced from various raw materials such as natural gas, coal or coal bed methane etc. through synthesis gas.

JFE has completed the development of Direct DME synthesis process and CO_2 recycle type ATR. DME is synthesized from synthesis gas of $H_2/CO=1$, which is produced by the ATR from natural gas with by-produced CO_2. This ratio is suitable for synthesis gas from gasification of coal ($H_2/CO=0.5\text{-}1.0$) or biomass ($H_2/CO\approx1$).

JFE has developed own efficient catalyst system and its mass production technology. Process technology has been confirmed by long term operation of 100tons/day demonstration plant. Catalyst performance in slurry phase reactor was stable. Conversion and DME selectivity were high. Scale-up technology has been established.

"JFE is ready for the licensing and catalyst supply".

Reference

1. Y.Ohno, T.Shikada, T.Ogawa, M.Ono and M.Mizuguchi, Preprints of Papers Presented at the 213th ACS National Meeting, Div. of Fuel Chemistry, (1997),705-709.
2. Y.Ohno, T.Ogawa, T.Shikada and Y.Ando, The 18th Congress of World Council, Buenos Aires, Argentine, (2001), DS6.
3. Y.Ohno, N.Inoue, K.Okuyama and T.Yajima, Proceedings of International Petroleum Technology Conference, Doha, Qatar, (2005), 10325.
4. S.N.Naqvi, SRI report No.245A,(2005)
5. R.Krishna and S.T.Sie, Fuel Processing Technology 64(2000), 73-105
6. W.D.Deckwer, Y.Serpemen, M.Ralek and B.Schmidt, Ind. Eng. Chem. Process Des. Dev.,21(1982),222-231.

Natural Gas Conversion VIII
F.B. Noronha, M. Schmal, E.F. Sousa-Aguiar (Editors)

The performance of $Pt/CeZrO_2/Al_2O_3$ catalysts on the partial oxidation and autothermal reforming of methane

Vanessa B. Mortola[a], Juan A. C. Ruiz[b], Diego da S. Martinez[a], Lisiane V. Mattos[b], Fábio B. Noronha[b], Carla E. Hori[a]

[a]*Faculdade de Engenharia Química - Universidade Federal de Uberlândia, Av. João Naves de Ávila, 2160/Bloco 1K, Uberlândia-MG 38400-902, Brazil*
[b]*Laboratório de Catálise, Instituto Nacional de Tecnologia/MCT, Av. Venezuela, 82/518, Centro, Rio de Janeiro-RJ 21081-312, , Brazil*

Abstract

The goal of this work was to evaluate the effect of the support calcination temperature (1073, 1173 and 1273 K) on the $Pt/CeZrO_2/Al_2O_3$ performance on the partial oxidation and autothermal reforming of methane. The sample calcined at 1273 K was less stable on the partial oxidation of methane probably due to the smaller contact between mixed oxide and Pt on the alumina surface, which favors the coke deposition. Moreover, for all samples, the addition of water led to a better activity and stability on autothermal reforming of methane.

1. Introduction

The partial oxidation and autothermal reforming of natural gas are advantageous alternatives in the syngas production when compared with steam reforming largely used today, since they are exothermic processes and, consequently the energy costs are minimized [1].
The transformation of methane in syngas, through reforming process, happens at high temperatures, usually above 1073 K, and still, coke can be formed in secondary reactions. Previous works showed that $Pt/CeZrO_2/Al_2O_3$ catalysts present higher activity and stability on partial oxidation of methane, due to the coke removal mechanism promoted by mixed oxide [2,3].

The aim of this work is to understand the influence of the support calcination temperatures (1073, 1173 and 1273 K) in the physical, chemical and catalytic properties of $Pt/Ce_{0.5}Zr_{0.5}O_2/Al_2O_3$ catalysts on the partial oxidation and autothermal reforming of methane reactions.

2. Experimental

The Al_2O_3 was calcined at 1073, 1173 and 1273 K, under air flow for 6 h. After that, $CeZrO_2$ was added to the alumina by incipient wetness impregnation method with an aqueous solution of cerium (IV) ammonium nitrate (Aldrich) and zirconium nitrate (MEL Chemicals) in order to obtain a 14wt.% of $Ce_{0.5}Zr_{0.5}O_2$. This loading was chosen because it is the theoretical value necessary to form a monolayer of ceria on the surface of the alumina. Finally, the supports were aged in a muffle furnace for 4 h at 1073, 1173 and 1273 K. Platinum was added to the supports by incipient wetness impregnation, using an aqueous solution of H_2PtCl_6 (Aldrich). After the impregnation with 1.5 wt.% of platinum, the samples were dried at 373 K, and calcined under air (50 ml/min) at 673 K for 2 h. The samples were characterized through BET surface area, X-ray diffraction (XRD), CO_2 infrared spectroscopy, oxygen storage capacity (OSC), temperature programmed reduction (TPR) and the metallic dispersion was evaluated using the cyclohexane dehydrogenation reaction. Details about experimental conditions are presented elsewhere [2,3]. Partial oxidation (POX) and autothermal reforming (ATR) of methane were performed in a quartz reactor at atmospheric pressure. Prior to the reaction, the catalyst was reduced under H_2 at 773 K for 1 h and then heated to 1073 K under N_2. The reactions were carried out at 1073 K and WHSV = 260 h^{-1}. For partial oxidation of methane, it was used a CH_4/O_2 ratio of 2.0. The autothermal reforming of methane was carried out using a CH_4/O_2 and H_2O/CH_4 ratio of 2.0 and 0.2, respectively. The exit gases were analyzed using a gas chromatograph equipped with a thermal conductivity detector.

3. Results and Discussion

Increasing calcination temperature from 1073 to 1273 K decreased the BET surface area of the alumina samples from 96 to 38 m^2/g. For $Pt/CeZrO_2/Al_2O_3$ catalysts, the results obtained in the BET analysis (Table 1) also showed that the increase of temperature caused a decrease in the surface area of the samples. This can be attributed to the alumina sintering processes. The OSC values (Table 1) strongly decreased when the calcination temperature increased from 1073 K to 1173 K, although there was not a significant change in the OSC value when the temperature was increased to 1273 K. The TPR profiles of all the samples (not shown) exhibited two reduction regions: one at lower temperature (440 - 670 K) that presented the highest H_2 consumption and is related to the reduction of Pt oxide and superficial $CeZrO_2$, and another region, at higher temperature (>1200 K), which may be attributed to the reduction of bulk $CeZrO_2$ [4]. The values of H_2 consumption during TPR experiments (Table 1) agree with BET and OSC results and show a decrease of reducibility with the increase of calcination temperature. This is probably associated to the sintering of $CeZrO_2$ particles, which makes harder to remove bulk oxygen from the inside of these particles [5]. Figure 1 shows the X-ray diffraction patterns for all the catalysts. For the samples calcined at 1073 and 1173 K, it was possible to identify the presence of γ-Al_2/O_3 phase. However, for the sample submitted to 1273 K, an additional phase, possibly α-Al_2/O_3, was also present.

Table 1. Results of BET, OSC, TPR, D_{CeO2} and Pt dispersion for $Pt/CeZrO_2/Al_2O$ catalysts.

$C_{alcinations}$ (K)	BET area (m^2/g)	OSC (μmoles de O_2/gcat)	TPR (μmoles de H_2/gcat)	D_{CeO2} (nm)	Pt Dispersion (%)
1073	85	458	700	8	47
1173	70	237	564	9	41
1273	37	279	421	12	39

The three XRD profiles also exhibit lines related to ceria cubic phase shifted to higher 2θ values. This could be verified with a slow scanning speed between 2θ = 27° and 32° (not shown) for all the samples. The peak characteristic of ceria cubic phase (JCPDS-4-0593) shifted from 28.6° to 28.8°, for the catalyst submitted to 1073 K, and to 2θ = 28.97° for sample calcined at 1173 K. This result indicates that zirconium was introduced into the ceria lattice [2,3]. On the other hand, the sample calcined at 1273 K showed peaks related to a tetragonal zirconia phase, which suggests the presence of two phases: one rich phase in ceria and another one rich in zirconia. The compositions of these solid solutions were calculated using the procedure described by Kozlov et al. [6]. The results showed that the samples treated at 1073 and 1173 K contain 25 and 46% of Zr incorporated into the cerium oxide lattice while the sample calcined at 1273 K presented two peaks with different compositions: $Ce_{0.875}Zr_{0.125}O_2$ (2θ = 28.8°) and $Ce_{0.125}Zr_{0.725}O_2$ (2θ = 30°). These results are in agreement with Yao et al. [7] who identified the formation of solid solutions with heterogeneous compositions for samples aged at 1273 K and zirconia loadings above 10%. CeO_2 average particles sizes (D_{CeO2}) were calculate through Scherrer equation using Ce(111) reflection (Table 1). The results showed that the increase in the calcination temperature from 1073 K to 1173 K did not cause a significant change in the particle size. However, the sample treated at 1273 K showed larger particles, which is consistent with the results reported by Yao et al. [7].

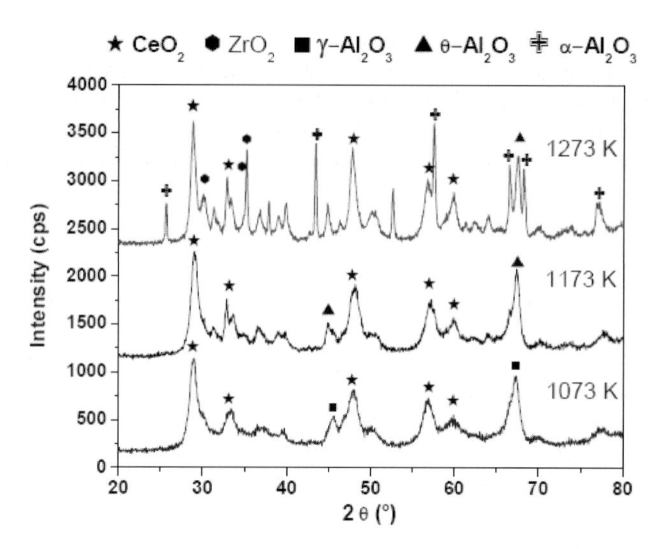

Figure 1: X-ray diffraction patterns of $Pt/CeZrO_2/Al_2O_3$ catalysts aged at 1073, 1173 and 1273 K.

The values of Pt dispersion (Table 1) are in agreement with those obtained in previous reports [2,3] and decreased with the increase of the calcination temperature. The degree of coverage of the alumina surface was evaluated through the optical density of the band at 1235 cm^{-1} calculated through adsorbed CO_2 infrared data [8]. In fact, the optical density of the band at 1235 cm^{-1} is a measurement of the interaction of CO_2 with the hydroxyl groups of the alumina. For catalysts treated at higher calcination temperature (1273 K) this measurement is inaccurate because of the removal of the hydroxyl groups after the exposure to this high temperature and then it was not performed. The results obtained by adsorbed CO_2 infrared data showed a decrease on the degree of alumina coverage by the ceria-based oxide from 80.7 to 66.2 % when the aging temperature is increased from 1073 to 1173 K.
Figure 2 shows the CH_4 conversion on POX reaction. The samples calcined at 1073 and 1173 K presented similar performance during the reaction. It was observed a slight deactivation during 24 h for both samples. However, $Pt/CeZrO_2/Al_2O_3$ calcined at 1273 K strongly deactivated along the reaction. The CO selectivity decreased while the CO_2 formation increased during the reaction (not shown) for all samples. However, this effect was more accentuated for the catalyst calcined at 1273 K. These results can be explained by the two-step mechanism of partial oxidation of methane [2]. The first step comprehends the methane combustion, producing CO_2 and H_2O. In the second step, syngas is produced through CO_2 and steam reforming of unreacted methane. The support takes part in the CO_2 dissociative adsorption close to the metallic particle, suppling oxygen to the metal surface.
When the support does not exhibit OSC, carbon deposits around the metal particle and CO_2 dissociation is inhibited. In the case of $Pt/CeZrO_2/Al_2O_3$

catalysts, a high degree of coverage of the alumina by the $CeZrO_2$ oxide provides a good interaction of the metal with the mixed oxide, promoting the carbon removal mechanism of the metallic surface [2]. In this work, the increase of CO_2 formation during the reaction indicates that the second step (CO_2 reforming) is being inhibited. This effect was more significant for the sample calcined at 1273 K, although the metal dispersion and OSC values for this catalyst are similar to the ones obtained for the sample calcined at 1173 K. Therefore, the smallest stability of the catalyst calcined at 1273 K could be related to a low degree of coverage of the alumina by the mixed oxide. Although it was impossible to measure the degree of coverage of the alumina for this sample by FTIR, the XRD analysis showed that the calcination of the support at 1273 K strongly increased average particle size of the ceria based oxide. This fact probably decreased the contact between platinum particles and this mixed oxide on the surface of alumina. Recently, it was shown that a Pt/Al_2O_3 catalyst suffers a strong deactivation in POX, due to the absence of the metallic particle cleaning mechanism [9]. Then, the coke removal mechanism of the metallic particle in the $Pt/CeZrO_2/Al_2O_3$ catalyst calcined at 1273 K is not favored, causing the catalyst deactivation.

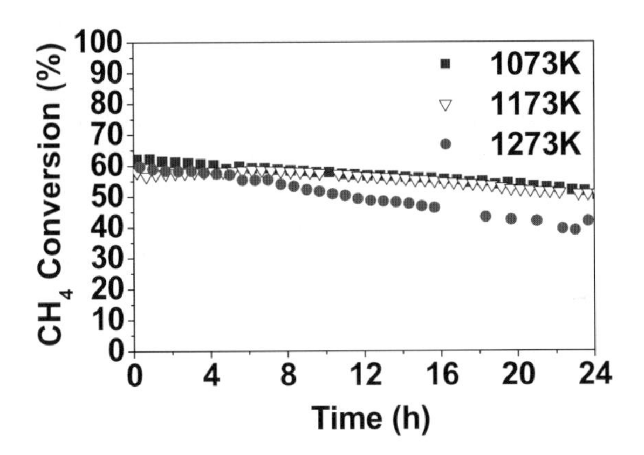

Figure 2: CH_4 conversion on partial oxidation of methane over $Pt/CeZrO_2/Al_2O_3$ catalysts calcined at 1073, 1173 and 1273 K.

Concerning the ATR reaction (Figure 3), the sample calcined at 1073 K presented a slight deactivation only at the beginning of the reaction, while the other samples were very stable during the reaction. Furthermore, the CH_4 conversions were higher than those observed on partial oxidation of methane for all the samples. The higher values of CH_4 conversion can be associated to the occurrence of steam reforming of methane due to the addition of water. $CeZrO_2$ support plays the same role on the stability of the catalysts observed on POX reaction. Moreover, the better stability of the catalysts on ATR is also related to the introduction of water, which avoids the carbon deposition.

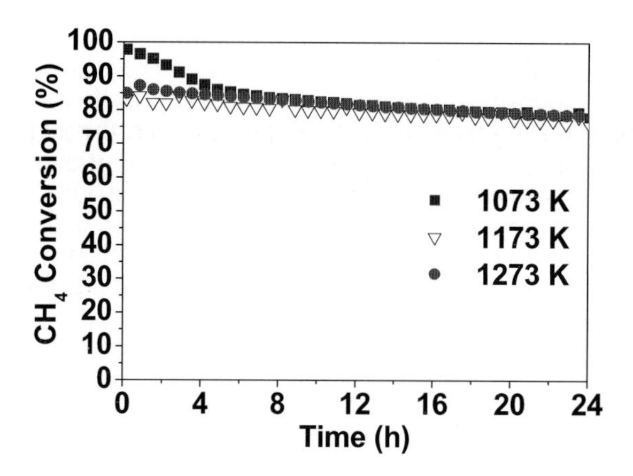

Figure 3: CH_4 conversion on autothermal reforming of methane over $Pt/CeZrO_2/Al_2O_3$ catalysts.

4. Conclusions

The $Pt/CeZrO_2/Al_2O_3$ catalysts calcined at 1073 and 1173 K presented a higher stability than the sample calcined at 1273 K on partial oxidation of methane. According to the XRD results, the sample calcined at 1273 K showed particle agglomeration. This probably caused a smaller contact between mixed oxide and Pt on the alumina surface and consequently the coke removal mechanism was less efficient for this sample. Furthermore, the activity and the stability of $Pt/CeZrO_2/Al_2O_3$ catalysts were higher on autothermal reforming of methane. The addition of water leads to the occurrence of steam reforming of methane and avoids the carbon deposition.

References

1. J.R. Rostrup-Nielsen, Catal Today 71 (2002) 243.
2. P.P. Silva, F.A. Silva, H.P. Souza, A.G. Lobo, L.V. Mattos, F.B. Noronha, C.E. Hori, Catal. Today 101 (2005) 31.
3. P.P. Silva, F.A. Silva, L.S. Portela, L.V. Mattos, F.B. Noronha, C.E. Hori, Catal. Today 107-108 (2005) 734.
4. H.C. Yao, Y.F. Yao, J. Catal. 86 (1984) 256.
5. J.P. Cuiff, G. Blanchard, O. Touret, A. Seigneurin, M. Marczi, E. Quéméré, SAE paper 970463 (1996).
6. A.I. Kozlov, D.H. Kim, A. Yezerets, P. Andersen, H.H. Kung, M.C. Kung, J. Catal. 209 (2002) 417.
7. M.H. Yao, R.J. Baird, F.W. Kunz, T.E. Hoost, J. Catal. 166 (1997) 67.
8. R.Frety, P.J.Lévy, V. Perrichon, V.Pitchon, M.M.Chevrier, C.Gauthier, F.Mathis, Stud.Surf.Sci.Catal. 96 (1995) 405.
9. L.V. Mattos; E.R. Oliveira; P.D. Resende; F.B. Noronha; F.B. Passos Catal. Today 77 (2002) 245.

Natural Gas Conversion VIII
F.B. Noronha, M. Schmal, E.F. Sousa-Aguiar (Editors)

Development of monolith catalyst for catalytic partial oxidation of methane

Kenichi Imagawa,[a*] Atsuo Nagumo,[a] Ryuichi Kanai,[a] Takeshi Minami,[a] Keiichi Tomishige,[b] Yoshifumi Suehiro[c]

[a]*R&D Center, Chiyoda Corp., Yokohama, Kanagawa, 221-0022, Japan,*
[b]*Institute of Materials Science, University of Tsukuba, Tsukuba, Ibaraki 305-8573, Japan,*
[c]*Oil & Gas Technology Research & Development, Japan Oil, Gas and Metals National Corp., Chiba, Chiba 261-0025, Japan*

1. ABSTRACT

In an attempt at developing of the catalysts for catalytic partial oxidation, the Rh/(CeO_2+ZrO_2+MgO) catalyst showed good activity. And the investigation for the effect of the residence time on product gas compositions suggested that the direct catalytic partial oxidation (D-CPOX: $CH_4+0.5O_2 \rightarrow CO+2H_2$) occurred at first, then steam reforming, CO_2 reforming or methanation reaction followed D-CPOX. The results of XPS analysis of these catalysts represented that the Rh species on the CeO_2+ZrO_2+MgO support was reduced more easily than that on the MgO support.

2. INTRODUCTION

Because of the trend toward larger capacity of syngas generation for methanol and GTL production, higher throughput has been required for syngas generation reactor. Chiyoda Corp., Tsukuba Univ. and JOGMEC (Japan Oil, Gas and Metals National Corporation) have jointly developed a catalytic partial oxidation process for this demand. In this report, we mentioned the development of novel catalyst which is Rh supported on the mixture of MgO, ZrO_2 and CeO_2.

* Corresponding author, E-mail: kimagawa@ykh.chiyoda.co.jp

3. EXPERIMENT

Preparation of granule catalyst

The metal oxide supports were prepared by the decomposition of corresponding hydroxides. Appropriate amounts of corresponding metal hydroxides (ex. $Ce(OH)_4$, $Zr(OH)_4$, and $Mg(OH)_2$) were completely mixed with 5wt% graphite, which was added as a gum. The obtained mixtures were then pressed into disks followed by calcination in flowing of air at 1200°C for 6 hours. Subsequently, the calcined disks were crushed and sieved. Particles in a size of 0.355–0.425 mm were chosen as supports for the loading.

The 2000wt-ppm Rh-supported catalysts were prepared by the incipient wetness method using an aqueous solution of $Rh(CH_3COO)_3$ and then dried at 60°C for 16 hours. The dried samples were finally calcined at 950°C for 3 hours.

Preparation of monolith catalyst

Corresponding metal oxide powder was dispersed into the water to prepare coating slurry. A ring-shaped ceramic foam (16 $mm^{O.D.}$ x 7 $mm^{I.D.}$ x 5 mm^{t}) was coated with this slurry and calcined. Monolithic catalysts were also prepared by the incipient wetness method. Calcined foam was impregnated with the Rh aqueous solution and calcined again.

Performance test of catalyst

The schematic diagram of a reactor is shown in Fig.1. The tubular reactor was made of Inconel® with an internal diameter of 16 mm. 0.84 cm^3 of the catalyst was employed in it. A thermocouple in a guard pipe with an outer diameter of 7 mm was inserted at the center of the reactor for the measurement of the temperature gradient.

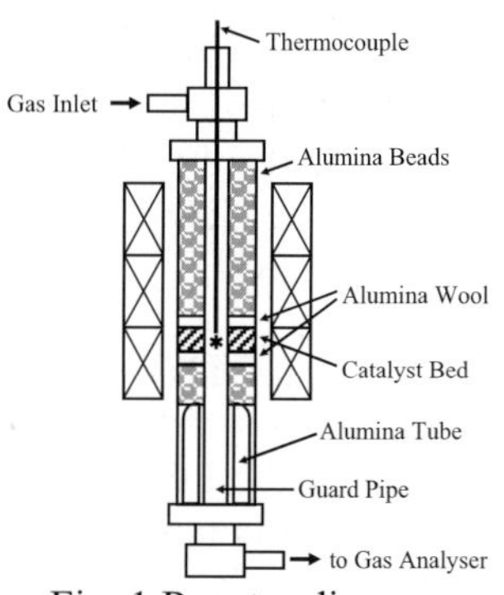

Fig. 1 Reactor diagram

Before activity tests, the catalysts were reduced in H_2 at around 900°C for 1 hour. The catalyst screening tests were carried out by feeding the mixture of methane, oxygen, and argon ($CH_4/O_2/Ar$=30/15/55 in mole ratio under normal pressures with GHSV=400, 000 h^{-1}). The temperature of the catalyst bed outlet was controlled at 650°C. The composition of the product gas was analyzed by a gas chromatograph equipped with a TCD. In all performance tests of this paper, the O_2 conversion was almost 100%.

The CH_4 conversion, H_2 selectivity and CO selectivity were calculated from product gas composition by the following expressions.

$$CH_4\ \text{conv.} = \frac{CO + CO_2}{CO + CO_2 + CH_4} \qquad H_2\ \text{sel.} = \frac{1/2H_2}{CO + CO_2} \qquad CO\ \text{sel.} = \frac{CO}{CO + CO_2}$$

4. RESULTS and DISCUSSION

Additive effect of ZrO_2 and CeO_2 to Rh/MgO

In order to improve the catalytic performance of the Rh/MgO catalyst, ZrO_2 having mobile lattice oxide ions and CeO_2 having storage capacity of oxygen were added into the MgO support. The results of a partial oxidation reaction over the obtained granular catalyst are shown in Figs. 2 and 3.

As shown in Fig. 2(a), in the range of ZrO_2 content is lower than 50%, the selectivity to H_2 and CO increased along with increment of ZrO_2 content. On the other hand, the selectivity to H_2 decreased in the range where ZrO_2 content is higher than 50%. The BET surface area and the amount of carbon formed during the activity tests are compared in Fig. 2(b). The catalyst with higher ZrO_2 content gave larger amount of carbon deposition and higher surface area. Based on the above results, the optimum ZrO_2 content determined to be 50%.

Fig. 3 shows the effect of additive amount of CeO_2 to the MgO support. Different from the case of ZrO_2, the methane conversion, H_2 selectivity, and CO

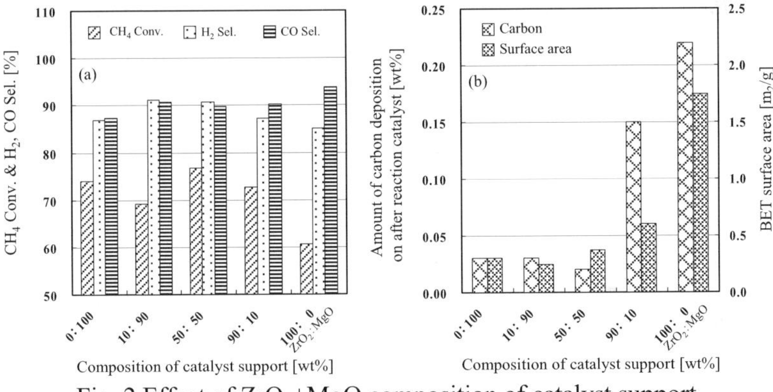

Fig. 2 Effect of ZrO_2+MgO composition of catalyst support

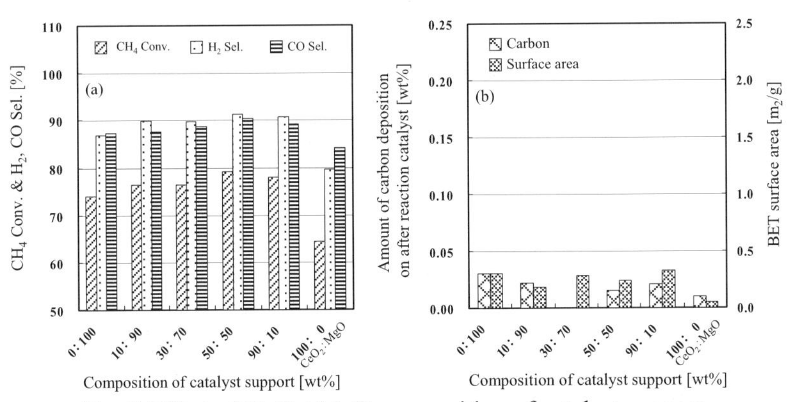

Fig. 3 Effect of CeO_2+MgO composition of catalyst support

selectivity increased with increment of CeO_2 content in the range of 10 to 90% (Fig. 3(a)). Nevertheless, the MgO-free Rh/CeO_2 sample shows much lower performance than the Rh/MgO catalyst. It is suggested that MgO plays a key role. In addition, carbon deposition on all the CeO_2-contained samples is smaller than that on the Rh/MgO catalyst (Fig. 3(b)). Presumably, the oxygen species on the CeO_2 surface is responsible for the resistance to carbon formation.

Effect of ZrO_2+CeO_2+MgO support

As discussed above, the addition of ZrO_2 or CeO_2 was effective for catalytic partial oxidation. Then, we tested the catalyst of which support consisted of CeO_2, ZrO_2 and MgO. Fig.4 shows that the effect of support composition on catalytic partial oxidation reaction. The Rh/(CeO_2+ZrO_2+MgO) catalyst showed higher activity than the Rh/(ZrO_2+MgO) catalyst, and slightly higher than the Rh/(CeO_2+MgO) catalyst. Fig. 5 shows that the influence of temperature on the catalytic performance. Clearly, the Rh/(CeO_2+ZrO_2+MgO) catalyst showed higher activity than the Rh/(CeO_2+MgO) catalyst at lower temperature. Especially, CH_4 conversion and H_2 selectivity were higher than the equilibrium values at lower temperature. These results suggest that the D-CPOX reaction proceeds.

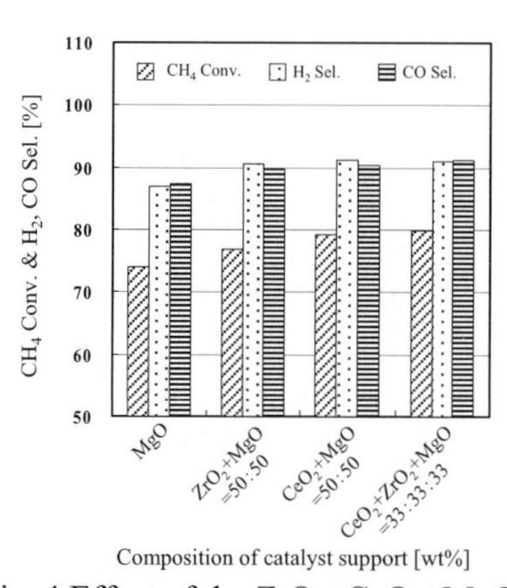

Fig. 4 Effect of the ZrO_2+CeO_2+MgO catalyst support

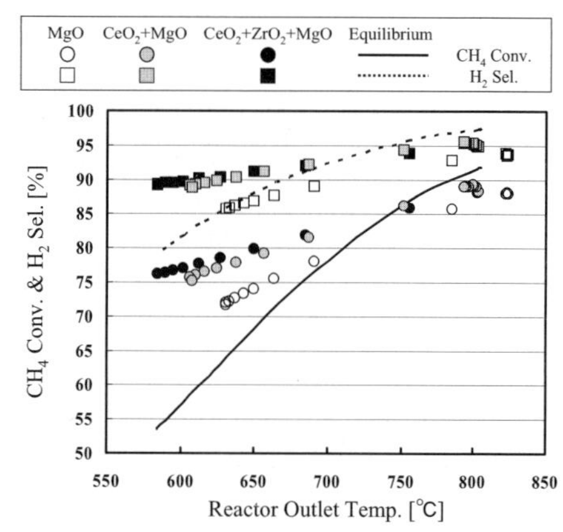

Fig. 5 Effect of temperature and catalyst supports

Effect of residence time

The residence time was defined by the division of volume of the catalytic zone by the volumetric total flow rate.

Fig. 6 shows the influence of residence time on the catalytic performance of the Rh/(CeO_2+ZrO_2+MgO) where the outlet temperature of the catalyst bed was 800°C and 650°C. At 800°C, CO selectivity was practically constant, and CH_4 conversion and H_2 selectivity increased according to increment of the residence time. But, at 650°C, CH_4 conversion, H_2 selectivity and CO selectivity decreased according to increment of the residence time. These results suggest that the D-CPOX reaction proceeds at first, then steam reforming and CO_2 reforming reactions can follow the D-CPOX at 800°C of outlet temperature, on the other hand, methanation reaction can follow the D-CPOX reaction at 650°C.

Recently, some literatures present the existence of D-CPOX route and complete oxidation followed by the reforming reactions on the same catalyst [1-2]. Our results can be explained by these two aspects of a catalytic partial oxidation.

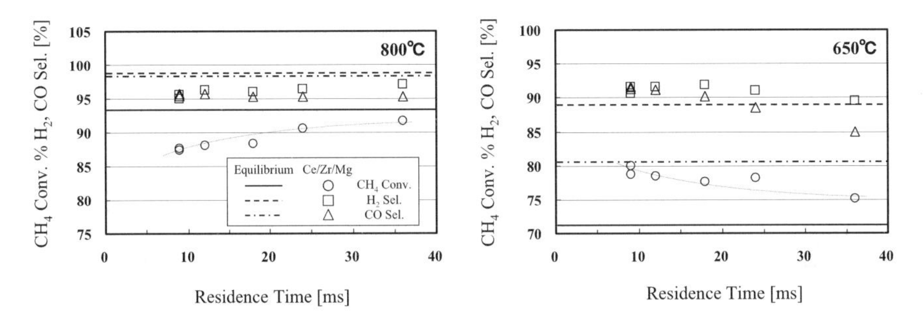

Fig.6 Influence of residence time on the Rh/CeO_2+ZrO_2+MgO catalyst

Catalytic stability of the monolith catalyst

The life test of Rh supported on the CeO_2+ZrO_2+MgO which is coated on the ceramic foam was carried out by feeding hydrocarbon (HC : a mixture of 89% CH_4, 6% C_2H_6, 3% C_3H_8, and 2% C_4H_{10}), oxygen and steam ($C/O_2/H_2O$ =2/1/0.1) under pressure of 1 MPa and GHSV=1,000,000 hr^{-1} (Fig.7). Under above condition, conversion of CH_4, C_2H_6, C_3H_8, C_4H_{10} were roughly 70%, 80%, 100%, 100% respectively. During the operation, the catalyst showed small decrease of conversion and selectivities, but it is expected that the practical use is promising if the stability is improved.

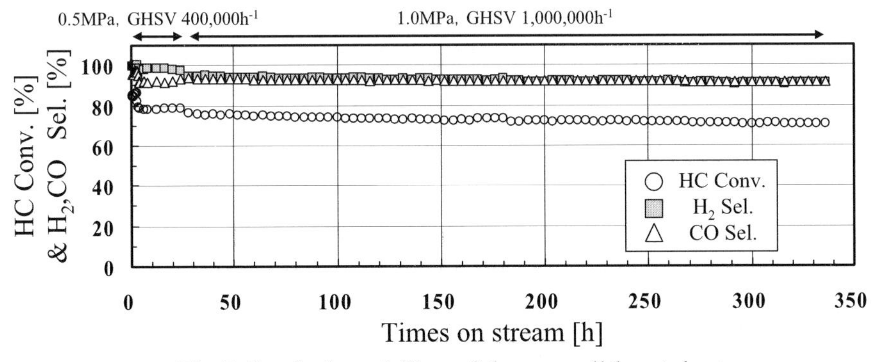

Fig.7 Catalytic stability of the monolith catalyst

Catalyst characterization

Generally, the reduced Rh species (Rh^0) has good ability for D-CPOX.

First, we studied the effect of temperature for the reduction of Rh on each support by XPS analysis (Fig. 8). From the comparison of the spectra of the samples after reduction at 650°C, most of Rh species on the CeO_2+ZrO_2+MgO can be reduced sufficiently, however only a part of Rh species on the MgO can be reduced at the same temperature.

Next, we measured the XPS of the Rh/MgO and the Rh/(CeO_2+ZrO_2+ MgO) which were used in the reaction test at the temperature of around 600°C. The activity and selectivity of the Rh/(CeO_2+ZrO_2+MgO) catalyst was higher than those of the Rh/MgO catalyst as shown in Fig. 5. More Rh^0 species were detected on the Rh/(CeO_2+ZrO_2+MgO) catalyst as compared to that of the Rh/MgO catalyst (Fig. 8).

These results show that the Rh species on the CeO_2+ZrO_2+MgO have higher reducibility than those on the MgO, and this can be related to higher catalyst performance of the D-CPOX.

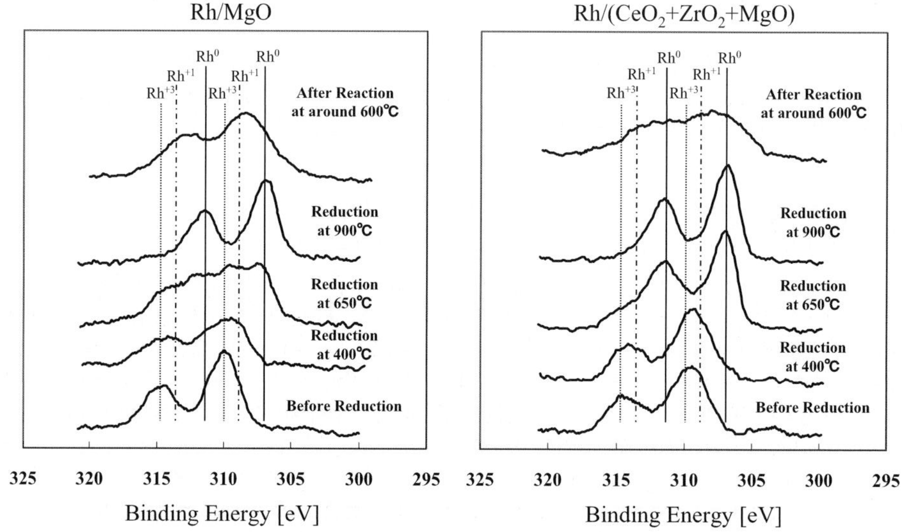

Fig.8 XPS of the Rh/MgO and the Rh/(CeO_2+ZrO_2+MgO) catalysts

5. ACKNOWLEDGEMENTS

Financial support was provided by the Japan Oil, Gas and Metals National Corporation (JOGMEC) and is gratefully acknowledged.

[1] R. Horn, , L. D. Schmidt, et al. *Journal of Catalysis*, Vol.242, p.92-102 (2006)
[2] M. Lyubovsky, S. Roychoudhury, et al., *Catalysis Letter*, Vol.99, p.113-117 (2005)

Natural Gas Conversion VIII
F.B. Noronha, M. Schmal, E.F. Sousa-Aguiar (Editors)

Promoter effect of Ag and La on stability of Ni/Al_2O_3 catalysts in reforming of methane processes

N,V. Parizotto,[a] R.F. Fernandez,[b] C.M.P. Marques,[a] J.M.C. Bueno[b]

[a]Department of Chemistry and [b]Department of Chemical Engineering, Universidade Federal de São Carlos, C.P. 676, 13565-905, São Carlos, SP, Brazil

Abstract: The promoter effect of Ag and La on the stability of Ni/Al_2O_3 catalysts in the steam reforming of methane (SRM), autothermal (ATRM) and partial oxidation of methane (POM) was investigated. The Ag addition leads to high resistance to carbon deposition during SRM, but at more oxidant conditions (ATRM and POM) the oxi-reduction process leads to NiO formation and catalyst deactivation by $NiAl_2O_4$ formation. La addition to the support leads to highly stable catalysts, even at oxygen-rich reaction conditions.

1. Introduction

The CH_4 reforming processes, with H_2O and O_2 on supported Ni catalysts [1] lead to H_2 production or H_2/CO syngas. Ni-based catalysts are the most used in SRM with high conversion and low catalyst cost in comparison to noble metal catalysts (Pt, Pd, Rh, Ir). But they are susceptible to carbon deposition and Ni oxidation, which depends on the feed composition and reaction temperature. To avoid carbon filaments formation, the reaction should be performed at high steam/CH_4 (S/C) ratio >3. The modification of the catalysts surface properties, changing Ni superficial structure through the formation of a superficial alloy, is also used to decrease carbon deposition. Metals such as Ag and Au on Ni substrate have shown thermodynamic parameters that lead to the formation surface Au-Ni and Ag-Ni alloys [2]. Promoters such as alkaline-earth (Ca, Mg) and rare-earth metal oxides (La, Ce) are used to retard the surface area loss of alumina at elevated temperatures required for high CH_4 conversion [3]. The catalysts composed by Ag-Ni/La-

Al_2O_3 resulted in high resistance to coke formation at SRM at low S/C ratio, but there are no studies about their application under more oxidant conditions as ATRM, where the reactor is fed with a CH_4:H_2O:O_2 mixture. The current study attempts to elucidate the effect of feed composition in the reforming of methane during stability tests of Al_2O_3 supported Ni catalysts promoted with Ag and La.

2. Experimental

La-coated alumina carrier (LaAl) was prepared by impregnation of a γ-Al_2O_3 (Strem Chemicals) with an aqueous solution of $La(NO_3)_2$ (Aldrich) in order to obtain a support modified with 12.0 wt. % of La, followed by a treatment at 1173K under air flow for 8 h. The samples with 15.0 wt. % of Ni and 0.3 wt.% of Ag loading were prepared by co-impregnation of a Al_2O_3 or La-Al_2O_3, with an aqueous solution of $Ni(NO_3)_2.6H_2O$ (Aldrich) and/or $AgNO_3$ (Merck). The samples were calcined under flowing air at 723 K for 2 h. Different sample compositions and their denominations are shown in Table 1.

Table 1. Reaction data for steam reforming of methane: activity and apparent activation energy (E_a^{app}) obtained at S/C ratio of 2; carbon deposition obtained at S/C ratio of 0.5 and 6 h on stream.

Sample	Loading Ag-Ni-La/Al_2O_3 (wt%)	Activity 773K (10^{-4} $mol_{CH4}/s.g_{Ni}$)	E_a^{app} (kJ/mol)	Carbon deposition ($mg/g_{cat}.h$)
NiAl	0-15-0-Al_2O_3	9.7	78	118
NiLaAl	0-15-12-Al_2O_3	5.5	80	96
AgNiAl	0.3-15-0-Al_2O_3	7.2	105	3.2
AgNiLaAl	0.3-15-12-Al_2O_3	7.5	104	5

XRD spectra were collected in a Rigaku Multiflex diffractometer with Cu-Kα radiation. FTIR of CO adsorbed spectra were recorded by a FT-IR THERMO NICOLET 4700 NEXUS spectrophotometer in diffuse reflectance mode. The samples were reduced *in situ* at 923K in a 25%H_2/N_2 mixture flow, for 2h and CO adsorption was made by CO pulses of 9 torr at 298K. XANES spectra at Ni K-edge of the samples were measured at the D06A - DXAS beam line at the National Synchrotron Radiation Laboratory (LNLS), Campinas, Brazil. The oxidized samples were heated under CH_4:O_2 (1:0.5) flow of 112.5 ml/min from 295 to 973 K at 10 $K.min^{-1}$ and held at 973 for 30 min. The XANES spectra were collected *in situ* in the transmission mode.

The catalytic tests were performed during methane reforming with various feed compositions of H_2:H_2O:CH_4, H_2O:O_2:CH_4 and O_2:CH_4, in fixed bed tubular quartz reactor. Before the reaction, the samples were reduced at 773 K under H_2 flow for 2h. The stability tests were performed for 6 or 24h on stream at 923 or 1073K. The temperature-programmed surface reaction (TPSR) tests were performed for POM with O_2:CH_4 ratio of 1:2 for the oxidized NiLaAl sample. The temperature was raised from 573 to 1073K at the rate of 10K.min^{-1}, and then decreased from 1073 to 573K. The effluents were analyzed on line by a VARIAN 4400 gas chromatograph.

3. Results

XRD spectra of all oxidized samples calcined at 723 K, which are not presented here, showed diffraction patterns that characterize NiO and γ-Al_2O_3 presence. The addition of Ag to NiAl catalyst does not seem to change the NiO particles size. No XRD patterns of LaO_x species were observed.

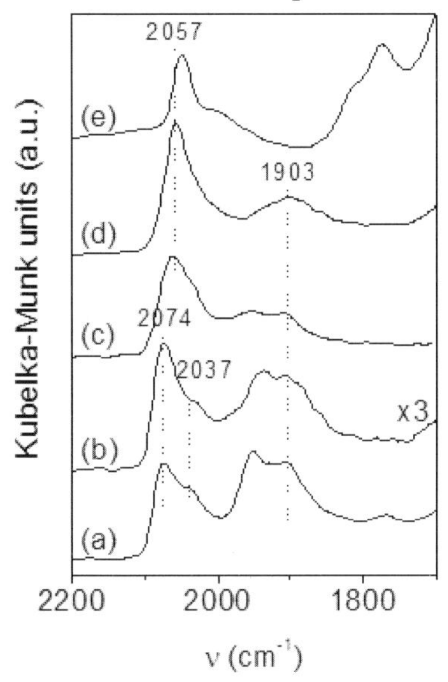

Figure 1 - DRIFTS spectra of adsorbed CO at 298 K on NiAl (a), NiLaAl (b), AgNiAl (c), AgNiLaAl (d) and AgAl (e) samples.

In CO adsorption spectra, presented in Fig. 1, the bands can be divided into two groups: high-frequency region (HF) above 2000 cm^{-1} and low-frequency region (LF) bellow 2000 cm^{-1}. The bands at HF region are related to linear adsorbed CO species at the Ni surface [4], and the bands at LF region are related to CO species adsorbed in the bridged form [5]. CO adsorbed on NiAl catalyst presents bands at LF, which are more intense than those at HF and this gives a HF/LF intensities ratio of 0.7. With the La addition to support alumina (Fig. 1b), for NiLaAl, the CO bands position and the HF/LF ratio are kept the same. Nevertheless, the band at the highest frequency 2073 cm^{-1}, which will be denominated HF1, has an increase in the relative intensity, while the other band at 2030 cm^{-1} presents a decrease if compared to the NiAl spectrum. AgNiAl spectrum (Fig. 1c) shows considerable modifications relative to the NiAl spectrum (Fig. 1a) at HF region with great increase in the intensity of HF1 bands. The HF/LF ratio becomes 1.3, suggesting that the surface structure is modified with Ag. AgNiLaAl CO adsorbed spectrum (Fig. 1d) is very similar to that of AgNiAl with HF/LF ratio of 1.1. The adsorbed CO spectrum in AgAl sample (Fig. 1e) shows that at the adopted conditions for CO adsorption, CO linearly adsorbs on Ag particles generating the band at 2050 cm^{-1} [6]. The increase in the HF/LF ratio caused by Ag addition may be directly related to the changes on Ni surface structure, which could generate Ni sites where CO molecules adsorb preferentially in the linear form or even, linear CO adsorption at Ag sites [6]. The promoter can also modify the Ni surface causing defects with very low Ni coordination, whose geometry allows the adsorption of two CO molecules at one Ni atom.

Table 1 shows the reaction rate and E_a^{app} to SRM on AgNiAl, NiAl, NiLaAl and AgNiLaAl catalysts obtained at 773 K and feed composition of H_2O/CH_4 =2 and H_2/CH_4=0.4. The results exhibit that the activity is strongly suppressed by the presence of Ag. NiAl and NiLaAl samples show an E_a^{app} around 78 kJ/mol, while Ag-containing catalysts show higher E_a^{app} around 105 kJ/mol.

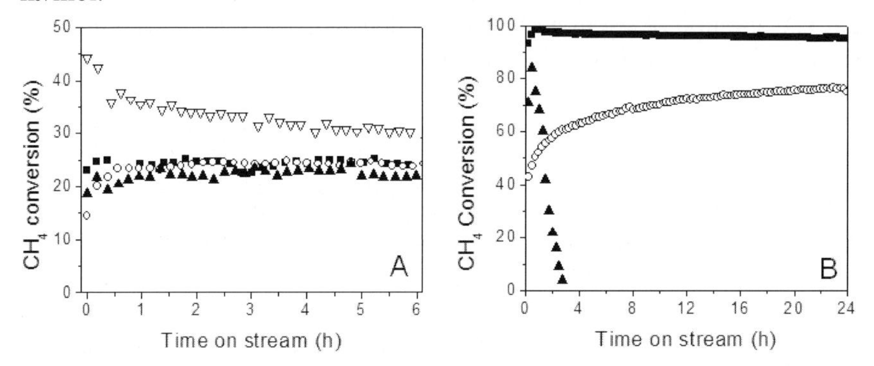

Figure 2. (A) CH_4 conversion with time on stream during SRM (T= 873K and S/C ratio of 0.5); (B) CH_4 conversion with time on stream, during ATRM (T=1073K and $H_2O:O_2:CH_4$=0.65:0.5:1). The samples are denoted by NiAl (▽), NiLaAl (■), AgNiAl (▲), and AgNiLaAl (○);

The results of methane conversion in SRM as a function of time on stream at 873 K and feed composition of H_2O/CH_4=0.5 and H_2/CH_4=0.4 are presented in Fig. 2A. NiAl and NiLaAl catalysts presented intense carbon deposition, as it can be verified by the results of thermo-gravimetric analysis (Table 1), while AgNiAl remained stable, with low carbon formation after 6 h on stream. The stability tests at 1073 K with feed composition of $H_2O:O_2:CH_4$ (0.65:0.5:1) in ATRM are shown in Fig. 2B. AgNiAl, similar to NiAl, showed strong deactivation and became inactive after 3h on stream. XRD results revealed a $NiAl_2O_4$ spinel formation, indicating that Ni is oxidized and reacts with Al_2O_3, forming $NiAl_2O_4$. On the other hand, NiLaAl was stable for 24 hours on stream and showed higher CH_4 conversion. These results suggest that La blocks the Al_2O_3 sites that are very reactive with NiO and the oxi-reduction process occurs with the net rate favoring reduced Ni. AgNiLaAl catalyst showed lower initial CH_4 conversion (Fig. 2A), which agrees with the lower activity in SRM (Table 1), but the conversion increased with time on stream.

In order to obtain an insight at the Ni state under an oxygen-rich atmosphere, additional catalytic tests were performed with NiLaAl oxidized catalysts at various temperatures during POM. Fig.3A shows the profiles of the effluents composition with increasing temperature. The CH_4 consumption starts at temperature >673 K, when NiO is reduced. Similar temperature was previously observed in the temperature-programmed reduction of oxidized NiAl by CH_4 [7]. O_2 is totally consumed at 973 K, and H_2 is present when CH_4 consumption starts. Differing from what was observed during the cooling, when the oxygen is totally consumed in all temperatures region (Fig.3B).

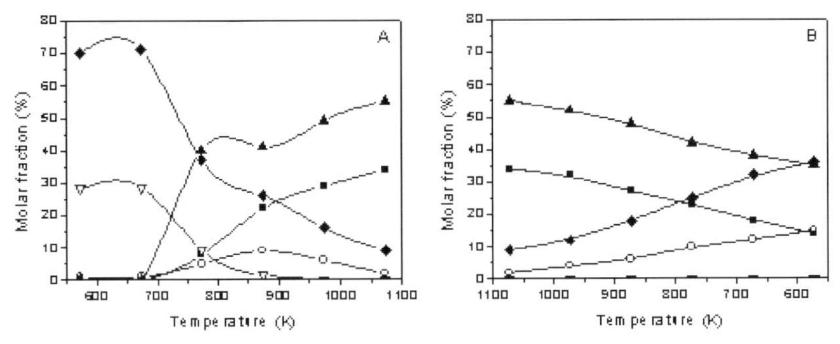

Figure 3 – TPSR for NiLaAl: Effluent composition in terms of molar fraction during partial oxidation of methane for the fresh sample in heating (A) and cooled (B) temperature. CO (■), CO_2 (○), H_2 (▲), O_2 (▽), and CH_4 (◆).

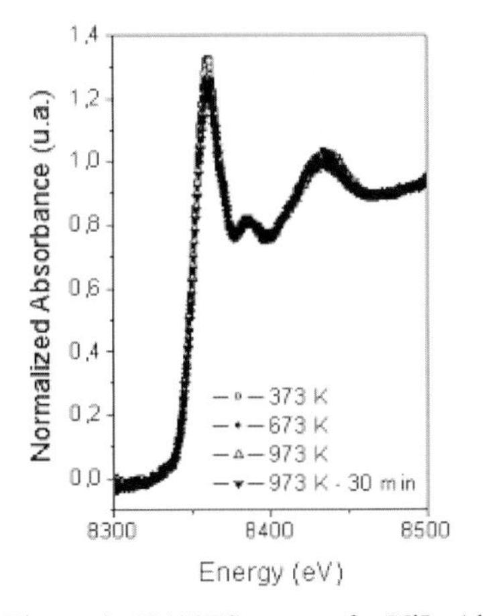

Figure 4 - XANES spectra for NiLaAl sample during heating at POM.

XANES spectra at Ni K-edge for NiLaAl catalyst were obtained in order to verify Ni oxidation state under POM atmosphere (Fig.4). The NiLaAl oxidized sample was analyzed under an O_2:CH_4 (1:2) flow, similar to the atmosphere at the entrance of the catalytic bed. In these conditions XANES spectra at Ni K-edge is characteristic of Ni^{2+}, which remained at the oxidized state (Ni^{2+}) when the temperature was increased from 473 to 973 K. This suggests that although NiO can be reduced by CH_4 at high temperatures, the oxidation is preferential at the beginning of catalytic bed, favoring NiO.

4. Discussion

FT-IR spectra of adsorbed CO show that Ag and La addition at alumina supported Ni catalysts may cause modifications on the Ni surface structure. This effect is more evident with the Ag addition, where the HF/LF ratio highly increases in relation to unpromoted NiAl catalysts. In addition, a significant shift to low frequency of HF2 band is observed to Ag-containing catalysts, which may indicate an increase of electron density.

The heating TPSR for the NiLaAl sample (Fig.3A) during POM reaction shows the coexistence of O_2 and H_2 at high temperatures and the

H_2/CO ratio is around 1.8. As demonstrated in a previous work [8], CH_4 dissociates to form hydrogen and surface carbon species on Ni^0. From this result, it is reasonable to suppose that CH_4 is activated in Ni^0 sites to produce H_2, which is fast desorbed avoiding oxidation by oxygen species. Carbon species at the surface would be oxidized by NiO and/or surface oxygen species, resulting in CO, CO_2 and Ni^0. There is a complete conversion of O_2 above 973K (Fig. 3A). On the other hand, when NiLaAl is cooled from 1073 to 573K, a hysteresis is found and the catalyst becomes very active to O_2 conversion even at low temperatures (Fig. 3B). As previously demonstrated, CH_4 reduces NiO ($4NiO + CH_4 \rightarrow CO_2 + 4H_2O + Ni^0$) [8]. Nevertheless, XANES spectra at Ni K-edge for unreduced NiLaAl sample under POM conditions show that Ni remains, mostly, at the oxidized state (Ni^{2+}) when heated from 573 to 1073 K (Fig. 4). At the entrance of catalytic bed, where O_2 concentration is high, Ni is predominantly in NiO form, and when the O_2 concentration decreases downstream the catalytic bed, Ni^0 state predominates. The initial low activity of AgNiLaAl for ATRM in relation to NiLaAl catalyst, agrees with the lower activity of Ag-promoted catalysts for SRM. Even if, from TPR results, it is expected that Ag is initially in reduced form, when the catalyst is exposed to ATRM, with O_2 in the entrance of reactor, Ag can be oxidized, resulting in changes at Ni surface structure. From the SRM results, the increasing activity during ATRM is expected if the promoter effect of Ag is unmade.

5. Conclusions

Reduced Ni catalysts are oxidized by O_2 and subsequently reduced by CH_4 in the entrance of the catalytic bed at POM conditions, but Ni is predominantly oxidized due to higher activity of O_2 relative to CH_4. Ag-Ni alloy at Ni surface is efficient to control carbon deposition in SRM. The stability of AgNiLaAl catalyst in SRM, ATRM and POM suggests that under O_2-rich feed the mobile oxygen species are dislocated to the catalytic bed, and the Ag may be oxidized and its promoter effect unmade, resulting in an increase in the AgNiLaAl catalyst activity with time on stream in ATRM and POM reactions.

References

1. D.L. Trimm Catal. Today 49 (1999) 3.
2. J.H. Larsen, I. Chorkendorff, Surf. Sci. Rep. 35 (1999) 163
3. A. Vázquez, et al. J. of Sol. State Chem. 128 (1997) 161.
4. C.W. Garland, et al, J. Phys. Chem. 69 (1965) 1195.
5. R.P. Eischens, S.A. Francis, W.A. Pliskin, J. Phys. Chem. 60 (1955) 194.
6. J. Müslehiddinoglu, M. A. Vannice, J. of Catal. 213 (2003) 305.
7. J. A. C. Dias, J. M. Assaf, J. Pow. Sour. 139 (2005) 176–181.
8. R. Jin, et al, App. Catal. A: Gen. 201 (2000) 71.

The Effect of Pt Loading and Space Velocity on the Performance of $Pt/CeZrO_2/Al_2O_3$ Catalysts for the Partial Oxidation of Methane

Fabiano de A. Silva[a], Diego da S. Martinez[a], Juan A. C. Ruiz[b], Lisiane V. Mattos[b], Fabio B. Noronha[a], Carla E. Hori[a]

[a]*Federal University of Uberlandia – School of Chemical Engineering, Av. João Naves de Ávila, 2160, bl. 1K, Campus Sta Mônica, CEP: 38408-902, Uberlândia-MG, Brazil*
[b] *National Institute of Technology, Av. Venezuela, 82,Centro, CEP: 20081-310, Rio de Janeiro - RJ, Brazil*

ABSTRACT

In this study, the perfomance of $Pt/CeZrO_2/Al_2O_3$ with Pt loadings of 0.5, 1.0 and 1.5 wt% on the partial oxidation of methane was evaluated. The results showed that the increase of metal content caused a decrease of the metal dispersion, but it did not affect the catalysts reducibility properties. The methane conversion and CO selectivity increased with the use of higher Pt content. The deactivation observed for the lower Pt contents could be due to the unbalance between methane decomposition and oxygen transfer process. The increase of space velocity decreased the catalyst stability and selectivity for H_2 and CO.

1. Introduction

The partial oxidation of methane (POM) has been considered as a potential process to generate syngas that can be then transformed into heavier hydrocarbons. Recently, our group has shown that the use of ceria-zirconia mixed oxides can improve the catalysts activity, stability and selectivities for CO and H_2 of Pt/Al_2O_3 samples for POM [1-3].
Besides those properties, another important issue that sometimes is overlooked is the space velocity at which the reactors operate. Schmidt and coworkers had shown that the use of very high space velocities for POM reaction might lead to a decrease of methane conversion and of syngas selectivity [4]. Therefore, the

goal of this work was to evaluate the effect of different space velocities on the performance of $Pt/CeZrO_2/Al_2O_3$ catalysts. In addition, the effect of the use of lower Pt loadings on the dispersion and on the catalytic performance was also investigated.

2. Experimental

The alumina was obtained by the calcination of aluminum hydroxide (Catapal-A - Sasol) in a muffle furnace at 873 K for 6 hours. After this, the sample was doped with 1 wt% of cerium through an incipient wetness impregnation using an aqueous solution of $(NH_4)_2Ce(NO_3)_6$ (Aldrich) and calcined at 1173 K under air flow (100 mL/min). $CeZrO_2/Al_2O_3$ supports were synthetized through wet impregnation of the doped alumina in a rotavapor during two hours. $(NH_4)_2Ce(NO_3)_6$ (Aldrich) and $ZrO_2(NO_3)_2$ (Aldrich) were used as precursors to obtain 19 wt% $CeZrO_2$ on alumina with a Ce/Zr ratio equals to 1. The platinum was added by incipient wetness impregnation (1.5, 1.0 and 0.5 wt%) using an aqueous solution of H_2PtCl_6 (Aldrich). After the impregnation, the samples were dried at 373 K and calcined at 673 K for two hours in air flow (50 mL/min). The samples were characterized through BET surface area, X-ray Diffraction (XRD), temperature programmed reduction (TPR) and oxygen storage capacity (OSC) measurements and the experimental details are given elsewhere [2,3]. The platinum dispersion was evaluated using a structure insensitive reaction, the dehydrogenation of cyclohexane [3]. This reaction was done at 543 K, with a H_2/Cyclohexane ratio equals to 13. The partial oxidation of methane (POM) was performed at different space velocities, 261, 522 and 1044 h^{-1} and the feed had a $CH_4:O_2$ ratio of 2:1. Before each catalytic test for the POM, 10 mg of sample diluted with 90 mg of SiC was reduced under H_2 at 773 K for 1 hour. Then, the sample was heated up to the reaction temperature, 1073 K, under N_2. The products of the reaction were analyzed using a gas chromatograph (Agilent 6890).

3. Results and Discussion

Table 1 shows BET surface areas obtained for all catalysts. The samples exhibited similar surface areas and this result was expected since the alumina impregnated with ceria-zirconia was the same and Pt loadings were low. The highest BET area was obtained for the sample with the lowest platinum loading. The XRD patterns are presented in Figure 1. It can be observed for all patterns that the peak corresponding to the cerium oxide with cubic structure moved from $2\theta = 28.6°$ to $2\theta = 29.4°$. Accordingly to the literature, this shift is an indication of the formation of a solid solution between the cerium and zirconium oxides [1-3, 5]. The average particle sizes of the ceria mixed oxide were estimated using the Scherrer equation and the values was around 6 nm for all catalysts.

Table 1: Results of BET, TPR and OSC for all catalysts.

Samples	BET (m^2/g)	H_2 uptake (μmoles H_2 / g_{cat})	OSC (μmoles O_2 / g_{cat})	Pt Dispersion (%)
1.5%Pt/$CeZrO_2$/Al_2O_3	110	730	242	57
1.0%Pt/$CeZrO_2$/Al_2O_3	106	741	245	77
0.5%Pt/$CeZrO_2$/Al_2O_3	124	740	246	100

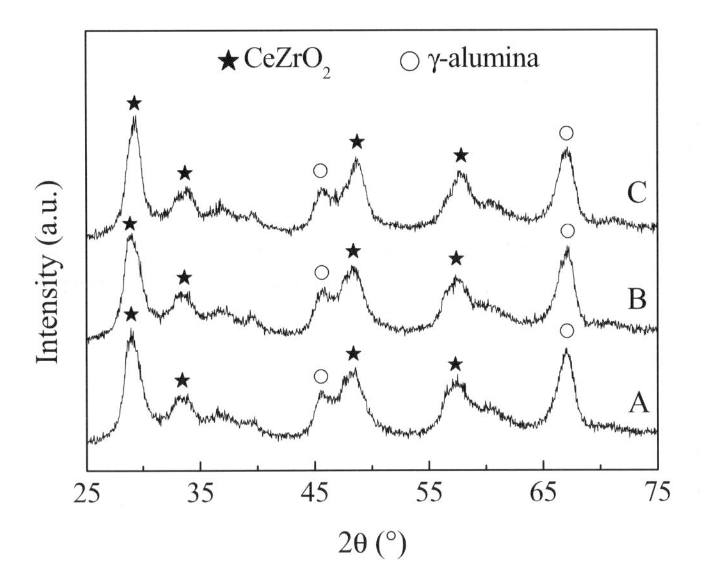

Figure 1: X-ray diffraction patterns of 0.5% Pt/$CeZrO_2$/Al_2O_3 (A), 1.0%Pt/$CeZrO_2$/Al_2O_3 (B) and 1.5% Pt/$CeZrO_2$/Al_2O_3 (C)

The reduction profiles of Pt/$CeZrO_2$/Al_2O_3 catalysts are shown in Figure 2. The majority of the H_2 consumption observed for the three samples was observed at low temperatures, around 550 K. The H_2 consumption can be attributed to the reduction of platinum and of the ceria-zirconia mixed oxide promoted by this metal [2, 3]. The increase of Pt content shifted the maximum to lower temperatures, starting from 550 K for the sample containing 0.5 wt% of Pt to 535 K for the one with 1.0 wt% Pt and to 505 K for the catalyst with 1.5 wt% Pt. For the catalysts containing 1.5 wt% of Pt, there are two peaks. This could be an indication that there is a part of the ceria-zirconia support with higher interaction with platinum and there is smaller portion of the ceria based oxide with lower degree of interaction with the metal. Since the theoretical value of H_2 consumption necessary to reduce the Pt is very small (77, 51 and 26 μmoles for the samples containing 1.5, 1.0 and 0.5 wt% of Pt) and the H_2 uptakes during TPR, shown in Table 1, are very similar, one may suggest that these catalysts have an equivalent degree of reducibility. These results were confirmed by the oxygen storage capacities that are also presented in Table 1. The O_2 consumption results show that the reducibility of the catalysts was not affected by the decrease on the metal content.

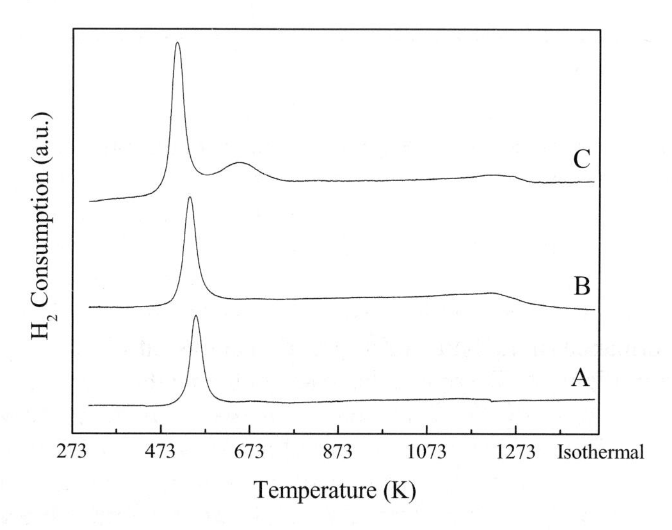

Figure 2: Temperature programmed reduction profiles of 0.5% $Pt/CeZrO_2/Al_2O_3$ (A), 1.0%$Pt/CeZrO_2/Al_2O_3$ (B) and 1.5% $Pt/CeZrO_2/Al_2O_3$ (C)

The platinum dispersion was evaluated by the dehydrogenation of cyclohexane and the results are presented in Table 1. The decrease of Pt content improved significantly the metallic dispersions, reaching 100% for the sample with the smaller loading. This result was expected since the use of less metal can lead to better dispersion on the surface of the support.

The methane conversions of the three samples during the partial oxidation of methane, as well as, the CO and CO_2 selectivities are presented in figure 3. The catalyst containing the lowest Pt loading exhibited higher deactivation during the catalytic test. In spite of the higher metal dispersion, this catalyst has approximately the same oxygen storage capacity than the others. In a recent study, we demonstrated that the stability of Pt supported catalysts on the partial

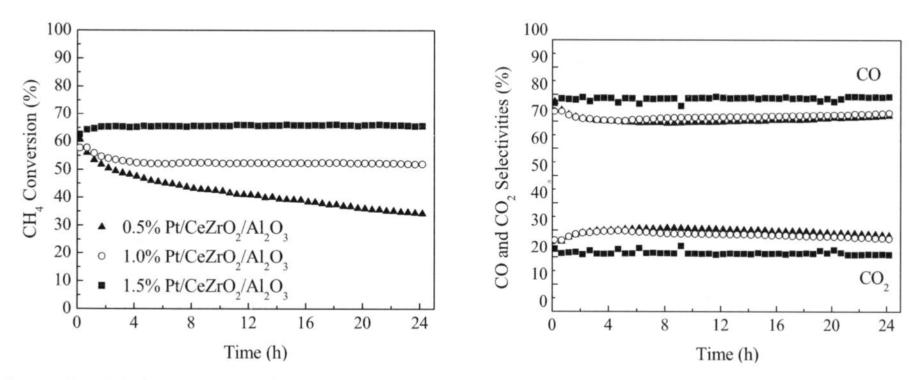

Figure 3 - Methane conversion (A) and CO and CO_2 selectivities (B) during methane partial oxidation reaction for $Pt/CeZrO_2/Al_2O_3$ catalysts with different Pt content (WHSV = 522 h^{-1}).

oxidation of methane is associated to the proper balance between oxygen storage capacity of the support and the metal dispersion [1]. Therefore, in this work, the strong deactivation observed for the lower Pt content could be due to the unbalance resulting from the methane decomposition and oxygen transfer. In this case, the oxygen transfer is probably not fast enough to keep up with the methane decomposition on the metal surface and then carbon deposition occurs. The performance of 1.5%$Pt/CeZrO_2/Al_2O_3$ catalyst at different space velocities is shown in Figure 4. There is a decrease on the methane conversion and on the catalyst stability as the space velocity is increased, particularly at 1044 h^{-1}. On the other hand, H_2 selectivity is stable around 90% for all the samples. Probably the use of high space velocities supply more methane and oxygen to the catalyst, leading to the formation of H_2 and carbon on the surface. The oxygen transfer processes that are responsible for the cleaning mechanism of the surface

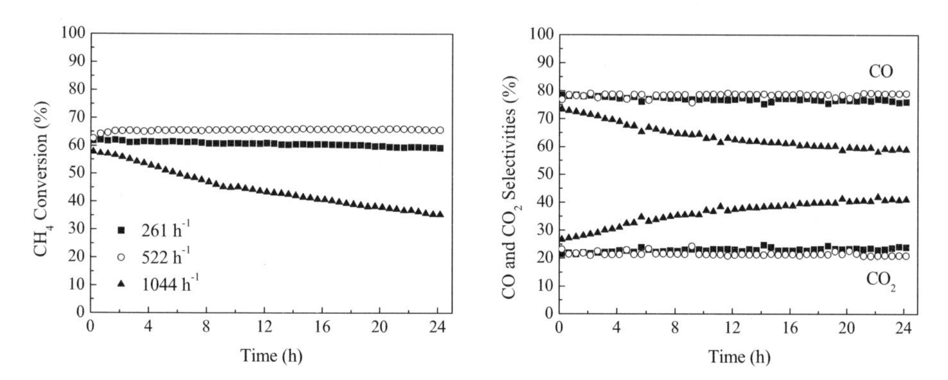

Figure 4 - Methane conversion (A) and CO and CO_2 selectivities (B) during methane partial oxidation reaction for 1.5%$Pt/CeZrO_2/Al_2O_3$ catalysts at different space velocities.

probably can not keep up with the elevated rates of carbon generation on the surface. This creates an unbalance between the two processes, elevating the CO_2 selectivity and decreasing the methane conversion.

4. Conclusions

Although the use of different platinum loadings changed the platinum dispersions of $Pt/CeZrO_2/Al_2O_3$ catalysts, it did not cause a significant difference on their redox properties. However, the performance of the samples for POM reaction was altered, specially the CO and CO_2 selectivites, as well as the methane conversion stability. The sample with higher Pt content was more active, stable and selective to the formation of CO than the others, probably due to a better balance between the methane decomposition and oxygen transfer process, which would keep the surface free of carbon deposits. The use of high space velocities supply more methane and oxygen to the catalyst and the oxygen transfer processes that are responsible for the cleaning mechanism of the surface probably can not keep up with the elevated rates of carbon generation on the surface. This creates an unbalance between the two processes, elevating the CO_2 selectivity and decreasing the methane conversion.

Acknowledgments

The authors would like to thank Capes and Fapemig for the financial support.

References

[1] F.B. Passos, E.R. de Oliveira, L.V. Mattos, F.B. Noronha, Catal. Today, 101 (2005) 23.
[2] P. P. Silva; F. A. Silva; H. P. Souza; A. G. Lobo; L. V. Mattos; F. B. Noronha; C. E. Hori, Catalysis Today, 2005, 101, 31.
[3] P.P. Silva, F.A. Silva, L.S. Portela, L.V. Mattos, F.B. Noronha, C.E. Hori, Catal. Today, 107-108 (2005) 734.
[4] K.L. Hohn, L.D. Schmidt, Appl.Catal. A: General, 211 (2001), 53-68.
[5] A. I. Kozlov; D. H. Kim; A. Yezerets; P. Andersen; H. H. Kung; M. C. Kung Journal of Catalysis 2002, 209, 417.

Natural Gas Conversion VIII
F.B. Noronha, M. Schmal, E.F. Sousa-Aguiar (Editors)

Promoter effect of CeO_2 on the stability of supported Pt catalysts for methane-reforming as revealed by *in-situ* XANES and TEM analysis

A.P. Ferreira[a], J. C. S. Araújo[a], J. W. C. Liberatori[a], S. Damyanova[b], D. Zanchet[c], F.B. Noronha[d], J.M.C. Bueno[a]

[a]*Universidade Federal de São Carlos-UFSCar, São Carlos, SP, Brazil*
[b]*Institute of Catalysis, Bulgarian Academy of Sciences, 1113 Sofia, Bulgaria*
[c]*Laboratório Nacional de Luz Síncrotron-LNLS, Campinas, SP, Brazil.*
[d] *Instituto Nacional de Tecnologia - INT, Rio de Janeiro, RJ, Brazil*

Abstract: The stability of Pt/Al_2O_3 and $Pt/12CeO_2$-Al_2O_3 catalysts during autothermal reforming of methane was investigated. The catalysts were characterized by XPS, TEM and *in situ* Pt L_3-edge XANES-TPR-H_2 and XANES-POM. On $Pt/12CeO_2$-Al_2O_3, Pt is easily reduced and shows higher stability against sintering, which is related to the Pt-O-Ce-support interaction.

1. INTRODUCTION

The reaction of CH_4 with H_2O and O_2 on supported transition metals produces H_2/CO mixtures that are useful for the synthesis of fuels and petrochemicals. The H_2/CO ratio can be modified by adjusting the composition of the reaction mixture, CH_4:H_2O:O_2. The mechanism proposed for the reaction consists in coupling the exothermic total oxidation of methane, which produces CO_2 and H_2O, with the endothermic reforming of the unreacted CH_4 with CO_2 and H_2O [1]. Hicks *et al.* [2] have demonstrated that methane oxidation reaction over Pt catalysts, the first step of adsorption of oxygen on Pt, one of the most active and stable metals for these reactions, is extremely fast. Wei and Iglesia [3] have demonstrated that the CH_4 turnover rates increase monotonically with increasing Pt dispersion on the support for CO_2 reforming, H_2O reforming, and CH_4 decomposition reactions, suggesting that the coordinative instauration increases the activation reactivity of the C-H bond [3]. Therefore, a high dispersion of Pt is desirable to produce an active catalyst. On the other hand, the thermodynamic properties of methane reforming demand a high reaction temperature (1073K) to obtain high methane conversion and, under these conditions, metal agglomeration occurs, leading to a decrease in activity during time on stream. Nagai et al. [4] reported that Pt did not sinter after aging treatment at 1073K in air on a Pt/ceria-based oxide catalyst, while sinterization occurs on Pt/Al_2O_3 catalyst.

The aim of this study is to elucidate the promoter effect of CeO_2 on the stability of Pt/12CeO_2-Al_2O_3 catalysts during the reforming of methane reactions by detailed morphological and structural characterization.

2. EXPERIMENTAL:

Catalyst preparation: The 12CeO_2-Al_2O_3 support, containing 12 wt. % of CeO_2 was prepared by the impregnation of γ-Al_2O_3 (S_{BET}=205 m^2/g) with an aqueous solution of $(NH_4)_2[Ce(NO_3)_6]$. γ-Al_2O_3 and 12CeO_2-Al_2O_3 supports were calcined in air at 1173 K. Pt/Al_2O_3 and Pt/12CeO_2-Al_2O_3 catalysts containing 1 wt. % of Pt were prepared by the impregnation of the support with a solution of $H_2PtCl_6.6H_2O$ in ethanol, followed by calcination at 773K for 2 h or at 1173 K for 24h.

Characterization: Surface areas (S_{BET}) and pore volumes (Vp) were measured by N_2 adsorption at 77 K using Micromeritics ASAP 2000 apparatus. The fresh and the used catalysts were investigated by transmission electron microscopy (TEM) using a JEM 3010 microscope, operating at 300 kV (1.7 Å point resolution). X-ray diffraction (XRD) patterns of the samples were collected with a Rigaku DMAX 2500 PC diffractometer using Cu-K_α radiation. X-ray photoelectron spectra (XPS) were obtained with an ESCALAB II VG Scientific spectrometer by using a monochromatic Al-K_α source (E = 1486.7 eV). X-ray absorption near edge structure (XANES) measurements in the Pt L_3-edge (11564 eV) were carried out at beamline D06A-DXAS of the Brazilian National Synchroton Latoratory (LNLS), using a Si (111) polychromator. Spectra were collected *in situ* during temperature-programmed reduction in H_2 and during heating of the reduced sample from room temperature up to 973K in a flow of the reaction mixture containing a CH_4:O_2 ratio of 2:1.

Catalytic Test: The partial oxidation and the autothermal reforming of CH_4 were conducted in a fixed-bed quartz reactor at atmospheric pressure. The catalyst (20 mg) was diluted with SiC (36 mg). The samples were reduced *in situ* at 773 K for 1h in a flow of 10%H_2/N_2 (at 50 mL/min) and then heated up to 1073 K under N_2. The reactant mixture contained a H_2O:CH_4:O_2 ratio of x:2:1 (where x = 0 or 0.64), and a CH_4 and O_2 flow rate of 100 cm^3/min was used.

3. RESULTS AND DISCUSSION

Surface areas (S_{BET}) and pore volumes of supported Pt catalysts are summarized in Table 1. Pt catalysts supported on Al_2O_3 and $12CeO_2$-Al_2O_3 showed surface areas and pore volumes similar to those of the supports. The dispersion of Pt, estimated by the rate of cyclohexane dehydrogenation, in different catalysts was: Pt/Al_2O_3 > Pt/$12CeO_2$-Al_2O_3.

Table 1: Chemical analysis, surface area (S_{BET}), pore volume (V_p), Pt dispersion (Pt-Disp.) and CeO_2 particle size (D_{XRD}) of the samples.

Sample	CeO_2 (wt%)	S_{BET} (m^2/g)	Vp (cm^3/g)	Pt Disp.	D_{XRD} (nm)
Al_2O_3	-	78	0.20		-
$12CeO_2$-Al_2O_3	11.2	75	0.19		17.2
Pt/Al_2O_3		76	0.24	0.66	-
Pt/$12CeO_2$-Al_2O_3	11.2	72	0.19	0.54	17.4

XRD patterns of the CeO_2-Al_2O_3-support after calcination at 1173 K showed diffraction lines characteristic of γ-Al_2O_3 and CeO_2 fluorite structure. The crystallite size of CeO_2 in CeO_2-Al_2O_3-supported Pt catalysts after calcination at 773 K is shown in Table 1. The lines associated with Pt oxide were not observed in the samples calcined at 773 K. In addition, the reduction of these samples in H_2 at 773 or 1073 K did not lead to the appearance of the lines characteristic of metallic Pt. However, when the Pt/Al_2O_3 and Pt/$12CeO_2$-Al_2O_3 catalysts were calcined in air at 1173 K for 24 h, the diffractogram of the reduced samples exhibited the patterns of metallic Pt.These results suggest that both Pt/Al_2O_3 and Pt/$12CeO_2$-Al_2O_3 catalysts are not resistant to sintering of Pt in O_2 at high temperatures.

The XPS parameters of oxidized and reduced samples are summarized in Table 2. For all calcined samples, the binding energy (BE) values of Pt $4d_{5/2}$ are in the range of 315.3-315.8 eV, characteristic of oxidized Pt [5]. The BE of Pt $4d_{5/2}$ increases with the CeO_2 content. This result may be attributed to different effects: (i) electron transfer between Pt and CeO_2 and (ii) interaction between Pt and Cl^- ions. For all reduced samples, the BE values of Pt $4d_{5/2}$ are in the range of 314.1-314.4 eV. These values are slightly higher than those of completely reduced Pt (313.5 eV), which means that the surface Pt maintains some δ^+ character. In Table 2, it can be seen that Pt is more easily reduced on alumina.

Table 2: XPS parameters of Pt/Al_2O_3 and Pt/$12CeO_2$-Al_2O_3 oxidized and reduced samples

Sample	BE (eV) Oxidized samples			BE (eV) Reduced samples	
	O1s	Pt $4d_{5/2}$	Ce $3d_{5/2}$	Ce$3d_{5/2}$	Pt $4d_{5/2}$
Pt/Al_2O_3	531.3	315.3			314.1
Pt/$12CeO_2$-Al_2O_3	529.5sh,531.1	315. 8	882.0	882.0	314.4

Fig. 1a shows the evolution of *in situ* Pt L_3-edge XANES-H_2 spectra of the oxidized PtO_x/Al_2O_3 as temperature was increased under H_2. The initial spectrum of the PtO_x/Al_2O_3 sample is characterized by a steeply rising absorption peak denominated "white line", which changes with the temperature. In the case of the Pt L_3-edge XANES, the absorption intensity of the white line reflects the vacancy in the 5d orbital of Pt atoms, i.e., a large white line is observed for the oxidized Pt, whereas a small white line is observed for reduced Pt. The results show that the intensity of the absorption at 11564 eV became slightly decreased at the temperature region between 400-450 K, indicating that a small fraction of Pt oxide is reduced, while the majority of the Pt oxide is reduced at temperatures higher than 773 K. The *in situ* Pt L_3-edge XANES-H_2 spectra of the oxidized $PtO_x/12CeO_2$-Al_2O_3 sample (data not shown) presented a similar profile to that obtained for PtO_x/Al_2O_3 in Fig. 1a. However, the Pt in $PtO_x/12CeO_2$-Al_2O_3 is reduced at temperatures about 80 K lower than in PtO_x/Al_2O_3.

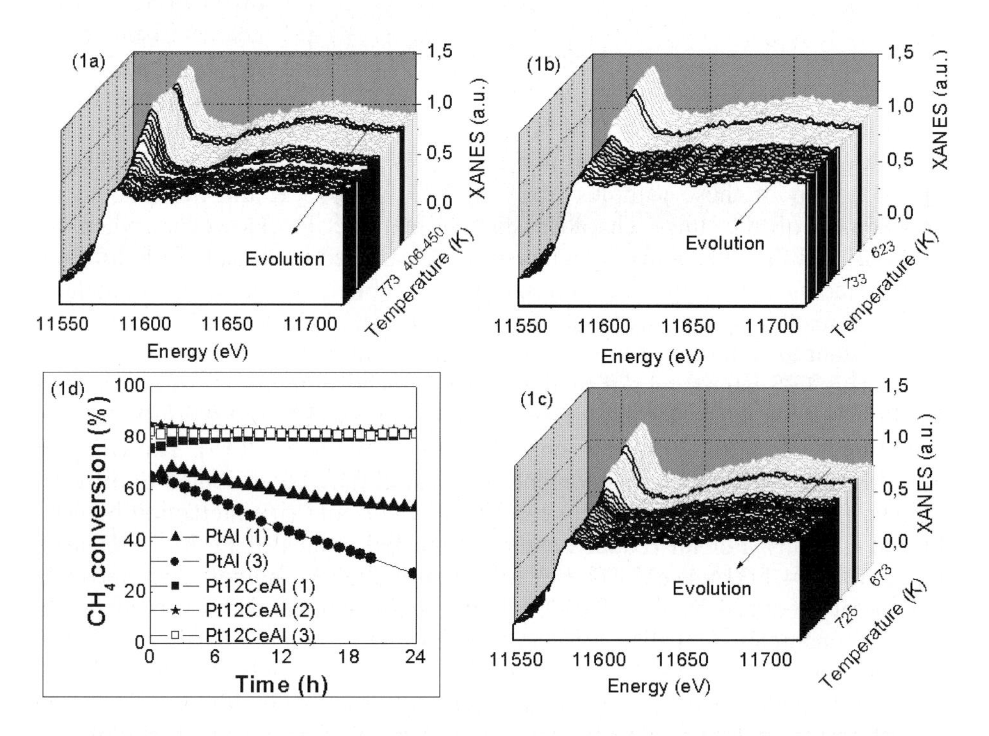

Figure 1. (a) Pt L_3-edge XANES-H_2 spectra of PtO_x/Al_2O_3. Pt L_3-edge XANES-POM spectra of (b) Pt/Al_2O_3 and (c) $Pt/12CeO_2$-Al_2O_3. (d) Stability test of Pt/Al_2O_3 (PtAl) and $Pt/12CeO_2$-Al_2O_3 (PtAl12CeAl) under reaction mixtures of various compositions (H_2O:O_2:CH_4): cond. (1)=(0.6:0.5:1.0); cond. (2)=(0.2:0.5:1.0) and cond. (3)=(0.0:0.5:1.0).

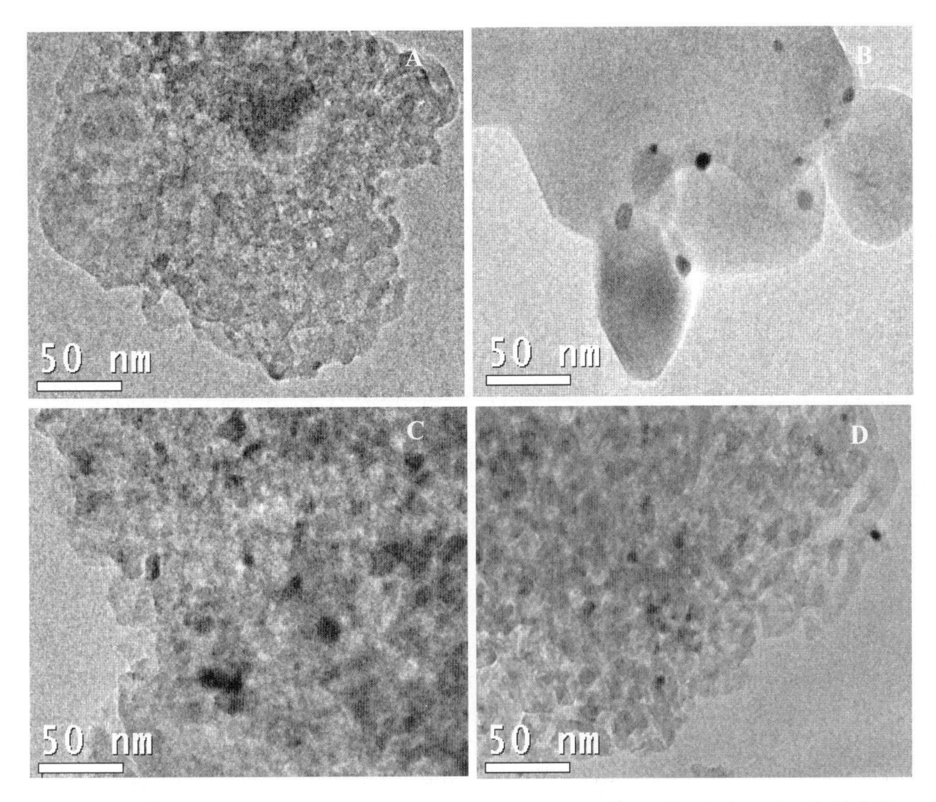

Fig. 2: TEM images of fresh and used catalysts in a reacton mixture of H_2O:CH_4:O_2 (0.64:2:1): Pt/Al_2O_3 (a) fresh and (b) used; $Pt/12CeO_2$-Al_2O_3 (c) fresh and (d) used.

Figures 1b and 1c show the *in situ* Pt L_3-edge XANES-POM spectra of the pre-reduced Pt/Al_2O_3 and $Pt/12CeO_2$-Al_2O_3 samples, respectively, as a function of temperature. The CH_4:O_2 ratio was kept at 2:1, which corresponds to reaction conditions of the POM. The samples were pre-reduced in H_2 at 1073 K, cooled down to room temperature and the H_2 flow was replaced by a mixture of CH_4:O_2. The white line became significantly more intense when the H_2 was changed to CH_4:O_2, showing that Pt was oxidized in both Pt/Al_2O_3 and $Pt/12CeO_2$-Al_2O_3 catalysts. The oxidation is more pronounced in the Pt/Al_2O_3, which shows a more intense white line in the same conditions. Therefore, it is reasonable to suppose that Ce favors the stabilization of reduced Pt species. For the Pt/Al_2O_3 catalyst (Fig. 1b), a slight decrease in the white line is observed at temperatures around 623-673 K, while an abrupt decrease occurs at around 733 K. A similar behavior was shown by the $Pt/12CeO_2$-Al_2O_3 catalyst (Fig. 1c). It is important to stress that the intensity of the white line at the reaction temperature (1073 K) became smaller than that observed for Pt/Al_2O_3. These results indicate that Pt is more easily reduced in $Pt/12CeO_2$-Al_2O_3.

The results of the stability test of the Pt/Al_2O_3 and $Pt/12CeO_2$-Al_2O_3 samples as a function of time on stream under reaction mixtures containing various compositions (H_2O:O_2:CH_4) are shown in Fig. 1d. It is clear that the stability increases greatly with the presence of CeO_2. The Pt/Al_2O_3 catalysts strongly deactivated with the time on stream, whereas the 12 wt. % CeO_2-loaded catalyst is quite stable.

To understand the difference in stability, TEM images were taken of fresh and used Pt/Al_2O_3 and $Pt/12CeO_2$-Al_2O_3 catalysts (Figure 2). In the fresh samples, both catalysts showed similar results, with Pt particles smaller than 5 nm in diameter. On the other hand, TEM images of the used Pt/Al_2O_3 catalyst confirmed its low thermal stability and the strong occurrence of Pt sintering, whereas no significant change of the Pt dispersion was detected in the used $Pt/12CeO_2$-Al_2O_3 catalyst. The higher stability of $Pt/12CeO_2$-Al_2O_3 catalysts is assigned to thermal stability of the support and to a possible lower mobility of Pt-O-Ce-support species.

4. CONCLUSIONS

TEM results showed that Pt particles strongly sintered in Pt/Al_2O_3 catalyst under reaction conditions of auto-thermal reforming of methane, while Pt in $Pt/12CeO_2$-Al_2O_3 showed higher stability. *In situ* XANES-POM spectra showed that PtO_x in $Pt/12CeO_2$-Al_2O_3 is more easily reduced than in Pt/Al_2O_3 on the POM reaction. The Pt-O-support interaction is reestablished by oxidation of Pt at a low temperature, suggesting that the Pt-support interaction stabilizes Pt dispersion.

ACKNOWLEDGEMENTS

The authors wish to acknowledge the financial support of the PETROBRAS (0050.0007696.04.2).

REFERENCES

1. H.Y. Wang, E. Ruckenstein, J. Catal. 186 (1999) 181.
2. H. C. Yao, M. Sieg, H. K. Plummer, J. Catal. 59 (1979) 365.
3. J. Wei, E. Iglesia, J.Catal. 224 (2004) 370.
4. Y. Nagai, T. Hirabayashi, K. Dohmae, N. Takai, T. Minami, H. Shinjoh, S. Matsumoto, J. Catal. 242 (2006) 103.
5. Shyu, J.Z.; Otto, K. Appl. Surf. Sci. 1988, *32,* 246.
6. Jackson, S.D.; Willis, J.; McLellan, G.D.; Webb, G.; Keegan, M.B.T.; Moyes, R.B.; Simpson, S. Wells, P.B.; Whyman, R. J. Catal. 1993, 139, 191.

Natural Gas Conversion VIII
F.B. Noronha, M. Schmal, E.F. Sousa-Aguiar (Editors)

Advanced Auto-thermal Gasification Process

Yoshiyuki Watanabe[a], Nobuhiro Yamada[a], Akira Sugimoto[a], Koichiro Ikeda[b], Naoki Inoue[b], Fuyuki Noguchi[b], Yoshifumi Suehiro[c]

[a]*JGC Corporation, 2-3-1, Minato Mirai, Nishi-ku, Yokohama, 220-6001, Japan*
[b]*Osaka Gas Co., Ltd., 6-19-9, Torishima, Konohana-ku, Osaka, 554-0051, Japan*
[c]*Japan Oil, Gas and Metals National Corporation, 1-2-2, Hamada, Mihama-ku, Chiba, 261-0025, Japan*

ABSTRACT

The Advanced Auto-thermal Gasification (A-ATG) Process, developed through a large-scale national project in Japan, is a synthesis gas production process that combines ultra-deep desulphurization of a feed gas (i.e., natural gas) with a new, high performance reforming catalyst.

The A-ATG catalyst allows the oxidation reaction and the reforming reaction to take place simultaneously and has a high activity and a high reaction rate. Therefore the amount of catalyst can be reduced to one fourth of that previously required by a conventional process using a pre-reformer.

A pilot plant has demonstrated that the A-ATG Process results in a compact plant and reduced construction costs, and operation of this pilot plant has provided a firm basis for the development of commercial plants.

1. INTRODUCTION

With increasingly stringent environmental requirements and rising oil prices, the world's attention is focusing on clean fuels produced from natural gas using gas-to-liquids (GTL) processes, and accordingly, a number of large GTL plants are currently being constructed in various countries that produce natural gas.

The Advanced Auto-thermal Gasification (A-ATG) Process is a new, high-efficiency process to produce synthesis gas from natural gas. It is hoped that the successful development of this process will lead to a substantial reduction in the construction costs for synthesis gas production systems, which account for a large percentage of the total construction costs of GTL plants [1], and will, thereby, contribute to the wider use of GTL fuels.

The development of the A-ATG process is being carried out jointly by Osaka Gas Co., Ltd., and JGC Corporation under the sponsorship of the Japan, Oil, Gas and Metals National Corporation (JOGMEC), an Independent Administrative Corporation in Japan.

The development project to verify the feasibility of the A-ATG process included the design and construction of a pilot plant with a synthesis gas production capacity of 2,000 Nm^3/hr (total of hydrogen and carbon monoxide), which is sufficient to produce 65 BPSD of GTL products. Additionally, this development project included collecting design data through a demonstration test using the pilot plant and developing a conceptual design for a commercial plant.

The features of the A-ATG Process are described below, together with information regarding its development status and pilot plant test results.

Fig. 1. Overview of A-ATG Pilot Plant

2. A-ATG PROCESS

A synthesis gas production process based on the combination of a pre-reformer and a secondary reformer is the state of the art for commercial plants at the present time [2,3]. Fig. 2 compares the main flows of the conventional process and the A-ATG process.

In the conventional process, desulfurized natural gas is reformed with steam at a low temperature in the pre-reformer and, then, further reformed at a high temperature in the secondary reformer, using burners. However, the pre-reformer must be relatively large because of the low reaction rate (GHSV 5,000 hr^{-1} or so), and there is a limit to how much the secondary reformer can be scaled up because of burner design issues.

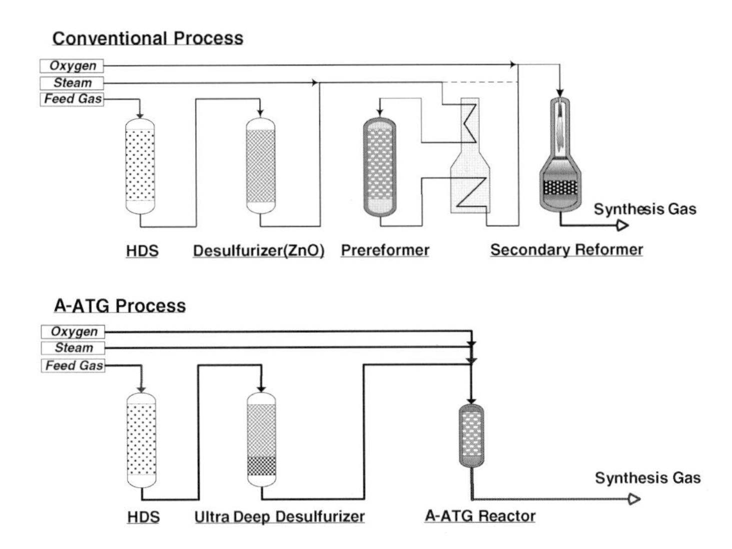

Fig. 2. Comparison between the Conventional Process and the A-ATG Process

The A-ATG Process combines ultra-deep desulfurization of the feed gas with a new, high performance reforming catalyst which allows simultaneous oxidation (exothermic reaction) and reforming (endothermic reaction), with an entry temperature below 300°C and without using burners. Moreover, the maximum reactor temperature is lower than that in the secondary reformer used in the conventional process.

The long-term operating stability of the catalyst is ensured by the ultra-deep desulfurization of the feed gas, which reduces the sulfur content of the feed gas to just a few parts per billion, far lower that the "parts per million" levels achieved in conventional plants. Specifically, by reducing the concentrations of sulfur compounds in the feed gas to levels less than 2.4 mol. ppb before the feed gas is admitted to the reforming, A-ATG reactor, it is possible to eliminate almost entirely the deleterious effects of sulfur compounds, i.e., a reduction in reforming catalyst activity over time and a concomitant rise in reforming reactor temperature [4].

The A-ATG catalyst, a supported precious metal catalyst, allows the oxidation reaction and the reforming reaction to take place simultaneously and has a high activity and a high reaction rate (GHSV: Over 20,000 hr^{-1}). Therefore the amount of catalyst can be reduced to one fourth of that previously required by a conventional process using a pre-reformer, and the reactor vessel can, therefore, be smaller. In addition, the feed gas heater can be eliminated because the feed gas supply temperature can be lower. Thus, by reducing the size of the reactor vessel and eliminating the feed gas heater, the space required for installation of the unit can be greatly reduced.

Therefore, in a commercial GTL plant, using the simpler and more compact A-ATG Process, it should be possible to reduce construction costs by more than 10%.

3. RESULTS AND DISCUSSION

The development of the A-ATG Process had the following objectives:

1. Development of a reactor in which oxygen and feed gas can be uniformly mixed and supplied to a catalyst bed without combustion of the gas
2. Development of a catalyst which can withstand high reaction temperatures
3. Development of a process with long-term operating stability

To achieve these objectives efficiently, a small experimental plant (1/50 scale of the pilot plant) was used, and design data was collected, scaled up and used for the design of the pilot plant.

Using this small experimental plant, the A-ATG process was first verified under mild operating conditions and with a relatively low oxygen feed rate (O_2/Carbon ratio = 0.2 mol/mol). Then, demonstration tests further verified operation at a high oxygen ratio (O_2/Carbon ratio = 0.5 to 0.7 mol/mol), during which, synthesis gas was produced using only the A-ATG reactor. Finally, in May, 2006, high-pressure operation (3 MPaG) at a high oxygen ratio was achieved, and synthesis gas was produced with a residual methane content of around 1%.

Fig. 3 shows the temperature profile in the reactor. Increasing the oxygen ratio, raised the maximum temperature in the catalyst bed. The maximum

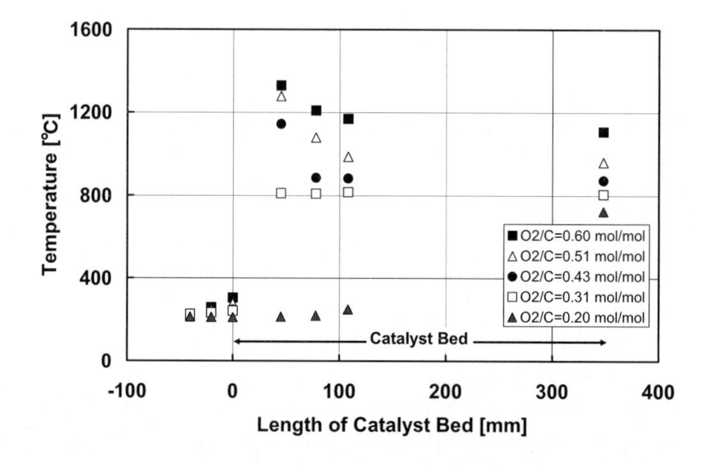

Fig. 3. Temperature Profile in the A-ATG Reactor in the Small Experimental Plant

temperature exceeded one thousand, several hundred degrees Celsius when the plant was operated at the desired high oxygen ratio. It was verified that the catalyst performance remained stable after 150 hours of operation.

The technical knowledge obtained with the small experimental plant was applied to the design of the larger, pilot plant, and in July, 2006, high-pressure operation at a high oxygen ratio was achieved for the first time in the pilot plant. Further, the performance of the pilot plant remained stable over time.

Fig. 4 shows how the composition of the synthesis gas changed over time during the high-temperature, high-oxygen-ratio demonstration test using the pilot plant.

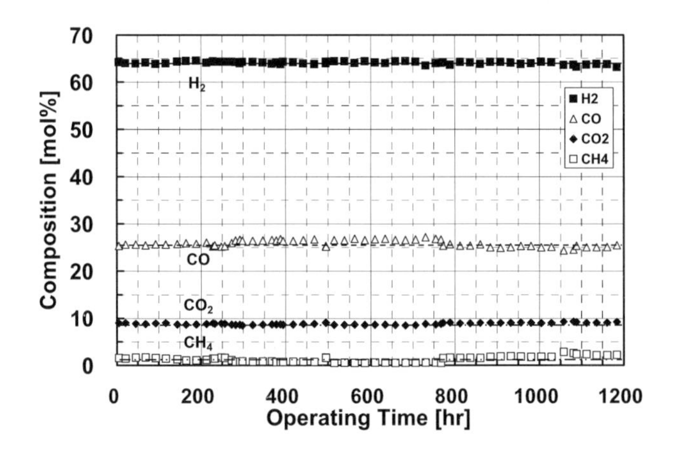

Fig. 4. Synthesis Gas Composition in the A-ATG Pilot Plant

During this demonstration test, there were no remarkable changes in the process conditions, such as reactor temperature or pressure (See Fig. 5), and it was confirmed that the H_2 and CO content of the synthesis gas obtained remained stable throughout the test and that a residual methane content of approximately 1% was maintained, as well.

The operating conditions and composition of the synthesis gas were changed slightly at the 230 hr, 500 hr, 760 hr and 1040 hr points to evaluate the performance under different reaction conditions. The residual methane content increased slightly at 760 hr and 1040 hr points due to a decrease in the oxygen feed rate. The composition of the synthesis gas achieved the equilibrium gas composition calculated for the operating conditions during the test, and no variation in the composition due to deactivation of the catalyst was observed.

This demonstration test verified that the catalyst maintained stable performance under high-pressure and high-oxygen-ratio conditions and made it possible to conduct an endurance test that extended over one thousand hours. In addition, the demonstration test provided information and data that could be used to design commercial plants, such as start-up and shutdown procedures, operating conditions under which auto-ignition of the premixed gas and backfiring were prevented, and data for the design of the reactor vessel, itself.

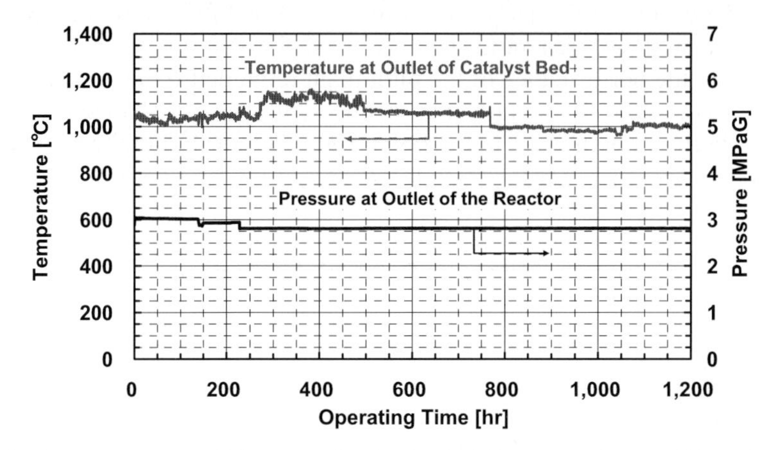

Fig. 5. Temperature and Pressure in the A-ATG Reactor in the Pilot Plant

4. CONCLUSION

The demonstration tests conducted using the pilot plant verified the performance of the A-ATG Process and provided the data and information needed for the design of a commercial plant.

The A-ATG Process is simple and compact and can be easily scaled up, and this process can, therefore, reduce the cost of the plants used to produce synthesis gas for the production of ammonia, methanol, DME, GTL, and other products.

REFERENCES

[1] S. C. Clarke and B. Ghaemmaghami, OFFSHORE WORLD, October-December (2003) 55.
[2] DME Handbook (Japanese version), Japan DME Forum, Ohmsha, p174 (2006). (English version will be printed.)
[3] S.Sreb and H.Gohna, World Methanol Conference, Copenhagen (Denmark), November 8-10 (2000).
[4] Osamu Okada, Kota Yokoyama, and Naoki Inoue, WO 2006/001438 A1, January 5, (2006).

Natural Gas Conversion VIII
F.B. Noronha, M. Schmal, E.F. Sousa-Aguiar (Editors)

Bifunctional Catalysts for Hydrogen Production from Dimethyl Ether

Galina Volkova, Sukhe Badmaev, Vladimir Belyaev, Lyudmila Plyasova, Anna Budneva, Evgeny Paukshtis, Vladimir Zaikovsky, Vladimir Sobyanin

Boreskov Institute of Catalysis, Pr. Akademika Lavrentieva 5, Novosibirsk, 630090, Russia

Abstract

The production of hydrogen directly from dimethyl ether (DME) was performed on bifunctional Cu-Ce/Al_2O_3 catalysts. The catalysts were characterized by XRD, HRTEM, EDX and IR spectroscopy of low-temperature CO adsorption. The high hydrogen productivity up to 600 mmol $g^{-1}h^{-1}$ may be explained by assuming that (1) DME dehydration occurs on acid sites of γ-Al_2O_3 and (2) methanol steam reforming takes place on mixed oxide phase CuO-CeO_2, solid solution of copper ions in cerium dioxide with ratio Cu/Ce from 12/86 to 33/67 at.%.

1. Introduction

Dimethyl Ether is expected to become an important energy resource as a raw material for hydrogen production for fuel cells. DME as well as methanol can be easily and selectively converted to hydrogen-rich gas at relatively low temperature (250-350°C) compared to other fuels such as natural gas, gasoline and LPG [1-3]. DME is relatively inert, non-corrosive, non-carcinogenic. DME is more favorable for steam reforming (SR) than methanol from the economical viewpoint because it can be produced from syn-gas more effectively [4-5]. Overall DME SR is expressed by the equations:

$$CH_3OCH_3 + H_2O = 2CH_3OH$$
$$CH_3OH + H_2O = 3H_2 + CO_2$$
$$CO_2 + H_2 = CO + H_2O$$

The first step of DME SR is hydration of DME into methanol the second step is methanol steam reforming into hydrogen-rich gas. Besides, during DME SR a reverse water gas shift reaction may occur to produce carbon monoxide. The mechanical mixtures of two catalysts: solid-acid for DME hydration and copper-containing oxide for methanol steam reforming, are usually used for DME SR [6-9]. The main disadvantage of such catalysts is their layering that leads to decrease of activity. The design of a bifunctional catalyst for DME SR seems to be a promising way to overcome this problem. Cu/Al_2O_3, Cu-Zn/Al_2O_3, Cu-Pd/Al_2O_3, Cu-Ru/Al_2O_3, Cu-Pt/Al_2O_3, Cu-Rh/Al_2O_3, Cu-Au/Al_2O_3 and Cu/$Ga_8Al_2O_{15}$ systems were studied as bifunctional catalysts for DME SR reaction [10-13] but demonstrated low hydrogen productivity. Researchers from the Boreskov Institute of Catalysis have developed the bifunctional Cu-Ce/Al_2O_3 catalysts for DME SR providing ten-fold increase of H_2 productivity [14]. Here we present the performance and physical characterization of Cu-Ce/Al_2O_3 catalysts.

2. Experimental

The Cu-Ce/Al_2O_3 catalysts were synthesized by treating γ-alumina in solutions of copper and cerium salts taken at the given ratio. Samples were dried at 100°C and calcined at 450°C for 3 hours. Catalysts were evaluated in fixed bed flow reactor with gas analysis on line. Reaction conditions: H_2O/DME/N_2=20/60/20, GHSV=10000 h^{-1}, pressure 1 atm, temperature 250-370°C. Prior to reaction, catalysts were reduced at 300°C in stream of H_2-N_2 for 1 hour.

HRTEM images were obtained on a JEM-2010 electron microscope (JEOL, Japan) with a lattice-fringe resolution of 0.14 nm at an accelerating voltage of 200 kV. The high-resolution images of periodic structures were analyzed by the Fourier method. Local energy-dispersive X-ray analysis (EDXA) was carried out on an EDAX spectrometer (EDAX Co.) fitted with a Si (Li) detector with a resolution of 130 eV. FT-IR spectroscopy was used to determine the acidity of the samples by monitoring the low temperature CO adsorption. The samples were degassed in the IR cell at 400°C and then cooled to -173 °C and treated with low doses of CO from 0.1 to 10 torr. The spectra were recorded with 4 cm^{-1} resolution using Shimadzu FT-IR-8300 spectrometer. The number of CO adsorption sites was calculated by fitting the data to equation: $N=A/\rho A_o$, where A (cm^{-1}) is the integrated intensity of band, ρ (g cm^{-2}) is the mass of material normalized on pellet surface, A_o is the molar integrated absorption coefficient equal to 0.8 cm μmol^{-1} for band 2185-2200 cm^{-1}. Procedure for measurement of A_o is described in [15]. XRD studies were carried out on a diffractometer URD-63 with CuKα radiation using graphite monochromator with reflected beam. The lattice parameters and dispersion of CeO_2 were determined from line (111), $\Delta\alpha \pm 0.003$ Å.

3. Results and Discussion

Table 1. Performance of bifunctional $Cu-Ce/Al_2O_3$ catalysts in DME SR at 350°C, pressure 1 atm, $H_2O/DME/N_2$=60/20/20; GHSV= 10000 h^{-1}

Catalyst	DME	Outlet gas composition, %					H_2
	conv. %	H_2	CO_2	CO	CH_4	CH_3OH	mmol/g h
$4Cu-4Ce/Al_2O_3$	71	50	17.3	0.3	0.08	0.64	473
$8Cu-4Ce/Al_2O_3$	80	55	18.5	0.7	0.13	0.35	523
$12Cu-4Ce/Al_2O_3$	81	57	18.5	0.7	0.25	0.6	520
$8Cu-2Ce/Al_2O_3$	98	59,5	19.3	0.7	1.6	0.07	610
Equilibrium	100	60	15.7	6.7	0	0.00013	

The high hydrogen productivity up to 600 mmol $g^{-1}h^{-1}$ and low CO concentration in hydrogen-rich gas were observed for all tested catalysts (Table 1). The H_2 productivity is more than one order of magnitude higher than the early reported (Table 2). The content of copper and cerium and their ratio slightly affected the catalyst activity, it reduced by 30% at the lowest concentration of $CuO+CeO_2$. Performance of an alumina–free catalyst $10Cu/CeO_2$ was very poor; the DME conversion was within 1%.

Table 2. Comparison of bifunctional catalysts in DME SR

Catalyst	T,°C	GHSV h^{-1}	Inlet gas $DME/H_2O/N_2$	DME conv.,%	H_2 mmol/g h	References
$Cu-Ce/Al_2O_3$	350	10000	20/60/20	71-98	470-610	Present work
$Cu-Zn/Al_2O_3$	350	180	25/75/0	97	11.3	[10]
$Cu-Zn/Al_2O_3$	350	1400	20/60/20	58	48.2	[11]
$Ga_8Al_2O_{15}$	350	20000	1/3/96	80	43.0	[12-13]

Equilibrium conversion of DME and concentration of H_2 over $8Cu-2Ce/Al_2O_3$ catalyst are reached at temperature 350°C (Figure 1). The CO concentration is lower than 1% and remains below the equilibrium value at 250-370°C.

FT-IR spectra of $\gamma-Al_2O_3$ and $Cu-Ce/Al_2O_3$ catalysts in the hydroxyl region (Figure 2a) showed that the bands at 3730 and 3770 cm^{-1} are decreased (curve 2) and disappeared (curve 3) under loading of Cu-Ce oxides. However, no change in bands intensity of the most acidic hydroxyl groups at 3675 and 3690 cm^{-1} was detected.

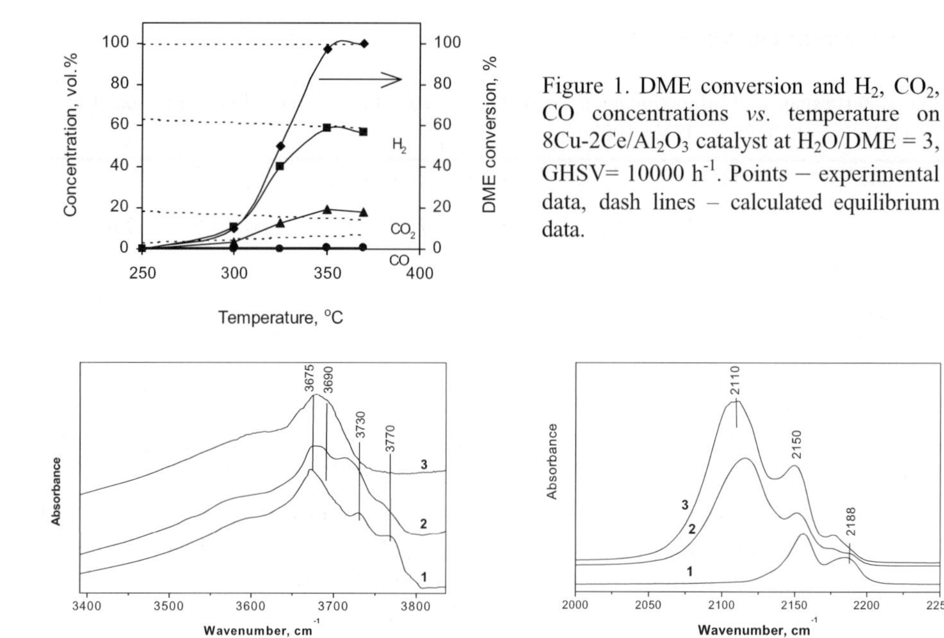

Figure 1. DME conversion and H_2, CO_2, CO concentrations *vs.* temperature on $8Cu\text{-}2Ce/Al_2O_3$ catalyst at $H_2O/DME = 3$, GHSV= 10000 h^{-1}. Points – experimental data, dash lines – calculated equilibrium data.

Figure 2. FT-IR spectra of (1) Al_2O_3, (2) $4Cu\text{-}4Ce/Al_2O_3$ and (3)$12Cu\text{-}4Ce/Al_2O_3$ catalysts in: (a) hydroxyl region and (b) adsorbed CO.

FT-IR spectra of adsorbed CO (Figure 2b) revealed bands characteristic of both Lewis (2185-2200 cm^{-1}) and Brønsted (2110-2175 cm^{-1}) forms of adsorption. Table 3 collects the acidity data calculated from bands 2185-2200 cm^{-1} for Lewis acid sites. It can be seen that amount of acid sites on the surface of $\gamma\text{-}Al_2O_3$ significantly reduced after loading of copper-cerium oxides. Nevertheless, the Lewis acid sites typical for γ-alumina were registered on the surface of all bifunctional $Cu\text{-}Ce/Al_2O_3$ catalysts.

Table 3. Acidity of the catalysts from FTIR spectra of CO adsorption

Catalyst	νCO, cm^{-1}	Lewis acid sites, μmol g^{-1}
$\gamma\text{-}Al_2O_3$	2184-2200	600
$4Cu\text{-}4Ce/Al_2O_3$	2190	110
$8Cu\text{-}2Ce/Al_2O_3$	2190	70
$12Cu\text{-}4Ce/Al_2O_3$	2190	50

Table 4. Characterization of $Cu\text{-}Ce/Al_2O_3$ catalysts by XRD and EDX analysis

Catalyst	Phase composition	a (111) CeO_2, Å	D (111) CeO_2, Å	Cu/Ce at. %
CeO_2	CeO_2	5.417	80	
$4Cu\text{-}4Ce/Al_2O_3$	$CeO_2+CuO+Al_2O_3$	5.381	40	33/67
$8Cu\text{-}2Ce/Al_2O_3$	$CeO_2+CuO+Al_2O_3$	5.378	35	no data
$12Cu\text{-}4Ce/Al_2O_3$	$CeO_2+CuO+Al_2O_3$	5.402	42	12/88

Results of XRD and EDX analysis for $Cu\text{-}Ce/Al_2O_3$ catalysts are presented in Table 4 and on Figure 3a. Three crystalline phases: CuO, CeO_2 and Al_2O_3 were identified in bifunctional $Cu\text{-}Ce/Al_2O_3$ catalysts. Content of CuO phase decreased with decreasing of copper concentration in the samples; in $4Cu\text{-}4Ce/Al_2O_3$ catalyst only traces of CuO were observed. It can be seen from Table 4 that the unit cell parameter of CeO_2 decreased from 5.417 Å to 5.402-5.378 Å. It indicates the formation of the solid solution of copper oxide in cerium dioxide. The exact atomic ratio of metals Cu/Ce in $CuO\text{-}CeO_2$ solid solutions was determined from the EDX spectra (Figure 3a). It was shown that the atomic ratio of Cu/Ce ranged from 12/88 to 33/67 at. % (Table 4).

The dispersion of CeO_2 in the range of 35-42 Å has been registered by XRD for all bifunctional catalysts (Table 4). TEM observations presented in Figure 3b show the formation of 50-100 nm aggregates of mixed $CuO\text{-}CeO_2$ oxide. It means that $CuO\text{-}CeO_2$ solid solution is composed of elementary particles (35-42 Å) agglomerated in large particles of 50-100 nm in size. These particles of mixed $CuO\text{-}CeO_2$ oxide are located within the large pores of γ-alumina.

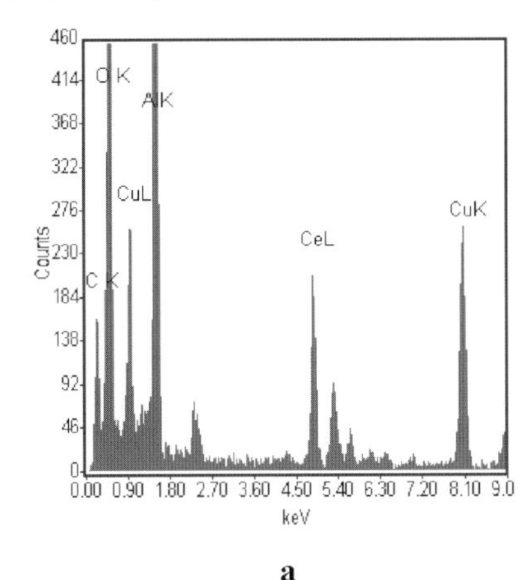

a

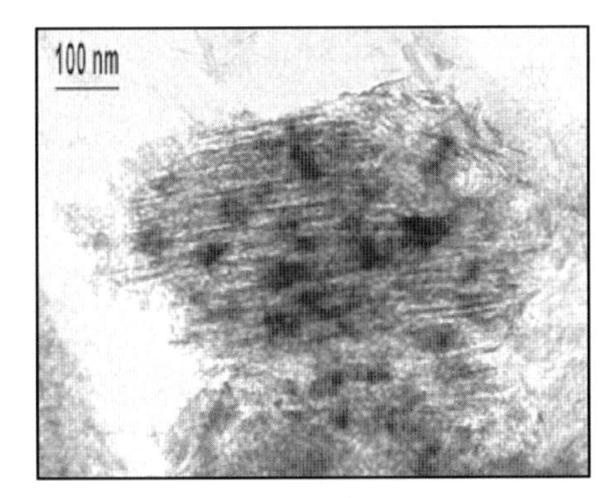

b

Figure 3. EDX spectra (a) and TEM micrograph (b) of $4Cu\text{-}4Ce/Al_2O_3$ catalyst

It is known that the activity of copper containing mixed oxides for methanol synthesis and for water gas shift reaction is determined by the formation of solid solution of copper ions in zinc-alumina mixed oxides [16-18]. These catalysts are also very active for methanol steam reforming. Our aim was to provide the formation of the CuO-CeO_2 solid solution on the surface of alumina and at the same time to prevent the blocking of alumina acid sites. We have succeeded for the first time in design of bifunctional catalysts for DME SR with two types of active sites on the surface of alumina: solid solution of copper-cerium oxides and acid sites.

4. Conclusion

It was shown that bifunctional Cu-Ce/Al_2O_3 catalysts (CuO+CeO_2 10-20 wt. %, Cu/Ce wt. ratio from 1/1 to 4/1) are novel and effective catalysts for hydrogen production from DME. Two types of active sites were registered on the surface of these catalysts: acid sites for DME dehydration into methanol and CuO-CeO_2 solid solutions for steam reforming of methanol to hydrogen rich gas. The formation of solid solutions CuO-CeO_2 is a key factor responsible for high performance of Cu-Ce/Al_2O_3 catalysts.

References

1. V.A. Sobyanin, S. Cavallaro, S. Freni, Energy Fuels 14 (2000) 1139.
2. T.H. Fleisch, R.A. Sills, M.D. Briscoe, J. Natural Gas Chem., 11 (2002) 1.
3. International DME Association Website: www.aboutdme.org
4. T. Shikada, Y. Ohno, T. Ogawa, M. Ono, M. Mizuguchi, K. Tomura, K. Fujimoto, Stud. Surf. Sci. Catal. 119 (1998) 515.
5. T. Ogawa, N. Inoue, T. Shikada, O. Inokoshi, Y.Ohno, Stud. Surf. Sci. Catal. 147 (2004) 379.
6. V.V. Galvita, G.L. Semin, V.D. Belyaev, T.M. Yurieva, V.A. Sobyanin, Appl. Catal. A: Gen. 216 (2001) 85.
7. Y. Tanaka, R. Kikuchi, T. Takeguchi, K. Eguchi, Appl. Catal. B: Env. 57 (2004) 211.
8. T. Nishiguchi, K. Oka, T. Matsumoto, H. Kanai, K. Utani, S. Imamura, Appl. Catal. A: Gen. 301 (2006) 66.
9. T.A. Semelsberger, K.C. Ott, R.L. Borup, H.L. Greene, Appl. Catal. B: Env. 65 (2006) 291.
10. K. Takeishi, H. Suzuki, Appl. Catal. A: Gen. 260 (2004) 111.
11. T.A. Semelsberger, K.C. Ott, R.L. Borup, H.L. Greene, Appl. Catal. A: Gen. 309 (2006) 210.
12. T. Mathew, Y. Yamada, A. Ueda, H. Shioyama, T. Kobayashi, Catal. Lett. 100 (2005) 247.
13. T. Mathew, Y.Yamada, A.Ueda, H. Shioyama, T. Kobayashi, Appl. Catal. A: Gen.286 (2005) 11.
14. S.D. Badmaev, G.G. Volkova, V.D. Belyaev, L.M. Plyasova, V.A. Sobyanin, Rassian Patent 2 286 210 (2006).
15. E.A. Paukshis, IR spectroscopy for heterogeneous acid-base catalysis. Nauka, Novosibirsk, 1992 (in Russian).
16. J.B. Hansen, in G.Ertl, H. Knozinger, J.Weitkamp (Eds.) Handbook of Heterogeneous Catalysis, Vol. 5, Wiley-VCH, Weinheim, 1997, p.1856.
17. K. Klier, Adv. Catal. 31 (1982) 243.
18. T.M. Yurieva, Catal. Today 51 (1999) 457.

Natural Gas Conversion VIII
F.B. Noronha, M. Schmal, E.F. Sousa-Aguiar (Editors)

Autothermal reforming of methane over nickel catalysts prepared from hydrotalcite-like compounds

Mariana M.V.M. Souza[*1,2], Nielson F. P. Ribeiro[2], Octávio R. Macedo Neto[2], Ivna O. Cruz[1] and Martin Schmal[2]

[1]*Escola de Química- Federal University of Rio de Janeiro, Centro de Tecnologia, Bloco E, sala 206, 21940-900, Rio de Janeiro, RJ, Brazil *mmattos@eq.ufrj.br*
[2]*NUCAT/PEQ/COPPE- Federal University of Rio de Janeiro, Centro de Tecnologia, Bloco G, sala 128, 21945-970, Rio de Janeiro, RJ, Brazil*

1. Introduction

There has been substantial interest in recent years in alternative routes for conversion of natural gas (methane) to synthesis gas, a mixture of CO and H_2, which can be used to produce chemical products with high added values, such as hydrocarbons and oxygenated compounds.

Autothermal reforming (ATR), a combination of steam reforming and partial oxidation reactions, is an advantageous route for syngas production for both economical and technical reasons. It has low-energy requirements due to the opposite contribution of the exothermic methane oxidation and endothermic steam reforming. The combination of these reactions can improve the reactor temperature control and reduce the formation of hot spots, avoiding catalyst deactivation by sintering or carbon deposition. Moreover, ATR allows the production of syngas with a wider range of H_2/CO ratio by manipulating the relative concentrations of H_2O and O_2 in the feed [1,2]. All these advantages indicate that ATR should be the technology of choice for large-scale GTL (gas-to-liquid) plants [3]. In addition, a fuel processor based on autothermal reforming of methane could provide a low cost and compact system, with fast start-up and capability to follow load variations, more adequate for fuel cell electric vehicles [4].

Supported metal catalysts have been used in the reforming reactions of hydrocarbons and are conventionally prepared by wet impregnation of different supports. This method is not fully reproducible and may give rise to some heterogeneity in the distribution of the metal on the surface. Moreover, the fine metal particles tend to sinter at high temperature, resulting in the catalyst deactivation.

Hydrotalcite-like compounds (HTLCs) are anionic clays, layered-double hydroxides with lamellar structure, showing interesting properties, such as high surface area, "memory effect" and basic character [5]. Upon heating, these compounds form a homogeneous mixture of oxides with very small crystal size, stable against thermal treatments, which by reduction result in high dispersion of the metallic crystallites.

In this context, the objective of this work is to study nickel catalysts prepared from hydrotalcite-like compounds, with different Ni compositions, for autothermal reforming of methane, comparing their performance with a conventional nickel impregnated catalyst.

2. Experimental

2.1. Catalyst Preparation

Three Ni-Mg-Al HTLCs (10NiHT, 20NiHT and 30NiHT) were prepared by co-precipitation from aqueous solutions, following the procedure described by Corma et al. [6]. Solution A, containing nitrates precursors (Al/(Ni+Mg+Al)=0,25 and [Ni+Mg+Al]=1.5M), was added dropwise (1 mL/min) into solution B, formed by Na_2CO_3 and NaOH (CO_3^{2-}/Al^{3+} = 0,375 and OH^-/Al^{3+} = 6,3). The gel formed was aged for 42h at 60°C. The solid obtained was filtered and washed with distilled water at 80°C until pH 7, dried at 100°C overnight and calcined in flowing air (120 mL/min) up to 500°C, for 30 min. After calcination, the nominal NiO loading was 10, 20 and 30wt%. The calcined catalysts were named XXNiHTc where XX is 10, 20 or 30, depending on the NiO content. For comparison, a sample without nickel (MgAlHT) was also prepared.

The reference catalyst (20NiAl) was prepared by wet impregnation of alumina with an aqueous solution of nickel nitrate, followed by drying and calcination at the same conditions. The nominal NiO loading was 20wt%.

2.2. Characterization techniques

The chemical composition of the synthesized samples was determined by X-ray fluorescence (XRF) using a Rigaku spectrometer. X-ray diffraction (XRD) patterns were recorded in a Rigaku miniflex X-ray diffractometer equipped with a graphite monochromator using CuKα radiation (40 kV).

The textural characteristics, such as BET specific area and pore volume (BJH method), were determined by N_2 adsorption-desorption at -196°C in a Micromeritcs ASAP 2000. Prior to the analysis the samples were outgassed for 20h at 200°C.

2.3. Catalytic tests

The reaction was carried out in a fixed-bed flow-type quartz reactor loaded with 20 mg of catalyst, under atmospheric pressure. The total feed flow rate was held constant at 200 cm^3/min (WHSV= $160h^{-1}$), with molar ratios of O_2/CH_4=0.25 and H_2O/CH_4= 0.5. The steam was added to the system by saturator with temperature control. The activity tests were performed at different temperatures, ranging from 400 to 900°C in steps of 50°C that were kept for 30 min at each temperature. The loss in catalyst activity at 800°C was monitored up to 48 h on stream. The reaction products were analyzed by on-line gas chromatograph (CHROMPACK CP9001), equipped with a Hayesep D column and a thermal conductivity detector.

3. Results and Discussion

3.1. Catalyst characterization

The chemical composition of the synthesized samples is presented in Tables 1 and 2. The composition of the as-synthesized samples is similar to those of the gel of synthesis, indicating an approximately complete incorporation of the cations in HTLC structure. The nickel loading of all samples is slightly lower than the nominal values due to loss of nickel during washing.

The XRD patterns of HTLC precursors (Figure 1A) exhibit sharp and symmetrical reflections for (003), (006), (110) and (113) planes and broad and asymmetric reflections for (102), (105) and (108) planes, characteristic of a well-crystallized HTLC in carbonate form. The absence of other phases suggests that both Ni^{2+} and Al^{3+} have isomorphically replaced Mg^{2+} cations in the brucite-like layers.

Based on the rhombohedral symmetry of hydrotalcites, the lattice parameters a (cation-cation distance in the brucite-like layer) and c=3c'(thickness of one brucite-like layer and one interlayer) were calculated, as shown in Table 1. These lattice parameters are almost the same for all samples, showing an isomorphous substitution of Mg by Ni cations in the brucite layers, without distortion of the lamellar structure, since ionic radius of Mg^{2+} and Ni^{2+} are very similar, 0.65 and 0.72 Å, respectively [5].

After calcination at 500°C, the lamellar structure disappears and XRD patterns show the presence of poorly crystallized MgO periclase phase, as illustrated in Figure 1B. These results indicate that both nickel and aluminum are well dispersed in MgO matrix without the segregation of a spinel-phase, in accordance to the literature [7]. The impregnated catalyst presented the characteristic peaks of cubic NiO and $NiAl_2O_4$.

Table 1. Chemical composition, structural and textural characteristics of the as-synthesized samples.

Sample	Ni/Mg* (molar)	Al/(Ni+Mg+Al)* (molar)	a (Å)	c (Å)	S_{BET} (m^2/g)	V_{pore} (cm^3/g)
MgAlHT	-	0.22 (0.25)	3.06	23.40	60.9	0.472
10NiHT	0.069 (0.088)	0.22 (0.25)	3.06	23.32	83.6	0.476
20NiHT	0.158 (0.20)	0.22 (0.25)	3.06	23.24	88.7	0.402
30NiHT	0.276 (0.36)	0.18 (0.25)	3.05	23.21	39.8	0.176

* The values in parentheses are nominal values.

Table 2. Chemical composition and textural characteristics of the calcined catalysts.

Sample	NiO* (wt%)	MgO (wt%)	Al_2O_3 (wt%)	S_{BET} (m^2/g)	V_{pore} (cm^3/g)
MgAlHTc	-	62.4	20.6	203.5	0.890
10NiHTc	7.45 (10)	58.5	21.8	164.7	0.694
20NiHTc	16.5 (20)	51.3	21.0	286.0	0.694
30NiHTc	23.1 (30)	44.3	15.9	183.3	0.344
20NiAl	24.5 (20)	-	75.5	189.2	-

* The values in parentheses are nominal values.

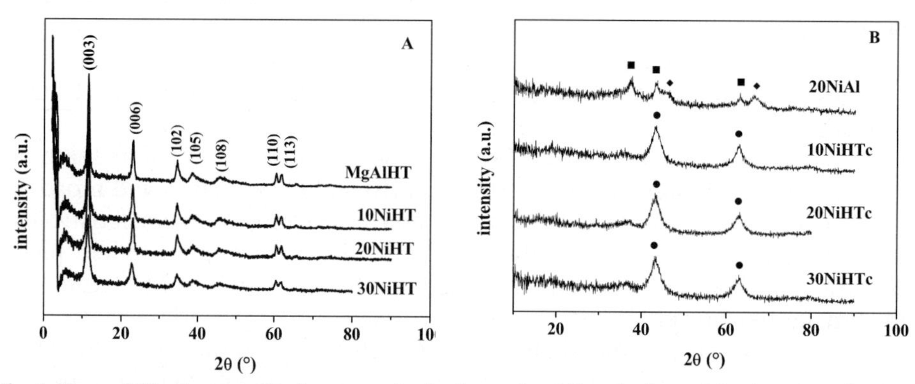

Fig. 1. X-ray diffractograms for the as-synthesized samples (A) and after calcination at 500°C (B).
● - MgO ■ - NiO ♦ - $NiAl_2O_4$

Textural characteristics (Table 1) of the 10NiHT and 20NiHT samples are very similar, with BET surface area of about 85 m^2/g and pore volume of 0.43 cm^3/g. Both surface area and pore volume of the 30NiHT are significantly lower. After calcination, it was observed an increase in surface area and pore volume for all samples, as shown in Table 2, which can be attributed to the elimination of carbonate anions as CO_2, leading to the destruction of the layered structure (as observed by XRD). The surface area of the impregnated catalyst is almost the same as that of pure alumina (200 m^2/g).

3.2. Catalytic tests

The catalytic activities were evaluated for autothermal reforming of methane, as a function of temperature, between 400 and 900°C, and the results are displayed in Figure 3 in terms of CH_4 conversion and H_2/CO product ratio. The hydrotalcite derived catalysts presented similar activities, with CH_4 conversion between 32% at 400°C and 94% at 900°C and O_2 conversion of 100% over the whole temperature range. The activity of the impregnated catalyst is much lower until 550°C (CH_4 conversion <10%), with similar conversions to those of HTLCs samples between 600 and 750°C, followed by a deactivation behavior at temperatures higher than 800°C, with decreasing CH_4 conversion.

The H_2/CO ratio varied between 5.1 at 400°C and 1.6 at 900°C for 30NiHTc and 20NiHTc catalysts, with 10NiHTc showing H_2/CO ratios slightly higher. The 20NiAl did not show any H_2 production until 550°C, with H_2/CO ratios close to those of 20NiHTc beyond 600°C.

The observed high H_2/CO ratios (> 4.0) at low temperatures suggests that water-gas shift (WGS) reaction occurs to a great extent with reforming of methane, as already reported in the literature [1,8]. At the same time, the decrease in H_2/CO ratio with increasing reaction temperature is consistent with the fact that WGS reaction is thermodynamically unfavorable at higher temperatures.

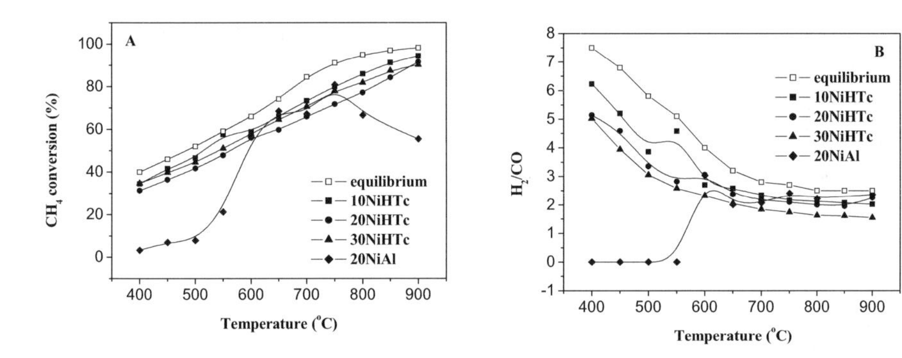

Fig. 3. CH_4 conversion (A) and H_2/CO product ratio (B) of Ni catalysts for autothermal reforming of methane as a function of temperature.

The catalyst stabilities were evaluated at 800°C up to 48h on stream. The equilibrium CH_4 conversion at this temperature is 95%. The performance of derived hydrotalcite catalysts was very similar, with CH_4 conversion of about 85% without any apparent deactivation during the whole experiment. The impregnated catalyst also presented a stable performance, close to the behavior of the HTLC catalysts.

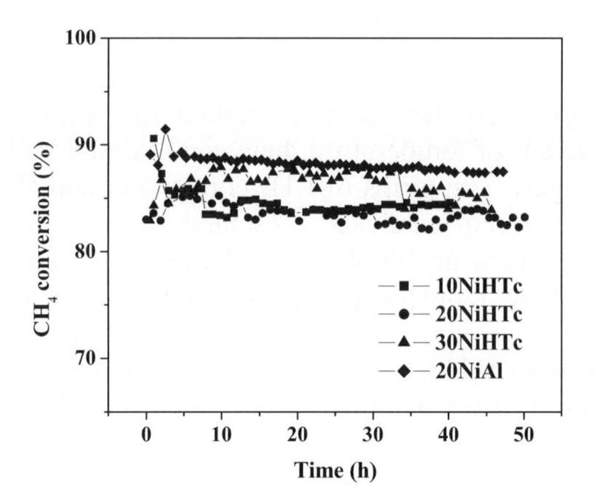

Fig. 4. CH_4 conversion as a function of time on stream at 800°C for autothermal reforming of methane.

4. CONCLUSIONS

Well-crystallized Ni-Mg-Al HTLCs were synthesized, showing an isomorphous substitution of Mg by Ni cations in the brucite layers. Upon calcination, the lamellar structure of HTLC disappeared and the formation of a poor crystallized MgO-periclase phase was observed. No other phases were detected, suggesting that both nickel and aluminum are well dispersed in MgO matrix. The derived hydrotalcite catalysts showed high activity for autothermal reforming of methane over the whole temperature range studied (400-900°C) and good stability at 800°C.

REFERENCES

1. S. Liu, G. Xiong, H. Dong, W. Yang, Appl. Catal. A 202 (2000) 141.
2. Z.-W. Liu, K.-W. Jun, H.-S. Roh, S.-E. Park, J. Power Sources 111 (2002) 283.
3. D.J. Wilhelm, D.R. Simbeck, A.D. Karp, R.L. Dickenson, Fuel Proc. Tech. 71 (2001) 139.
4. S.H. Chan, H.M. Wang, Int. J. Hydrogen Energy 25 (2000) 441.
5. F. Cavani, F. Trifiró, A. Vaccari, Catal. Today 11 (1991) 173.
6. A. Corma, V. Fornes, F.Rey, J.Catalysis 148 (1994) 205.
7. A.C.C. Rodrigues, C.A. Henriques, J.L.F. Monteiro, Mater. Research 6 (2003) 563.
8. M.E.S. Hegarty, A.M. O'Connor, J.R.H. Ross, Catal. Today 42 (1998) 225.

Natural Gas Conversion VIII
F.B. Noronha, M. Schmal, E.F. Sousa-Aguiar (Editors)

A high effective catalytic system for the production of synthesis gas for gas-to-liquid

Shizhong (Jason) Zhao[a], XD Hu[b], David Rogers[a], and David Tolle[a]

[a]*Sud-Chemie Inc., 1600 West Hill Street, Louisville, KY40210, USA*
[b]*Süd-Chemie China Operation, Shanghai, China*

1. Introduction

The synthesis gas generation section accounts for a significant portion of the capital and operational costs of a Gas-To-Liquid (GTL) plant. Therefore, the economics of the synthesis gas process is vital for the commercial success of a GTL process. Fischer-Tropsch synthesis requires the feed gas with a H_2/CO ratio of 2.1 for a typical Co-based catalyst and 1.7 for an Fe-based catalyst [1]. However, the conventional natural gas steam reforming process generates synthesis gas with a H_2/CO ratio greater than 3. The GTL processes need synthesis gas with a lower H_2/CO ratio than the traditional steam reforming can make, which energizes the study of other synthesis gas generation processes, such as catalytic partial oxidization and the mixture of steam reforming, oxidization, and CO_2 reforming. The ultimate goal is to make synthesis gas for the Fischer-Tropsch process as cost-effective as possible. This goal can be accomplished by directly making synthesis gas with desired H_2/CO ratio at high selectivity and throughput.

Steam and CO_2 mixed reforming has been commercially practiced in producing reducing gas for the steel industry for more than 30 years. Depending on the ratio of CO_2 and steam to the hydrocarbons, the H_2/CO ratio of the synthesis gas can be changed anywhere from less than 1 (CO_2-reforming) to greater than 3 (steam reforming). Since both steam reforming and CO_2-reforming are strongly endothermic, large and expensive tubular reactors are usually employed for the consideration of mass and heat transfer. To make a GTL process more economically attractive, reducing the size of the synthesis gas unit by including partial oxidization reaction could be a solution since it is a much faster exothermic reaction. The process would be more thermally neutral, and therefore, can be operated at very high space velocities in a vessel that reduces the size and the complexity of the reactor. In order for the high gas space velocity process to work properly, catalysts with high selectivity, activity, resistance to carbon formation, and high thermal stability are required.

Rhodium has been widely accepted as the most active and stable metal for synthesis gas generation via partial oxidization and autothemal reforming at high space velocities. Both Pt and Ir are reported to be either less stable, active, selective, or coking resistant [2]. Ruthenium is considered to be good for steam reforming but less suitable when O_2 is present in the feedstock. Palladium deactivates rapidly and forms carbon [3]. Nickel is active but forms carbon if significant amounts of steam are not present. A group of modified non-Rh precious metal catalysts is reported in this paper. It was demonstrated that, at a significantly lower metal loading, the new bimetallic catalysts are more active, stable, and selective than the 4wt% Rh catalyst in partial oxidization and low steam/carbon ratio ATR due to the stabilization of the precious metal particles and the promotion of coking resistance. The catalysts are applied to make synthesis gas with easily adjustable H_2/CO ratios.

2. Experimental

The catalysts employed in this study were made by a proprietary method. Alpha-Al_2O_3 with surface area of 4.7 m^2/g was used as the support. For catalyst evaluation, 30x40 mesh catalyst particles were loaded in a 10mm ID tubular reactor with 2 times dilution of similar size quartz. The catalyst was reduced in H_2 at 850°C for 1 hour. All the tests were carried out using CH_4 as the hydrocarbon source and pure O_2 as the oxygen source. The flow rates of CH_4 and O_2 were controlled by mass flow controllers and H_2O by a pump. The reaction of CH_4 and O_2 was at stoichiometric. After reaction, the effluent was cooled and demoisturized before being analyzed by a gas chromatograph equipped with two columns, a 5Å molecular sieve column and a TDX-01 column. The gas hourly space velocity (GHSV) was controlled at 500,000/h unless specified. The catalysts are characterized by TPR, TEM, CO chemisorption, and other conventional methods.

3. Results and Discussion

Eight samples were prepared with total metal (Pt + Promoter) loading of 2.5wt% and at different promoter/Pt ratios. Chemisorption data shown in Figure 1 indicate that with the promoter only, the catalyst has about a half the CO uptake of the Pt-containing catalysts. All the Pt-containing catalysts exhibit similar CO-uptake even through the absolute amount of Pt on the catalyst increases from about 0.25 wt% to 2.5 wt%. In other words, a small amount of

Pt addition boosts the chemisorption of CO to almost maximum. These catalysts are tested for initial activity at 700°C, 800°C, 900°C, and 1000°C at a CH_4/O_2 ratio of 2 and GHSV of 500,000/h as illustrated in Figure 2. The initial activity does not change appreciably with the increase in Pt loading as long as Pt is present in the catalyst. This result is similar to what was noticed on CO uptake in chemisorption. When Pt is not present, including the sample with no metal and the sample containing promoter only, the catalysts have insignificant activity at low temperatures. At high temperatures, a small amount of CH_4 is converted that is most likely due to non-catalytic reactions.

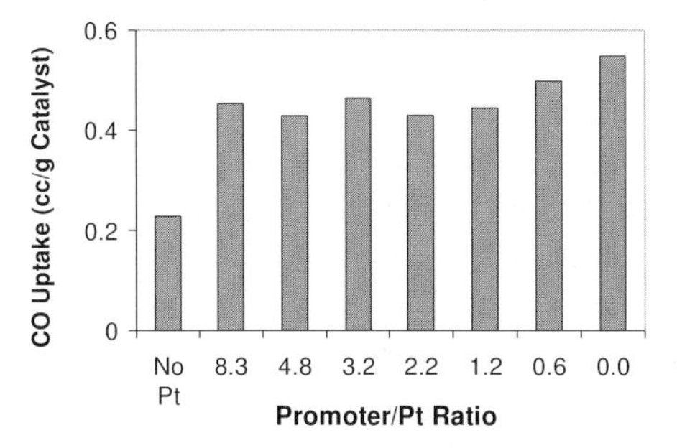

Figure 1: CO Chemisorption of Pt bimetallic catalysts

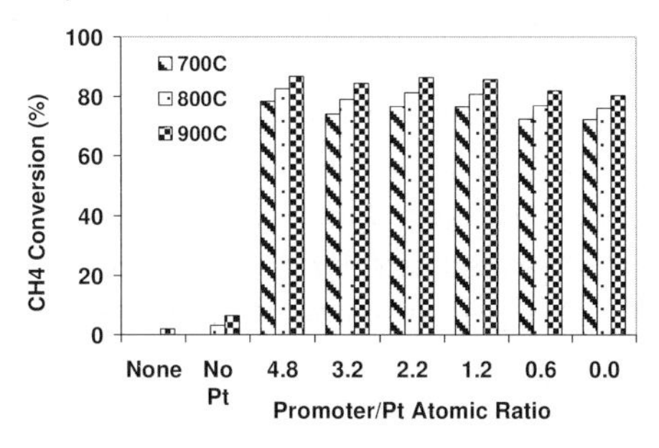

Figure 2: Initial activity of Pt bimetallic catalysts

Some of the TPR data are presented in Figure 3. There are two peaks for the sample containing promoter only indicating two reduction steps of the promoter. The promoter is reduced to a lower valence that peaks at around 266°C and finally to metal around 345°C. With a small amount of Pt addition, a peak at low temperature region around 150°C appears. This temperature is higher than the reduction temperature of Pt that peaks at about 100°C, which suggests the reduction of Pt and part of the promoter around 150°C. On the high temperature region, there is only one peak remaining when Pt is present. The reduction peaks at about 290°C for the catalyst with a promoter/Pt ratio of 4.8. The peak moves toward lower temperature region when Pt loading increases and the promoter loading decreases. It is evident that the presence of Pt helps the reduction of the promoter.

Catalyst stability was initially evaluated by scanning at temperatures from 700°C to 1000°C at an increment of 100°C and then coming back to the lower temperatures. The results of the Pt only catalyst and the one with promoter/Pt at a ratio of 4.8 are summarized in Figure 4. Although only insignificant difference between these two catalysts is noticed with increase in temperature from 700°C to 1000°C, substantial discrepancy in performance can be easily identified after the catalysts have undergone temperature ramping, which implies that the bimetallic catalyst is more thermally stable. Long-term stability tests were carried out at 900°C and a CH_4/O_2 ratio of 2. The results for the Pt only catalyst (Catalyst C) and the bimetallic catalyst with 0.5 wt% Pt and 2 wt% promoter (Catalyst D) are compared with these of 2 wt% Rh (Catalyst A) and 4 wt% Rh (Catalyst B) as summarized in Table 1. Catalyst A showed good initial CH_4 conversion at 93.5% and H_2 selectivity at 94.6% but deactivated constantly during the course of 119.6 hours on stream. It took 4.6 hours for CH_4 conversion to decrease 1% on this catalyst. When the catalyst deactivates, the H_2 selectivity declines as well, however, the CO selectivity does not change appreciably. Consequently, the H_2/CO ratio of the synthesis gas decreased to 1.63 at the end of the run. To try to make the Rh catalyst more stable, Catalyst B contains twice as much Rh as Catalyst A. The catalyst was tested at 500,000/h GHSV for 46.5 hours first and then at 250,000/h. At GHSV 500,000/h Catalyst B performed similar to Catalyst A. The higher Rh loading does not give a significant improvement on the performance of the catalyst. After reducing the space velocity to half, the decline in catalyst activity slowed.

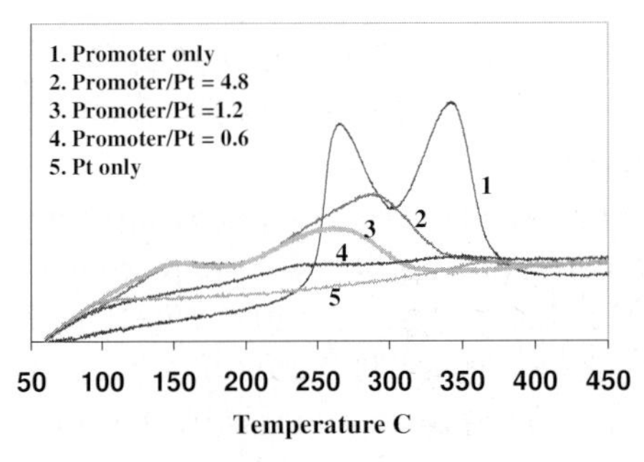

Figure 3: TPR results of the Pt bimetallic catalysts

Figure 4: Catalyst activity temperature scans

Table 1: Long-term catalyst performance results summary

Catalyst	A	B	C	D
Precious Metal Loading	2wt% Rh	4wt% Rh	2.5wt% Pt	0.5wt% Pt Modified
Initial X_{CH4} (%)	93.5	92.9	92.7	93.9
Initial S_{H2} (%)	94.6%	92.4	100	100
Hour on Stream (h)	119.6	46.5/100	7.4	235
End of Run X_{CH4} (%)	67.5%	81.8	85.3	91.3
End of Run S_{H2} (%)	79.3%	87.3	96.5	99.6
End of Run S_{CO} (%)	97.2%	98.2	98.6	99.0
End of Run H_2/CO	1.63	1.78	1.98	2.0
Stability (h/∇1% X_{CH4})	4.6	4.2	1.0	90.3

Catalyst C showed initial activity comparable to that of the Rh catalysts at 92.7% CH_4 conversion and better initial H_2 selectivity at 100%, but it deactivated very rapidly. The test was stopped after only 7.4 hours on stream. With only 0.5wt% of Pt, Catalyst D demonstrated equivalent initial activity to the Rh catalysts but much better stability and selectivity during 235 hours on stream. The stability is in agreement with what was found in temperature scanning tests. It took 90.3 hours for the CH_4 conversion to fall 1% on this catalyst. It had 10 - 20% higher H_2 selectivity than the Rh catalysts at the end of run and the synthesis gas generated from this catalyst had a H_2/CO ratio close to 2 during the run. Catalyst B, C and D before and after reaction were examined by high resolution TEM as presented in Figure 5. The fresh catalysts were reduced at same conditions before TEM tests.

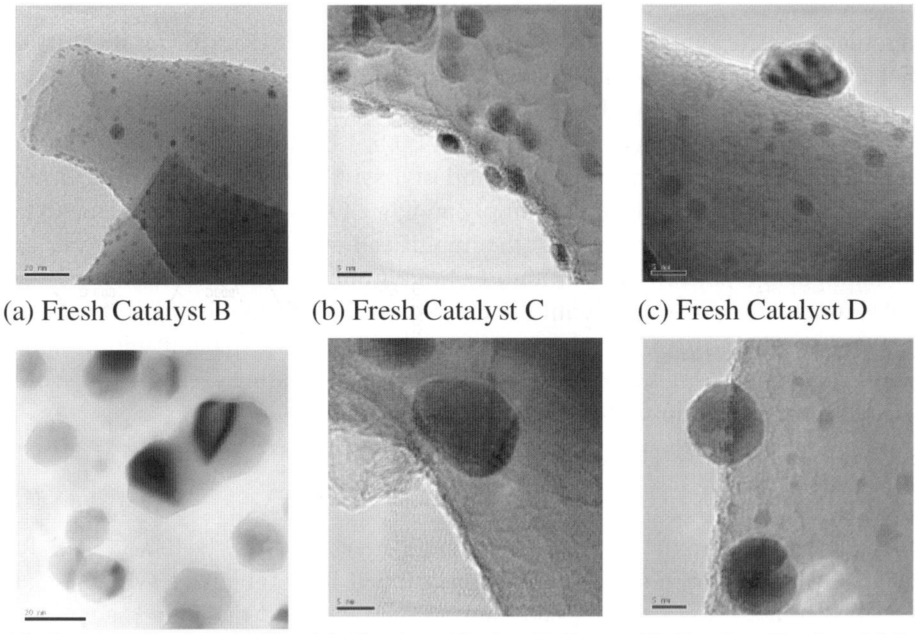

(a) Fresh Catalyst B (b) Fresh Catalyst C (c) Fresh Catalyst D

(d) Catalyst B after 100h (e) Catalyst C after 7.4h (f) Catalyst D after 235h

Figure 5: TEM images of the Rh, Pt, and bimetallic catalyst

It can be seen that Pt particles on catalyst D were mostly smaller than 10nms and their size did not change appreciably after 235 hours on stream. On the other hand, Pt particles grew noticeable on Catalyst C after only 7.4 hours on stream. The rhodium particle size on Catalyst B increased considerably during the course of reaction. The catalyst particles sintered obviously, which is the reason of deactivation. It was evident that the promoter stabilized the Pt catalyst, and it contributes to the high stability of the bimetallic catalyst. With the promoter, only a small amount of precious metal is needed to achieve high activity and stability. Actually, the precious metal-lean catalysts performed better than the precious metal-rich ones in terms of stability and this makes the catalyst more cost effective. In addition to the Pt bimetallic catalysts, other precious metal bimetallic catalysts were also investigated. All the catalysts contain the same amount of precious metal and promoter in weight. The stability of the Ru and Ir bimetallic catalysts is slightly lower than that of the Pt bimetallic catalyst but is much better than the Rh monometallic catalysts in Table 1. It takes 48.4 and 65.8 hours for the Ir and Ru bimetallic catalyst to decrease 1% in CH_4 conversion respectively. The addition of the promoter to Rh also improved the stability but the effect is less significant compared to Pt, Ir, and Ru. On the other hand bimetallic Pd catalyst is still very unstable.

In order to make synthesis gas with H_2/CO ratio greater than 2.0 for a Co-based F-T catalyst, a small amount of steam can be added to the feedstock. Catalyst D was tested under steam/carbon ratios from 0 to 1.5. Keeping CH_4/O_2 ratio at 2, synthesis gas generated at H_2O/CH_4 ratio of 0.5, 1.0 and 1.5 had H_2/CO ratios of 2.12, 2.38, and 2.62 respectively. Similarly, synthesis gas with a H_2/CO ratio less than 2.0 can be produced by simply adding CO_2 in the feedstock. Due to the improved coking resistance by the modification, the new Pt catalyst can maintain stable performance in high CO_2 environment and in conditions where coke formation is thermodynamically favorable. With this improved catalytic system, required H_2/CO synthesis gas can be directly generated with a single unit operated at a high space velocity by simply adjusting feedstock based on the requirements of downstream systems.

4. Conclusions

A group of bimetallic catalysts, that are better than the well accepted Rh catalyst, allows one not only to design a much more compact synthesis gas generation unit for a GTL plant, but also to operate the unit much more economically. The bimetallic catalyst is about 20 times more stable, 10% to 20% more selective than the Rh catalyst that contains eight times more precious metal. The addition of the promoter to the precious metal catalysts stabilizes the precious metal particles and results in high activity and more stable performance in partial oxidization and autothermal reforming of CH_4.

References:

1. M. E. Dry, *Catalysis Today* **71,** 227-241(2002).
2. D.A. Hickman and L.D. Schmidt, *Science* **259**, 343-246 (January 15, 1993)
3. P. Torniainen, X. Chu, and L. D. Schmidt, *J. of Catalysis* **146**, 1-10 (1994).

Natural Gas Conversion VIII
F.B. Noronha, M. Schmal, E.F. Sousa-Aguiar (Editors)

Conversion of methanol to hydrocarbons: Hints to rational catalyst design from fundamental mechanistic studies on H-ZSM-5

Morten Bjørgen,[a] Karl-Petter Lillerud,[b] Unni Olsbye,[b] Stian Svelle[b]

[a]*Haldor Topsøe A/S, Nymøllevej 55, DK-2800 Lyngby, Denmark*
[b]*Centre for Materials Science and Nanotechnology, Department of Chemistry, University of Oslo, P.O. Box 1033 Blindern, N-0315 Oslo, Norway*

In this study, the mechanism of the methanol-to-hydrocarbons (MTH) reaction over H-ZSM-5, which is the archetype MTH catalyst, has been pursued. From isotopic labeling experiments, it is demonstrated that the reaction scheme for alkene formation from methanol over H-ZSM-5 is essentially different from those previously drawn for H-beta and H-SAPO-34. In addition to a modified hydrocarbon pool mechanism, wherein ethene is formed from the lower methylbenzenes, a parallel C_{3+} alkene methylation/cracking cycle is operative. The results showing that ethene and propene are formed through different catalytic cycles, are of utmost importance for understanding and possibly controlling the ethene/propene selectivity in methanol-to-propene catalysis.

1. Introduction

Methanol can be made from carbonaceous feedstocks via synthesis gas and further converted into gasoline range hydrocarbons or alternatively light alkenes. The zeolite catalysed conversion of **m**ethanol **t**o **h**ydrocarbons is commonly denotet the MTH reaction, but as process conditions and catalyst choice alter the product selectivities, the abbreviations MTO (**m**ethanol **t**o **o**lefins) and MTG (**m**ethanol **t**o **g**asoline), are frequently used. Throughout the years, the MTH reaction has challenged the catalysis community by its complex mechanism [1]. The long-standing question has been the mechanism behind the step where C-C bonds are formed from oxygenates (methanol/dimethylether). There are presently strong evidences disfavoring direct mechanisms where methanol/dimethylether molecules are combined to form e.g. ethene during steady state conversion [1,2]. On the other hand, an all-dominating indirect route known as the "hydrocarbon pool mechanism" has been gradually accepted over the last years [1-5]. The hydrocarbon pool is now described as a catalytic scaffold, constituted by larger organic molecules adsorbed in the zeolite, to which methanol/dimethylether is added and from which alkenes and water are formed in a closed catalytic cycle. The identity and the operation of the hydrocarbon pool are now fairly well described for some specific catalysts, in particular zeolite H-beta [6,8] and H-SAPO-34 [7]. For these particular systems, the hydrocarbon pool has been identified as highly methyl substituted benzene rings ("polymethylbenzenes") or the benzenium cations derived thereof. In the wide pore beta zeolite, the terminal benzene methylation product, i.e. the heptamethylbenzenium cation, is unquestionably the species responsible for the major part of alkene formation. According to the hydrocarbon pool mechanism over the beta zeolite, the heptamethylbenzenium cation splits off propene and butene under a concomitant formation of the tetra-, and trimethylbenzenium ions, respectively . The heptamethylbenzenium cation is then regenerated by subsequent methylations of these lower polymethylbenzenium ions (after deprotonation) by methanol/dimethylether, and the cycle is thus completed. The most extensive fundamental mechanistic understanding has until now been attained for the H-SAPO-34 and H-beta catalysts, but, as will be shown, many aspects may vary with the zeolite topology.

2. Experimental

A commercially available H-ZSM-5 (Si/Al = 140) from Zeolyst International has been used in this study. The catalytic reactions were carried out in a fixed bed reactor using 60 mg catalyst. The reaction temperature was in the range 290-390 °C. Methanol, either ^{13}C enriched (Cambridge Isotope Laboratories, 99% ^{13}C isotopic purity) or ordinary ^{12}C methanol (BDH Laboratory Supplies, > 99.8% chemical purity) was fed by passing a He stream (35 $mLmin^{-1}$) through a saturation evaporator kept at 20 °C. Resulting feed rate (WHSV) was 7.0 $gg^{-1}h^{-1}$. Three types of experiments were performed: i) Ordinary ^{12}C methanol was reacted for predetermined times at given reaction temperatures. GC-FID effluent analyses were performed during the reaction and analyses of the organic material retained in the zeolite (see below) were carried out after each run. ii) Alternatively, the methanol feed was stopped after a given time by bypassing the evaporator and flushing the catalyst with carrier gas at the reaction temperature for a set time followed by analysis of the retained material. iii) By using two separate and identical feed lines, it was possible to switch from ^{12}C to ^{13}C methanol without otherwise disrupting the conditions. The isotopic compositions of both effluent compounds and retained material were determined at increasing ^{13}C methanol reaction times. Analysis of the organic material that is confined within the zeolite pores during the reaction was performed by thermally quenching the reaction, transferring a part of the catalyst mass (40 mg) to a screw-cap Teflon vial, dissolving the catalyst in 1.0 mL 15 % HF, extracting the liberated organic molecules from the aqueous phase with 1.0 mL CH_2Cl_2, and finally analyzing the organic phase using a GC-MS.

3. Results and discussion

At 370 °C, the conversion was 85 % (dimethylether included in reactants) over the fresh zeolite (time on stream = 20 min) and the main products were propene, butenes, and a variety of C_5-C_{10} hydrocarbons. The conversion level and product selectivities remained stable for days of operation under the given set of conditions. As the reaction intermediates of the MTH reaction are known to be fairly large organic molecules confined in the zeolite channels, a main interest has been the build-up and stability of such species during the reaction. Fig. 1 presents GC-MS chromatograms of the organics that were present inside the zeolite after 10 min, 30 min, 1 h, and 20 h of methanol reaction at 370 °C. As the figure shows, a few compounds are dominating, and virtually all peaks represent polymethylbenzenes (polyMBs). Hexamethylbenzene (hexaMB) is most prominent, followed by pentamethylbenzene (pentaMB), tetramethylbenzenes (tetraMBs), trimethylbenzenes (triMBs) and p/m-xylene. Hydrocarbons with molecular mass higher than hexaMB were not present in significant amounts. Previous studies on the organics formed inside wide pore zeolites/zeotypes (H-beta and H-SAPO-34) under MTH conditions have shown similar results, but the well defined cut-off in the distribution

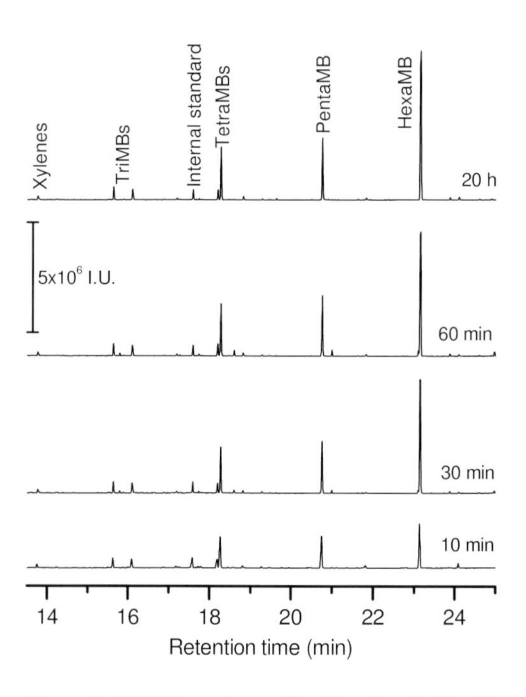

Figure 1. GC-MS analyses of retained material after methanol conversion at 370 °C. Time on stream 10 min to 20 h.

of confined organics at hexaMB (Fig. 1) must be a result of the restricted space in the channels of ZSM-5, as far larger polyaromatics are found in copious amounts in zeolites/zeotypes providing larger cavities [7-9]. The accumulation of hexaMB in the zeolite at this rather high temperature (and conversion) indicates a low reactivity of this species. This is in strong contrast to the above mentioned investigations where hexaMB (and heptaMB+) has been found to be highly unstable and only present in minute amounts at temperatures above 325 °C. In an earlier study it was essential to lower the reaction temperature to 300 °C in order to observe the most reactive retained species in the beta zeolite during the MTH reaction [8]. In that case, heptaMB+ was the dominant compound in the zeolite pores at the lower temperatures, but this compound was hardly observable above 300 °C due to a fast decomposition into alkenes and lower polyMBs. We therefore conducted a temperature series also for H-ZSM-5 in the range 290-390 °C (not shown) in order to trace possible species that are too reactive to be seen at more realistic temperatures. At 290 °C the concentration of retained organics was very low, and no new possible intermediates or heptaMB+ appeared. Thus, a reaction route similar to the previously suggested mechanism based on heptaMB+ is not probable for H-ZSM-5. It is seems reasonable that the ZSM-5 channels are too narrow to allow the terminal methylation step where heptaMB+ is formed from hexaMB. In order to obtain more information about the stability of the confined hydrocarbons, an experiment was carried out where the feed was stopped after 20 min of reaction, and the catalyst flushed with carrier gas for 2 and 8 min at the reaction temperature. In Fig. 2, the effect of catalyst flushing on the hydrocarbon deposits in H-ZSM-5 is shown. Upon flushing, the polyMBs disappear from the zeolite channels at rates according to the following order: p/m-xylene > 1,3,5-triMB > 1,2,4-triMB > 1,2,4,5-tetraMB > 1,2,3,5-tetraMB > pentaMB = hexaMB. Based on what we have seen for other zeolites, the result is rather surprising. Similar experiments on H-beta and H-SAPO-34, but at a considerably lower reaction temperature (325 °C), leads to a total disappearance of hexaMB already 1-2 min after methanol feed has been terminated [6,9]. In the present case penta- and hexaMB show a very slow decomposition rate and this observation makes these species less likely reaction intermadiates in H-ZSM-5. It

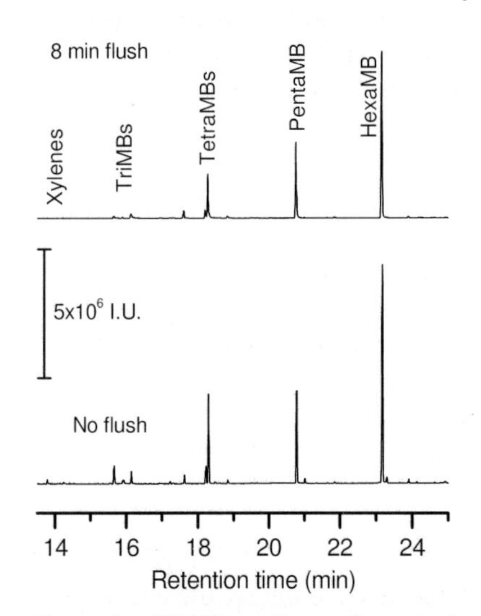

Figure 2. GC-MS analyses of retained material after 20 min of methanol conversion at 370 °C (bottom) and 20 min of reaction followed by 8 min of flushing with carrier gas at 370 °C.

should indeed be mentioned that the observed slow decomposition of penta- and hexaMB during the flushing experiment does not necessarily demonstrate that they have a low reactivity in the working zeolite: After several minutes of flushing, the concentration of methanol in the pores is low and the slow decomposition of penta- and hexaMB may not reflect their actual reactivity in the presence of methanol. The reactivity of the different species in the presence of methanol was therefore further investigated by isotopic labeling experiments. The reactivity patterns of the confined organics towards methanol in the working catalyst were assessed by a series of experiments conducted by reacting ordinary ^{12}C methanol for 18 min followed by a rapid switch to ^{13}C methanol feed for a shorter time (0.5, 1, or 2 min). Isotopic analysis of both the retained material and the gas phase products was performed after 0.5, 1, and 2 min of 13C methanol reaction. The time evolutions of total ^{13}C content for the retained hydrocarbons and gas phase components are shown in Fig. 3a and 3b, respectively. The apparently low reactivity of the higher polymethylbenzenes suggested by the flushing experiment described above is clearly confirmed by this experiment: Fig. 3a shows that the incorporation of ^{13}C is consistently slower as the number of methyl groups on the aromatic ring increases. The low rate of ^{13}C incorporation of hexaMB in particular, is in stark contrast to data obtained in a similar manner for H-beta and H-SAPO-34 where hexaMB unquestionably has the highest reactivity [7,8]. *As a direct implication of this, it may be stated that a hydrocarbon pool mechanism involving the highest methylbenzenes, proven to be all dominating for H-beta and H-SAPO-34, cannot be applicable to the H-ZSM-5 catalyst.* In addition, a rather important implication regarding the different mechanisms contributing to formation of ethene and the higher alkenes can also be inferred from this experiment. Fig. 3b

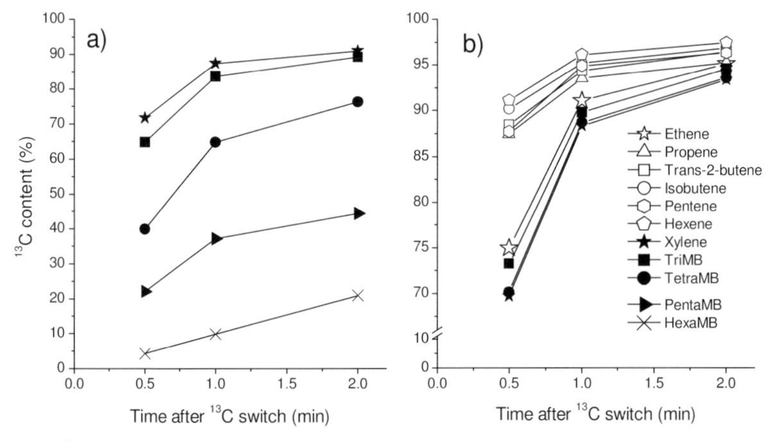

Figure 3. Total ^{13}C content in the retained hydrocarbons (part a) and the effluent compounds (part b) after 18 min of ^{12}C methanol reaction followed by a switch to ^{13}C methanol and further reaction for 0.5, 1.0 and 2.0 min at 350 °C. Note the scale break in part b.

shows that the rates of ^{13}C incorporation are significantly slower for ethene and the aromatics compared to the C_3-C_6 alkenes: At 0.5 min of ^{13}C methanol reaction ethene and the aromatics contain 70-75 % ^{13}C, whereas the C_3-C_6 alkenes contain 87-91 % ^{13}C. This suggests a mechanistic link between ethene and the C_8-C_{10} aromatics. Moreover, the significantly higher ^{13}C content in the C_3-C_6 alkenes suggests a different or additional route for methanol incorporation into these compounds. It seems very probable that additional ^{13}C is introduced into the C_3-C_6 alkenes via repeated methylations, and that the carbon atoms are distributed among the product fractions through subsequent cracking steps. Kinetic investigations have shown that ethene methylation is between one and two orders of magnitude slower than for propene and linear butenes [10]. Hence, ethene methylation is a very minor reaction under the current set of conditions and ethene is virtually excluded from this route.
It can now be established that two mechanistic cycles run simultaneously during the MTH reaction over H-ZSM-5: Ethene formation from the lower methylbenzenes, and a methylation/cracking cycle involving only the C_3-C_6 alkenes. A highly relevant question regarding selectivity control is whether these two cycles operate completely separately or if they are entangled in some way. A totally independent operation is probably not the case of H-ZSM-5: There is a continuous production of aromatics during the reaction; hence there must be a corresponding formation of aromatics from the higher alkenes formed in the C_3-C_6 alkene cycle. This implies that the ethene-aromatics cycle needs a constant supply of aromatics from the C_3-C_6 alkene cycle over H-ZSM-5. However, the two cycles do not have be mutually dependent. Based on the low reactivity of ethene towards methanol relative to that of propene and butenes mentioned above, the contribution to the C_3-C_6 alkenes involving ethene is very small and may not be required. Thus, for a given catalyst topology or for a given set of conditions we may imagine that methanol can be converted solely according to the C_3-C_6 alkene cycle and ethene formation can then be avoided.

References

1. M. Stöcker, Microporous Mesoporous Mater. 29, (1999), 3-48.
2. W. Song, D. M. Marcus, H. Fu, J. O. Ehresmann, J. F. Haw, J. Am. Chem. Soc. 124, (2002), 3844-3845.
3. D. M. Marcus, K. A. McLachlan, M. A. Wildman, J. O. Ehresmann, P. W. Kletnieks, J. F. Haw, Angew. Chem., Int. Ed. 45, (2006), 3133-3136.
4. I. M. Dahl, S. Kolboe, J. Catal. 149, (1994), 458-464.
5. U. Olsbye,M. Bjørgen, S. Svelle, K. P. Lillerud, S. Kolboe, Catal. Today, 106(1-4), (2005), 108-111.
6. M. Bjørgen, U. Olsbye, S. Kolboe, J. Catal., 215, (2003), 30-44.
7. B. Arstad, S. Kolboe, J. Am. Chem. Soc., 123, (2001), 8137-8138.
8. M. Bjørgen, U. Olsbye, D. Petersen, S. Kolboe, J. Catal. 221 (1), (2004), 1-10.
9. B. Arstad, S. Kolboe, Catal. Lett. 71, (2001), 209-212.
10. S. Svelle, P.O. Rønning, U. Olsbye, S. Kolboe, J. Catal. 234, (2005), 385-400.

Natural Gas Conversion VIII
F.B. Noronha, M. Schmal, E.F. Sousa-Aguiar (Editors)

Dry Reforming and Partial Oxidation of Natural Gas to Syngas Production

Leonardo José Lins Maciel,[a] Aleksándros El Áurens Meira de Souza,[a] Solange Maria de Vasconcelos,[a] Augusto Knoechelmann,[a] Cesar Augusto Moraes de Abreu[a]

[a]DEQ, CTG, UFPE, Av. Prof. Artur de Sá, s/nº - Cidade Universitária, Recife-PE - 50740-521, Brazil

Abstract

Dry reforming with partial oxidation of natural gas to syngas process with nickel catalysts supported on γ-alumina was compared with a dry reforming process, where methane conversion and the hydrogen and carbon monoxide yields at 1023 K, 1073 K and 1123 K, in both processes, were evaluated. This work presents the results of the evaluation of the dry reforming process when O_2 was added to the reactor with the methane and CO_2 feed. The syngas obtained from methane reforming, with and without partial oxidation, may be employed as a raw material for oxo-alcohol, polycarbonate and formaldehyde syntheses.

Keywords: catalysis, nickel, aluminum oxide, natural gas, syngas, dry reform.

1. Introduction

Natural gas (NG) conversion and its employment as fuel are important options as new energy sources. Syngas from NG, a mixture of CO and H_2, may be used to produce high added value chemical products such as hydrocarbons, fuels [1], and oxygenated compounds [2]. In gas-to-liquid (GTL) processes [3], where natural gas is firstly converted to syngas, 60-70% of the costs of the overall process is associated with the syngas production. The availability of natural gas and CO_2 (from thermoelectric plants, refineries, alcoholic fermentation) indicates that, instead of the traditional steam reforming of methane (SRM), the

catalytic reforming with carbon dioxide may be employed. However, the main obstacle of the process is given by the catalyst deactivation due to carbon deposition during the reaction. The catalytic partial oxidation of methane (POM) may drastically decrease or even eliminate the carbon deposition. Furthermore, as the catalytic partial oxidation of methane is exothermic, a process with that reaction may be more efficient due to the high energy demands of the endothermic steps. The introduction of O_2 in the reaction generates heat *in situ* that can be used to increase the energy efficiency. The excess of O_2 probably favors the reoxidation of carbonaceous residues formed on the catalytic surface, avoiding the catalyst deactivation. The increase in CH_4 conversion with the O_2 addition can be related with two distinct effects: one, a fast combustion of a part of methane with the O_2, and the other, an increase in the reforming reaction rate due to the temperature increase caused by the exothermicity of combustion reaction [4].
A recent work [5] presented a detailed mechanism of the thermal reforming of methane with CO_2, and another [6] presented the thermodynamically favorable conditions for the hydrogen production in the catalytic partial oxidation of methane.

2. Experimental

The catalyst with a nominal composition of Ni(5%)/γ-Al_2O_3 was prepared by impregnating the support with a precursor salt solution of $Ni(NO_3)_2.6H_2O$ in excess. The process was evaluated with 50 mg of catalyst packed in a tubular fixed-bed quartz reactor. The feed gas was a mixture with a molar ratio of CH_4:CO_2:O_2 = 1:0.5:0.08, diluted in argon at 75%, and a total flow rate of 1800 cm^3/min where the spatial time was 1.7 $kg_{cat}.s/m^3$ (STP). The catalytic reforming was performed at 1023 K, 1073 K and 1123 K under atmospheric pressure. The reactor effluents were analyzed by gas chromatography.
The data of this work were compared with the dry reforming of methane data in a previous work performed by the same research group. These data were obtained from a dry reforming reaction with a Ni(4.82% wt)/γ-Al_2O_3 catalyst. Considering that the performance of the catalyst does not vary perceptively when its Ni content is in the range of 3.5 to 7.0% [7], the comparison between CH_4-CO_2 and CH_4-CO_2/O_2 operations performed in this work is acceptable.

3. Results

3.1. Characterization of catalyst

The nickel content of the catalyst was analyzed via atomic absorption spectrometry (AAS) indicating 3.92% of Ni. Thermogravimetry (TG) and differential thermal analysis (DTA) indicated that the loss of water from the impregnated precursor was 13.0% in weight and NO_2 loss 10.0%. The loss of NO_2 is referred to nitrate decomposition, which occurs at a maximum of 541.4°C.

Characterization by X-ray diffraction (XRD) allowed to identify the presence of three phases of alumina: alpha, delta and theta-alumina. After impregnation, nickel nitrate, NiO (calcination) and Ni (reduction) were identified.

Infrared spectrophotometry (FT-IR) identified the presence of crystalline water (3451 cm^{-1}) and hydrogen bridges (3451 cm^{-1} and 1637 cm^{-1}) in the catalyst structure in all the preparation steps. Bands indicated the presence of NO_2 groups (1631 cm^{-1}) and nitrates (NO_3^-) (1383 cm^{-1} and 827 cm^{-1}) after the precursor salt impregnation (Fig. 1).

Thermal analysis (TG/DTG), allowed to quantify the coke formed in the CH_4-CO_2/O_2 process. Therefore, 0.323 mg, 3.250 mg and 1.223 mg of coke were formed at the 1023 K, 1073 K, and 1123 K, respectively, showing that, the coke formation increased with the temperature due to favorable conditions for the precocious catalyst poisoning. Besides the amount of coke formed at the catalyst for the operation at 1123 K, there were also evidences of Ni_3C formation in greater quantity, resultant of the interaction between the coke and the metal on the surface. The presence of coke, as amorphous carbon and Ni_3C, especially after the reaction at 1123 K, is also indicated in the diffractograms of the catalyst employed in the three operations in Fig. 2.

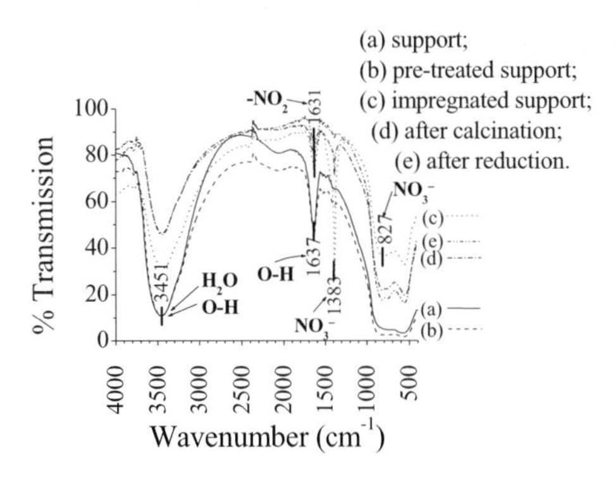

Fig. 1. Spectra of the steps of Ni(3.92% wt)/γ-Al_2O_3 catalyst preparation.

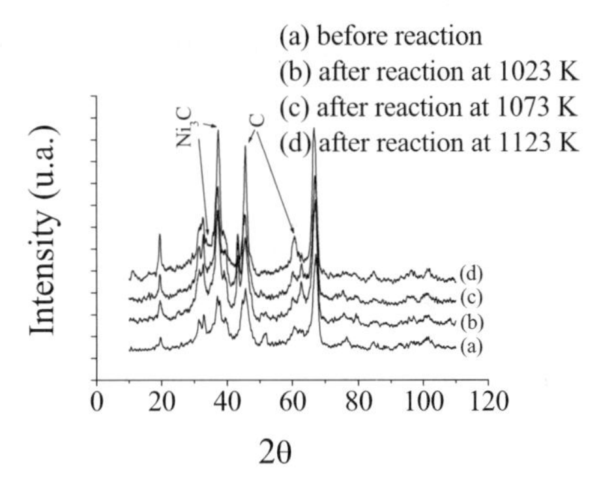

Fig. 2. X-ray diffractograms of a catalyst submitted to CH_4-CO_2/O_2 processes, 1 bar, Ni(3.92% wt)/γ-Al_2O_3.

Under the operating conditions, the reaction occurred without any limitations for mass transfer. At 1023 K and 1073 K, steady state was reached after 240 min of operation, whereas at 1123 K it was reached after 740 min. The average conversions of CH_4 at 1023 K, 1073 K and 1123 K were 40.7%, 55.1% and 5.1% at the steady state, respectively. The results are presented in Fig. 3 and 4.

3.2. Dry reforming of methane and partial oxidation of methane operations

Methane-CO_2 and methane-CO_2 in presence of oxygen processes with nickel catalyst were compared through methane conversion, hydrogen and carbon monoxide yields obtained at 1023 K, 1073 K and 1123 K. Methane conversions (Fig.3) were higher at 1023 K and 1073 K (40-60%), while at 1123 K the level was reduced to 5%. The highest carbon deposition was observed at 1123 K, resulting in catalyst deactivation. With temperature increase, dry reforming of methane was favored (endothermic), in spite of methane partial oxidation (exothermic). The carbon deposited is usually reduced, regenerating the catalyst through its reaction with oxygen (exothermic) or with carbon dioxide (endothermic). In Fig. 4, methane conversions are compared at 1073 K for the two processes.

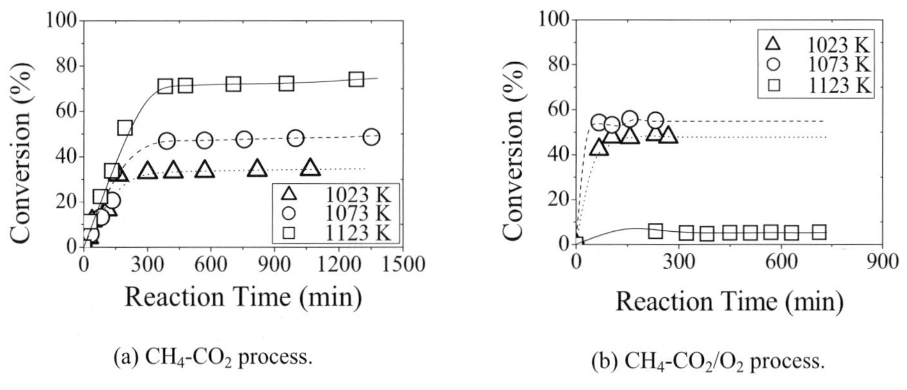

(a) CH_4-CO_2 process. (b) CH_4-CO_2/O_2 process.

Fig. 3. Methane conversion in function of the reaction time. Influence of temperature, 1 bar.

The main reaction steps (4-5) representing the methane-CO_2 oxidation process and the mechanism described for the methane-CO_2 reforming process (1-3) were employed to analyze the experimental results.

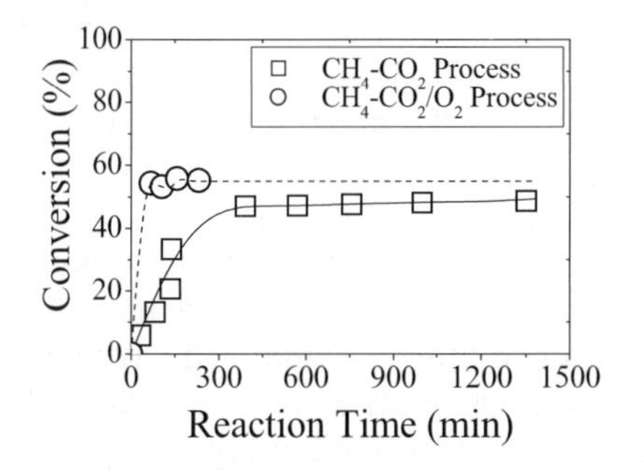

Fig. 4. Methane conversion in function of the reaction time, 1073 K, 1 bar.

$$CH_{4(g)} \rightleftarrows C_{(s)} + 2H_{2(g)} \quad (1)$$

$$C_{(s)} + CO_{2(g)} \rightleftarrows 2CO_{(g)} \quad (2)$$

$$CO_{2(g)} + H_{2(g)} \rightleftarrows CO_{(g)} + H_2O_{(g)} \quad (3)$$

$$CH_{4(g)} + 0.5O_{2(g)} \rightarrow CO_{(g)} + 2H_{2(g)} \quad (4)$$

$$C_{(s)} + O_{2(g)} \rightarrow CO_{2(g)} \quad (5)$$

In Fig. 5, H_2 and CO yields of the process are represented in function of the reaction time for the three temperatures. Hydrogen and carbon monoxide were the major products obtained experimentally what can be justified by the reaction steps of the employed mechanism. At the operating conditions of 1023 K and 1073 K, H_2 and CO yields were similar, approximately 30-40%. Reduced yields were obtained at 1123 K. To compare the processes, H_2 and CO yields (1073 K) in function of the reaction time are presented in Fig. 6.

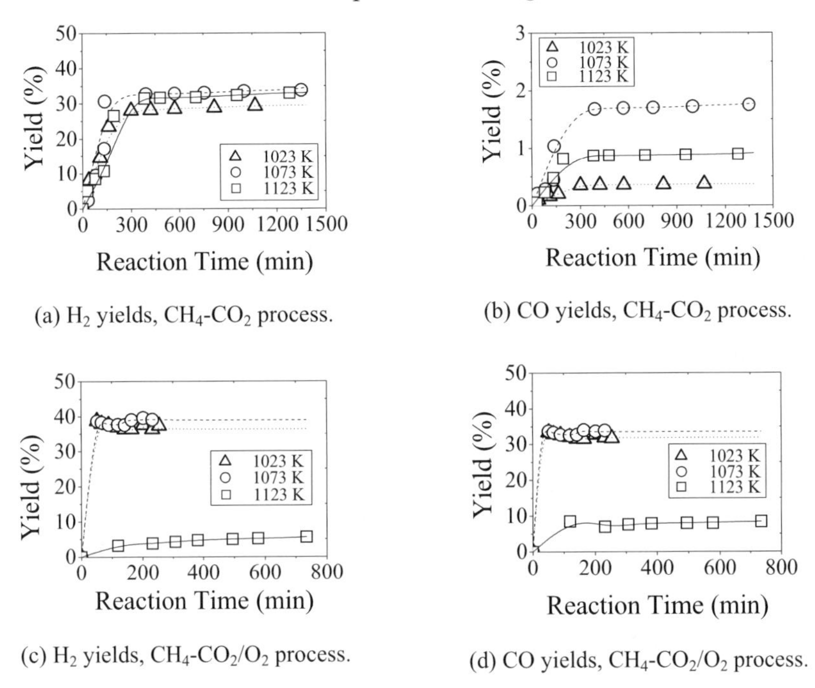

Fig. 5. Yields in function of the reaction time. Temperature influence, 1 bar.

Higher yield levels for hydrogen (39.0%) and carbon monoxide (33.6%) were observed when oxygen was used in the process. Carbon oxidation (Eq. 5) followed by Boudouard reverse reaction (Eq. 2) promotes the catalyst regeneration during the deactivation reaction and increases the CO production. Taking into account the potential applications of the produced syngas, evaluations of H_2/CO ratio were made considering the composition of this product, whose profiles in function of the reaction time can be visualized in Fig. 7. In the methane-CO_2 reforming processes, associated or not with the partial methane oxidation, the H_2/CO ratios are in the range of 1. Through the evaluation of H_2/CO, the obtained syngas of methane reforming, with or without the partial oxidation, presents great potential for oxo-alcohol, polycarbonate and formaldehyde syntheses.

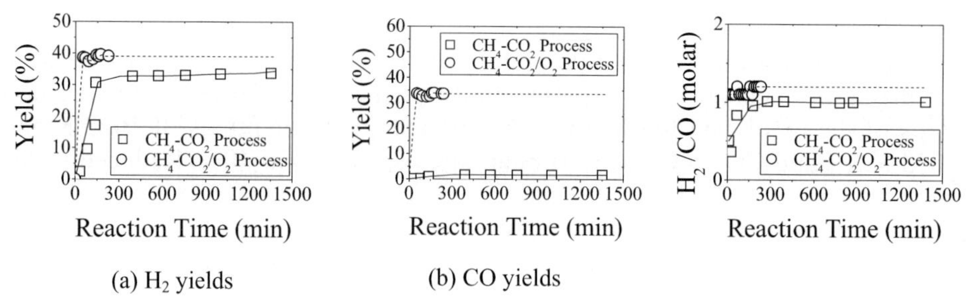

Fig. 6. Yields in function of the reaction time, 1073 K, 1 bar.

Fig. 7. H_2/CO molar in function of the reaction time, 1073 K, 1 bar.

4. Conclusions

Dry reforming of methane presented a conversion of 49.3%, and the yields of 34.4% and 1.8% for hydrogen and carbon monoxide, respectively. The dry reforming process associated with the partial oxidation of methane presented an average conversion of 54.9% and average yields of 39.0% and 33.6% for hydrogen and carbon monoxide, respectively. The influence of the oxygen presence as a feed component was observed, promoting the partial oxidation of methane. Besides, the oxygen promoted the catalyst reactivation through the oxidation of the formed coke. This process was also improved through coke-CO_2 reaction with carbon monoxide production.
The obtained syngas of methane reforming (H_2/CO about 1), with or without its partial oxidation, can be used for oxo-alcohol, polycarbonate and formaldehyde productions.

Acknowledgements

We would like to acknowledge UFPE, FINEP and PETROBRAS, for financial support, and Mr. J. A. Pacífico, for the dry reforming data employed for comparison in this work.

References

[1] J.P. Van Berge, J. Van De Loosdrecht and S. Barradas, Production Of Hydrocarbons From A Synthesis Gas, WO Patent No. 0 207 883 (2002).
[2] C.D. Long, A.M. Daage and J.R. Koveal Jr., Fischer-Tropsch Process, WO Patent No. 02/20 439 (2002).
[3] R.L. Ferreira, J.V. Bomtempo and E.L.F. Almeida. Estudo das inovações tecnológicas em GTL com base em patentes: o caso Shell. 2º Congresso Brasileiro de P&D em Petróleo & Gás. (2003).
[4] M.M.V.M. Souza and M. Schmal, Appl. Catal. (A), No. 281 (2005) 19.
[5] S-G. Wang, Y-W. Li, J-X. Lu, M-Y. He and H. Jiao, J. Mol. Struct. (Theochem), No. 673 (2004) 181.
[6] Y-S. Seo, A. Shirley and S. T. Kolaczkowski, J. Power Sources, No. 108 (2002) 213.
[7] C. Li and Y-W. Chen, Thermochim. Acta, No. 256 (1995) 457.

Natural Gas Conversion VIII
F.B. Noronha, M. Schmal, E.F. Sousa-Aguiar (Editors)

Preparation of Ni/SiO_2, Ni/SiO_2-CaO and Ni/SiO_2-MgO catalysts for methane steam reforming

Alano V. da Silva Neto[a], Patrícia P. C. Sartoratto[a], Maria do Carmo Rangel[b]

[a]*Instituto de Química, Universidade Federal de Goiás, Campus Samambaia 74001-970, Goiânia, Goiás, Brazil*
[b]*Instituto de Química, Universidade Federal da Bahia. Campus Universitário de Ondina, Federação. 40 170-280, Salvador, Bahia, Brazil*

The preparation of mesoporous silica-supported nickel, by adding nickel nitrate and citric acid into a colloidal dispersion of sphere-shaped silica nanoparticles, was described in this work. The promotion with calcium and magnesium caused a decrease in specific surface area but led to an improvement of both activity in methane steam reforming and selectivity to hydrogen, making the catalysts suitable for hydrogen production. The magnesium-doped catalyst is the most promising one.

1. Introduction

Steam reforming of methane, the main component of natural gas, is the most used and economical route to produce hydrogen for industrial purposes. The catalysts are often nickel-based ones, which go on severe deactivation by coke deposition. During the reaction, coke is formed at high temperatures on both nickel and on the metal-support interface, causing the loss of small metallic particles from the support [1]. Therefore, there is a demand for catalysts that can stabilize nickel particles on the support. In this sense, amorphous silica prepared by the sol-gel method is a versatile material to be used as catalyst support, since its porosity can be modulated and the stability of metal can be improved by the addition of basic promoters to the silica framework [2]. Besides, the use of organic additives in the sol-gel precursor mixtures, such as citric acid may also enhance the metal distribution in the silica support due to their ability to coordinate metal ions [3].

Takahashi et al. have recently reported the preparation of Ni/SiO_2 catalysts using the sol-gel process and citric acid as a non-surfactant template, under acidic reaction condition [4]. The catalysts obtained showed NiO nanoparticles from 6 to 9 nm and improved textural properties for samples containing up to 30% Ni. However, the drawback of using the sol-gel acid condition is a poor structural stability of the mesoporous silica at high temperatures [5]. More recently, Lee et al. synthesized mesoporous silica framed by sphere-shaped silica nanoparticles using a base-catalyzed sol-gel reaction and citric acid as template agent [6]. The pore size of this mesoporous silica was easily controlled up to 15 nm and the materials maintained a narrow pore size distribution, high specific surface area and high pore volume at high temperatures. The highly branched structure of the colloidal silica nanoparticles gave rise to a more rigid silica framework that shrunk less during calcination. These characteristics make the material interesting to be used in the preparation of supported nickel catalysts for steam reforming of methane at high temperature conditions.

In this study, nickel supported on amorphous mesoporous silica was prepared, by incorporating nickel nitrate and citric acid into a colloidal dispersion of sphere-shaped silica nanoparticles. Calcium and magnesium promoters were also introduced into the catalyst and the structural and textural characteristics were evaluated. The methane steam reforming was selected as a model reaction for evaluating the catalysts.

2. Experimental

The catalysts were prepared by the sol-gel method [6], using nickel nitrate hexahydrate and calcium or magnesium nitrate. The synthesis was performed using tetraethoxysilane (TEOS) in alkaline conditions at a TEOS:NH_3:H_2O:ethanol molar ratio of 1:0.086:53.6: 40.7. The TEOS/ethanol solution was stirred vigorously at 50 °C and a NH_3/H_2O solution was added dropwise mantaining at this temperature for 3 h, resulting in a stable colloidal silica sol. Citric acid, CA, was then added to the colloidal silica sol at TEOS:CA molar ratio of 1:1, followed by vigorous stirring at room temperature, for 10 min. Then, nickel nitrate was added to the mixture that was stirred at room temperature, for additional 15 min. Alternatively, calcium or magnesium nitrate was added to the previous mixture and stirred for 15 min. The resulting sols were poured into glass recipients and dried at 70 °C for 24 h. The xerogels were calcined at 500 ^{o}C, for 2 h to decompose citric acid and nitrate ions and to crystallize nickel oxide. The SiNi, SiNiCa and SiNMg samples produced were calcined at 900 ^{o}C, for 2 h. A reference SiO_2 sample was also prepared. The nickel, calcium and magnesium contents in the samples were determined by

atomic absorption analysis, using a Perkin Elmer 5000 instrument. Nitrogen adsorption-desorption isotherms were taken in a Micromeritics ASAP 2010 instrument, on samples previously degassed at 120 °C for 24 h, under vacuum. The specific surface areas were calculated from the adsorption curve (BET method) and the pore size distribution curves were obtained from the desorption branch by using the Barrett–Joyner–Halenda (BJH) method. The powder X-ray diffractograms (XRD) of the samples were recorded in a Shimadzu/XRD 6000, using CuKα radiation (λ=1.54056). The crystalline phase was identified using the JCPDS (Join Committee on Powder Diffraction Standards) data.
Fourier transform infrared spectra (FTIR) of non-calcined and calcined samples were recorded in a Bomem/MB100 spectrometer in the 4000 to 400 cm^{-1} region, using a diffuse reflectance accessory. Potassium bromide was used to record the background spectra and to dilute the samples.
The TPR experiments were carried out in a Micromeritics model TPD/TPO 2900 equipment with samples previously heated under nitrogen (150 °C, 2 h). The measurements were performed from room temperature to 1000 °C at a rate of 10 °C min^{-1} using a 5% H_2/N_2 mixture.
The catalytic tests were carried out in a stainless steel fixed bed microreactor containing 0.15 g of catalyst (100 mesh size), for 6 h. The sample was heated up to 700 °C at a rate of 10 °C min^{-1}, under nitrogen flow (60 mL min^{-1}) and then reduced at this temperature under hydrogen flow (40 mL min^{-1}), for 2 h. The runs were performed at atmospheric pressure and 600 °C using a steam to methane molar ratio of 4. Methane (60 mL.min^{-1}) was mixed with nitrogen and steam to produce a mixture of 10% CH_4 and fed to reactor. The reaction conditions were chosen to achieve a conversion of 10% with a commercial catalyst (alumina-supported nickel) and to avoid any diffusion effect. The products were analyzed by on line gas chromatography.

3. Results and Discussion

The nickel content (% wt) in the solids calcined at 500 °C was 12.0% for SiNi and SiNiCa samples and 11.5% for SiNiMg. The calcium or magnesium contents were around 3.8%.These values are close to the expected ones. The X-ray diffraction patterns of the samples are shown in Fig. 1(a). The typical X-ray peaks of the cubic structure of nickel oxide (111, 200, 220, 311, 222) can be observed for all samples (JCPDF 78-0643). The amorphous nature of the silica matrix is evident from the broad peak at 2θ between 15° and 35°. Furthermore, minor peaks are observed in the SiNiMg sample (900 °C), indicating the crystallization of magnesium silicate (JCPDF 87-0061). The XRD data did not indicate the presence of nickel and calcium silicate in the samples. The X-ray line broadening of the most intense peak (200) provided the average diameter of the nickel oxide nanocrystalline domains which varied from 6.1

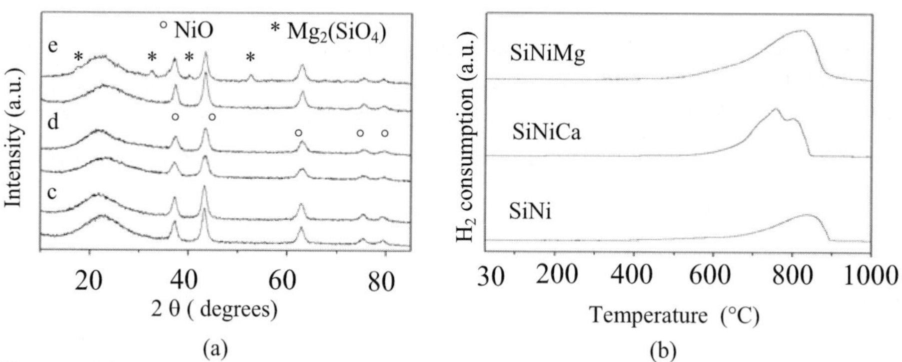

Figure 1. (a) X-ray diffractograms and (b) TPR profiles of the samples. SiNi (c), SiNiCa (d), SiNiMg (e) calcined at 500 °C (bottom) and 900 °C (top).

to 8.2 nm, as shown in Table 1. The average nanoparticle size was estimated using the Scherrer's equation [7]. The thermal treatment at 900 °C resulted in a little increase of the average diameter of the nickel oxide nanoparticles.

Before calcination, all samples showed FTIR spectra with an absorption band at 1733 cm^{-1}, assigned to carbonyl stretching (ν C=O) of citric acid and two broad absorptions (1600-1400 cm^{-1}) due to carboxylate stretching modes, indicating that citrate ions are bonded to nickel by coordinative linkages [8]. Other bands at 1100, 800 and 460 cm^{-1} due to Si-O-Si stretching (asymmetric and symmetric) and deformation modes were also noted, besides, a band at 955 cm^{-1} due to the Si-O stretching of silanol groups [9]. After calcination, the samples did not show the nitrate and carbonyl/carboxylate absorptions, indicating that these species were decomposed upon heating. The SiNiMg sample (900 °C) showed two extra bands (900, 605 cm^{-1}) attributed to Si-O-Mg stretching of magnesium silicate [10,11].

The textural properties of the samples are summarized in Table 1. The specific surface area and the pore volume of the catalyst heated at 500 °C were lower than that of silica; the SiNiCa and SiNiMg samples showed the lowest values. They decreased upon heating at 900 °C, while the average pore size and the average size of NiO crystallites did not change significantly. It means that nickel oxide, in the pores of the silica matrix, inhibits the pore collapse at high temperatures, when magnesium and calcium are absent. This can be explained by considering strong interactions between the nickel oxide nanoparticles and silanol moieties, including Si-O-Ni bonds. Furthermore, when calcium and magnesium ions were incorporated into the silica framework, the mobility of silica arrays might be enhanced, favoring the pore collapse at high temperatures. The pore size distribution was sharp regardless the calcination temperature. The presence of particle-like silica structure and citric acid before the addition of the guest ions, into the sol-gel mixture, seems to be a procedure that assures the preservation of silica mesoporous structure, for samples with up to 12% Ni.

Table 1.
Textural properties of the catalysts. hydrogen uptake in TPR experiments and coke produced during methane steam reforming.

Sample [a]	Sg [b] (m^2g^{-1})	PV [c] (cm^3g^{-1})	PS [d] (nm)	NiO CS [e] (nm)	H_2 uptake in TPR (μmol)	Coke (%)
SiO_2/500	593	1.5	8.9	-	--	--
SiO_2/900	184	0.4	7.1	-	--	--
SiNi/500	498	1.4	10.4	7.2	--	--
SiNi/900	330	1.1	10.3	7.4	44.2	1.2
SiNiCa/500	324	0.8	8.1	6.1	--	--
SiNiCa/900	132	0.4	8.1	6.8	46.3	0.31
SiNiMg/500	346	1.0	9.0	7.9	--	--
SiNiMg/900	183	0.5	9.3	8.2	47.5	0.06

[a]The number indicates the heating temperature in °C; [b]Specific surface area; [c]Total pore volume; [d]Average pore size; [e]CS, crystallite size, estimated from XRD according to Scherrer's equation.

Fig. 1(b) shows the TPR profiles of the catalysts calcined at 900 °C. All samples showed a reduction peak at high temperatures (700-800 °C), indicating that nickel oxide nanoparticles are interacting strongly with silica [12]. This is confirmed by the total hydrogen consumed in each case (Table 1) which was lower than the theoretical value (around 118 μmol) calculated considering the total reduction of nickel. The addition of calcium and magnesium slightly shifted the curve to lower temperature and also decreased the hydrogen uptake, indicating that both metals decreased the interaction of nickel with the support. Fig. 2(a) shows the methane conversion as a function of time over the catalysts. The SiNi and SiNiCa samples were quite stable during reaction while the SiNiMg showed a light decrease. These results are in accordance with the low amounts of coke, produced on the catalysts as shown in Table 1. The most active catalyst was the magnesium-containing sample (70%) followed by the calcium-doped one (45%) which is more active than the undoped sample (30%). The selectivity to hydrogen followed the same tendency (Fig. 2). High amounts of hydrogen were produced over the doped samples, especially on the magnesium-doped one; however, these solids produced low amounts of carbon monoxide, showing that they are suitable for producing hydrogen rather than to produce syngas. These results are in accordance with previous works [13,14], using alumina-supported nickel, according to which the alkaline earth oxides are able to provide lower support acidity, higher nickel dispersions and better steam activation. This causes a high amount of the active sites available during reaction, by the inhibition of metal surface blocking by coke deposition, increasing the methane conversion.The magnesium-doped catalyst was the most suitable one for methane reforming, due to the high ability of magnesium in preventing coke, as stated earlier [15].

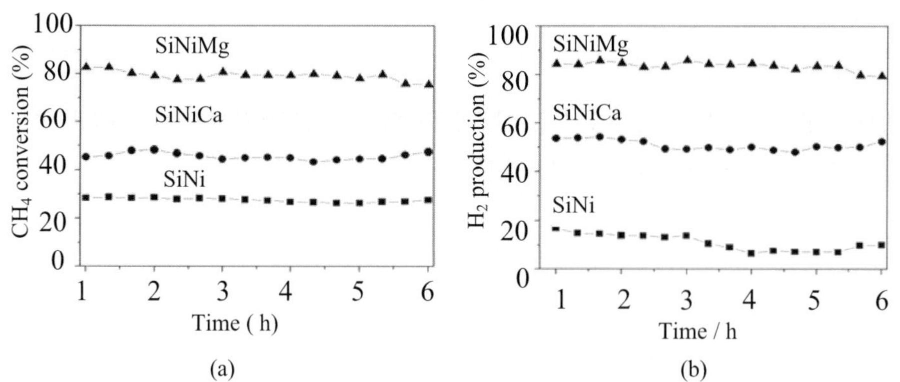

Figure 2. (a) Methane conversion and (b) hydrogen produced on the samples during the methane steam reforming. SiNi (square), SiNiCa (circle) and SiNiMg (triangle).

4. Conclusions

Silica-supported nickel prepared by the sol-gel method modified with citric acid is a promising catalyst to methane steam reforming. The addition of calcium and magnesium decreases the specific surface area but increases the activity and the selectivity to hydrogen. Magnesium-based catalyst showed the highest activity and selectivity of hydrogen and produced low amounts of carbon monoxide being suitable for hydrogen production.

5. References

[1] A. Fonseca, E. M. Assaf, J. Power Sour, 142 (2005) 154.
[2] R. Takahashi, S. Sato, T. Sodesawa, S. Tomiyama, Appl. Catal. A Gen. 286 (2005) 142.
[3] S. Tomiyama R. Takahashi, S. Sato, T. Sodesawa, S. Yoshida, Appl. Catal. A Gen. 241 (2003) 349.
[4] R. Takahashi, S. Sato, T. Sodesawa, M. Suzuki, N. Ichikumi, Micropor. Mesopor. Mater 66 (2003) 197.
[5] R.Takahashi,S.Sato,T.Sodesawa,M. Kawakita, K.Ogura, J. Phys. Chem. B 104 (2000) 12184.
[6] D. W. Lee, S K. Ihm. K. H. Lee, Micropor. Mesopor. Mater. 83 (2005) 262.
[7] B. D. Cullity, Introduction to Magnetic Materials, Addison-Wesley, Canada, 1972, p.181.
[8] K. Nakamoto, Infrared and Raman Spectra of Inorganic and Coordination Compounds, John Wiley & Sons: Canada, 1986.
[9] J. L. Bellamy, The infrared spectra of complex molecules, Chapman and Hall, London, 1975.
[10] F.F. Castillon, N.Bodganchikava, S. Fuentes, M..A.Borja, Appl. Catal. A Gen.175 (1998) 55.
[11] J. B. E. Caillerie, M. Kermarec, O. Clause, J. Phys. Chem. 99 (1995) 17273.
[12] B. Mile, D. Stiling , M.A Zammitt, , A. Lovell, M. Webb, J. Catal. 144 (1988) 217.
[13] J. T. Richardson, B. Turk, M. V. Twigg, Appl.Catal. A: Gen (1996) 97.
[14] J. de Lisboa, D.C.R.M Santos, F.B.Passos, F.B.Noronha, Catal. Today 101 (2005) 15.
[15] G. Xu, K. Shi, Y. Gao, H. Xu, Y. Wei, J. Molc. Catal. A: Chemical 147 (1999) 47.

Natural Gas Conversion VIII
F.B. Noronha, M. Schmal, E.F. Sousa-Aguiar (Editors)

Perovskites as catalyst precursors: partial oxidation of methane on $La_{1-x}Ca_xNiO_3$

S. M. de Lima[a], M. A. Peña[b], J. L. G. Fierro[b], J. M. Assaf[a]

[a]*Universidade Federal de São Carlos, Laboratório de Catálise, Departamento de Engenharia Química, São Carlos, Brasil, Fax:(016) 3351-8266, psania@iris.ufscar.br.*
[b]*Instituto de Catálisis y Petroleoquímica, CSIC, Cantoblanco, 28049 Madrid, Spain*

Perovskite-type oxides such as $La_{1-x}Ca_xNiO_3$ (x = 0, 0.05, 0.1, 0.3, 0.5 and 0.8) were prepared from citrate precursors and characterized by XRD, TPR and XPS. Catalytic tests in the partial oxidation of CH_4 were carried out in a tubular reactor at 973 K. After the catalytic tests the catalysts were studied by TPO. Partial substitution of La by Ca was performed to stabilize Ni particles and to prevent carbon deposition. The XRD profiles showed that the perovskite structure was the only compound identified within the $0 \leq x \leq 0.05$ range, whereas for $x \geq 0.1$ compounds such as La_2NiO_4, NiO and CaO were observed in addition to the perovskite oxide. On the other hand, segregation of NiO was confirmed by TRP and XPS, even in the unsubstituted perovskite (x = 0). The partial replacement of La by Ca decreased the catalytic activity in methane partial oxidation at x = 0.05, but as the Ca content rose further, activity increased.

1. Introduction

The partial oxidation of methane (POM) to produce CO and H_2 has received increasing attention in recent years, mainly because of its potential application as a commercial source of synthesis gas [1–3]. The reaction is exothermic, yields H_2/CO ratios suitable for use in the production of methanol and Fischer-Tropsch synthesis, and operates at temperatures where no NOx emissions are produced. The catalysts reported to be active in the POM reaction to syngas are either noble metals [3-5] or Ni-based catalysts [6-8]. Despite the high activity of noble metal-based catalysts, the high cost of such systems limits their widespread industrial application. It has observed in nickel catalysts that the degree of reducibility of nickel [9] and the basicity of the support substrate [10-11] are key requirements to achieve very high hydrogen yields. Over the last few decades, much attention has been given to lanthanum-transition metal-based perovskite oxides, which were introduced into catalysis some years ago [12]. Represented by the formula $La_{1-x}A_xMO_3$, where M stands for a transition metal such as Co, Mn, Fe, Ni or a combination of these, La-M perovskite oxides are characterized by a unique structure capable of accepting a wide range of defects [12]. Since the properties of the perovskite precursor can be modified, according to the choice and stoichiometry of A- and B- site cations, the aim in this paper was to examine the influence of the addition of Ca to the $LaNiO_3$ structure, to form the $La_{1-x}Ca_xNiO_3$ structure, on the properties and catalytic performance in the partial oxidation of methane.

2. Experimental

2.1. Catalysts preparation

The perovskite-type oxides $La_{1-x}Ca_xNiO_3$ were synthesized by a modification of the citrate method [13-14]. A concentrated solution of citric acid (Merck) was prepared and added to a solution of the precursor of the B-site cation ($Ni(NO_3)_2.6H_2O$ (Aldrich)) with an excess of ethylene glycol, and the mixture was kept at 333 K for 1 h under constant stirring. Next, solutions of lanthanum and calcium nitrates (Aldrich) were added, the whole remaining at 333 K for a further hour. The resulting solution was slowly evaporated at 383 K for 48 h until a spongy material was obtained. This citrate precursor was crushed and decomposed at 823 K for 3 h and finally calcined at 1173 K for 10 h.

2.2. Characterization techniques

X-ray diffraction (XRD) patterns of all calcined samples were obtained in a Siemens D-5005 diffractometer with CuK_α radiation in the 2θ scanning mode (range 5-80°).
Temperature-programmed reduction (TPR) experiments were carried out on a Micromeritics TPD/TPR 2900 apparatus interfaced with a microcomputer. Samples of about 30 mg were placed in a U-shaped quartz tube and initially purged in an air stream of 50 mL/min at 773 K for 1 h and then cooled to ambient temperature. Reduction profiles were then recorded by passing a stream of 10% H_2/Ar at a rate of 50 mL/min, while heating the sample at a rate of 10 K/min from room temperature to 1173 K. A cold-trap was placed just before the thermal conductivity detector (TCD) of the instrument to remove the water from the exit stream.
The oxidation state of Ni on the catalyst surface was studied by X-ray photoelectron spectroscopy (XPS) with a VG Escalab 200R spectrometer equipped with a hemispherical electron analyzer and Mg K_α ($h.\nu = 1253.6$ eV) X-ray source. Peak intensities were estimated by integrating the area under each peak after smoothing and subtraction of Shirley background and fitting of the experimental peak by a least-squares routine to Gaussian and Lorentzian lines.

Temperature-programmed oxidation was performed on catalysts after the catalytic tests in a TA Instruments SDT 2960 Simultaneous TGA analyzer, at a heating rate of 10 K/min, from room temperature up to 1173 K in an oxidizing atmosphere, to verify possible carbon formation.

2.3. Catalytic activity measurements

The catalytic tests were carried out in a tubular quartz fixed-bed reactor (i.d. 10 mm) under atmospheric pressure with CH_4/O_2 = 2:1 at 973 K. The samples of catalyst (100 mg) were first reduced *in situ* for 5 h at 973 K in a hydrogen flow (40 mL/min) and then tested at 973 K for approximately 10 h. The reactants and products were analyzed by on-line Varian GC-3800 gas chromatograph provided with two thermal conductivity detectors.

3. Results and Discussion

Fig. 1 revealed that the conditions used during calcination were sufficient to produce the perovskite as the main phase. However, in the catalysts with $x \geq 0.1$, NiO and La_2NiO_4 spinel, were observed too. An important observation is that at high Ca contents the dominant phases are NiO and CaO, but the spinel and perovskite phases still exist.

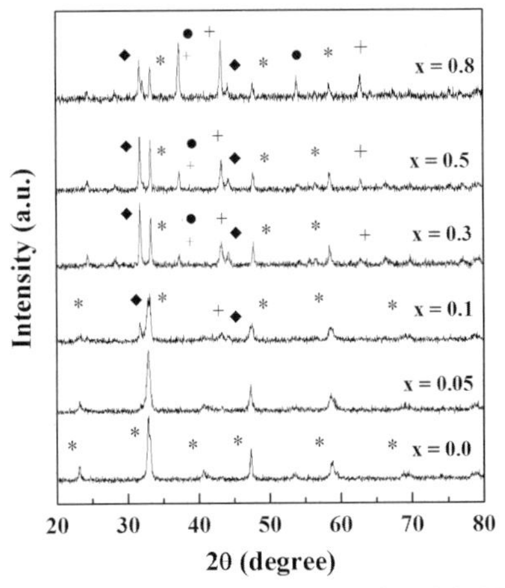

Fig. 1. XRD of the calcined $La_{1-x}Ca_xNiO_3$ - $LaNiO_3$ (*), La_2NiO_4 (♦), NiO (+), CaO (•).

TPR plots (Fig. 2a) exhibited two reduction peaks for $LaNiO_3$, the first corresponding to the formation of $La_2Ni_2O_5$ and the second to the formation of metallic Ni [15]. When La was partially substituted by Ca, the NiO and La_2NiO_4 phases were observed in the XRD patterns together with the perovskite structure at x = 0.1. Therefore, based on this analysis and observing, in Fig. 2a, that at values of $0 \leq x \leq 0.1$ an additional intermediate reduction peak appears, it can be suggested that the first reduction peak at 615 K corresponds to reduction of

Ni^{3+} to Ni^{2+} in the perovskite structure, whereas the peak at 653 K is due to the reduction of Ni^{2+} in the NiO to Ni^0, which happen almost simultaneously, so that it is not possible to visualize the separation of these peaks in all the samples.

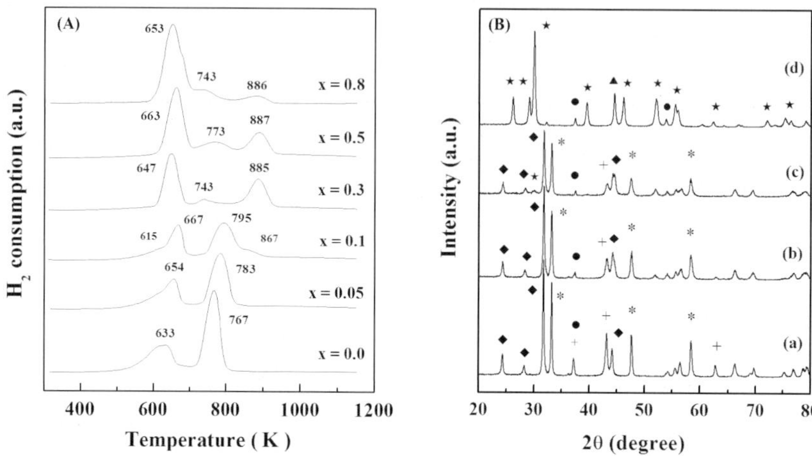

Fig. 2. (A) TPR of the calcined $La_{1-x}Ca_xNiO_3$ and (B) XRD of the sample with x=0.3 after successive steps of reduction. Curve (a) before reduction; (b) reduced at 693 K; (c) reduced at 823 K and (d) reduced at 973 K. $LaNiO_3$ (*), La_2NiO_4 (♦), NiO (+), CaO (•), La_2O_3 (★) and Ni^0 (▲).

Another notable feature of Fig. 2a is the leftward shift in the second reduction peak and the appearance of a third one, at high temperatures, when x = 0.1. On the basis of the XRD patterns and the fact that the third reduction peak grows to a maximum at x = 0.3, this can be assigned provisionally to reduction of Ni in the spinel phase. To verify that this TPR peak refers in fact to reduction of the spinel phase, XRD patterns were recorded after each step of the sample reduction with x = 0.3 (Fig. 2b). Proceeding through successive steps of reduction, it was observed that the lines assigned to both the perovskite structure and the spinel phase declined, but continued to be present in the first two reduction steps. Only after the final step does La_2O_3 appear, as all the Ni is reduced to metal, confirming the hypothesis regarding the TPR results.

The binding energies of O 1s and Ni $2p_{1/2}$ core-levels of calcined $La_{1-x}Ca_xNiO_3$ samples, are summarized in Table 1. Three photoemission lines can be seen for O 1s, which correspond to three distinct oxygen species [16] that are present in all the samples. The line with lowest binding energy (528.9-529.1 eV) is attributed to oxygen ions in the crystal lattice (O^{2-}), the intermediate energy line (530.8-531.1 eV) to oxygen in hydroxyl and/or carbonate groups and that with highest binding energy (532.0-532.4 eV) to strongly adsorbed molecular water [17].

The Ni 2p profiles are very complex because of the overlapping of the Ni $2p_{3/2}$ and La $3d_{3/2}$ peaks (not shown). Owing to the proximity of the binding energies of Ni^{2+} and Ni^{3+}, it is difficult, if not impossible, to distinguish each of these

Table 1. Binding energies (eV) of core levels of $La_{1-x}Ca_xNiO_3$ samples (area % in parenthesis)

Sample	x = 0.05	x = 0.1	x = 0.3	x = 0.5	x = 0.8
O 1s	528.9 (32) 530.9 (35) 532.0 (33)	529.1 (31) 531.1 (42) 532.4 (27)	528.9 (28) 530.9 (49) 532.3 (23)	528.9 (34) 530.8 (44) 532.2 (22)	529.0 (30) 530.8 (47) 532.0 (23)
Ni $2p_{1/2}$	872.8	872.9	872.8	872.7	872.7

species by photoelectron spectroscopy alone. The most intense peak for Ni $2p_{3/2}$, which appears around 855 eV, is typical of Ni^{2+}/Ni^{3+} ions surrounded by oxide ions [18]. This peak is accompanied by a satellite line, positioned about 6 eV to the high binding-energy side of the peak (ca. 861 eV), which constitutes a "fingerprint" of Ni^{2+} ions and provides conclusive evidence of their presence in significant numbers at the sample surface, together with the Ni^{3+} ions. These results agree with the XRD data that indicated the presence of NiO and La_2NiO_4, in both of which the Ni occurs as the divalent cation.

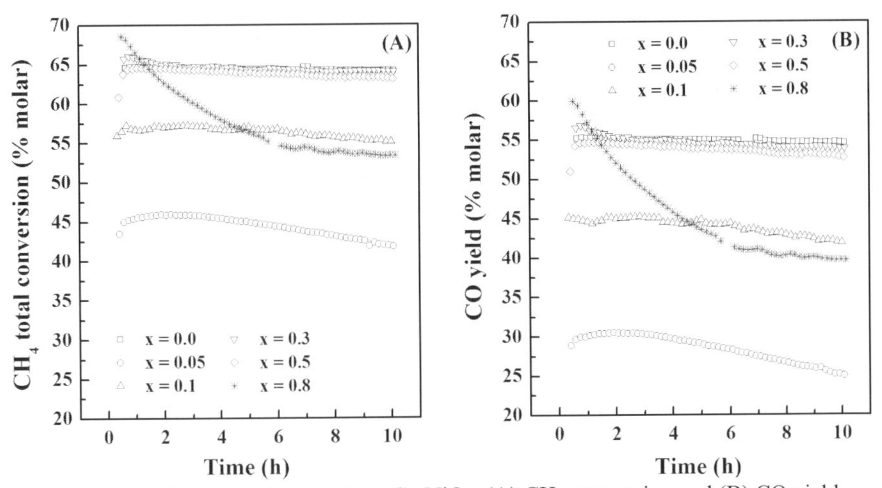

Fig. 3. Partial oxidation of methane on $La_{1-x}Ca_xNiO_3$: (A) CH_4 conversion and (B) CO yield versus reaction time.

Fig. 3a shows that the Ca-free sample was highly active for partial oxidation of methane, with 68% CH_4 total conversion and 55% CO yield (Fig. 2b). When La was partially replaced by Ca (x = 0.05), a drastic fall in methane conversion was observed, but the activity increased again for $x \geq 0.1$. After 6 hours of reaction, the activity decreased by more than 15% in the catalyst with x = 0.8. This decrease is generally attributed to carbon formation on the catalyst surface. However, carbon deposition was not observed on these catalysts (TPO not shown), so the deactivation must be attributed to oxidation of a part of the Ni metallic sites that loses activity during the partial oxidation of methane. For all catalysts, the H_2/CO ratio is in the range 1.6-1.8. The addition of Ca improved the activity of the reaction, but this improvement has not a simple dependency on the Ca content, presenting a minimum of activity for x=0.05.

4. Conclusions

According to the XRD patterns obtained, the $La_{1-x}Ca_xNiO_3$ oxides exhibited a single phase with a perovskite structure, in samples with $0 \leq x \leq 0.05$, while a mixture of this structure with La_2NiO_4, NiO and CaO phases was detected at $x \geq 0.1$. Nonetheless, when the samples were analyzed by TPR and XPS, Ni^{2+} was observed together with the perovskite, even at x = 0.0. The partial replacement of La by Ca firstly decreased catalytic activity in the partial oxidation of methane at x = 0.05, increasing it again with further rises in the Ca content.

The authors gratefully acknowledge CNPq Project 473598/04-3 for funding this work.

References

1. M. Prettre, Ch. Eichner, M. Perrin, Trans. Faraday Soc. 42 (1946) 335.
2. D.A. Hickman, L.D. Schmidt, Science 259 (1993) 343.
3. D.A. Hickman, L.D. Schmidt, J. Catal. 138 (1992) 267.
4. E.P.J. Mallens, J.H.B.J. Hoebnik, G.B. Marin, Catal. Lett. 33 (1995) 291.
5. M.A. Peña, J.P. Gomez, J.L.G. Fierro, Appl. Catal. A: Gen. 144 (1996) 7.
6. A. Slagtern, U. Olsbye, Appl. Catal. A: Gen. 110 (1994) 99.
7. Y.H. Hu, E. Ruckenstein, J. Catal. 158 (1996) 260.
8. J. Barbero, M.A. Peña, J.M. Campos-Martin, J.L.G. Fierro, P.L. Arias, Catal. Lett. 87 (2003) 211.
9. F. van Looij, J.W. Geus, J. Catal. 168 (1997) 154.
10. V.A. Tsipouriari, Z. Zhang, X.E. Verykios, J. Catal. 179 (1998) 283.
11. Y. Lu, Y. Liu, S. Shen, J. Catal. 177 (1998) 386.
12. J. Kirchnerova, M. Alifanti, B. Delmon, Appl. Catal. A: Gen. 231 (2002) 66.
13. S.M. Lima, Preparação e aplicação de óxidos tipo perovskita $La_{1-x}Ce_xNiO_3$ e $La_{1-x}Ca_xNiO_3$ para obtenção de gás de síntese a partir do metano, (D. Thesis), São Carlos - SP, UFSCar, 2006.
14. S.M. Lima, M.A. Peña, J.L.G. Fierro, J.M. Assaf, Appl. Catal. A: Gen. 311 (2006) 95.
15. S.M. Lima and J.M. Assaf, Catal. Lett. 108 (2006) 63-70.
16. M.A. Peña and J.L.G. Fierro, Chem. Rev. 101 (2001) 1981.
17. M. Crespin and W.K. Hall, J. Catal. 69 (1981) 359.
18. D. Briggs and M.P. Seah, Practical Surface Analysis by Auger and and X-Ray Photoelectron Spectroscopy, 2nd Ed., Wiley, Chichester, 1990.

Natural Gas Conversion VIII
F.B. Noronha, M. Schmal, E.F. Sousa-Aguiar (Editors)

Ni/$CeZrO_2$-based catalysts for H_2 production

Sandra C. Dantas, Janaína C. Escritori, Ricardo R. Soares, Carla E. Hori

School of Chemical Engineering, Federal University of Uberlandia, Av. Joao Naves de Avila 2121, Bloco 1K, Campus Santa Monica, Uberlandia MG, 38400-902, Brazil

Abstract

Ni/$CeZrO_2$-based catalysts were investigated in autothermal methane reforming. The use of alumina propitiated a higher surface area for the samples, which favored the formation of smaller nickel oxide, and metallic nickel particles as observed by XRD and metal dispersion experiments. XRD analysis showed the formation of a ceria zirconia solid solution. In general, the catalytic performance of Ni/$CeZrO_2$/Al_2O_3 was better than the ones observed for the other samples, which may be correlated to the higher reducibility and better redox properties verified for this catalyst.

1. Introduction

Among several processes used to obtain hydrogen, the steam methane reforming (SMR) shows the highest H_2-yield; however, it requires high energy input. On the other hand, catalytic partial oxidation (CPOX) releases energy, despite presenting a lower H_2/CO ratio. Since the autothermal reforming (ATR) integrates both processes, it has attracted more attention in the last years.
Nickel supported on alumina system is used traditionally due to its low cost. However, it suffers a severe deactivation during the first hours on stream. Several promoters, such as CeO_2 and $CeZrO_2$ have been used in order to improve its stability. Several researchers [1,2] have correlated the stability increase to the presence of redox sites, using CeO_2 or $CeZrO_2$-promoted catalysts, respectively. One of the drawbacks of the use of ceria as a support is its higher cost and lower thermal stability compared to alumina. Therefore, the objective of this study was to evaluate the properties and the performances of Ni/$CeZrO_2$ and Ni/$CeZrO_2$/Al_2O_3 for autothermal reforming of methane.

2. Experimental

$CeZrO_2$ bulk with atomic Ce/Zr ratio equals to 1 (102 m^2/g) was prepared by co-precipitation technique, using $(NH_4)_2Ce(NO_3)_6$ and $ZrO(NO_3)_2$ as precursors and calcination at 773 K during 4 hours. $CeZrO_2$ (same Ce/Zr ratio) was also supported on γ-Al_2O_3 (Catapal- Sasol) pre-calcined at 1173 K during 6 hours (127 m^2/g), by wet co-impregnation of the same aqueous solution precursors in order to obtain a 12 wt% $CeZrO_2/Al_2O_3$ carrier. This support was also calcined at 1173 K during 6 hours. 10 wt% of nickel was added by incipient wetness impregnation using an aqueous nickel nitrate solution. In order to avoid the formation of nickel aluminate, which is very difficult to reduce, the samples were calcined in air flow at 723 K by 4 hours.
The catalysts were characterized using BET, X-ray diffraction (XRD), temperature programmed reduction (H_2-TPR), and temperature programmed desorption (CO_2-TPD) measurements. The nickel dispersion was evaluated using the dehydrogenation of cyclohexane, a structure insensitive reaction [3]. The reaction mixture was obtained by bubbling hydrogen through a saturator containing cyclohexane (99.9%) at 285 K (H_2/cyclohexane = 13.2). The reaction temperatures varied from 520 to 570 K. At these conditions, no mass transfer or equilibrium limitations were observed. The composition of effluent gas phase was measured by online gas chromatograph (Shimatzu) equipped with a thermal conductivity detector and a Chrompack CP-WAX 57 CB column.
Before the catalytic tests, the dispersion measurements or CO_2-TPD experiments, the samples were reduced under H_2 flow (50 mL/min), increasing the temperature up to 773 K at 10 K/min. The temperature remained 3 hours at 773 K and then, He replaced the H_2 flow and the sample was heated up to the reaction temperature (1073 K). The catalytic tests were carried out using 12 mg of catalyst and a total flow of 180 mL/min. The feed composition was $2CH_4:1H_2O:0.5O_2$. Effluent gases from the reactor were analyzed by a gas chromatograph (Shimadzu) equipped with a Hayesep column.

3. Results and Discussion

Table 1 summarizes the results obtained during the characterization of the samples. As expected the sample supported on $CeZrO_2$ presented the lowest BET area, 88 m^2/g, similar to values previously reported in the literature Hori et al. [4]. The addition of $CeZrO_2$ to the alumina did not change significantly the total area and both catalysts had areas above 100 m^2/g.
The XRD patterns obtained for the three samples between 25 and 75° are shown in figure 1. For all catalysts, peaks relative to the NiO (37.2°, 43.2° and 63°) may be observed [2,5,6] and for the alumina containing samples, profiles A and C, it is possible to identify peaks characteristic γ-Al_2O_3 [3]. On the other hand, for the catalysts containing $CeZrO_2$, it was observed a shift of peaks relative to

Table 1 – Characterization Results

Catalysts	BET area (m^2/g_{cat})	Dp_{NiO} (nm)	Ni dispersion (%)	TPR consumption H_2:Ni	TPD (μmols/g_{cat}) CO_2	TPD (μmols/g_{cat}) CO
Ni/Al_2O_3	111	6	6	0.92	290	-
Ni/$CeZrO_2$	86	24	1	1.31	722	330
Ni/$CeZrO_2$/Al_2O_3	105	7	4.5	1.34	215	-

the cubic ceria phase to higher 2θ positions. The peak with higher intensity shifted from 2θ = 28.6° to 29.3°, which indicates that zirconia was incorporated into CeO_2 lattice and formed a solid solution. Since the literature reports a peak position of 29.4° for ceria zirconia solid solution with a Ce/Zr ratio equals to 1, it is possible that a highly dispersed zirconia phase is present and it was not detected [4].

The NiO particle size was estimated using the Scherrer equation and the results are also presented in Table 1. It may be observed that the sample supported on ceria-zirconia had a NiO particle size more than three times higher than the alumina containing samples. Probably the higher BET area propitiated a better dispersion of the nickel on the surface of the carrier. The metal dispersion was estimated through the cyclohexane dehydrogenation (Table 1). These results are in agreement with the NiO particle diameters calculated by XRD, although the dispersion of Ni is somewhat smaller than the value that one would obtain through Dp_{NiO}. Accordingly to the literature, the reduction process may cause some sintering of the metal [6].

Figure 2 presents the H_2 consumption profiles during TPR experiments. The sample Ni/Al_2O_3 presented a wide reduction region, between 673 K and 1273 K

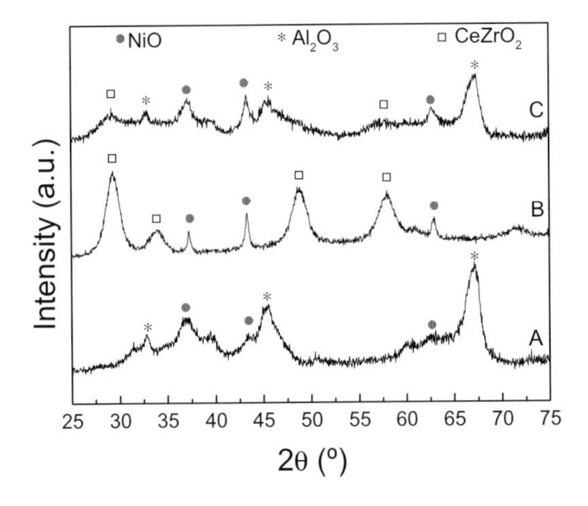

Figure 1 – X-ray diffraction profiles for Ni/Al_2O_3 (A), Ni/$CeZrO_2$ (B) and Ni/$CeZrO_2$/Al_2O_3 (C)

with a maximum around 1000 K, which could indicate the possible formation of nickel aluminate [7]. The TPR profile of the sample $Ni/CeZrO_2$ showed a reduction peak around 720 K and a shoulder at 830 K. This TPR profile is in agreement with previous literature reports [8,9], the lower temperature peak can be attributed to the reduction of the relatively free NiO particles, while the higher temperature peak are assigned to the reduction of complex NiO species in intimate contact with the $CeZrO_2$ [9].

Table 1 also presents the values of hydrogen consumption during the TPR. The amount of H_2 necessary to complete NiO reduction is 1704 $\mu mols/g_{cat}$. The sample Ni/Al_2O_3 showed reduction higher than 90%. However, the other samples presented H_2 consumption above the theoretical values for the reduction of nickel. This result indicates that $CeZrO_2$ was also reduced during the analysis, which is in agreement with previous studies in the literature [2,8,10].

The desorption profiles of CO_2 and CO are presented in Figure 3. All profiles showed at least two peaks of CO_2 desorption, one at lower temperature and another at higher temperature. The CO_2 desorption is continuous between these two regions and after the peak at higher temperature, the desorption continued until 800 K. The reference sample, Ni/Al_2O_3, showed two peaks, around 420 and 550 K. This result is in agreement with the ones obtained by Roh et al. [9]. The CO_2-TPD profile for the sample $Ni/CeZrO_2/Al_2O_3$ is similar to the one obtained for Ni/Al_2O_3, but it has smaller intensities. Since $Ni/CeZrO_2/Al_2O_3$ has just 12 wt% of ceria zirconia oxide, in a per gram of ceria basis, this sample has the highest amount of CO_2 desorbed. Hou et al. [11] observed that the increase

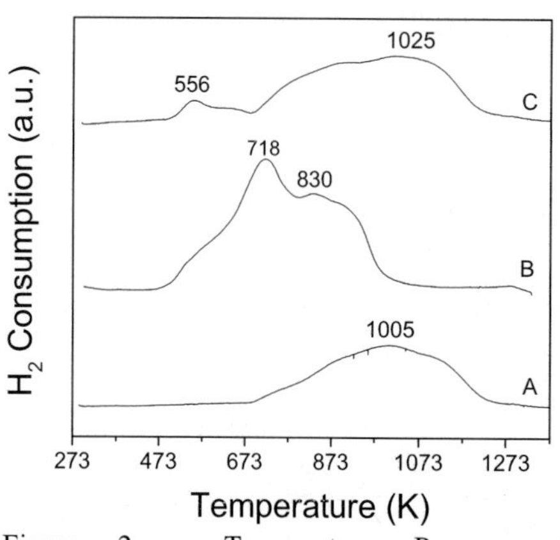

Figure 2 – Temperature Programmed Reduction profiles:
Ni/Al_2O_3 (A) $Ni/CeZrO_2$ (B) $Ni/CeZrO_2/Al_2O_3$ (C)

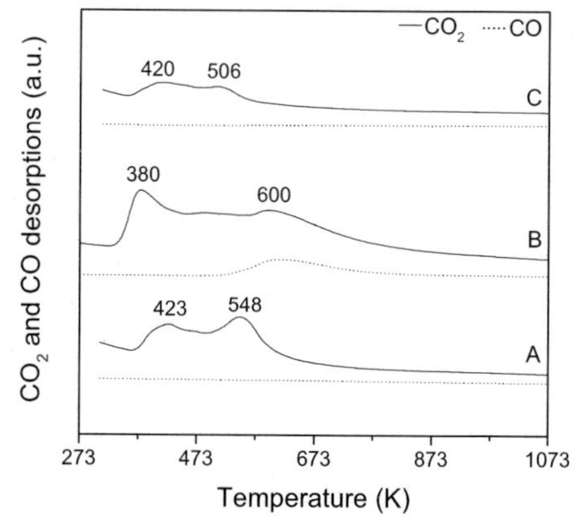

Figure 3 – Temperature Programmed Desorption of CO_2 profiles:
Ni/Al_2O_3 (A) $Ni/CeZrO_2$ (B) $Ni/CeZrO_2/Al_2O_3$ (C)

of the CO_2 adsorption could be correlated to the production of higher quantities of oxygen species on the catalyst surface leading to higher activity and stability and low coke formation during the reforming of methane.

Figure 4 presents the methane conversion as a function of time during the autothermal reforming catalytic tests. For Ni/Al_2O_3, it is possible to observe a slight deactivation, probably due to coke formation. After 24 hours, the methane conversion was around 40%. On the other hand, $Ni/CeZrO_2$ maintained its activity in this level during the 24 hours of reaction, around 50% and for $Ni/CeZrO_2/Al_2O_3$ the methane conversion was approximately 65%, after 24 hours of reaction. This higher activity observed for $Ni/CeZrO_2/Al_2O_3$ may be correlated to good nickel dispersion, higher reducibility observed during the TPR experiments and better redox properties observed during the CO_2-TPD measurements. CO and CO_2 selectivities are presented in figure 5. All the samples had a high selectivity to CO formation, Ni/Al_2O_3 and $Ni/CeZrO_2$ around 80% and $Ni/CeZrO_2/Al_2O_3$ above 90%.

Notice that CO and CO_2 selectivities add up to 100%, indicating a low carbon deposition for Ni/Al_2O_3 or no coke formation in the case of ceria containing catalysts, which is consistent with the stability observed for the methane conversion data for these samples. Moreover, the carbon balance indicated that there was almost no carbon deposition on the catalysts surface. The presence of ceria-zirconia helped to keep the surface free of carbon deposits, since it provides an additional path for the adsorption and dissociation of O_2 and H_2O, forming oxygen and hydroxyl groups on the surface of the support. These groups may be transferred to the metal through the metal-support interface and then react with the carbon containing species to form CO or CO_2. According to

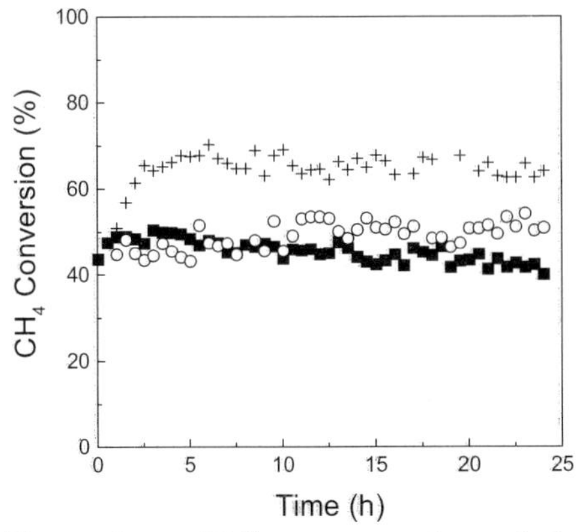

Figure 4 – Methane conversions during autothermal methane reforming at 1073 K
■ Ni/Al_2O_3 O $Ni/CeZrO_2$
+ $Ni/CeZrO_2/Al_2O_3$

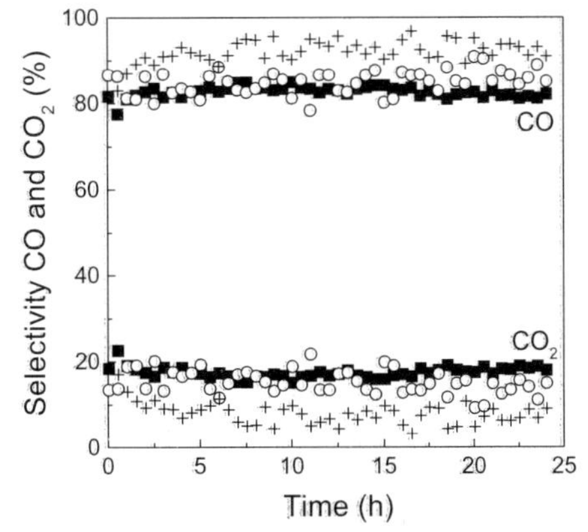

Figure 5 – CO and CO_2 selectivities during autothermal methane reforming at 1073 K
■ Ni/Al_2O_3
O $Ni/CeZrO_2$ + $Ni/CeZrO_2/Al_2O_3$

Wang et al. [12], at low oxygen concentrations adsorbed on the metal, the formation of CO and H_2 is favored. All the samples presented high H_2 selectivity which was around 90%. H_2/CO ratios were around 2.77 for Ni/Al_2O_3, 2.7 for $Ni/CeZrO_2$ and 2.64 for $Ni/CeZrO_2/Al_2O_3$ sample. These different values were more due to the slight variation of the CO selectivities than a disparity on the H_2 selectivity.

4. Conclusions

BET surface area results showed that the use of alumina propitiated a higher surface area for the samples, which favored the formation of smaller nickel particles as observed by XRD and metal dispersion experiments. XRD patterns also revealed the formation of a ceria zirconia solid solution, although it is possible to have a small amount of a highly dispersed zirconia phase. The use of ceria based mixed oxides propitiated better reducibilities and CO_2 desorption properties than the Ni/Al_2O_3 reference sample. In general, the catalytic performance of $Ni/CeZrO_2/Al_2O_3$ was better than the ones observed for the other samples, which may be correlated to the higher reducibility and better redox properties verified for this catalyst.

Acknowledgments

The authors would like to thank Fapemig and Finep for the financial support.

References

[1] N. Laosiripojana; S. Assabumrungrat, *Applied Catal.A: General*, 290 (2005) 200.
[2] H.S. Roh ; K.W. Jun ; S.E. Park, *Applied Catal.A: General*, 251 (2003) 275.
[3] P.P. Silva, F.A. Silva, L.S. Portela, L.V. Mattos, F.B. Noronha, C.E. Hori, *Catal. Today,* 101 (2005) 31.
[4] C.E. Hori; H. Permana; K.Y.S. Ng; A. Brenner; K. More; K.M. Rahmoeller; D. Belton, *Applied Catal. B: Environ.* 16 (1998) 105.
[5] S.M. Stagg-Willians; F.B. Noronha; G. Fendley; D.E. Resasco J. *of Catalysis* 194 (2000) 240
[6] H.S. Roh; H.S.Potdar; K.W. Jun, *Catal. Today*, 39 (2004) 93.
[7] R. Molina; G. Poncelet, *J. of Catalysis* 173 (1998) 257.
[8] W.S. Dong; H.S. Roh; K.W. Jun; S.E. Park; Y.S. Oh, *Applied Catal. A: General* 226 (2002) 63.
[9] H.S. Roh; K.W. Jun; W.S. Dong; J.S. Chang; S.E. Park; Y.I. Joe *Journal of Molecular Catal. A: Chemical* 181 (2002) 137.
[10] F.B. Passos; E.R. Oliveira; L.V. Mattos; F.B. Noronha *Catal.Today* 101 (2005) 23.
[11] Z. Hou; T. Yashima *Applied Catalysis A: General* 261 (2004) 205.
[12] D. Wang; O.Dewaele, A.M.Groote, G.F. Froment, *J. of Catalysis* 159 (1996) 418.

Natural Gas Conversion VIII
F.B. Noronha, M. Schmal, E.F. Sousa-Aguiar (Editors)

Effect of aluminum content on the properties of lanthana-supported nickel catalysts to WGSR

Manuela de S. Santos[a], Guillermo Paternina Berrocal[a], José Luís Garcia Fierro[b], Maria do Carmo Rangel[a*]

[a]*Instituto de Química, Universidade Federal da Bahia. Campus Universitário de Ondina, Federação. 40 170-280, Salvador, Bahia, Brazil. *E-mail:mcarmov@ufba.br*
[b]*Instituto de Catálisis y Petroquímica, CSIC, Cantoblanco, 28049, Madrid, Spain*

The effect of aluminum content on the properties of lanthana-supported nickel catalysts was evaluated in this work, in order do find new catalysts for water gas shift reaction (WGSR). It was noted that small amounts of aluminum increased the specific surface and the activity and thus is benefic to the catalyst. However, high amounts of aluminum causes a decrease on both properties and this was assigned to a strong interaction between nickel and aluminum, which stabilizes the Ni^{2+} species on the surface and makes the production of the active phase (metallic nickel) more difficult.

1. Introduction

The search for processes to obtain high pure hydrogen has increased in recent years, due to its large importance as fuel and for the production of other high value products. From the commercial point of view, the most important route to produce hydrogen is the steam reforming of natural gas. However, this reaction also produces carbon monoxide that can poison most of metallic catalysts. In order to remove these products from the gaseous stream, and also to increase the hydrogen production, most of industrial plants have another unit in which carbon monoxide is converted to carbon dioxide, which is removed from the stream in a further step [1]. The oxidation of carbon monoxide to carbon dioxide is carried out in the presence of steam and is known as water gas shift reaction (WGSR). In order to achieve rates for commercial purposes, this

reaction is often performed in two steps in industrial processes. The first one, called high temperature shift (HTS), occurs in the range of 320 at 450°C in favorable kinetic conditions, while the second step (low temperature shift, LTS) at 200 to 250°C is favored by thermodynamics [1-3].
The commercial catalysts used in the HTS reaction comprise hematite (α-Fe_2O_3), containing chromium oxide among other dopants. In commercial processes, this solid is reduced in situ, producing magnetite (Fe_3O_4), which is believed to be the active phase. This catalyst is very active and selective and is resistant against several poisoning; besides, it has a low cost. However, they have the inconvenience of being toxic, due to the chromium compounds and also of deactivating with time, due to the decrease of the specific surface area [2-5]. These features have motivated several studies addressed to new catalysts.
In this work, lanthana-supported nickel catalysts (with aluminum or not) were studied with the aim of developing new catalysts to HTS reaction. These solids have the advantage of being chromium-free and the support can also prevent the deactivation of the active phase.

2. Experimental

The support (lanthana) was prepared by adding a lanthanum nitrate solution (250 mL, 1 M) and 250 mL of 8.5% (v/v) of an ammonium hydroxide solution to a beaker with water, at room temperature. After the addition of the reactants, the system was kept under stirring for 24 h and then centrifuged (2500 rpm, 4 min). The gel was dried in oven at 120 °C, for 24 h, ground and sieved in 80 mesh and heated under air flow at 550 °C, for 4 h. This solid was named L sample. In the preparation of the LA10 sample (La/Al (molar)= 10), 250 mL of aluminum nitrate solution (0.1 M) and lanthanum nitrate (1 M) were used while to get the sample with La/Al = 1, an aluminum nitrate solution 1 M was used. The catalysts with 10% of nickel were prepared by wet impregnation of a nickel nitrate solution on pure lanthanum oxide or aluminum-doped ones, using 1.4 mL of the nickel nitrate solution (3.4 $mol.L^{-1}$) per gram of the support. The solution was kept in contact with the support for 24 h, at room temperature and then filtered and dried in oven at 120 °C, for 24 h. The samples were heated under air flow, for 3 h at 600 °C, producing NL, NLA10 and NLA1 samples.
The support precursors were characterized by thermogravimetry (TG) and the supports and the catalysts was analyzed by specific surface area measurements, X-ray diffraction (XRD), temperature programmed reduction (TPR), chemical analysis and X-ray photoelectron spectroscopy (XPS). After the reaction, the catalysts were analyzed by XRD, specific surface area measurements and XPS.
The elemental analysis of the solids was determined by flame atomic absorption spectrometry using a SpectrAA 220 Varian equipment. The sample (0.005 g)

was previously dissolved in a mixture of hydrochloridric acid and nitric acid (3:1). The XRD was performed at room temperature with a Shimadzu model XRD 6000 instrument, using CuKα radiation generated at 40 kV and a nickel filter. The specific surface areas were measured in a Micromeritics model TPR/TPO 2900 equipment on samples previously heated under nitrogen (160 °C, 1 h). The samples were analyzed with a 30% N_2/He mixture. The temperature programmed reduction was performed in the same equipment, using a 5% H_2/N_2 mixture (60 $mL.min^{-1}$) at a heating rate of 10 $^{o}.min^{-1}$. The solid was heated up to 1000 °C and the consumption of hydrogen was measured by a thermal conductivity detector. X-ray photoelectron spectra were obtained with a VG Escalab 220R spectrometer equipped with a MgKα X-ray radiation source (hν= 1253.6 eV) and a hemispherical electron analyzer.
The catalyst performance was evaluated using 0.15 g of catalyst powder (100 mesh) and a fixed-bed microreactor. The experiments were carried out under isothermal condition (370 °C) and at atmospheric pressure, employing a steam to process gas molar ratio of 0.6. These conditions were chosen to get 10% of conversion, using a commercial catalyst. A standard mixture containing 9.82% CO, 9.70% CO_2, 19.67% N_2 and H_2 (balance) was used as process gas. The reaction products were analyzed by gas chromatography, using a CG-35 instrument with Porapak Q and molecular sieve columns.

3. Results and Discussion

The X-ray diffractograms of the supports (not shown) displayed peaks of the lanthanum oxide, La_2O_3 (JCPDS 05-0602) and of the orthorhombic lanthanum nitrate hydroxide, $La(OH)_2NO_3$ (JCPDS 26-1144), except for the solid with La/Al= 1 that showed an amorphous halo. The production of lanthana was confirmed by DTA and TG curves, which showed a peak at about 500 °C, followed by a weight loss. No aluminum-containing phase was detected for the LA1 sample, a fact which can be assigned to its poor crystallinity [6]. Figure 1 shows the X-ray diffractograms of the catalysts before and after the WGSR. In fresh catalysts, a mixed phase of lanthanum and nickel oxide, La_2NiO_4 (JCPDS 80-1346) and lanthanum oxide, La_2O_3 (JCPDS 05-0602), besides nickel oxide, NiO (JCPDS 44-1159) and lanthanum nitrate hydroxide, $La(OH)_2NO_3$ (JCPDS 26-1146), were detected. During the WGSR, nickel oxide changed to metallic nickel (JCPDS 87-0712), while the other phases remained in the solids. The sample with the highest amount of aluminum (NLA1) remained amorphous to X-ray even after the reaction.
The specific surface areas are shown in Table 1. The addition of small amounts of aluminum to lanthana caused an increase of specific surface area from 12 (L) to 28 $m^2.g^{-1}$ (LA10), indicating that it acts as a textural promoter. However, the

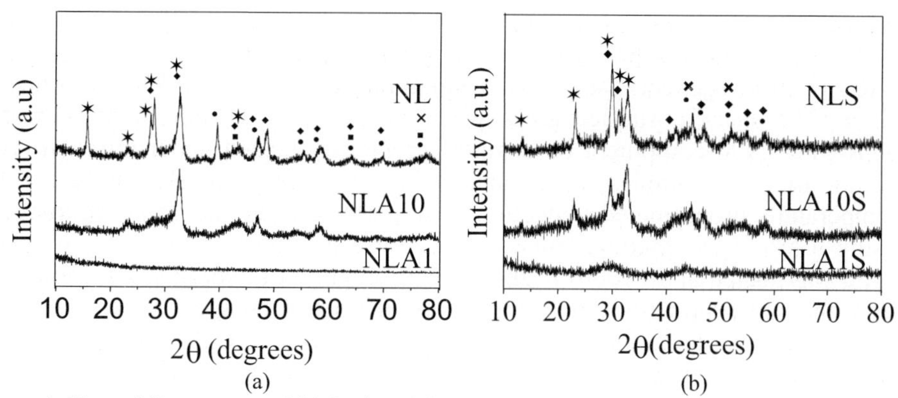

(a) (b)

Figure 1. X-ray diffractograms of (a) fresh and (b) spent catalysts. NL sample: lanthana-supported nickel. LA10 and LA1 samples: nickel on aluminum-doped lanthana with La/Al (molar)=10 and 1, respectively. S= spent catalysts. • La_2O_3; ▪ NiO; • La_2NiO_4; ✶ $La(OH)_2NO_3$; × Ni^0.

Table1.
Specific surface areas of the catalysts (Sg), binding energies of characteristic core levels of lanthanum, aluminum and nickel and surface composition of fresh catalysts. NL sample: lanthana-supported nickel; NLA10 and NLA1: nickel supported on aluminum-doped lanthana with La/Al (molar)= 10 and 1.

Samples	Sg ($m^2.g^{-1}$)	La3d5/2	Al2p	Ni2p 1/2	Ni / (Al+La) (surface)	Ni / (Al+La) (bulk)
NL	11	834.8	--	871.5	1.69	0.28
NLA10	15	834.8	73.7	871.8	1.15	0.28
NLA1	6.0	834.9	73.7	871.9	2.92	0.28

addition of higher amounts of aluminum (LA1) caused a decrease (9.0 $m^2.g^{-1}$). The addition of nickel did not change the specific surface area for the sample without aluminum but decreased for the other samples; this can be related to the production of amorphous aluminum compounds, not detectable by XRD, but related to a large peak in DTA curves in the range of 600-950 °C.
The binding energies (BE) of some characteristic core levels of lanthanum, nickel, aluminum and nickel are displayed in Table 1.The BE for lanthanum (834.8 eV) is characteristic of La^{3+} species [7] and was the same in all species, indicating they are in the same chemical environment. The same behaviour was noted regarding the BE values for aluminium and nickel, which are typical of Al^{3+} and Ni^{2+} species, respectively [7]. The solid surface composition, expressed

as atomic ratio, is also shown in Table 1.It can be noted that the amount of nickel was higher on the surface than in the bulk for all samples. The NLA1 sample is the richest in nickel, among the samples.

The TPR profiles of the supports (not shown) displayed two peaks, which were shifted to higher temperatures, due to the addition of small amounts of aluminum (LA10). By adding higher amounts of this metal (LA1), they overlapped at an intermediate temperature, suggesting a strong interaction between aluminum and the support. These peaks are also assigned to nitrate species and lanthana reduction [8,9]. The presence of nitrate species in the solids was confirmed by XPS. As expected, the reduction profiles of the catalysts are quite different as compared to the support ones (Figure 2a). A low temperature peak (around 230 °C) appeared, which was assigned to the reduction of nickel oxide particles in weak interaction with the support [10]; this peak is absent in the sample with the highest amount of aluminum (NLA1), indicating that all nickel are in medium or strong interaction with the support. The NL and NLA1 curves also showed three other peaks above 400 °C, the first related to reduction of nickel in strong interaction with the support and the others associated with the reduction of La_2NiO_4 compound. These peaks are also related to the reduction of nitrate species and of lanthanum oxide [8,9]. The addition of more aluminum (NLA1 sample) caused the overlapping of these peaks and shifted them to higher temperature, indicating that this dopant made all these reduction processes more difficult, probably due to the reduction of aluminum-based compounds (not detectable by XRD). It means that there is a strong interaction among the metals for the NLA1 sample.

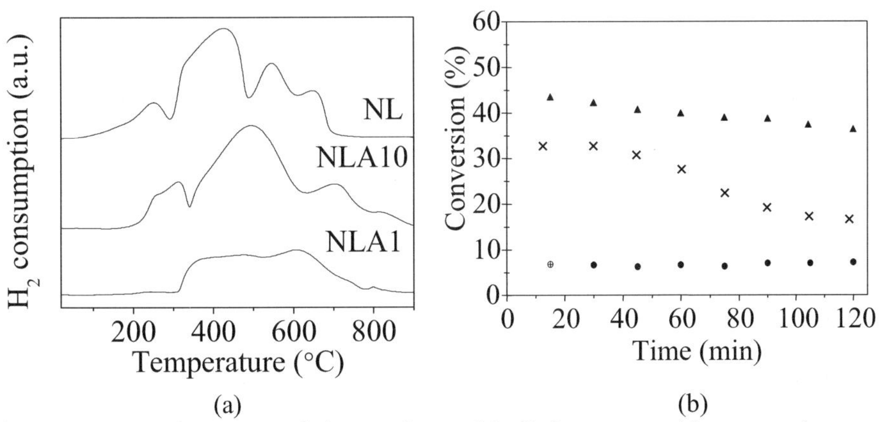

Figure 2. (a) TPR curves of the catalysts. (b) Carbon monoxide conversion as a function of time. × NL sample (lanthana-supported nickel); ▴ NLA10 and • NLA1: with La/Al (molar)= 10 and 1, respectively.

All catalysts were active in HTS reaction, as shown in Figure 2b. The pure and doped-lanthana did not show any activity, indicating that nickel is the active phase. When small amounts of aluminum were added to the catalysts, the activity strongly increased reaching a value close to the equilibrium value (46%). However, it decreased sharply when larger amounts of aluminum were added. This can be assigned to the strong interaction between nickel and the support [11-13],since the surface is rich in aluminum for this sample. Therefore, although this sample contains the largest amount of nickel on the surface most of the species are stabilized as Ni^{2+} and can not go into reduction to produce the active phase (metallic nickel).The presence of aluminum also made the catalysts more stable under the reaction condition, as shown in Figure 2b.

4. Conclusions

Lanthana-supported nickel is an efficient catalyst for WGSR. When small amounts of aluminum are added to this solid, no other phase is noted, besides those detected in undoped solid, but the specific surface area and the activity in WGSR increase. However, the further addition of aluminum causes a sharp decrease of specific surface area and in activity, a fact which can be assigned to strong interaction among nickel, lanthana and aluminum, as detected by TPR. This stabilizes the Ni^{2+} species and makes the production of active phase (metallic nickel) more difficult, causing a decrease in catalytic activity.

References

1. C. Fukuhara, H. Ohkura, K. Gonohe, A. Igarashi, Appl. Catal A: Gen. 279 (2005) 195.
2. V. Twigg, M. V. Loyd, D. E. Ridler, Catalyst Handbook, Wolfe Publishing, 1989.
3. D. S. Newsome, Catal Rev-Sci Eng, 21(1980) 275.
4. M. I. Temkim, Adv. Catal.,28 (1979) 263.
5. I. J. Lima, J. Millet, M. Aouine, M. C. Rangel, Appl. Catal A: Gen. 283 (2005) 91.
6. H.P.Klug,L.E. Alexander, X-Ray Diffraction Procedures,A Wiley-Interscience Publication.
7. C. D. Wagner, W. M. Riggs, L. E. Davis, J. F. Moulder, G. E. Muilenberg, Handbook of X-Ray Photoelectron Spectroscopy, Perkin-Elmer Coorporation Eden Prairie, 1978.
8. S. Ho, T. Chou, Ind. Eng. Chem. Res. 34 (1995) 2279.
9. A. Jones, B. McNicol, Temperature-Programmed Reduction for Solid Materials Characterization, Marcel Dekker, Inc. New York, 1986.
10. E. Ruckenstein, Y. H. Hu, J. Catal., 161 (1996) 55.
11. M. Parvary, S. H. Jazayeri, A. Taeb, C. Petit, A. Kiennemann, Catalysis Communications, 2 (2001) 357.
12. Z. Xu, Y. Li, J. Zhang, L. Chang, R. Zhou, Z. Duan, Appl. Catal A: Gen, 210 (2001) 45.
13. S. Wang, G. Q. Lu, Appl. Catal A: Gen, 169 (1998) 271.

Natural Gas Conversion VIII
F.B. Noronha, M. Schmal, E.F. Sousa-Aguiar (Editors)

Evaluation of Pd/La_2O_3 catalysts for dry reforming of methane

Luiz Carlos P. Fernandes Júnior[a], Sergio de Miguel[b], José Luís Garcia Fierro[c] and Maria do Carmo Rangel[a]*

*[a]GECCAT, Instituto de Química, Universidade Federal da Bahia, Campus Universitário de Ondina, 40170-290, Salvador, Bahia, Brazil. E-mail: *mcamov@ufba.br*
[b]Instituto de Investigaciones en Catálisis y Petroquímica - INCAPE, Santiago del Estero 2654, 3000 Santa Fe, Argentina
[c]Instituto de Catálisis y Petroquímica, CSIC, Cantoblanco, 28049, Madrid, Spain

Lanthana-supported palladium catalysts with different metal loads were evaluated in carbon dioxide of methane (dry reforming). A strong interaction between palladium and lanthana was noted. The catalysts produced low hydrogen to carbon monoxide molar ratio and thus the reaction can be combined with methane steam reforming to adjust this ratio for several purposes. The catalyst with 1% Pd was the most active leading to conversion above 80% and hydrogen to carbon monoxide molar ratio of 0.4.

1. Introduction

The interest for methane reforming with carbon dioxide (dry reforming) has been growing in the last years, because of its commercial and environmental importance [1,2]. The reaction converts two of the cheapest carbon-containing gases (methane and carbon dioxide) into useful chemical products, such as syngas, a mixture of hydrogen and carbon monoxide [1,3]. In addition, in certain applications, dry reforming has a number of advantages over steam reforming and is likely becoming an increasingly important industrial reaction in the future [1]. For instance, the products carbon monoxide and hydrogen with equal molar ratio are suitable for the Fischer-Tropsch synthesis, as compared to the products of the steam reforming or partial oxidation of methane [4].
Several metals are able to catalyze the carbon dioxide reforming of methane, such as nickel, ruthenium, rhodium and palladium, among others [5, 6]. Due to their low cost and availability, nickel-based catalysts are the most convenient for

the use in industry [7]. However, nickel is the most susceptible metal to deactivation by coke deposition, the main drawback of dry reforming, while noble metal-based catalysts are less sensitive to carbon deposition [5,6,8]. Therefore, the use of noble metal continues to be extensively investigated [6]. In the present work, lanthana-supported palladium catalysts were prepared and evaluated in carbon dioxide reforming of methane in order to find alternative catalysts to the reaction.

2. Experimental

The support (L) was prepared by adding simultaneously solutions of hexahydrated lanthanum nitrate ($La(NO_3)_3.6H_2O$) and ammonium nitrate (25 % v/v) to a beaker with water, under stirring at room temperature. The pH value of the mixture was kept at 9.0. After the addition of reagents, the precipitated was kept for 24 h, under stirring at room temperature and then centrifuged. The gel was washed with water, dried at 120 °C for 24 h, sieved and calcined under air flow, at 650 °C for 4 h. The support thus obtained was impregnated with a solution containing palladium chloride ($PdCl_2$) and chloridric acid (1 % v/v), which was used as competitor agent to improve the dispersion of palladium on the support. After 12 h in a evaporator, the sample was dried at 120 °C for 12 h, calcined under air flow the 650 °C, for 4 h and reduced under hydrogen flow at 400 °C, for 3 h. Catalysts with 0.5 % and 1.0 % (weight) of palladium were prepared (PL05 and PL1). The samples were characterized by thermogravimetry (TG), differential thermal analysis (DTA), X-ray diffraction (DRX), specific surface area and dispersion measurements, temperature programmed reduction (TPR) , Fourier transform infrared spectroscopy (FTIR) and X-ray photoelectron spectroscopy (XPS). The FTIR experiments were performed using a Perkin Elmer equipment, model Spectrum One, in the range of 400 to 4000 cm^{-1}. The DTA and TG experiments were carried out in a Mettler Toledo model TGA/SDTA851 equipment, under air flow (50 mL.min-1) with a heating rate of 10 °C.min^{-1} from room temperature to 1000 °C. X-ray diffractograms were obtained by a Shimadzu model XD3A equipment, using CuKα radiation (λ=1.5420 A) and a nickel filter, in the range of 2θ= 10-80°. The specific surface areas were measured by nitrogen adsorption-desorption experiments, using the BET method and a Micromeritics model ASAP 2020 equipment. The palladium dispersion was measured in the same equipment. The TPR experiments were performed using a Micromeritics model TPD/TPR 2900 equipment, heating the samples from room temperature up to 900 °C, at a rate of 10 °C.min^{-1}, under a 5%H_2/N_2 mixture. The XPS spectra were acquired with a VG Scientific spectrometer, Escalab model 220i-XL, with source of X-rays, MgKα (1253 eV) anode and 400 W power and hemispheric electron analyzer. The Al2p peak (BE = 74.5 eV) was chosen as an internal reference which was in all cases in good agreement with the BE of the C 1s peak. The powder samples were pressed into small stainless steel cylinders and mounted onto a manipulator which allowed the

transfer from the preparation chamber into the spectrometer. Before the analysis, they were outgassed (10^{-9} mbar) and reduced under hydrogen at 500 °C (1 h) .
The different catalysts were tested in the methane reforming with carbon dioxide, in a flow equipment at 1 bar. The reproducibility of the test was assured by carrying out the experiments three times. The samples (0.2 g) were first reduced under flowing hydrogen at 823 K, for 3 h and then were heated under flowing helium up to the reaction temperature 1023 K. After reaching the reaction temperature, the reaction mixture (CH_4/CO_2 (molar)=1) with a flowing rate of 20 $ml.min^{-1}$ was fed. In order to avoid diffusional effects, the catalyst particle sizes were very small (< 80 mesh). Each run took 300 min and the products were analyzed by using an on line chromatography (TCD) system containing a column Supelco Carboxen 1006 plot (30m x 0.53 mm).

3. Results and Discussion

The X-ray diffractograms of the catalysts are shown in Figure 1. The support showed peaks of the hexagonal lanthanum hydroxide (JCPDS 06-0585) and of the orthorhombic lanthanum nitrate hydroxide ($La(OH)_2NO_3$) (JCPDS 26-1144). After the addition of palladium and further calcination, the diffractograms showed the pattern of the cubic lanthanum oxide nitrate ($LaONO_3$) (JCPDS 23-1149) and of the hexagonal and cubic lanthanum oxide (La_2O_3) (JCPDS 05-0602 and JCPDS 22-0369), indicating that some nitrate species remained in the final solid, probably delaying the lanthanum oxide formation. The presence of nitrate in solids was confirmed by an absorption band at 1430 cm^{-1} in the spectra (not shown) [9]. In the palladium-doped solids, bands at around 1460 and 846 cm^{-1} confirmed the presence of chloride species [9]. These findings were confirmed by TG analysis, which showed the loss of these species at high temperatures.

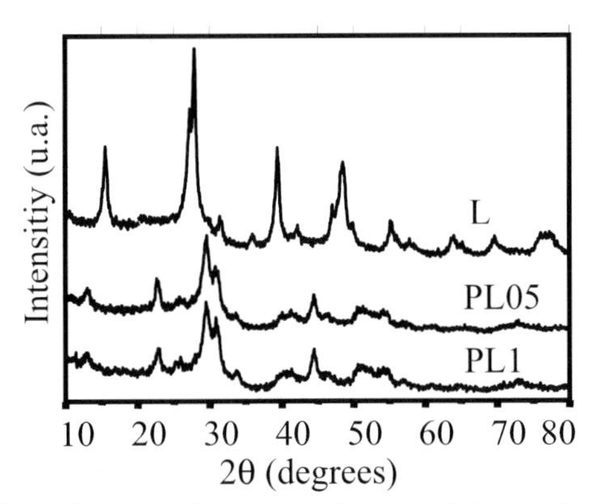

Figure 1. (a) X-ray diffractograms of the support (L) and of the catalysts. L:support; PL 05 and PL1: lanthana-supported palladium with 0.5 and 1% Pd, respectively.

The reduction profiles of the catalysts are shown in Figure 2. The support (L) showed a low temperature peak which can be assigned to the reduction of nitrate species in the solids [10] and a shoulder at around 700 °C, related to the partial reduction of lanthana [11].After the addition of palladium and further calcination, the TPR curves showed a broad peak in the range of 300-600 °C, which can be assigned to the reduction of nitrate species and a high temperature peak related to the reduction of palladium and lanthana in strong interaction [12,13]. It can be seen that a large amount of hydrogen was taken by nitrate species; this was confirmed by the amount of the hydrogen consumed by the support, which was higher than the palladium-containing catalysts (Table 1).

The binding energies (BE) of some characteristic core levels of lanthanum and palladium are displayed in Table 1. The binding energy for the La $3d_{5/2}$ peak was in close agreement with that for La^{3+} species in lanthanum oxide (La_2O_3) [14]. The binding energies for the Pd $3d_{5/2}$ peak are related to Pd^{2+} species (337.2 eV) and to metallic palladium (337.3 eV) [14]. A typical Pd 2p core-level spectrum of the catalysts is shown in Figure 2(b). It can be noted that the catalysts had the same proportion of theses species, regardless the metal load. On the other hand, they showed different amounts of palladium on the surface, being the PL1 sample the richest one in palladium.

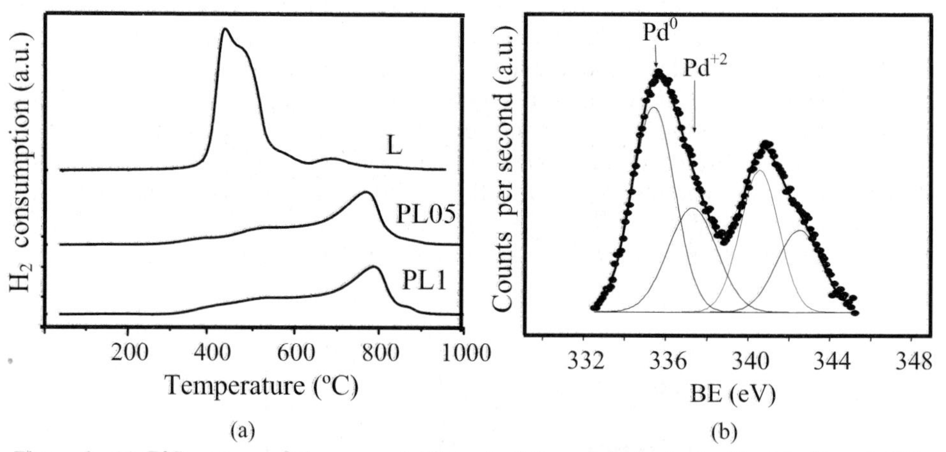

Figure 2. (a) TPR curves of the support (L) and of the catalysts. L: La_2O_3, PL 05 and PL1: lanthana-supported palladium with 0.5 and 1% Pd, respectively. (b) Typical Pd 2p core-level spectrum of the catalysts.

The specific surface areas of the catalysts decreased due to palladium addition (Table 1) and this was ascribed to the calcination of the solids after the metal impregnation. During this heating, lanthanum hydroxide changed to lanthanum oxide and the coalescence of pores and/or particles took place causing a decrease in specific surface areas. No difference was found in the specific surface areas of the samples due to different metal loads. On the other hand, the metallic dispersion decreased with the increase of the metal amount, as shown in Table 1.

Table 1. Binding energies (eV), surface composition and specific surface area of the catalysts. L: support, PL 05 and PL1: lanthana-supported palladium with 0.5 and 1% Pd, respectively.

Samples	H_2 uptake in TPR (mmol)	La 3d$_{5/2}$	Pd 3d$_{5/2}$	Pd/La (atom)	Sg ($m^2.g^{-1}$)	Dispersion (%)
L	2.88	835.0	-	-	27	--
PL05	1.79	835.2	335.4 (63) 337.3 (37)	0.052	10	19.9
PL1	1.49	834.8	335.3 (63) 337.2 (37)	0.301	9	10.6

The catalysts were active to dry reforming of methane and were stable during the time reaction (Figure 3). The activity increased with the amount of palladium in solids, a fact which can be assigned to the highest amount of palladium on the surface as noted by XPS experiments. Both catalysts showed low hydrogen to carbon monoxide molar ratio (Figure 4) and then they can be used in conjunction with the widely methane steam reforming when a ratio of the produced gas required should be less than that generated by steam reforming alone, as suggested previously [1]. The catalysts with the highest amount of palladium led to the highest H_2/CO ratio. It is probable that the reaction over palladium catalysts is a sensitive-structure one and then it seems that it would be necessary a determined ensemble of palladium atoms to be carried out. Hence the catalyst with the higher metal loading (1%) has a lower metallic dispersion, as observed by XPS (that is higher metallic particles) and a higher activity in dry reforming.

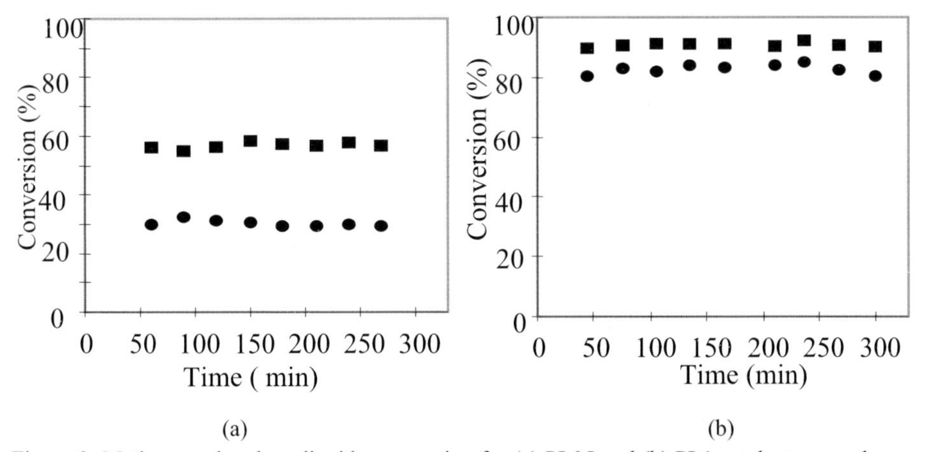

Figure 3. Methane and carbon dioxide conversion for (a) PL05 and (b) PL1 catalysts. • methane; ▪ carbon dioxide. PL05 and PL1 samples: lanthana-supported palladium with 0.5 and 1% Pd, respectively. The methane and carbon dioxide conversions at thermodynamic equilibrium (at the experimental conditions) were 88% and 98%, respectively.

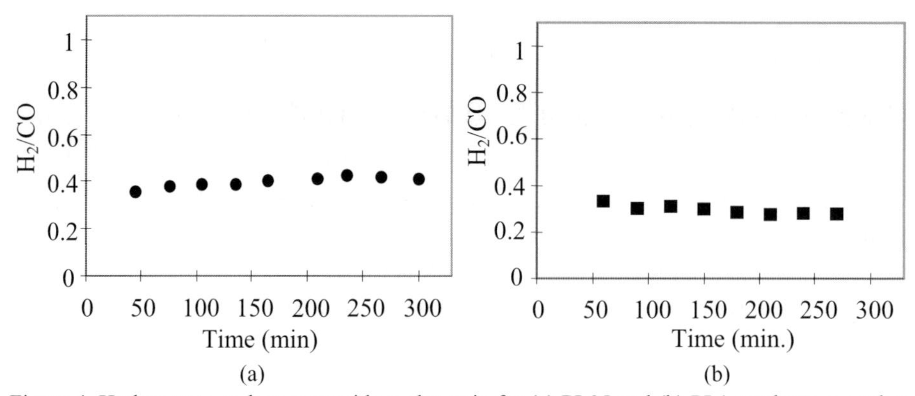

Figure 4. Hydrogen to carbon monoxide molar ratio for (a) PL05 and (b) PL1 catalysts. • methane; ▪ carbon monoxide. PL05 and PL1: lanthana-supported palladium with 0.5 and 1%Pd, respectively.

4. Conclusions

Lanthana-supported palladium is a promising catalyst to carbon dioxide reforming of methane. In solids with 0.5 and 1% of palladium, this metal strongly interacts with the support. The catalyst with 1% Pd is the most active leading to methane conversions above 80% and hydrogen to carbon monoxide molar ratio of 0.4. It can thus be combined with methane steam reforming to adjust the hydrogen to carbon monoxide molar ratio to several purposes.

References

[1] X. E. Verykios, Int. J. Hydrogen Energy 28 (2003) 1045.
[2] Y. H. Hu, E. Ruckenstein, Adv. Catal. 48 (2004) 297.
[3] C. Li, Y. Fu, G. Bian, Y. Xie, T. Hu, J. Zhang, Kinet. Catal., 45 (2004) 719.
[4] A. Slagtern, Y. Schuurma, C. Leclerg, X. E. Verykios, C. Mirodatos, J. Catal. 172 (1997) 118.
[5] J. R. Rostrup-Nielsen, Cat. Rev. Sci. Tech. 46 (2004) 247.
[6] Z. Hou, O. Yokota, T. Tanaka, T. Yashima, Appl. Catal. A: Gen 253 (2003) 381.
[7] L. Pelletier, D. D. S. Liu, Appl. Catal. A: Gen 317 (2007) 293.
[8] S. Wang and G. Q. Lu, Appl. Catal. B: Environm. 19 (1998) 267.
[9] A.Niquist, R. O. Kagel, Infrared Spectra of Inorganic Compounds, Academic Press, Orlando, 1971, p. 3.
[10] S. Ho, T. Chou, Ind. Eng. Chem. Res. 1995, 34, 2279;
[11] A. Jones and B. McNicol. Temperature-Programmed Reduction for Solid Materials Characterization. Marcel Dekker, Inc. New York. 1986. p. 137.
[12] S. Fuentes,N. E.Bogdanchikova, G.Diaz, M.Peraaza, G. C.Sandoval, Catal. Lett. 47 (1997) 27.
[13] A. Barrera, M. Viniegra, P. Bosch, V. H. Lara, S. Fuentes, Appl. Catal. B: Environm. 34 (2001) 97.
[14] C.D. Wagner, W.M. Riggs, L.E. Davis, J.F. Moulder and G.E. Muilenberg, Handbook of X-Ray Photoelectron Spectroscopy, Perkin-Elmer Coorporation, Eden Prairie, 1978, p. 110, 132.

Natural Gas Conversion VIII
F.B. Noronha, M. Schmal, E.F. Sousa-Aguiar (Editors)

SNOW: Styrene from Ethane and Benzene

Domenico Sanfilippo[a], Guido Capone[a], Alberto Cipelli[a], Richard Pierce[b], Howard Clark[b], Matt Pretz[b]
[a]*Snamprogetti S.p.A. Viale DeGasperi 16 - S. Donato Milanese, 20097, Italy*
[b]*The Dow Chemical Company, 2301 N. Brazosport Blvd., B-250 Freeport, TX 77541-3257 U.S.A.*

1. Introduction

Styrene is one of the most important monomers for the production of polymers, resins and rubbers. Styrene production exceeds 25 Million MT/y. The biggest consumer of Styrene monomer (SM) is Polystyrene (PS). Other major derivatives are expanded polystyrene (EPS), SB Latex, SB Rubber, styrene block co-polymers (eg ABS, MBS, SBS) and unsaturated polyester resins. SM is a commodity and a lower production cost is a critical factor in the value chain of its derivatives. The styrene market has been growing at a rate of 4.4% yearly for the last several decades. It is expected that worldwide capacity will expand by 5.5 million MT over the next five years
In the current art there are two commercial routes for the synthesis of SM:
a) The classic and largely applied EB/SM route: in a first step ethylbenzene (EB) is formed by alkylation of benzene with ethylene in the presence of acid catalyst (typically a zeolite). EB is catalytically dehydrogenated to SM in the second step. Unconverted EB is recycled back to the dehydrogenation section.

$$C_6H_6 + CH_2{=}CH_2 \Leftrightarrow C_6H_5\text{-}CH_2\text{-}CH_3 \Leftrightarrow C_6H_5\text{-}CH{=}CH_2 + H_2 \quad (1)$$

b) Alternatively and to a lower extent (<20% of the SM market) SM is co-produced together with propylene oxide (PO/SM route). The EB synthesis is followed by the EB peroxidation with the formation of the hydroperoxide. This latter reacts in a further step with propylene forming propylene oxide (PO) and ethylbenzyl alcohol that is dehydrated to styrene in a final step.

$$EB + O_2 \Rightarrow C_6H_5\text{-}CHOOH\text{-}CH_3 + C_3H_6 \Leftrightarrow PO + C_6H_5\text{-}CHOH\text{-}CH_3 \Rightarrow SM + H_2O \quad (2)$$

The conventional EB/SM route utilizes **benzene** and **ethylene** as raw materials and the core is represented by the EB catalytic dehydrogenation step. The continuous effort during several decades for improving this process in the areas of reactor technology and catalyst technology has brought it to a high degree of maturity and there is today little room for further improvement.

2. The innovative SNOW technology

The SNOW technology has been jointly developed by Snamprogetti and Dow (*SNOW* = *SN*amprogetti + D*OW*) and represents a technological and economical breakthrough in the Styrene Industry.

The SNOW technology is innovative as concerns many factors: raw materials, reactor design, heat supply system, catalyst, feed composition.

a) **Raw materials and reaction scheme**

The SNOW complex is fed with benzene and **ethane**, which is dehydrogenated in the same reaction system for EB dehydrogenation to produce the stoichiometric amount of ethylene necessary for the benzene alkylation (Fig. 1). In an alternative version the SNOW Unit can be fed with **ethylene** and **benzene**, similarly to the conventional technology (Fig 2: Ethylene Option).

The two Options (ethane or ethylene feed), allow the maximization of profit according to the location, price and availability of the raw materials.

The Ethane Option can add value to stranded or limited use gas streams. Ethane is a significant component of Natural Gas, and is also contained in some refinery and petrochemical streams (FCC and crackers off gas, by-product of liquid feed crackers). SNOW provides an opportunity to monetize Natural Gas without the need of an associated Ethylene Project, which may be attractive in some locations. Additionally, since this Option decouples SM production from Steam Cracker it becomes possible to build a new SM unit (or retrofit an existing conventional one) in refinery/petrochemical complexes without the need of debottlenecking the Steam Cracking plant.

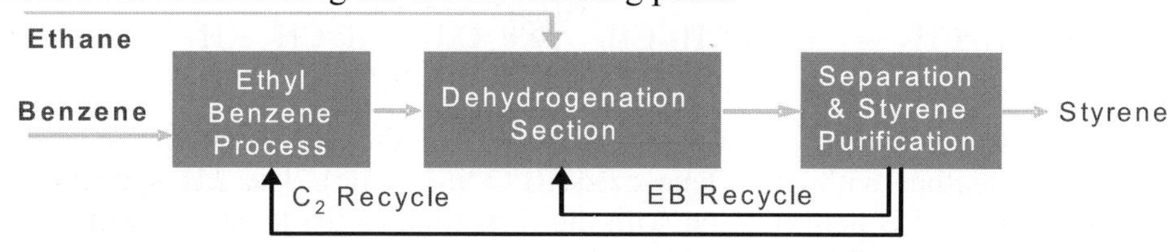

Fig. 1: Conceptual block diagram of SNOW technology (Ethane Option)

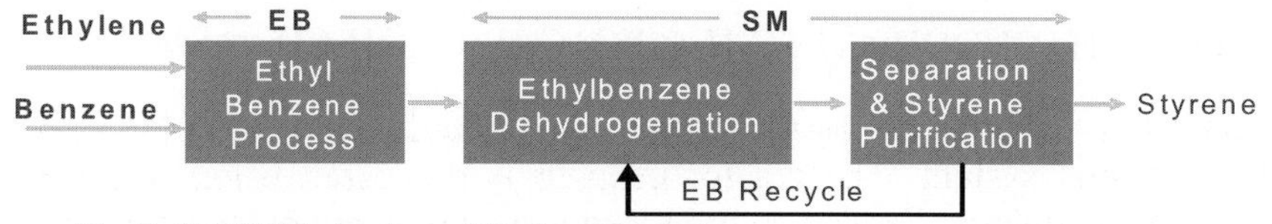

Fig. 2: SNOW Technology (Ethylene Option)

b) **Reactor Design**

Dehydrogenation of ethylbenzene to styrene and as well of ethane to ethylene is an endothermic, equilibrium-controlled reaction. Maximum conversion increases with temperature. In order to achieve reasonable economic conversion per pass, temperatures exceeding 600°C are a prerequisite.

Since the equilibrium conversion of ethane and EB are significantly different for a specific set of conditions, a suitable process design allows SNOW Ethane Option to produce per pass as much ethylene as EB converted.

The core of the SNOW technology is the reaction system, characterized by the use of fluidized beds with circulating catalyst, similarly to the well known and largely applied FCC technology (Fig. 3). The SNOW reactor is of Riser type, where the catalyst is fast moving upwards, entrained by the co-current hydrocarbons stream at a gas velocity of 4-20 m/s. Catalytic reactions are performed rapidly (1-5 seconds) in the riser. Temperature ranges between 590 and 700°C according to the feed type and riser level. Conventional technologies operate at EB partial pressures lower than one atmosphere (vacuum and steam dilution) to get a higher driving force and better selectivities. This is not required with SNOW.

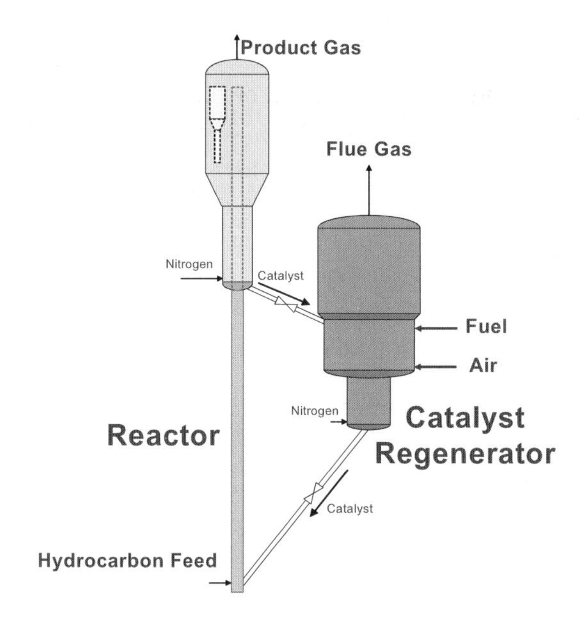

Fig. 3: SNOW reaction section

The products leave the disengaging zone after complete removal of the catalyst with a cyclone system. The reactor effluent is processed using conventional separation technology. EB can be produced using conventional technology.

c) **Heat Supply System**

The catalyst exiting the reactor is conveyed to the Regenerator (fig. 3), a bubbling fluidized bed where the initial catalyst activity is restored by burning the minor carbonaceous deposits built up on the catalyst surface during the reaction time and restructuring the active sites. Air is fed to the regenerator bottom countercurrently to the down-flowing catalyst. Flue gas leaves the disengaging zone of the regenerator through a cyclones system.

As already mentioned, the dehydrogenation of EB and ethane are highly endothermic:

$$C_6H_5\text{-}CH_2\text{-}CH_3 \Leftrightarrow C_6H_5\text{-}CH{=}CH_2{+}H_2 \quad \Delta H_r = 28.1 \text{ kcal/mole, } 264 \text{ kcal/kg} \quad (1)$$

$$CH_3\text{-}CH_3 \quad \Leftrightarrow \quad CH_2{=}CH_2 + H_2 \quad \Delta H_r = 32.7 \text{ kcal/mole, } 1086 \text{ kcal/kg} \quad (2)$$

The endothermicity results in a high heat demand and requires high heat fluxes at temperatures higher than 590°C along the reaction coordinates. In the SNOW technology this heat is supplied by the heat capacity of the circulating catalyst: "hot" catalyst, heated up in the regenerator at a temperature higher than that required for reaction, goes to the reactor. The catalyst cools down in the riser reactor during the reaction and flows back "cold" to the regenerator. Indeed in order to satisfy the heat balance and generate the heat necessary to the reaction loop in the regenerator some fuel is catalytically burnt, preferentially the hydrogen rich off gas of the process is used. It increases significantly the carbon efficiency of the process because of the low CO_2 emissions in the flue gas.

The high surface of catalyst ensures a very efficient heat exchange directly in the process side without any intermediate wall. Of course heat is recovered from the reactor/regenerator effluents. Since the heat generation in the reaction loop takes place through catalytic combustion of fuel gas at relatively low temperature, SNOW is characterized by very low CO/NO_x emissions.

d) **Catalytic System Design**

Conventional dehydrogenation technologies in the prior art have selected promoted iron oxide as a unique catalyst to be operated under low EB partial pressure and a significant dilution with steam.

For the SNOW project several candidates as active phase have been considered and tested. The chosen formulation is based on promoted Gallium oxide. Gallium systems are little known in scientific and patent literature. A new chemistry and catalysis has been developed for the SNOW technology.

When duly promoted, gallium catalyst performance is excellent for ethane as well as EB dehydrogenation. The catalyst is so active that it is possible to operate it with very short contact time, typical of a fast riser reactor. The selectivities to ethylene and styrene respectively are so high that no steam dilution is required. Thus the SNOW technology has the unique advantage of avoiding the steam cycle (generation, condensation, heat exchangers etc.). This

aspect is particularly important because it allows significant benefits in energy saving and CO_2 emission reduction. Coke formation is very low and the catalyst circulation provides a continuous regeneration from coke build up.
The commercial catalyst is a microspheroidal, attrition resistant, high heat capacity solid, synergistically designed and developed together with the reactor engineering. Catalyst PSD (Particle Size Distribution) and mechanical properties have been optimized for the use in a fluidized bed and make this catalyst more resistant than the typical FCC catalysts.
The SNOW technology allows, like in the other fluidized bed processes, the continuous make up of fresh catalyst, maintaining its "equilibrium" activity stable over the time. No turndown for catalyst substitution is required and no aging from Start-Of-Run to End-of-Run conditions is present.

3. SNOW Technology development steps

Snamprogetti and Dow joined their effort for developing the SNOW technology in early 2000. The initial technology development and engineering skills of Snamprogetti [1-3], and the complementary styrene production and marketing expertise of Dow have been integrated[4-7]. R&D activity has been carried out initially in microreactors and bench scale units. Results have been validated at pilot plant level utilizing a DCR (Davison Cracking Reactor), typically adopted for FCC catalyst testing.
The next step has been the construction and operation on a Dow site of a PDU (Process Demonstration Unit, fig. 8) that has allowed testing all aspects of the new technology including process yields, conversion and economics, SM product quality for derivatives application, reactor and catalyst reliability and so on. The PDU is on a scale intended to prove critical unit operations in both performance and reliability. The capacity (feed rates in the 500+ kg/hr range) is of sufficient size to minimize the risk of scale-up of critical unit operations. The PDU operates at the same conditions as a world scale unit, but has allowed enough variation around process parameters in order to optimize the process. Hydrodynamics have been modeled and demonstrated in large scale mock ups (cold flow models). Catalyst production has been successfully scaled up from laboratory formulations to commercial scale manufacture. Several tens of tons of catalyst have been produced in full scale equipment. The PDU operation has validated the catalyst performances, the scale up criteria and the whole process.

4. Final considerations

The Dow Chemical Company and Snamprogetti S.p.A. have synergistically developed a new route to produce SM from ethane and benzene (SNOW "Ethane Option") or ethylene and benzene (SNOW "Ethylene Option") which Dow/Snamprogetti will be uniquely offering in the market to styrene producers.

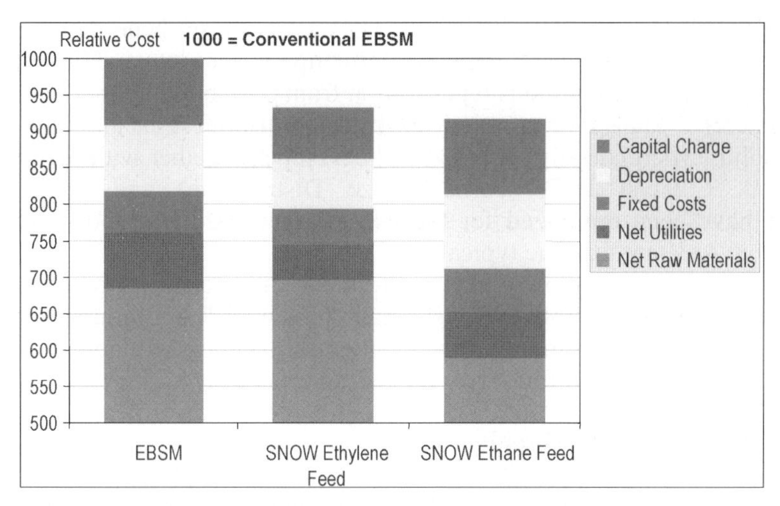

Fig. 4: Example of economics of SNOW (Ethane/Ethylene Options) relative to conventional

The innovative SNOW process is expected to enable significant cost savings in both versions from benzene-ethane and from benzene-ethylene.
Fig. 4 reports an example of the SNOW advantage versus the conventional EB/SM route. Of course the advantage may change with the plant location due to the different valorization of raw material, energy and equipment cost.
The "Ethane Option" moves away from the conventional Styrene production process and raw materials, captures ethane-ethylene price differential and eliminates the need for upstream investment in ethylene production at a steam cracker or ethylene purchases and offers greater location flexibility.
The "Ethylene Option" offers significant advantage in investment cost versus conventional approach.
SNOW will use less energy (no dilution steam) and emit less CO_2 than competing processes (no fuel gas, instead burns H_2 produced in the process).

Acknowledgements
The mentioned authors are the iceberg tip of an impressive number of colleagues in Dow and Snamprogetti and EniTecnologie that have been working in this project around the World with competence, passion, and creativity.

References:

1. F. Buonomo, D. Sanfilippo, R. Iezzi, E. Micheli USP 5,994,258 and USP 6,242,660 (1997) to Snamprogetti
2. F. Buonomo, G. Donati, E. Micheli, L. Tagliabue USP 6,031,143 (1997) to Snamprogetti
3. D. Sanfilippo, R. Iezzi USP 6,841,712 (1999) to Snamprogetti
4. I. Miracca, G. Capone USP 7,094,940 (2001) to Snamprogetti
5. D. Sanfilippo, A. Bartolini, R. Iezzi US Appl. 2004/0259727 to Snamprogetti
6. D. Sanfilippo, I. Miracca, G. Capone, V. Fantinuoli US Appl. 2005/01776016
7. W. Castor, S. Hamper, M. Pretz, S. Domke WO 2005/077867 to Dow

Natural Gas Conversion VIII
F.B. Noronha, M. Schmal, E.F. Sousa-Aguiar (Editors)

Designing Pt catalysts by sol-gel chemistry: influence of the Pt addition methods on catalyst stability in the partial oxidation of methane

A.P. Ferreira[a], V. B. Mortola[a], R. Rinaldi[b], U. Schuchardt[b], J.M.C. Bueno[a]

[a] *DEQ, Universidade Federal de São Carlos, PO Box 676, 13565-905 São Carlos, SP, Brazil. * e-mail: jmcb@power.ufscar.br.*
[b] *Instituto de Quimica, Universidade Estadual de Campinas, PO Box 6154, 13083-970 Campinas, SP, Brasil.*

Abstract: The effect of the preparation method of Pt/Al_2O_3 and Pt/CeO_2-Al_2O_3 catalysts, containing 12 wt % of CeO_2, on stability in the partial oxidation of methane was evaluated. Both supports and Pt catalysts were obtained by a sol-gel method. The preparation method and support composition have a strong influence on the agglomeration of Pt. Ce-containing catalysts, obtained by the one-step sol-gel method, show high stability in the partial oxidation of methane at 1073 K.

1. INTRODUCTION

Partial oxidation of methane (POM) is a promising process for producing synthesis gas from methane. In a previous work [1] it was shown that the activity and stability of Pt/CeO_2-Al_2O_3 catalysts for POM strongly depended on the Ce loading. The catalysts with high CeO_2 loading (12 wt %) showed both higher activity and higher stability.

The sol-gel method is an interesting route for synthesis of materials with controlled textural and structural properties [2, 3]. The sol-gel preparation of catalytic materials allows to tailor several properties of materials for catalysis, such as: (i) purity; (ii) textural properties, including samples with different physical forms; (iii) compositional homogeneity at the molecular level and (iv) addition of several components in a single-step synthesis, which provides unique oxide-oxide and/or metal-oxide interactions [4].

The so called one-step preparation of supported metal catalysts improved the catalytic performance of the metallic phase [5]. Pt sol formation, after adding the Pt precursor to the alumina or alumina-ceria sol, results in Pt nanoparticles with the same crystallite size, independent of Pt loading. The metals incorporated in the supports also have a modest effect on the structural properties.

This work attempts to show the influence of the Pt addition method on the properties and stability of CeO_2-Al_2O_3-supported Pt catalysts obtained by the sol-gel method. The stability of the incorporated Pt is studied in the partial oxidation of methane at 1073 K.

2. EXPERIMENTAL

The Al_2O_3 (Al) and 12 wt % CeO_2-Al_2O_3 (12CeAl) supports were prepared by the hydrolysis of Al(O-*s*Bu)$_3$ in ethanol, with addition of an aqueous solution of $Ce(NO_3)_3.6H_2O$, as described elsewhere [2]. The gels were dried at room temperature for 48 h. These gels were heated flowing air at 10 K min^{-1} from room temperature to 1223 K and maintain at 1223 K for 6 h.

The catalysts were prepared by two methods: (i) impregnation of the supports, calcined at 1223 K, by ethanolic solutions of the Pt-precursor [H_2PtCl_6 or Pt(acac)$_2$]; (ii) one-step preparation by the addition of the Pt-precursor during the Al(O-*s*Bu)$_3$ hydrolysis step. The different catalysts prepared are shown in Table 1. The Pt catalysts supported on alumina and on mixed CeO_2-Al_2O_3 are denominated Pt/Al and Pt/12CeAl, respectively.

For all samples the amount of Pt was around 1 wt % and before catalytic test the samples were treated under flowing air at 390 K for 1 h and then at 773 K for 4 h and finally reduced in H_2 at 773 K for 2h. The catalysts samples were aged flowing air at 1073 K for 24 h. The catalysts were characterized by adsorption-desorption of N_2 at 77 K, X-ray diffraction (XRD) and temperature programmed reduction (TPR-H_2). CO adsorption was investigated using Fourier transform infrared spectroscopy (FTIR). The catalyst Pt-12CeAl obtained by the one-step preparation was characterized under temperature-programmed reduction and under POM conditions using X-ray absorption near-edge structure (XANES) of the Pt L3-edge. POM and the stability evaluation of the Pt-containing catalyst were carried out at 1073K using a reactant mixture of CH_4:O_2 (2:1 mol/mol) and a flow rate of 100 cm^3 min^{-1}.

3. RESULTS AND DISCUSSION

The textural properties and the crystallite size of CeO_2 on the catalytic supports (Al and 12CeAl) and Pt-supported catalysts (Pt/Al, Pt/12CeAl) obtained by wet-impregnation and one-step sol-gel synthesis are summarized in Table 1.

Table 1. Properties of the supports and the Pt-catalysts.

Series	Sample	Pt-precursor	Preparation method	S_{BET} ($m^2 g^{-1}$)	V_p ($cm^3 g^{-1}$)	D_p (nm)	CeO_2 (nm)
-	Al	-	-	325 (1223)	0.5	9.6	-
-	12CeAl	-	-	135 (773) 87 (1223)	0.2	5.8 (1223)	6.0 12.4
A	A-Pt/Al	H_2PtCl_6	Wet-impregnation	90 (773)	0.4	6.7	-
	A-Pt/12CeAl			85 (773)	0.2	5.8	9.3
B	B-Pt/Al	$Pt(acac)_2$		87 (773)	0.2	6.7	-
	B-Pt/12CeAl			67 (773)	0.1	3.9	11.0
C	C-Pt/Al	H_2PtCl_6	One-step sol-gel synthesis	276 (773)	0.3	3.6	-
	C-Pt/12CeAl			252 (773)	0.2	3.6	6.0
D	D-Pt/Al	$Pt(acac)_2$		277 (773)	0.3	3.9	-
	D-Pt/12CeAl			220 (773)	0.2	3.6	5.0

Values in parenthesis correspond to the calcination temperatures (K); n.d. – not determined.

The porosity of the materials is non-structural as indicated by the type V isotherm with H2 hysteresis. The pore systems are formed by intraparticular slits and have narrow pore size distributions, which depends on the preparation method. The addition of Ce causes the decrease in the specific surface area (S_{BET}), pore volume (V_p) and average pore size (D_p). The decrease in S_{BET} and V_p is also seen after the impregnation of the supports Al and 12CeAl with the Pt-precursor in the catalysts prepared by wet-impregnation (series A and B). On the other hand, a remarkable increase of S_{BET} and V_p, relative to the unloaded support, 12CeAl, is observed for Pt12CeAl catalysts obtained in the one-step sol-gel synthesis (series C and D).

Uncalcined Al, 12CeAl and Pt/12CeAl (series C and D) materials have a pseudo-boehmite-like structure. The Al precursor after calcination at 773 K results in the γ-Al_2O_3 phase. The 12CeAl precursor after calcination at 773 K results in a mixture of γ-Al_2O_3 and CeO_2 with fluorite-like structure. The crystallite size of CeO_2 in the 12CeAl supports and Pt/12CeAl catalysts is shown in Table 1. XRD patterns of Pt/12CeAl catalysts obtained by calcination at 773 K of the wet-impregnation precursors (series A and B) show characteristic diffraction patterns of γ-Al_2O_3 and CeO_2 in a fluorite-like structure, which are similar to the unloaded supports. However, the diffraction patterns of Pt/12CeAl catalysts obtained by the one-step sol-gel method (series

C and D) have lower intensities compared to 12CeAl unloaded supports for the corresponding phases. The reflection of Pt species (2θ = 39°) was detected for Pt/Al catalyst calcined at 773 K (Fig. 1a). Nevertheless, lower Pt agglomeration and lower crystallinity of Al_2O_3 are observed in sample C (Fig. 1a), obtained by the one-step method. The reflection for Pt decreases significantly in Ce-containing catalysts (Fig. 1b). The reflection for Pt is also detected for A- and B-Pt/12CeAl catalysts obtained by impregnation and calcined at 773 K (Fig 1b); however, it is not observed in the C- and D-Pt/12CeAl catalysts obtained by the one-step sol-gel method and calcined at 773 K (Fig. 1b). The aging of the Pt/12CeAl catalysts at 1073 K for 24 h results in agglomeration of Pt as revealed by the reflection at 2θ = 39°. The process of agglomeration of Pt at higher temperatures seems to be stronger for the Pt/12CeAl catalyst obtained by wet-impregnation.

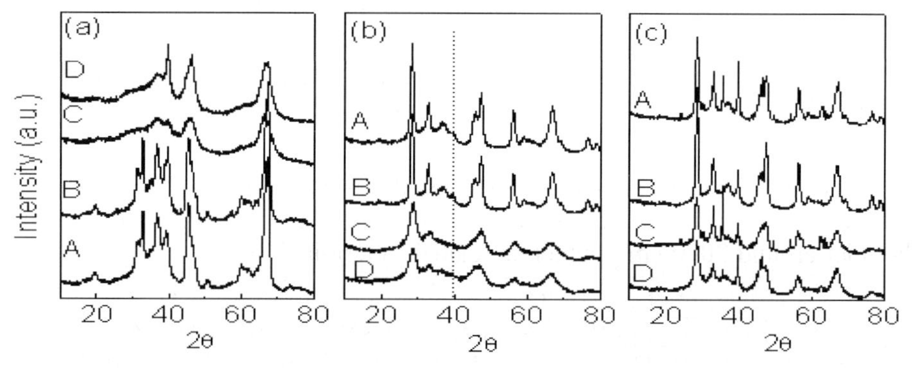

Fig. 1 XRD spectra of Pt/Al and Pt/12CeAl catalysts: a) corresponds to Pt/Al catalysts after calcination at 773 K of precursors obtained by the different methods, A, B, C and D, shown in Table 1; b) corresponds to Pt/12CeAl samples after calcination at 773 K and c) corresponds to Pt/12CeAl catalysts aged for 24 h in air at 1073 K. The dotted line in Fig 1b indicates a characteristic reflection for Pt.

TPR profiles of Pt/Al catalyst obtained by the one-step sol-gel method reveal that H_2 consumption is lower than that required for the reduction of the entire Pt loading, suggesting that, after calcination at 773 K, Pt is mainly in the reduced form. Similar results are observed for Pt/12CeAl samples. The oxidized contents of Pt are slightly higher for A- and B-Pt/12CeAl catalysts, which were obtained by the impregnation method. The TPR profile of CeO_2 shows peaks around 720 K and 1200 K, which are assigned, respectively, to the reduction of exposed oxygens and to the reduction of bulk ceria by the elimination of O^{2-} anions of the lattice and the formation of Ce_2O_3 [1]. The intensities of each H_2 consumption peak for reduction of CeO_2 depends strongly on the crystallite sizes of this phase. The A- and B-Pt/12CeAl catalysts obtained by impregnation show only one peak of H_2 consumption at 1200 K, which is attributed to the reduction of bulk CeO_2 to Ce_2O_3. For the C- and D-Pt/12CeAl catalysts these peaks around 1200 K have lower intensities than for A- and B-Pt/12CeAl catalysts, suggesting a smaller crystallite size of CeO_2, which is in agreement with XRD results (Table 1).

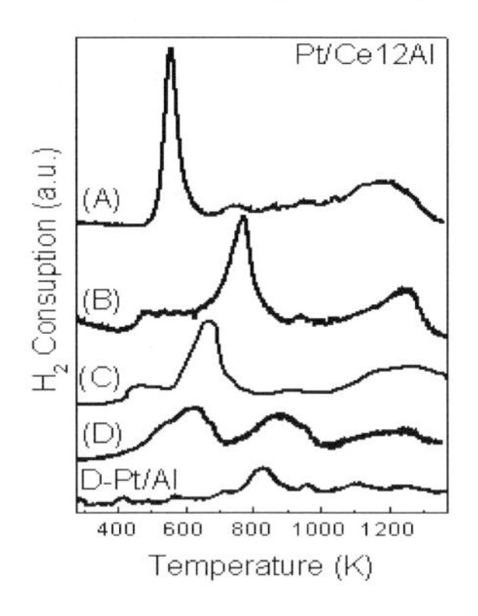

Fig. 2. TPR of D-Pt/Al and A-,B-,C- and D-Pt/12CeAl.

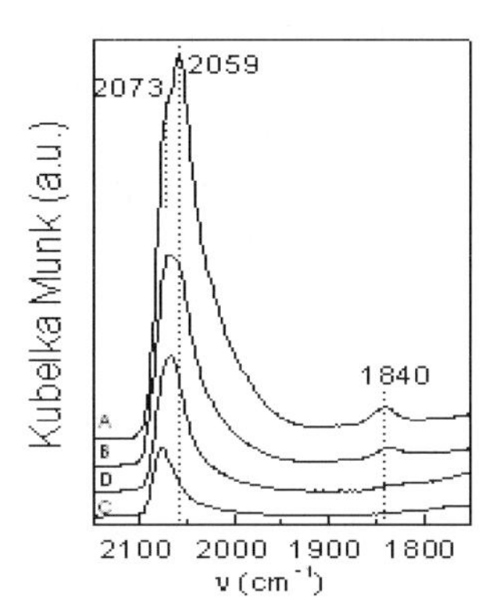

Fig. 3. FTIR of CO adsorbed on A-,B-,C- and D-Pt/12CeAl catalysts.

Fig. 3 shows the CO adsorption spectra for Pt/12CeAl catalysts calcined and reduced at 773 K. The spectra show bands around 2060 cm^{-1} with a shoulder around 2070 cm^{-1}, which is related to terminal-bonded CO and a band at 1840 cm^{-1} due to bridged-bonded CO [6]. The shoulder around 2070 cm^{-1} is related to CO adsorption on the most densely-packed planes [6]. The band at 1840 cm^{-1}, which is related to CO adsorption on the terrace sites [6], is stronger in the spectra of A- and B-Pt/12CeAl catalysts, but is practically absent in the C- and D-Pt/12CeAl catalysts. The total intensity of CO adsorption bands decrease in the order: A->B->D->C-Pt/12CeAl.

Fig. 4 shows the profiles of the catalyst stability test for B-Pt/Al and A-, B-, C- and D-Pt/12CeAl during the POM at 1073 K. A-Pt/Al catalyst shows a deactivation profile with time on stream in POM similar to that seen for B-PtAl catalyst. The A- and B-Pt/Al catalysts, obtained by the impregnation method, show a more pronounced deactivation compared to Pt/Al catalysts obtained by the one-step sol-gel method. The Ce-containing catalysts obtained by the impregnation methods show significant deactivation, while higher stability is found for Pt/12CeAl catalysts obtained by the one-step sol-gel method (Fig. 4). The TEM results (data not shown) suggest that the deactivation of the Pt/Al catalysts in POM reactions at 1073 K is due to the agglomeration of Pt and to carbon deposition on the Pt surface, while for C- and D-Pt/12CeAl catalyst the deactivation is likely to occur only by Pt agglomeration.

Fig. 5 shows the Pt L_3-edge XANES spectra of D-Pt/12CeAl calcined in air at 773 K, which demonstrates that the absorption at 11594 eV is significantly more intense than for the sample activated in H_2 at 773 K. This absorption becomes less intense at 1073 K for the sample used in POM. The white line intensity of Pt in POM is similar to that of Pt foil, suggesting that Pt is in the Pt^0 state. These results show that one fraction of PtO_x is reduced by H_2 at 773 K and another fraction of PtO_x, which is strongly interacting with the support, is reduced in POM at 1073 K. The high stability of D-Pt/12CeAl is assigned to the high interaction of PtO_x with the support, which is obtained when the catalyst is prepared by the one-step sol-gel method.

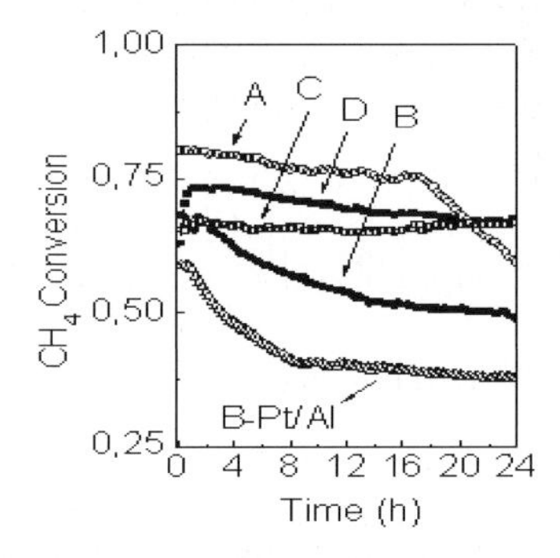

Fig. 4. Stability test of B-Pt/Al, and A-,B-,C-, D-Pt/12CeAl for POM at 1073 K.

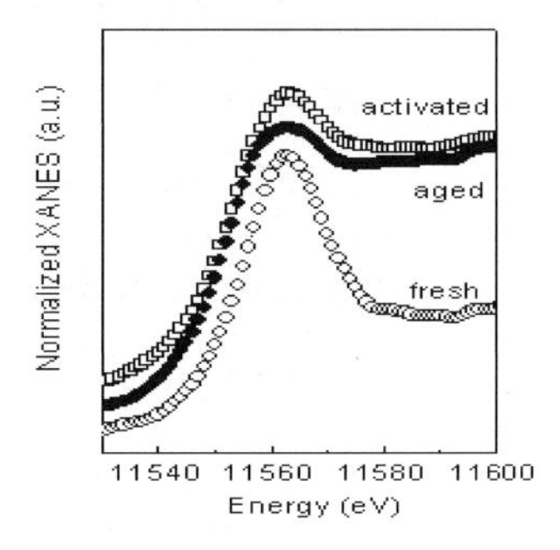

Fig. 5. Pt L_3-edge XANES spectra of the D-Pt/12CeAl sample: fresh (calcined in air at 773 K); activated (reduced in H_2 at 773 K) and aged (*in situ* POM at 1073 K).

4. CONCLUSION

The results reveal that the preparation method and composition of the support have strong influence on the agglomeration of Pt. Pt agglomeration on the Pt/12CeAl catalysts is favored by high temperatures in an oxidant atmosphere. Higher stability is obtained for Pt/12CeAl catalysts that are prepared by the one-step method. Ce-containing catalysts obtained by the one-step method show high stability in POM at 1073 K.

5. REFERENCES

1. A.C.S.F. Santos, S. Damyanova, G.N.R. Teixeira, L.V. Mattos, F.B. Noronha, F.B. Passos, J.M.C. Bueno. Appl. Catal. A: Gen. 290, 2005, 123.
2. X. Bokhimi; J. Sanchez-Valente; F. Pedraza J. Solid State Chem. 2002, 166, 182.
3. R. Rinaldi, F.Y.Fujiwara, W. Hoelderich, U. Schuchardt, J. Catal. 244, 2006, 92.
4. D. A. Ward, E. I., Ko, Ind. Eng. Chem. Res. 34, 1995, 421.
5. C.K., Lambert, R.D. Gonzalez. Appl. Catal. A: Gen. 172, 1998, 253.
6. B. A. Riguetto, S. Damianova, G. Gouliev, C. M. P. Marques, A. Petrov, J.M.C. Bueno. J. Phys. Chem. 108, 2004, 5349.

STUDIES IN SURFACE SCIENCE AND CATALYSIS

Volume 1 **Preparation of Catalysts I.** Scientific Bases for the Preparation of Heterogeneous Catalysts. Proceedings of the First International Symposium, Brussels, October 14–17,1975
edited by **B. Delmon, P.A. Jacobs and G. Poncelet**

Volume 2 **The Control of the Reactivity of Solids.** A Critical Survey of the Factors that Influence the Reactivity of Solids, with Special Emphasis on the Control of the Chemical Processes in Relation to Practical Applications
by **V.V. Boldyrev, M. Bulens and B. Delmon**

Volume 3 **Preparation of Catalysts II.** Scientific Bases for the Preparation of Heterogeneous Catalysts. Proceedings of the Second International Symposium, Louvain-la-Neuve, September 4–7, 1978
edited by **B. Delmon, P. Grange, P. Jacobs and G. Poncelet**

Volume 4 **Growth and Properties of Metal Clusters.** Applications to Catalysis and the Photographic Process. Proceedings of the 32nd International Meeting of the Société de Chimie Physique, Villeurbanne, September 24–28, 1979
edited by **J. Bourdon**

Volume 5 **Catalysis by Zeolites.** Proceedings of an International Symposium, Ecully (Lyon), September 9–11, 1980
edited by **B. Imelik, C. Naccache, Y. BenTaarit, J.C. Vedrine, G. Coudurier and H. Praliaud**

Volume 6 **Catalyst Deactivation.** Proceedings of an International Symposium, Antwerp, October 13–15,1980
edited by **B. Delmon and G.F. Froment**

Volume 7 **New Horizons in Catalysis.** Proceedings of the 7th International Congress on Catalysis, Tokyo, June 30–July 4, 1980. Parts A and B
edited by **T. Seiyama and K. Tanabe**

Volume 8 **Catalysis by Supported Complexes**
by **Yu.I. Yermakov, B.N. Kuznetsov and V.A. Zakharov**

Volume 9 **Physics of Solid Surfaces.** Proceedings of a Symposium, Bechyňe, September 29–October 3,1980
edited by **M. Láznička**

Volume 10 **Adsorption at the Gas-Solid and Liquid-Solid Interface.** Proceedings of an International Symposium, Aix-en-Provence, September 21–23, 1981
edited by **J. Rouquerol and K.S.W. Sing**

Volume 11 **Metal-Support and Metal-Additive Effects in Catalysis.** Proceedings of an International Symposium, Ecully (Lyon), September 14–16, 1982
edited by **B. Imelik, C. Naccache, G. Coudurier, H. Praliaud, P. Meriaudeau, P. Gallezot, G.A. Martin and J.C. Vedrine**

Volume 12 **Metal Microstructures in Zeolites.** Preparation–Properties–Applications. Proceedings of a Workshop, Bremen, September 22–24, 1982
edited by **P.A. Jacobs, N.I. Jaeger, P. Jirů and G. Schulz-Ekloff**

Volume 13 **Adsorption on Metal Surfaces.** An Integrated Approach
edited by **J. Bénard**

Volume 14 **Vibrations at Surfaces.** Proceedings of the Third International Conference, Asilomar, CA, September 1–4, 1982
edited by **C.R. Brundle and H. Morawitz**

Volume 15 **Heterogeneous Catalytic Reactions Involving Molecular Oxygen**
by **G.I. Golodets**

Volume 16 **Preparation of Catalysts III.** Scientific Bases for the Preparation of Heterogeneous Catalysts. Proceedings of the Third International Symposium, Louvain-la-Neuve, September 6–9, 1982
edited by **G. Poncelet, P. Grange and P.A. Jacobs**

Volume 17 **Spillover of Adsorbed Species.** Proceedings of an International Symposium, Lyon-Villeurbanne, September 12–16, 1983
edited by **G.M. Pajonk, S.J. Teichner and J.E. Germain**

Volume 18 **Structure and Reactivity of Modified Zeolites.** Proceedings of an International Conference, Prague, July 9–13, 1984
edited by **P.A. Jacobs, N.I. Jaeger, P. Jirů, V.B. Kazansky and G. Schulz-Ekloff**

Volume 19 **Catalysis on the Energy Scene.** Proceedings of the 9th Canadian Symposium on Catalysis, Quebec, P.Q., September 30–October 3, 1984
edited by **S. Kaliaguine and A. Mahay**

Volume 126 **Catalyst Deactivation 1999** Proceedings of the 8th International Symposium, Brugge, Belgium, October 10–13, 1999
edited by **B. Delmon and G.F. Froment**

Volume 127 **Hydrotreatment and Hydrocracking of Oil Fractions** Proceedings of the 2nd International Symposium/7th European Workshop, Antwerpen, Belgium, November 14–17, 1999
edited by **B. Delmon, G.F. Froment and P. Grange**

Volume 128 **Characterisation of Porous Solids V** Proceedings of the 5th International Symposium on the Characterisation of Porous Solids (COPS-V), Heidelberg, Germany, May 30–June 2, 1999
edited by **K.K. Unger, G. Kreysa and P. Baselt**

Volume 129 **Nanoporous Materials II** Proceedings of the 2nd Conference on Access in Nanoporous Materials, Banff, Alberta, Canada, May 25–30, 2000
edited by **A. Sayari, M. Jaroniec and T.J. Pinnavaia**

Volume 130 **12th International Congress on Catalysis** Proceedings of the 12th ICC, Granada, Spain, July 9–14, 2000
edited by **A. Corma, F.V. Melo, S. Mendioroz and J.L.G. Fierro**

Volume 131 **Catalytic Polymerization of Cycloolefins** Ionic, Ziegler-Natta and Ring-Opening Metathesis Polymerization
by **V. Dragutan and R. Streck**

Volume 132 **Proceedings of the International Conference on Colloid and Surface Science, Tokyo, Japan, November 5–8, 2000**
25th Anniversary of the Division of Colloid and Surface Chemistry, The Chemical Society of Japan
edited by **Y. Iwasawa, N. Oyama and H. Kunieda**

Volume 133 **Reaction Kinetics and the Development and Operation of Catalytic Processes** Proceedings of the 3rd International Symposium, Oostende, Belgium, April 22–25, 2001
edited by **G.F. Froment and K.C. Waugh**

Volume 134 **Fluid Catalytic Cracking V** Materials and Technological Innovations
edited by **M.L. Occelli and P. O'Connor**

Volume 135 **Zeolites and Mesoporous Materials at the Dawn of the 21st Century** Proceedings of the 13th International Zeolite Conference, Montpellier, France, July 8–13, 2001
edited by **A. Galameau, F. di Renso, F. Fajula and J. Vedrine**

Volume 136 **Natural Gas Conversion VI** Proceedings of the 6th Natural Gas Conversion Symposium, June 17–22, 2001, Alaska, U.S.A.
edited by **J.J. Spivey, E. Iglesia and T.H. Fleisch**

Volume 137 **Introduction to Zeolite Science and Practice.** 2nd completely revised and expanded edition
edited by **H. van Bekkum, E.M. Flanigen, P.A. Jacobs and J.C. Jansen**

Volume 138 **Spillover and Mobility of Species on Solid Surfaces**
edited by **A. Guerrero-Ruiz and I. Rodriquez-Ramos**

Volume 139 **Catalyst Deactivation 2001.** Proceedings of the 9th International Symposium, Lexington, KY, U.S.A, October 2001
edited by **J.J. Spivey, G.W. Roberts and B.H. Davis**

Volume 140 **Oxide-based Systems at the Crossroads of Chemistry.** Second International Workshop, October 8–11, 2000, Como, Italy.
Edited by **A. Gamba, C. Colella and S. Coluccia**

Volume 141 **Nanoporous Materials III** Proceedings of the 3rd International Symposium on Nanoporous Materials, Ottawa, Ontario, Canada, June 12–15, 2002
edited by **A. Sayari and M. Jaroniec**

Volume 142 **Impact of Zeolites and Other Porous Materials on the New Technologies at the Beginning of the New Millennium** Proceedings of the 2nd International FEZA (Federation of the European Zeolite Associations) Conference, Taormina, Italy, September 1–5, 2002
edited by **R. Aiello, G. Giordano and F. Testa**

Volume 143 **Scientific Bases for the Preparation of Heterogeneous Catalysts** Proceedings of the 8th International Symposium, Louvain-la-Neuve, Leuven, Belgium, September 9–12, 2002
edited by **E. Gaigneaux, D.E. De Vos, P. Grange, P.A. Jacobs, J.A. Martens, P. Ruiz and G. Poncelet**

Volume 144 **Characterization of Porous Solids VI** Proceedings of the 6th International Symposium on the Characterization of Porous Solids (COPS-VI), Alicante, Spain, May 8–11, 2002
edited by **F. Rodriguez-Reinoso, B. McEnaney, J. Rouquerol and K. Unger**

Volume 145 **Science and Technology in Catalysis 2002** Proceedings of the Fourth Tokyo Conference on Advanced Catalytic Science and Technology, Tokyo, July 14–19, 2002
edited by **M. Anpo, M. Onaka and H. Yamashita**

Volume 157 **Zeolites and Ordered Mesoporous Materials** Progress and Prospects
edited by **J. Čejka and H. van Bekkum**

Volume 158 **Molecular Sieves: From Basic Research to Industrial Applications.** Proceedings of the 3rd International Zeolite Symposium (3rd FEZA), Prague, Czech Republic, August 23–26, 2005
edited by **J. Čejka, N. Žilková and P. Nachtigall**

Volume 159 **New Developments and Application in Chemical Reaction Engineering** Proceedings of the 4th Asia-Pacific Chemical Reaction Engineering Symposium (APCRE'05), Syeongju, Korea, June 12–15, 2005
edited by **H.-K. Rhee, I.-S. Nam and J.M. Park**

Volume 160 **Characterization of Porous Solids VII** Proceedings of the 7th International Symposium on the Characterization of Porous Solids (COPS-VII), Aix-en-Provence, France, May 26–28, 2005
edited by **Ph.L. Llewellyn, F. Rodríquez-Reinoso, J. Rouqerol and N. Seaton**

Volume 161 **Progress in Olefin Polymerization Catalysts and Polyolefin Materials** Proceedings of the First Asian Polyolefin Workshop, Nara, Japan, December 7–9, 2005
edited by **T. Shiono, K. Nomura and M. Terano**

Volume 162 **Scientific Bases for the Preparation of Heterogeneous Catalysts** Proceedings of the 9th International Symposium, Louvain-la-Neuve, Belgium, September 10–14, 2006
edited by **E.M. Gaigneaux, M. Devillers, D.E. De Vos, S. Hermans, P.A. Jacobs, J.A. Martens and P. Ruiz**

Volume 163 **Fischer-Tropsch Synthesis, Catalysts and Catalysis**
edited by **B.H. Davis and M.L. Occelli**

Volume 164 **Biocatalysis in Oil Refining.**
edited by **M.M. Ramirez-Corredores and Abhijeet P. Borole**

Volume 165 **Recent Progress in Mesostructured Materials.** Proceedings of the 5th International Mesostructured Materials Symposium (IMMS2006), Shanghai, P.R China, August 5–7, 2006
edited by **D. Zhao, S. Qiu, Y. Tang and C. Yu**

Volume 166 **Fluid Catalytic cracking VII: Materials, Methods and process Innovations**. Studies in Surface Science and Catalysis
edited by **M.L. Occelli**

Volume 167 **Natural Gas Conversion VIII.** Proceedings of the 8th Natural Gas Conversion Symposium, Natal, Brazil, May 27–31, 2007
Edited by **F.B. Noronha, M. Schmal and E.F. Sousa-Aguiar**